INTRODUCTION TO DIGITAL MICROELECTRONIC CIRCUITS

K. "Gopal" Gopalan
Purdue University Calumet

IRWIN

Chicago • *Bogotá* • *Boston* • *Buenos Aires* • *Caracas*
London • *Madrid* • *Mexico City* • *Sydney* • *Toronto*

Irwin Book Team

Sponsoring editor: Scott Isenberg
Marketing manager: Brian Kibby
Project editor: Paula M. Buschman
Production supervisor: Bob Lange
Designer: Laurie J. Entringer
Manager, graphics and desktop services: Kim Meriwether
Compositor: Interactive Composition Corporation
Typeface: 10/12 Times Roman
Printer: R. R. Donnelley & Sons Company

Library of Congress Cataloging-in-Publication Data

Gopalan, K. Gopal
    Introduction to digital electronic circuits / K. Gopal Gopalan.
      p. cm.
    Includes index.
    ISBN 0-256-12089-7
    1. Digital electronics. 2. Semiconductors. I. Title.
TK7868.D5G664 1996
621.39'5–dc20   95–49218

Printed in the United States of America

1 2 3 4 5 6 7 8 9 0 DO 2 1 0 9 8 7 6 5

To
Kalaichelvi,

Ilango,
and
Elil
for their
love, understanding, and support

எப்பொருள் யார்யார்வாய்க் கேட்பினும் அப்பொருள்
மெய்ப்பொருள் காண்ப தறிவு.

To discern the truth in everything heard from whomsoever is knowledge.
Thirukkural, Chapter 42, Verse 3
Thiruvalluvar, Tamil Sage, 35 B.C.

PREFACE

Of all the new technologies that have evolved in the last few decades, perhaps the digital integrated circuit (IC) technology is the one that continues to experience a phenomenal growth in terms of overall circuit complexity, switching speed, and power dissipation. This growth has created a pivotal place for teaching digital electronics in the undergraduate electrical and computer engineering curricula. The vast amount of material arising from innovative circuit designs and newer device technologies, however, requires that the circuit analysis aspects of digital electronics be covered in a first course, separated from device design and chip layout considerations. While the chip level design and layout are important in the design of Very Large Scale Integration (VLSI) systems and Application-Specific Integrated Circuits (ASIC), clear understanding of the performance characteristics of available devices is required for designers of systems using 'off the shelf' ICs. Therefore, the pedagogical approach taken in this book is to cover the analysis and performance comparison of different gate level logic circuits. Since the logic design course covers the building block implementation of a digital system, it is appropriate that the digital electronics course consider the analysis aspects of these building blocks arising from different technologies, primarily at the circuit level. A strong background in the analysis and comparative strengths of available technologies, from the circuits point of view, is required to make practical design trade-offs. For a systems architect interested in developing noncustomized systems by interconnecting standard ICs, such a background can be readily developed in a course without the chip or the physical level of design. Furthermore, with newer IC technologies appearing every few years, a thorough treatment cannot be given in a single course covering both the technologies and the circuit designs. Finally, the availability of computer-aided VLSI design tools still requires the user of these tools to choose the appropriate technology based on the requirements of a given application. Thus, the circuit level analysis provides an appreciation of the circuit design techniques and equips students for the efficient design of digital systems.

Subscribing to this philosophy of analyzing digital circuits in a single course, *Introduction to Digital Microelectronic Circuits* covers the basic gates in all of the presently available logic families. In addition, circuit configurations for VLSI implementation, interfacing of logic families, regenerative logic circuits, analog-digital interfacing, semiconductor memories, and programmable logic devices are discussed. Where applicable, design examples based on logic level requirements are presented.

MicroSim™ PSpice® simulation of the logic families is emphasized throughout the book. PSpice is chosen because of its availability and convenience compared to other simulation tools. Since the basic logic circuits in each family typically have no more than 10 transistors, the student version of MicroSim PSpice, available at no cost from MicroSim, can readily handle the analysis. It has been the author's experience that with PSpice and personal computers, students tend to complete the simulation of a circuit and analyze the results more conveniently, and also better appreciate the importance of simulation.

Emphasis is placed on the analysis of IC gates available in the market in each logic family, while theoretical circuit configurations are considered only as possible examples.

v

With this emphasis and the laboratory experiments using IC and discrete (simulated) versions of gates from each family, students gain insight into the relative merits of different circuit configurations in each of the logic families studied.

KEY FEATURES

Every attempt has been made to offer a distinctive perspective on the subject of digital microelectronic circuits. In particular, this book:

- Develops the study of semiconductor devices and digital electronics for students with a background in basic circuit analysis and some exposure to physics and electronics.
- Presents a complete treatment of the analysis of bipolar logic gates, from the early RTL to the popular TTL families and the advanced Schottky TTL families.
- Provides comprehensive coverage of the basic, 10K, 10KH and 100K series of ECL gates, and the I^2L gates.
- Explains thoroughly the implementation of logic gates using different configurations of MOS devices.
- Extends the analysis of digital IC families to cover the more recent BiCMOS and GaAs technologies.
- Gives a balanced treatment of regenerative logic circuits using bipolar and MOS discrete and integrated circuits.
- Includes coverage of popular methods of analog-digital data conversions.
- Introduces LSI and VLSI systems with memories and gate arrays.
- Incorporates MicroSim PSpice modeling and simulation throughout.

ORGANIZATION AND OUTLINE OF CHAPTERS

The book is organized into 10 chapters. Each chapter begins with an introduction and ends with a summary of key points covered, references, review questions, problems, and experiments. End-of-chapter problems serve as exercises, and are also used frequently to illustrate and/or extend some of the concepts presented in the text. Problems requiring computer simulation for analysis or design verification and those involving lengthy calculations are indicated by *. Experiments at the end of each chapter are used to extract device parameters for readily available bipolar and MOS devices and to provide an understanding of the performance characteristics of basic logic circuits using these devices.

Chapter 1 outlines the basic steps in the design of a digital system, and the importance of analyzing a system at various levels of design. Ideal and practical logic inverter characteristics are presented.

Fundamentals of semiconductors and current conduction mechanisms are described in Chapter 2. Operation and modeling of junction diodes are discussed.

Chapter 3 gives a brief description of the structure and operation of bipolar junction transistors (BJTs). Ebers-Moll, hybrid-π, charge-control, and MicroSim PSpice models are presented.

Static and dynamic characteristics of BJT saturation logic families are analyzed in Chapter 4. Performance improvements of different TTL families are compared.

Chapter 5 presents the analyses of different current mode logic families and their implementations in large-scale integration systems. Interfacing of saturation and current mode logic families is studied.

Chapter 6 provides a brief description of the structure and operation of MOSFETs and MESFETs. Simplified models for hand calculations and MicroSim PSpice models are presented for these devices.

Analyses of different structures of MOSFETs, BiCMOS, and gallium arsenide MES-FET logic circuits are treated in Chapter 7. Interfacing of BJT and MOS logic families is discussed.

Multivibrator circuits as a class of sequential circuits are analyzed in Chapter 8. Both discrete and integrated circuit implementations using bipolar and MOSFET devices are considered.

Chapter 9 presents various analog-digital conversion techniques.

Chapter 10 provides an introduction to the implementation of bipolar and MOS memories. Different programmable logic devices are discussed as examples of VLSI systems.

AUDIENCE

This text is intended for a one-semester, upper-level undergraduate course in electrical and computer engineering.

Basic knowledge of circuit analysis at the level of a first engineering circuit analysis course is assumed. Introductory level of knowledge in semiconductors and electronics is helpful, but not required. Enough material, however, is included to cover logic device characteristics, currents in semiconductors, and the structure, characteristics, and models of diodes, BJTs, and the FETs. Additionally, a concurrent (or previous) course in digital logic fundamentals is helpful.

Most of the material in this text has been used at Purdue University Calumet in a one-semester, upper-level course. The course is required for computer engineering and optional for electrical engineering students with a background in basic analog electronic circuits at the diode, BJT, and FET level. With two hours of lecture and three hours of laboratory per week, all the chapters are covered at least partially. A minimum of 12 laboratory experiments covers the characteristics of devices, logic families, multivibrators, and data converters. Most of the experiments require students to determine the performance characteristics of logic families in the lab and compare them with calculated and simulated results.

ACKNOWLEDGMENTS

The author wishes to acknowledge the reviewers for their comments at various stages of the manuscript.

The encouragement and support of the Department of Engineering, Purdue University Calumet are greatly appreciated.

CONTENTS

Chapter 4

BIPOLAR JUNCTION TRANSISTOR
SATURATION LOGIC FAMILIES   138

Chapter 5

CURRENT-MODE LOGIC
FAMILIES   220

Chapter 10

SEMICONDUCTOR MEMORIES AND VLSI SYSTEMS 555

INTRODUCTION

This chapter provides the motivation for the analysis and design of digital microelectronic circuits. Digital systems are used extensively in all realms of modern life. We find them in applications ranging from home appliances, entertainment systems, and palmtop computers to health care products, high-speed computers, and communication systems. More and more applications using digital techniques appear every year, with high precision, small size, and low power dissipation. Analysis of digital electronic circuits is vital to understanding present technologies of microelectronic circuits and to designing these digital systems at all levels of integration. This chapter outlines the design steps and emphasizes the use of computer-aided tools for the analysis and design of complex digital systems. As a first step in the analysis of digital electronic technologies, we examine the performance characteristics of general inverters.

1.1 DIGITAL SYSTEMS

Digital systems operate on information, or data, represented in discrete form. The most common discrete form used is the *binary,* with two disjoint sets of voltage levels representing binary low (0) and high (1) states. With each voltage level constrained to vary within a specified range, the output of a digital system is predictable over a wide range of operating conditions. Other advantages of digital systems over analog, or linear, systems (in which information is represented by continuously varying voltages or currents) include low cost, easy extension of data size, long-time storage capability, and programmability.

Digital systems use electronic circuits that operate, most commonly, as switches, with open switch position designated as logic, or binary, 1 (or high), and closed position as 0 (or low). Alternatively, the output of a digital electronic circuit may be one of two well-defined ranges of voltages (or currents) for the two logic states. Semiconductor diodes and transistors are used as switching devices in digital systems, also called *logic* or *switching* systems. *Microelectronics* refers to the technology of fabricating a large number of electronic devices on a single chip of silicon or a compound semiconductor material such as gallium arsenide. The size of the active transistor area in chips has progressively decreased to about 0.1 μm^2 at present, while a density of over a million transistors is achieved in an overall chip size of less than 100 mm^2 [1, 2]. Compare this area and density with those of the chips available at the beginning of the integrated circuit era, circa 1966: a chip area of approximately 5 mm^2, for example, contained 50 active devices with areas of about 0.025 mm^2 each [3]. Simultaneous with high density, high operating speeds of close to a billion operations per second, and high data transfer rates of nearly 10 Gbits/s have recently been achieved. This remarkable increase in performance along with decrease in size is due primarily to advances in the technology of the semiconductor device fabrication process, and to the development of innovative circuit configurations.

In the following section we consider the steps in the design of a digital system.

1.2   DESIGN OF DIGITAL SYSTEMS

Design of a digital system, be it a simple traffic light control system or a complex high-speed computer, proceeds with the following general steps: *specification, functional design,* detailed *logic design,* and *fabrication and testing.* As we will see, for the more complex of these applications, two more steps may also be needed before the final fabrication step.

As with any system, the first step in the design is the detailed specification of the requirements of the system. In this step, the design engineer determines the required number and voltage levels of inputs and outputs, speed of performance, range of power supply, physical size, and operating environment.

The next step is to create a functional model of the system. At this phase, the system is described in terms of abstract blocks, which, when interconnected, simulate the intended behavior for the given input and initial conditions. The goal in this phase is to establish the required building blocks and their interconnections to meet the gross operational specifications of the system. This step is also called the *architecture,* or *register level* design, particularly when referring to computer design. The design at this step represents the behavioral, or input-output, model of the system. Currently, we describe and specify the behavioral model in an abstract language such as Verilog or VHDL.[1]

In the next step, the *logic,* or *structural,* design considers how to implement the blocks identified in the functional design stage. While some of the blocks may be available as off-the-shelf components, others must be realized from basic elements. A divide-by-N counter, or an N bit sequence detector, for example, may not be available directly for any given value of N. A logic designer carries out the logic design of such blocks in this phase, as well as the interfacing of each block with others, if necessary. This is the primitive level of design, where one chooses the applicable technology for each functional as well as logic block, based on such considerations as power, size, and speed.

---

[1] Verilog is a registered trademark of Cadence Design Systems, Inc. VHDL, which stands for VHSIC (very high speed integrated circuits) Hardware Description Language, was developed with the sponsorship of the U.S. Department of Defense.

The final step is the fabrication of the blocks in printed circuit boards and performance-testing of the assembled system. In this step, all the blocks identified in the previous steps are interconnected with appropriate power and signal sources. A test engineer validates the completed system by supplying or simulating the specified inputs and monitoring the outputs from the system.

The above process for a system design assumes the use of readily available, off-the-shelf components: at the logic design level, small-scale integration (SSI) circuits for gates and flip-flops, and, at the functional level, medium-scale integration (MSI) circuits such as counters and shift-registers, and, in some cases, large-scale integration (LSI) circuits, such as memory and logic arrays.[2] For a number of applications with low volume, and/or those that will need changes in specifications with time, this design methodology is sufficient. However, the process has several limitations for use in applications where size and power dissipation are also primary considerations. When there are number of different circuits at different integration levels, sizes, power, and cost, these add to the overall cost, power, size, and assembly and testing time. For example, a digital time/temperature display unit or a marquee sign can be built efficiently using SSI, MSI, and LSI components; a wristwatch or a notebook computer, on the other hand, becomes costly and prohibitively large in size, and dissipates a disproportionate amount of power for its application when built with off-the-shelf integrated circuits. In addition, the advantages of high-speed technology at the gate and the functional levels would be lost in the system due to the extensive wiring and interconnection needed. To achieve a more appropriate design for such applications the designer extends the previous design steps from the logic (or *gate*) level of design to the *circuit* level, then the *chip (device) layout* level, and, finally, arrives at chip fabrication. Figure 1.1 shows the complete design steps. Because the final product here is a single chip for a given application, it is an application-specific integrated circuit (ASIC). It is also a very large scale integration (VLSI) circuit if its size exceeds 10,000 equivalent gates.

Unlike in MSI/SSI-level design, VLSI and ASIC circuits have stringent constraints while offering flexibility at the circuit and chip level of design. For example, because of its small size, the number of external pins (connections) in an ASIC or VLSI circuit is limited to a few hundred. As a result, some of the input signals may need to be generated internally, and the number of output lines may be limited. This, in turn, could result in lower performance, increased size, more components, and more design changes. Size is also a significant factor in determining maximum power dissipation within the chip and the operating supply voltage. The functional level design must, therefore, consider size with other specifications. Design at the logic level in ASIC and VLSI circuits must minimize redundancy, to ensure the minimum number of gates and to minimize size. In addition, design at the circuit level—which is the primitive level for a given technology of ASIC or VLSI circuits—must obtain optimum gate circuit

---

[2] Generally accepted definitions for scale of integration are: 1 to 10 gates per chip: SSI; 10 to 100 equivalent gates per chip: MSI; 100 to 10,000 equivalent gates, or memory bits per chip: LSI; and over 10,000 equivalent gates: VLSI (very large scale integration). For chips with a million or more equivalent gates, the term ULSI (ultra large scale integration) is sometimes used. In all these cases, a gate represents the basic block—an inverter, typically—in the given technology.

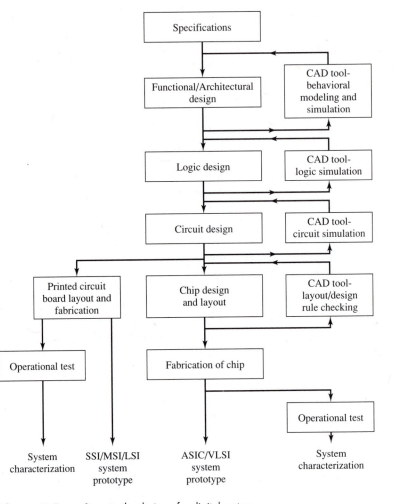

Figure 1.1   Steps in the design of a digital system

configuration in terms of speed, power, and size. Finally, careful layout at the
chip level ensures that a compact chip results, with the architecture dsigned at
the functional level. Testing of the chip after fabrication verifies that the final
circuit meets all of the performance specifications: static, dynamic, power, and
environmental.

Because of the interrelationship among the design phases, a digital system
design engineer must have expertise in one or more levels and must know enough
in all other levels to design and fabricate efficient systems. An architecture for
adding *N* bits of data, for example, may be more efficient using serial addition
than parallel addition if the source bits are given serially at a high rate. Therefore, an understanding of the adder implementation at the register level helps the
design engineer achieve an efficient design. Similarly, at the logic level, a combi-

nation of multiplexers or a programmable logic device might cut down design time and costly layout of different circuit patterns for realizing multiple logic functions in many variables. An innovative circuit design might, on the other hand, reduce size and power or increase speed of operation. With device technology advancing rapidly, the choices of available technology are crucial in determining overall performance and cost. If a design objective is the operation of a digital system at data transfer rates exceeding hundreds of Mbits/s, then a good choice might be Emitter-Coupled Logic (ECL), Bipolar-Complementary Metal-Oxide Semiconductor (BiCMOS), or gallium arsenide Metal-Semiconductor Field Effect Transistor (GaAs MESFET) technology. If, in addition to high speed, the system must dissipate low power, the choice is narrowed to BiCMOS technology, which has operating speeds rivaling that of ECL but at significantly reduced power dissipation. At present, ECL has the highest power dissipation; GaAs MESFET technology, on the other hand, exceeds ECL in speed but is more expensive due to its low yield and has not attained the same level of maturity as other technologies. However, with advances in GaAs device fabrication technology, data rates exceeding Gbits/s have been achieved. Additionally, we have recently seen the development of high density MOS (Metal-Oxide-Semiconductor) implementation of memory devices with storage capacities in the Gbit range. Clearly, these types of developments in technology must be considered in the design of high-speed, high-density, and low-power digital systems.

1.3   ANALYSIS OF DIGITAL SYSTEMS

To develop an error-free digital system in the form of a VLSI circuit, one must verify design at every step. Analysis of the functional level design blocks and their interconnections is necessary to check the input-output logic functionality of the system. Logic level analysis verifies the truth table, state table and/or flow table for each block and identifies timing errors (glitches, races, and hazards).[3] Analysis of basic gate circuits in a given technology presents static and dynamic performance characteristics and helps in devising novel configurations to meet overall design objectives. Verification of chip design is essential in a complex circuit to avoid costly and time-consuming errors when laying out active devices and making interconnections. In addition, layout must be checked for parasitic devices, noise sources and, more frequently, capacitances. Clearly, the top-down design approach has local feedback paths at each level so that problems identified during analysis can be corrected. This type of iterative design based on analysis and feedback is indispensable to meet all the performance specifications of the system and to prevent malfunctioning under all operating conditions. Finally, characterization of the final product by testing under

---

[3] For a detailed discussion of these topics, the reader is referred to logic design textbooks such as Roth [7] or Wakerly [8].

different conditions is warranted, to ensure satisfactory operation of the system and to document deviations from specifications.

As well as being part of the design process, analysis is essential for efficient use of a product. And understanding the technology and the characteristics of a digital device or circuit motivates us to develop refinements or newer technology.

1.4   COMPUTER-AIDED ANALYSIS AND DESIGN OF DIGITAL SYSTEMS

Analysis at each design level may be accomplished by paper and pencil (and perhaps, with the use of a calculator) for relatively simple systems or for those requiring only gross approximations to actual results. At the functional level, for example, timing diagrams, state tables, and truth tables can be derived, and the functionality verified, for systems with no more than a few sequential and combinational devices. Systems that have complex configuration or a large number of devices, however, require computer simulation programs to perform tedious and involved performance analyses. Simulation programs can handle large amounts of data and can perform rapid analyses under different operating conditions. These programs use built-in or user-defined device models and parameters, and initial state of the system. Simulation programs and other computer-aided design tools are available for all the levels of design shown in Figure 1.1. Verilog, for example, is a hardware description language that supports simulation and analysis of a digital system at the behavioral (functional), structural (logic) and switch (MOS circuit) levels. MicroSim™ PSpice® is a circuit simulation program that can analyze circuits consisting of active and passive devices as well as most of the commonly available analog and digital integrated circuits.[4] Simulation programs such as BDSYN and a host of others ("Tool suite") available from Massachusetts Microelectronics Center (M2C) can be used to analyze a system at the logic functional level. Finally, interactive systems are available for creating and modifying VLSI circuit layouts using CMOS. Magic by M2C, for example, is a system of software packages with built-in cell libraries for primitive functions as well as tested complex logic functions in the CMOS technology, and graphic design tools for capturing VLSI layout. The system can check circuit layout based on common design rules and connectivity, and it can generate graphic artwork to be used for fabrication. In addition, users can extract model parameters for the devices in the circuit for PSpice simulation. Simulation programs also aid in determining the speed of a system that is limited by the wiring and loading delays. Programs capable of extracting wiring delays from a specified layout enable the system designer to modify the design to meet the speed specifications. Hardware description lan-

---

[4] MicroSim™ PSpice® available from MicroSim Corporation, is an enhanced version of the public domain program, SPICE (Simulation Program with Integrated Circuit Emphasis), developed by the Electronics Research Laboratory of the University of California, Berkeley.

guages enable a designer to describe a complex system at a high level of abstraction. After satisfactory analysis using the logic and timing simulation tools available, the description can be converted to standard data formats ready for use in the fabrication stage.

It is clear from the brief description of simulation programs that some, such as PSpice, perform analysis mostly at one level of design; others, such as Magic, offer chip layout design as well as circuit analysis capability. Computer-aided analysis and design is, therefore, an integral part of designing a complex digital system at the ASIC and VLSI levels.

Despite the availability of automated systems for simulation at the functional, logic, and circuit levels, hand calculations are still needed for understanding a system at the circuit level. Rapid but grossly approximated hand calculations enable verification of the circuit performance during the design, and provide practical ranges of variables that control the operation of the circuit. Use of the estimated range of parameters from hand calculations reduces the analysis time of more detailed computer simulation. Additionally, this practice helps in obtaining accurate and refined results by restricting simulation to the useful range of variables. More importantly, the "quick-and-dirty" hand analysis, however approximate, provides insight into the performance of the circuit and enables the designer to compare alternative designs rapidly. It is, therefore, essential that design engineers develop skills to analyze circuits using simple models, and that they be able to extend the analysis with more elaborate models.

1.5 TEXT OBJECTIVE

The goal of this text is to assist the reader in developing the skill to understand and design digital systems at the circuit level. As seen above, digital system design is based on the understanding of the available technologies of semiconductor devices for implementing logic functions. Logic devices can be fabricated in different technologies to perform the same function. Gates and flip-flops, for example, can be built in the bipolar technologies (TTL and ECL), MOS technologies (NMOS, CMOS and BiCMOS), and GaAs MESFET technology. As a design engineer, the reader must understand the characteristics of the basic logic devices in each technology (or series, or family) and how these devices can be interconnected to realize complex functions. This analysis enables the designer to evaluate the relative merits of a family in terms of speed of operation, power dissipation, performance in a noisy environment, input signal levels, and loading conditions. Further analysis with chip layout brings out information regarding the size and the value of parasitic capacitance. Study of the basic gate structure and characteristics is also needed for designing systems using off-the-shelf components at the SSI/MSI/LSI level. In addition, an understanding of basic gate structure aids in analyzing complex internal structures of VLSI systems and in devising low-power, high-density, and high-speed logic devices.

This text first provides a review of digital microelectronic devices: semiconductor diodes, bipolar junction transistors (BJTs), and field-effect transistors using silicon and gallium arsenide. Next, we consider the gate level analysis of logic families using these devices. The student version of PSpice is used throughout for simulation and verification of hand calculations. Following these topics, we look at implementation of a class of sequential circuits, namely, regenerative logic circuits, using BJTs and MOSFETs, and the cover analog-digital data conversion. Finally, the text presents an introduction to memory devices and LSI/VLSI circuits.

We begin our analysis of basic logic devices with the inverter, which is an essential element in any digital integrated circuit series. Characteristics of an inverter are important in the selection of a logic family for a given application.

1.6   BASIC LOGIC INVERTER CHARACTERISTICS

First we study the properties of an ideal inverter. This study will enable us to compare the characteristics of practical inverters in each family.

1.6.1   IDEAL INVERTER CHARACTERISTICS

An ideal inverter, whose circuit symbol is shown in Figure 1.2a, operates with a single power (voltage) source, V_{CC}, and exhibits the input-output voltage transfer characteristics (VTC) shown in Figure 1.2b.

The two logic states are identified in the static voltage characteristics. As we see in Figure 1.2b, the logic levels are symmetric with respect to $V_{CC}/2$. Adopting positive logic,[5] we have, at the input, $0 < V_i < V_{CC}/2$ corresponding to logic low, or Boolean zero (0) state, and $V_{CC}/2 < V_i < V_{CC}$, to logic high, or Boolean one (1) state. At the output, the logic low has a voltage of $V_o = 0$, while the logic

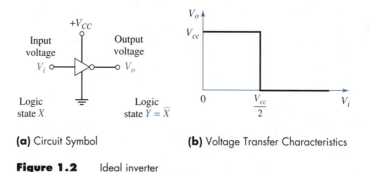

(a) Circuit Symbol   **(b)** Voltage Transfer Characteristics

Figure 1.2   Ideal inverter

[5] That is, the voltage corresponding to logic high (Boolean 1) > the voltage corresponding to logic low (Boolean 0).

high corresponds to $V_o = V_{CC}$. The transition from one state to the other occurs sharply at $V_i = V_{CC}/2$ so that there is no ambiguity in the logic state of input or output; that is, the input voltage V_i can be in one of only two ranges of voltages, while the output voltage V_o takes on one of two distinct levels.

Observe that a range of voltage at the input, $0 < V_i < V_{CC}/2$, is transformed to a single output voltage of $V_o = V_{CC}$; a second range of $V_{CC}/2 < V_i < V_{CC}$ is transformed to another single voltage of $V_o = 0$. Because of this many-to-one transformation from the input to the output, it is clear that the inverter exhibits a nonlinear relationship between V_i and V_o. Also, the incremental voltage gain, or slope, dV_o/dV_i is zero for $V_i < V_{CC}/2$ and for $V_i > V_{CC}/2$, and is infinity at $V_i = V_{CC}/2$.

If two inverters (ideal or otherwise) are cascaded together as shown in Figure 1.3, the output voltage of the first (*driver*) must correspond to one of the two well-defined ranges of the input of the second (*load*); that is, V_{o1} must be such that $0 < V_{o1} < V_{CC}/2$ at logic low, or $V_{CC}/2 < V_{o1} < V_{CC}$ at logic high. The second inverter can then regenerate at its output the logic state applied at the input of the first. Evidently, the ideal inverter possesses this property of regeneration of logic levels with $V_o = 0$ and $V_o = V_{CC}$. Note that the two *ranges* of input voltages, rather than the individual levels, allow for the variation of voltage levels at the input due to noise or loading. We will examine the effect of input voltage variation in the following section.

Another property of an ideal inverter is that it must be capable of driving (or sourcing) any amount of current into any type of load without altering its output logic state; it must also be able to sink any amount of current from any load. This implies that the output resistance of the ideal inverter is zero. Combined with the regeneration property, therefore, an ideal inverter can drive any number of identical circuits (loads) while maintaining its logic state at the output.

The *dynamic, or switching, response* of an ideal inverter is such that there is no delay between the input and the response. That is, the transmission, or *propagation delay* of the device for a change in input logic state to reflect at output is zero. Also, the rise time and the fall time of the output are the same as those of the input.[6]

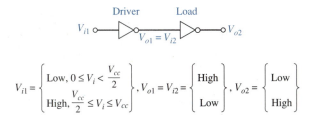

$$V_{i1} = \left\{ \begin{array}{l} \text{Low}, 0 \le V_i < \dfrac{V_{cc}}{2} \\ \text{High}, \dfrac{V_{cc}}{2} \le V_i \le V_{cc} \end{array} \right\}, V_{o1} = V_{i2} = \left\{ \begin{array}{l} \text{High} \\ \text{Low} \end{array} \right\}, V_{o2} = \left\{ \begin{array}{l} \text{Low} \\ \text{High} \end{array} \right\}$$

Figure 1.3   Logic regeneration in cascaded inverters

---

[6] These refer to the finite slopes with which the waveforms rise and fall between their logic high and low levels. See the following section.

In addition, an ideal inverter draws little or no power from the source or the input; its circuitry is simple, and occupies the smallest area. Of course, no logic family has an inverter that has all the characteristics of an ideal inverter. Instead, we use the ideal inverter as a reference for performance comparison of practical inverters of different logic families.

1.6.2   PRACTICAL INVERTER CHARACTERISTICS

Figure 1.4a depicts the "generic" voltage transfer characteristics of a practical inverter. There are two regions in the characteristics—the first near the origin and the second for large V_i—where V_o is relatively constant for a range of V_i. These regions, therefore, are defined to correspond to the two logic states. Lack of sharp transition from one reigon to the other, however, results in a transition, or undefined, band of voltages at input for which the output does not correspond to either of the logic states. The width of the transition region of V_i is determined using the slope criterion of $dV_o/dV_i = -1$. Note that the slope, or incremental gain, dV_o/dV_i is zero near the origin and V_{CC}, and negative in the transition band. Hence, the pairs of voltages V_i and V_o at which $dV_o/dV_i = -1$ define the extreme voltages, V_{IL}, V_{IH}, V_{OL}, and V_{OH} for logic low and high states. V_{IL} is the maximum allowable input voltage at logic low state, and V_{IH} is the minimum input voltage needed at logic high state. Clearly, an input voltage V_i in the range $V_{IL} < V_i < V_{IH}$ is in the transition region for which the output state is undefined. The *width of transition* region, which is a measure of performance, is

$$V_{tw} = V_{IH} - V_{IL} \tag{1.1}$$

Low V_{tw} reduces ambiguity in the input logic state. Additionally, a small transition band helps prevent an erroneous logic state (i.e., exchange of low and high) at input, which can arise from small changes due to noise voltage, as we will discuss next.

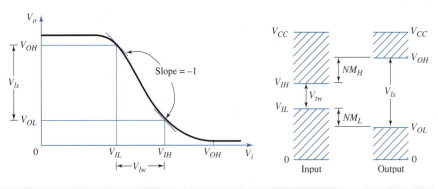

(a) Voltage Transfer Characteristics

(b) Logic Levels

Figure 1.4   Practical inverter characteristics

In general, the logic high and low voltages at the output, V_{OH} and V_{OL}, are also defined by the slope criterion. This is because in a system of cascaded circuits, the lowest input voltage is $V_i = 0$; the corresponding output voltage $V_o = V_{OH}$ is the maximum possible input voltage to the succeeding circuit. In practice, output voltage for $V_i = 0$ is used as V_{OH}, and V_O for $V_i = V_{OH}$ is considered V_{OL}. If $V_i = 0$ is not in the valid range of input, $V_i = V_{OL}$ is used to determine $V_o = V_{OH}$. Usually, V_O is constant and close to the supply voltage near the lower end of V_i, and V_{OL} can be determined from the given circuit configuration. Note that V_{OH} defined in both cases must satisfy $V_{OH} \geq V_{IH}$ for logic regeneration.

Logic swing is another, performance measure used to compare different logic families. It is defined by

$$V_{ls} = V_{OH} - V_{OL} \qquad \textbf{(1.2)}$$

Higher V_{ls} reduces ambiguity in the logic state and increases noise immunity, which we discuss further.

Figure 1.4b shows an example of the range of input and output voltages defined for proper logic operation. As we can see in this figure, uncertainty in the output logic state occurs for $V_{IL} < V_i < V_{IH}$. In addition, output voltage levels $V_o \geq V_{OH}$ and $V_o \leq V_{OL}$ are contained within the input levels, $0 \leq V_i \leq V_{IL}$ and $V_{IH} \leq V_i \leq V_{CC}$, respectively. This inclusion of output logic levels ensures that the input logic states are regenerated at the output of a succeeding inverter.

The ability of a digital circuit to maintain its logic state under varying voltage levels is measured by the *noise margins*. Logic high voltage of a practical inverter, for instance, can drop from its "nominal" value of V_{OH} because of *noise,* or undesirable voltages. Noise may arise from inadequate regulation or decoupling of power supply, electromagnetic radiation, inductive and capacitive coupling from neighboring circuits, and line drops. For an inverter driven by the output of another identical inverter, changes at the input logic voltage can also occur because of the temperature-dependent variations at the output voltage of the driving gate.

From the logic levels defined earlier, input voltage V_{i2} at logic high for the load (driven) gate in Figure 1.3 must be such that $V_{i2} \geq V_{IH}$; this allows the output of the driving gate (V_{o1}) to drop from its nominal high of V_{OH} to as low as V_{IH} (see Figure 1.4b). Hence, the margin of safety at logic high output is given by

$$NM_H = V_{OH} - V_{IH} \qquad \textbf{(1.3)}$$

NM_H is defined as the noise margin at logic high output. The noise margin at logic low level output is defined similarly by

$$NM_L = V_{IL} - V_{OL} \qquad \textbf{(1.4)}$$

The *absolute noise margin NM* for a gate is the smaller of the two noise margins,

$$NM = \min\{NM_L, NM_H\}$$

From Figure 1.4b we can see clearly that an inverter with large logic swing and small transition width has good noise immunity. Immunity to changes in input logic levels is also required when the characteristics of driving gates change due to temperature. The VTC shown in Figure 1.4a, for example, represent the nominal static behavior; variations in voltage levels with change in operating temperature (and load, as we see later) must be accommodated by the driven circuits. This is especially important in circuits such as ECL series (Chapter 5) which have low logic swing and transition width.

The following example illustrates the calculation of logic levels and noise margins.

Example 1.1 │ The voltage transfer characteristics of an inverter are shown in Figure 1.5. The input-output voltage relationship may be analytically expressed as

$$V_o = \begin{cases} 5\,V, & 0 \le V_i \le 1\,V \\ 5 - 2(V_{i-1})^2, & 1 \le V_i \le 2.35\,V \\ 5 - 2[2(V_i - 1)V_o - V_o^2], & 2.35 \le V_i \le 5\,V \end{cases}$$ **(1.5)**

Determine the logic levels V_{IL}, V_{IH}, V_{OL} and V_{OH}, and noise margins. What are the logic swing and transition width for the circuit?

Solution

Using the definitions of the logic levels, we can determine V_{IL}, V_{IH}, V_{OL}, and V_{OH} graphically from Figure 1.5. With the analytical relationship between V_i and V_o, the logic levels can be calculated as follows.

From Figure 1.5 and Equation (1.5), it is clear that V_o is constant at 5 V for $0 \le V_i \le 1$ V; hence, $V_{OH} = 5$ V.

The slope $dV_o/dV_i = -1$ occurs after the initial constant region. Therefore, using

$$V_o = 5 - 2(V_i - 1)^2, \qquad 1 \le V_i \le 2.35\,V$$ **(1.6)**

which is applicable at the beginning of the transition region, we have

$$dV_o/dV_i = 4 - 4V_i$$ **(1.7)**

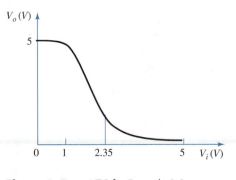

Figure 1.5   VTC for Example 1.1

Equating the slope to -1 in Equation (1.7) results in

$$V_{IL} = 1.25 \, \text{V}$$

and, using Equation (1.6), the corresponding output voltage is

$$V_o = 4.88 \, \text{V}$$

Note that there is no region of constant V_o for $V_i > V_{IL} = 1.25$ V; hence, the slope condition gives the minimum acceptable output voltage at logic high (without load or noise).

Taking the derivative of

$$V_o = 5 - 2[2(V_i - 1)V_o - V_o^2] \tag{1.8}$$

at the higher end of V_i and equating $dV_o/dV_i = -1$ results in

$$8V_o = 4V_i - 3 \tag{1.9}$$

Using Equations (1.8) and (1.9), we have, for input logic high level,

$$V_{IH} = 2.58 \, \text{V}$$

and the corresponding output voltage is

$$V_o = 0.91 \, \text{V}$$

Now, using $V_i = V_{OH} = 5$ V in Equation (1.5), we find

$$V_{OL} = 0.31 \, \text{V}$$

Hence the noise margins are

$$NM_L = V_{IL} - V_{OL} = 1.25 - 0.31 = 0.94 \, \text{V}$$

and

$$NM_H = V_{OH} - V_{IH} = 5 - 2.58 = 2.42 \, \text{V}$$

The logic swing is

$$V_{ls} = V_{OH} - V_{OL} = 3.97 \, \text{V}$$

and the transition width is $V_{tw} = V_{IH} - V_{IL} = 1.33 \, \text{V}$

A high noise margin is also essential in practical inverters and other digital devices so that their outputs can drive loads. Since practical load devices do draw power from their input sources, a noise margin ensures that the output of the driving gate is in the valid voltage range while supplying power to a load. This is also true for the load devices that supply power to the driving devices. Because of the variety of the loads that can be connected to a gate output, the load-driving capability of an inverter is defined in terms of the number of identical basic inverters of the same type or family. This number, called the *fan-out* of the driver, is the maximum number of identical inverters that can be driven by the output without altering its logic low or high state. Since inverters and other loads draw different amount of currents in logic low and high states, the circuit fan-out is the smaller of the number for logic low and high states. Fan-outs of other logic devices are also usually expressed in terms of the number of basic inverters they can drive. Standardizing the drive capability and input requirements of any device of a logic family in terms of the number of its basic inverters makes it simple to determine the loading effects in an interconnected system of different devices of the same family.

Example 1.2

The output of the inverter in Example 1.1 is modeled under static conditions by a voltage source of approximately 5 V in series with a resistance of 2 kΩ at logic high, and a source of approximately 0.9 V in series with 500 Ω at logic low. If the input currents drawn by the inverter are zero at logic low level and 2.25 μA at logic high, calculate the fan-out.

Solution

Since the input (sink) current for the load inverter (Figure 1.3) is zero at logic low, there is no change in voltage at the output of the driver. The fan-out of the driver, therefore, is virtually unlimited at logic low output.

At logic high output, each identical inverter connected to the driving gate draws a current of $I_i = 2.25\ \mu A$ as shown in Figure 1.6. Because of the finite output resistance R_o, this causes a drop in V_o. With N loads, therefore, output voltage reduces to

$$V_o = 5 - R_o N I_i = 5 - (2\,k\Omega)N(2.25\,\mu A) \qquad \text{(1.10)}$$

The lowest voltage permissible for V_o at logic high is

$$V_{OH\,min} = V_{IH} = 2.58\ \text{V}$$

Hence

$$N = (5 - 2.58)\frac{\text{V}}{4.5\ \text{mV}} = 537.8$$

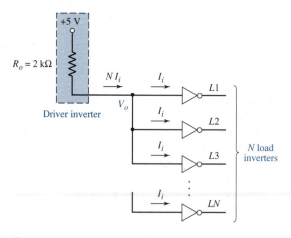

Figure 1.6   Logic high fan-out

Since fractional loads are not possible, we have a fan-out of $N = 537$ at logic high, which is also the fan-out of the inverter.(Notice that if we round N to 538, V_o goes into the transition region.)

If we drive all of the 537 gates using a single gate in the above example, the output of the driving gate drops to slightly above 2.58 V. Since this is the minimum voltage required at logic high level, it is susceptible to go into the uncertain region in the presence of noise. In practice, therefore, manufacturers specify a lower than maximum fan-out to guarantee a small noise margin even when loaded.

Another point to note is that the large value of $N = 537$ is applicable only when the input logic states are changing slowly. For this reason, the calculated fan-out is called the *static* or *dc fan-out* . If the input signal is switching between states at a fast rate, the actual number of loads that can be connected to the output, called the *dynamic fan-out,* is significantly lower than the dc fan-out. The reason for this is the stray, or unwanted, capacitance present in any load and in the interconnection paths. When the logic state at input changes, output cannot switch its state until the load capacitance is charged or discharged. Hence, sufficient time must be allowed for the loaded output to change state. Clearly, then, to operate the inverter or other logic devices at high switching frequencies, load capacitance must be minimized by reducing the number of load devices. We consider the effects of load capacitance following the definition of switching delays.

The above example of voltage transfer characteristics (and loading behavior, to some extent) is typical of ECL and NMOSFET logic families (Chapters 5 and 7). Other logic families, such as transistor-transistor-logic (TTL) and Schottky transistor-Logic (STL) families (Chapter 4), are usually described in terms of piecewise linear transfer characteristics as shown in Figure 1.7.

It is clear from Figure 1.7 that permissible logic levels correspond to the breakpoints terminating the zero-slope regions. These points are indicated as *a* and *b* in the figure.

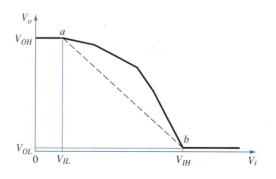

Figure 1.7   Piecewise linear voltage transfer characteristics

Although all logic levels considered so far are positive voltages, the analysis is valid for negative as well as mixed logic levels. In ECL circuits, for example, depending on the supply voltage and circuit configuration used, logic levels can be positive, negative, or bipolar.

Along with fan-out, *fan-in* for a general logic gate other than an inverter is defined as the number of independent inputs the particular gate family can have. Fan-in refers to the ability of the logic family to extend the number of inputs without adversely affecting the circuit performance, size, or complexity.

Propagation Delay   The speed of operation of a practical switching circuit depends on how fast a change at the input propagates through the circuit and causes a change at the output. The logic high-to-low and low-to-high propagation (transition) times are denoted by t_{PHL} and t_{PLH}, respectively. The two delays are defined between the 50 percent points of input and output voltage levels as shown in Figure 1.8. Note that input levels are shown as V_L and V_H, where $V_L \leq V_{IL}$ and $V_H \geq V_{IH}$, and both must be within allowable limits of the circuit.

The average of t_{PHL} and t_{PLH} is the average propagation delay t_P, given by

$$t_P = \frac{(t_{PHL} + t_{PLH})}{2} \tag{1.11}$$

When the input waveform is fairly steep, indicating negligible *rise and fall* times, output rise and fall times are used to describe the transient performance of the circuit. The *rise time* t_r is the time required for the waveform to rise from 10 percent to 90 percent of its final (high) value. The *fall time* t_f, similarly, is the time for the waveform to decrease form 90 percent to 10 percent of its final (low) value.

Propagation delays are measured by cascading an odd number N of inverters and connecting the final inverter output to the input of the first inverter as shown in Figure 1.9. The period of the output waveform that results in this "ring oscillator" corresponds to $2Nt_P$, where t_P is the average propagation delay. With a large number of cascaded inverters, laboratory oscilloscopes can be used to measure very small t_P (in the picosecond range). This method of measuring t_P

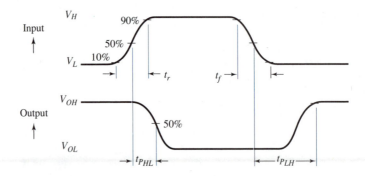

Figure 1.8   Propagation delays

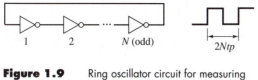

Figure 1.9   Ring oscillator circuit for measuring average propagation delay

by fabricating a ring oscillator in a chip is used to study the high-speed switching capabilities in developing technologies.

As stated earlier, output propagation delay and rise and fall times depend mainly on the lumped (load) capacitance at the output node. The following illustrates this.

The inverter in Example 1.2 presents a total capacitance to ground of 0.5 pF at its input. If 10 identical inverters are driven by a single inverter, calculate the delays at the output of the driver when switching: (a) from $V_o = 0.9$ V to $V_o = 4$ V, and (b) from $V_o = 4.9$ V to $V_o = 1$ V.

Example 1.3

Solution

The circuit for switching from low to high output is shown in Figure 1.10a where the load capacitances are lumped as $C_L = 5$ pF. C_L charges from the initial voltage of 0.9 V toward the final voltage of 5 V with a time-constant of (2 kΩ) (5 pF)= 10 ns. (We assume that the input resistance of the load is much larger than 2 kΩ so that its effect on the time-constant is negligible.) The charging equation is given by

$$V_o(t) = V_o(\infty) + [V_o(0) - V_o(\infty)]e^{-t/\tau} = 5 - 4.1e^{-t/10\text{ns}} \qquad \textbf{(1.12)}$$

In the absence of the propagation delay of the circuit, we shall assume that the only delay at the output is due to charging of the capacitor according to Equation (1.12).

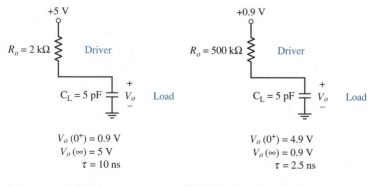

(a) Low-to-High Switching   **(b)** High-to-Low Switching

Figure 1.10   Calculation of delay times

Solving for $t = t_1$ at which instant $V_o(t_1) = 4\,\text{V}$, we obtain the low-to-high switching delay of

$$t_1 = 14.11\,\text{ns}$$

For switching from $V_o = 4.9\,\text{V}$ to $V_o = 1\,\text{V}$, referring to Figure 1.10b, we have

$$V_o(t) = 0.9 + 4\,\exp\!\left(\frac{-t}{2.5\,\text{ns}}\right) \tag{1.13}$$

Solving for $t = t_2$ when $V_o(t_2) = 1\,\text{V}$, we get the high-to-low switching delay of

$$t_2 = 9.22\,\text{ns}$$

Note that the voltages $V_o(t_1)$ and $V_o(t_2)$ are within the range of allowed voltages at the input of load gates. Clearly, the delays will be longer if V_o must reach V_{OL} or V_{OH}.

As assumed, response time analysis in the above example considers only the delays arising from charging or discharging of the output capacitance. In addition to these delays, practical digital circuits will also have delays in switching active devices from one mode of operation to another. Hence, to achieve increased speed of switching, design engineers must take steps to minimize (1) passive circuit delay, by reducing load capacitance, and (2) active device delay, by using appropriate circuit configuration, chip layout, and fabrication technology. Careful layout and circuit design, and minimization of wiring and interconnections are also essential in integrated circuits to reduce parasitic capacitance and hence overall switching delays.

The product of average propagation delay and power dissipation in a digital circuit is called the *delay-power product*. This is a figure of merit for circuit efficiency—a small value indicates that the circuit has a fast response time while dissipating very little power. In practice, however, there is a trade-off between propagation delay and power dissipation. It is possible to achieve a higher switching speed within a given logic family and circuit configuration at increased power dissipation; a case in point is the integrated injection logic family, which we consider in Chapter 5. Additionally, two different logic families may have nearly the same delay-power product, with one having smaller delay and greater dissipation than the other. ECL and BiCMOS technologies, which we examine in Chapters 5 and 7, fall in this category.

Another measure of speed of digital circuit is the *cycle time* t_{cyc}, which is the time between two successive low-to-high or high-to-low transitions at the same voltage at input or output. The *clock frequency* f_{cyc}, which is the reciprocal of t_{cyc}, is often used to denote the speed of the circuit. Typical values of t_{cyc} are on the order of 20 to 50 times the average propagation delay t_p.

SUMMARY

The design of a digital system begins with specification and ends with fabrication and testing. In between are the actual design phases, whose number may vary. Design of a digital system in the form of an ASIC or VLSI circuit is carried out in four phases for a given set of specifications: functional design, logic design, circuit design, and layout design. For error-free operation of the final system, design must be analyzed at each phase.

Computer-aided design tools are available to facilitate the analysis of a digital system at every level of design.

Simulation and analysis of digital systems at the circuit level is essential for (1) understanding the performance characteristics of different technologies and the complex internal structure of VLSI circuits, and (2) efficient implementation of logic level and chip level designs.

The logic inverter is a basic element in all digital technologies. Input logic levels, V_{IL} and V_{IH}, in practical inverters are defined by the slope criterion applied to the voltage transfer characteristic as $dV_o/dV_i = -1$; output logic levels V_{OL} and V_{OH} are determined from the circuit configuration. Input voltage in the transition region, $V_{IH} > V_i > V_{IL}$, corresponds to uncertainty in the logic state.

Width of transition region is given by $V_{IH} - V_{IL}$, and logic swing is given by $V_{OH} - V_{OL}$.

Immunity of a logic device to noise at its input is defined by noise margins: $NM_H = V_{OH} - V_{IH}$ and $NM_L = V_{IL} - V_{OL}$. With higher noise margins, a circuit can operate without logic error in noisy evironments.

Fan-out of an inverter refers to the number of identical inverters that it can drive while maintaining its present output logic state.

The static fan-out is high because of low power associated with the inputs of driven gates; dynamic fan-out is low due to increased load capacitance and, hence, propagation delays.

Propagation delays when switching output logic states are defined as time delays in going from 50 percent of input level to 50 percent of output level. These delays arise mainly from parasitic capacitances. Careful layout and circuit design minimizes propagation delays.

Delay-power product is a measure of the dynamic performance of logic circuits.

REFERENCES

1. *IEEE Spectrum,* January 1995, special issue on Technology.

2. *The Institute.* A News Supplement to *IEEE Spectrum* 16, no. 52 (September/October 1992).

3. Altman, L. (ed.). *Large Scale Integration.* Electronic Book Series, New York: McGraw-Hill Publications, 1976.

4. Muroga, S. *VLSI System Design, When and How to Design Very-Large-Scale Integrated Circuits.* New York: John Wiley & Sons, 1982.

5. Smith, M. J. "More Logic Synthesis for ASICs" *IEEE Spectrum* (November, 1992), pp. 44–48.

6. Hodges, D. A., and H. G. Jackson. *Analysis and Design of Digital Integrated Circuits.* New York: McGraw-Hill, 1988.

7. Roth, C. H., Jr. *Fundamentals of Logic Design.* St. Paul, MN: West Publishing, 1992.

8. Wakerly, J. F. *Digital Design Principles and Practices.* Englewood Cliffs, NJ: Prentice Hall, 1994.

REVIEW QUESTIONS

1. What are the steps in the design of a digital system?

2. What is the importance of analysis at each step of design?

3. How do computer simulation and analysis tools aid in the design of digital systems?

4. What causes performance deterioration when many basic logic gates are interconnected to realize a function, as opposed to a single block realization of a function?

5. How does overall chip size affect the design of VLSI and ASIC systems?

6. What is regeneration of logic levels? Why is it necessary?

7. How are input and output logic levels determined for an inverter?

8. How must the input and output logic levels be related to ensure logic regeneration?

9. Define logic swing and transition width of an inverter.

10. What is the noise margin? How are low and high noise margins defined?

11. What are some of the possible sources of noise in a large digital circuit?

12. How is the static fan-out of a circuit determined?

13. How does the noise margin affect static fan-out?

14. What is dynamic fan-out? Why is it, in general, less than static fan-out?

15. In general, what causes switching delays in digital circuits?

16. State ways of reducing switching delays.

17. Define t_{PHL} and t_{PLH}.

18. Define delay-power product. What is its importance?

PROBLEMS

1. If 100 inverters are connected at the output in Example 1.2, calculate the driver voltage at logic high.

2. Determine the fan-out in Example 1.2 if the logic high noise margin must be at least the same as the logic low noise margin when loaded.

*3. The transfer characteristics of a noninverting buffer in the ECL family are described by the following equation:

$$V_o = \begin{cases} 0, & V_i \leq 0.85 \text{ V} \\ 2.6[1 - 1/\{1 + \exp((V_i - 1)/V_T)\}], & 0.85 \text{ V} \leq V_i \leq 1.2 \text{ V} \\ 2.6 \text{ V}, & V_i \geq 1.2 \text{ V} \end{cases}$$

Calculate the logic levels and noise margins. Use $V_T = 26 \text{ mV}$.
Note: Since gain is positive in the transition regions for noninverting circuits, use slope = 1 at logic transitions.

4. Output of a TTL inverter driving five identical inverters may be modeled by a 3.6 V source in series with a 500 Ω resistance at logic high output, and by a 0 V source in series with a 20 Ω resistance at logic low output. If the total load capacitance is 20 pF, determine the time required for the output to change from (*a*) 0 to 3 V, and (*b*) 3.6 V to 0.4 V.

5. *Experiment:* Characteristics of an inverter from the 7400 family of TTL gates are studied in this experiment.
   a. Connect the input of one of the six inverters in the 7400, 74C00, or 74LS00 TTL family of hex inverters to a variable supply. Measure the inverter output voltage as the input is varied from 0 to 5 V. Obtain a graph of the static voltage transfer characteristic of the inverter. (For a quick look at the characteristic, apply a repetitive ramp voltage (0 to 5 V with 1 ms period) at the input and display the output on an oscilloscope using the *X-Y* mode.) Determine the logic levels and the noise margins.
   b. By cascading an odd number of inverters to form a ring oscillator as shown in Fig. 1.9, determine the average propagation delays of the inverters.
   c. Compare your experimental results of logic levels, noise margins, and propagation delays with the values specified by the manufacturer in the data sheets, (Note the test conditions under which the manufacturer's results are specified.)

2

INTRODUCTION TO SEMICONDUCTORS AND JUNCTION DIODES

INTRODUCTION

We begin the study of digital microelectronic circuits by examining the electrical behavior of semiconductors. In this chapter we analyze the characteristics of the most fundamental electronic device, the junction diode. The chapter also provides an introduction to the computer modeling of the diode, and discusses the applications of different forms of diodes in switching circuits.

2.1 CONDUCTION IN SOLIDS

The flow of current in solids is due to the motion of electrons. Although electrons are negatively charged particles, by convention, current flow is considered positive in the direction of positive charge flow. Hence, positive current direction in solids is in the direction opposite to the direction of the flow of electrons. By definition, current I in amperes (A) in a given direction is the rate of positive charge flow per unit time, given by

$$I = \frac{\Delta Q}{\Delta t} \qquad\qquad (2.1)$$

where ΔQ is the charge, in coulomb (C), transferred in time Δt seconds (s). With the charge of an electron at approximately -1.6×10^{-19} C, therefore, it takes the flow of approximately 6.24×10^{15} electrons per second to constitute a current of 1 mA in a given direction.

Recall from basic chemistry that matter is composed of atoms. In the Bohr model, each atom consists of positively charged protons and neutrons (along with other subatomic particles) in the nucleus, surrounded by negatively charged, orbiting electrons. The number of electrons orbiting the nucleus is the same as the number of protons. With the charge of a proton equal in magnitude to that of an electron, the atom as a whole is electrically neutral.

Atoms in metals such as aluminum and silver are arranged spatially in a systematic array to form a crystal structure. Each of the orbiting electrons in the atom possesses kinetic energy due to motion and potential energy due to Coulomb force between the positive nucleus and the negative electron. In a single atom in the Bohr model, electrons can only have distinct and well-separated energies called *energy states*. However, close proximity of other atoms in a solid affects the potential energy of each orbiting electron. In a crystalline solid, the orderly arrangement of atoms and their interactions cause a shift in energy levels; that is, each energy state in the isolated atom is split as many times as the number of atoms. With the multitudinous numbers of atoms in crystalline solids (on the order of $10^{23}/cm^3$), the split energy levels form two *energy bands* separated by an energy gap, or *bandgap,* as shown in Figure 2.1a. Electrons are not allowed to possess energy in the bandgap; hence, the bandgap is known as the forbidden band. The energy bands, which are essentially continuous because of the large number of closely spaced levels, belong to the entire combination of atoms in the solid instead of any particular individual atom. Since electrons in each atom exist in definite orbits at definite distances from the nucleus, those that are farthest from the nucleus have the highest energy; hence, they require the least amount of external energy to be freed from the influence of the nucleus.

The highest energy band containing electrons for a given atom is the *valence band*. Electrons with energies in the valence band, that is, those in the outermost orbit, are bound to neighboring host atoms and are called *valence electrons*. Valence electrons strongly determine the chemical properties of the atom. If an electron in the bound valence band acquires sufficient energy to overcome the bandgap, it moves into the next higher energy level, which is the *conduction band*. Since an electron in the conduction band is no longer bound to any particular atom, it can move freely in the crystal and contribute to current flow. In metals, the discrete energy levels of electrons are so closely spaced that they appear to be continuous and the bandgap is small, or the two bands overlap (Figure 2.1b). Hence, a small amount of energy from an applied electric field can cause movement of electrons from the valence band, or electrons from the conduction band, and these excited electrons contribute to current flow. The bandgap energy, which is the energy needed to free electrons from the valence

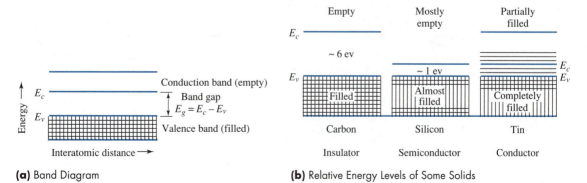

(a) Band Diagram

(b) Relative Energy Levels of Some Solids

Figure 2.1   Energy bands in crystalline solids

band and move them into the conduction band, is on the order of an electron-volt in metals.[1] In the case of insulators, the valence band is full (i.e., electrons occupy all allowable orbital distances and hence allowable energy states), while the conduction band is empty. The bandgap is so large, typically above 3 eV at room temperature, that high levels of external energy must be applied to move electrons from the valence band and cause conduction. At high applied energy, however, irreversible dielectric breakdown may occur during conduction in the insulator. Table 2.1 gives comparative bandgap energies for some common materials.

Table 2.1   Bandgap energy at 300 K

Material	Bandgap, eV
Silicon	1.12
Germanium	0.67
Gallium arsenide	1.42
Indium phosphide	1.35
Carbon (crystalline diamond)	6
Silicon dioxide	9

Electric current in a metal depends on the velocity of electrons. At room temperature, many electrons acquire sufficient energy to enter the conduction band. These conduction electrons are in constant random motion within the crystal lattice. Using the kinetic theory of gases, the mean velocity of these electrons—the thermal velocity—under thermal equilibrium is approximately 10^5 m/s at room temperature. Tightly bound atoms that are freed of their valence electrons are called *ions;* these ions vibrate about their neutral positions. As a result, according to the electron gas model, there are elastic and inelastic collisions between the sea of free electrons and the almost stationary ions. Because there is no net flow of electrons per unit time in any given direction, the bouncing off of the electrons does not contribute to current. When we apply a uniform electric field of intensity E (measured in volts/meter) in addition to the random motion the electrons acquire a component of steady velocity in the direction opposite to the field. The additional velocity due to the applied field results in a net flow of electrons, which constitutes positive current in the direction of the field. The velocity in the direction of $-E$ is called the *drift velocity,* and is given by

$$v = \mu E \qquad \qquad \textbf{(2.2)}$$

[1] An electron-volt (eV) represents an energy of 1.6×10^{-19} joule(J) which is the energy associated with the movement of an electron (of charge -1.6×10^{-19} C) resulting in a potential of 1 V.

where the proportionality constant μ measured in m^2/V-s is the *mobility* of electrons. Mobility depends on the physical structure of the material. Because of the statistical nature of the random motion of the electrons and the energy transfer during collisions with the ions, only average drift velocity and mobility are defined. Mobility for copper, for example, is 3.2×10^{-3} m^2/V-s; for aluminum, it is 4.15×10^{-3} m^2/V-s.

If n is the number of net flow of free electrons passing a unit volume at a drift velocity of v, the *drift current density* J is given by

$$J = nev \text{ A/m}^2 \qquad (2.3)$$

where e is the charge of an electron. Using Equation (2.2), therefore,

$$J = ne\mu E \qquad (2.4)$$

Equation (2.4) is a description of Ohm's law, $J = \sigma E$, where

$$\sigma = ne\mu \qquad (2.5)$$

in siemens/meter (S/m or $(\Omega\text{-m})^{-1}$) is the *conductivity,* and its reciprocal, $1/\sigma$ in m/S (or Ω-m) is the *resistivity* of the material. Metals are characterized by large conductivity and hence low resistivity. Electrical insulators such as silicon dioxide, on the other hand, have large resistivities. Table 2.2 shows typical conductivity and resistivity values for some common materials.

Table 2.2   Electrical conductivity and resistivity at 300 K

Material	Conductivity (S/m)	Resistivity (Ω-m)
Conductors		
Silver	62×10^6	1.6×10^{-8}
Copper	59×10^6	1.7×10^{-8}
Gold	43×10^6	2.3×10^{-8}
Aluminum	38×10^6	2.6×10^{-8}
Insulators		
Bakelite	10^{-12} to 10^{-11}	10^{11} to 10^{12}
Glass	10^{-14} to 10^{-11}	10^{11} to 10^{14}
Silicon dioxide	10^{-14} to 10^{-12}	10^{12} to 10^{14}
Quartz	10^{-18} to 10^{-17}	10^{17} to 10^{18}
Semiconductors		
Intrinsic (pure) germanium	2.2	0.45
Intrinsic (pure) silicon	4.3×10^{-4}	2.3×10^3
Gallium arsenide	1.3×10^{-6}	7.7×10^5

2.2   CONDUCTION IN SEMICONDUCTORS

A semiconductor is a solid whose electronic properties lie in the intermediate range between conductors and insulators (see Table 2.2). Silicon (Si) and germanium (Ge) are the most commonly used single-element semiconductors. Silicon is the second most abundant element present on the earth's crust (after oxygen). In addition to its abundance, silicon has certain advantages over germanium in terms of electronic properties which, in turn, lead to advantages in the fabrication of microelectronic circuits. For these reasons, silicon is the most widely used semiconductor in all integrated circuits. Usage of germanium, which paralleled that of silicon in the early semiconductor devices, is now limited primarily to discrete devices, which are individually packaged diodes and bipolar junction transistors.

Two-element compound semiconductors such as gallium arsenide (GaAs) and indium phosphide (InP) are used increasingly in high-speed/high-frequency and optoelectronic applications. Note from Table 2.1 that these compound semiconductors have energy gaps comparable to those of silicon and germanium. More importantly, electron mobilities in compound semiconductors are much higher than those in silicon (Table 2.4); this is advantageous for high-speed operation. In addition, in a compound semiconductor, energy is converted in the form of light (instead of heat as in silicon or germanium) when electrons return from the conduction band to the valence band; conversely, carriers for current conduction can be generated in compound semiconductors by absorption of light energy. Hence, compound semiconductors are also used in light-emitting and light-detecting devices, as we see in Section 2.9.

2.2.1   CHARGE CARRIERS IN SEMICONDUCTORS

Pure form of silicon, known as *intrinsic silicon,* has impurity atoms of fewer than one per 10^{10} silicon atoms. It is obtained by purifying silica (SiO_2). A silicon atom has four valence electrons and is located in the fourth column of the periodic table (Table 2.3). As illustrated by Figure 2.2, it can exist in three different forms: the amorphous form has irregular atomic structure, or short-range order; the crystalline form has perfect long-range order, with repeated and

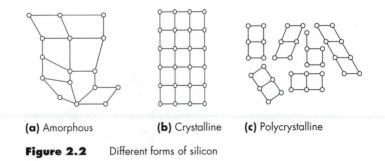

(a) Amorphous   **(b)** Crystalline   **(c)** Polycrystalline

Figure 2.2   Different forms of silicon

predictable structure; and the polycrystalline form has small crystalline grains separated by disordered zones. Each form gives rise to different electrical properties, and all three forms are used in microelectronic devices. Amorphous silicon is used in specialized applications such as solar cells and liquid-crystal displays. With electrical properties similar to a metal, polycrystalline silicon is used for making ohmic interconnections and as gate electrodes in MOSFET devices (Chapter 6). A vast majority of devices, however, are made of crystalline silicon because, by adding impurity atoms, we have excellent ability to control its electrical properties. Hence we focus our attention on the crystalline silicon.

Table 2.3   Semiconductor elements in the periodic table

	Column III	Column IV	Column V
	+3	+4	+5
Element Symbol	Boron B	Carbon C	Nitrogen N
Atomic Atomic	5 10.81	6 12.01	7 14.01
Number Weight			
	Aluminum Al	Silicon Si	Phosphorus P
	13 26.98	14 28.09	15 30.97
	Gallium Ga	Germanium Ge	Arsenic As
	31 69.72	32 72.59	33 74.92
	Indium In	Tin Sn	Antimony Sb
	49 114.82	50 118.69	51 121.75

With four valence electrons, every silicon atom forms a strong bond with four neighboring atoms at equal angular position. Figure 2.3a shows this tetrahedral bonding; its structure is the same as the diamond lattice of crystalline carbon. The *covalent bond*, or sharing of a valence electron with each adjacent atom, is represented in two dimensions in Figure 2.3b. The core with +4 denotes the immobile ion, having a positive charge of four times the electronic charge magnitude. At temperatures close to absolute zero, the valence electrons in each silicon atom are tightly held by the neighboring silicon atoms and, hence, no free carriers are available for charge transport. Consequently, a pure silicon crystal is an insulator at low temperatures. The bandgap, or the energy needed to break a covalent bond, is about 1.1 eV at 300 K.

The compound semiconductor gallium arsenide also forms a crystal lattice similar to silicon, with gallium (a trivalent element) and arsenic (a pentavalent

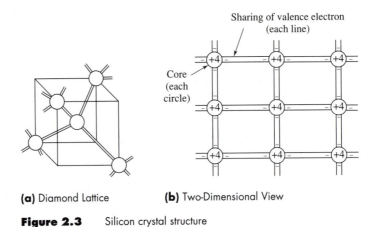

(a) Diamond Lattice   **(b)** Two-Dimensional View

Figure 2.3   Silicon crystal structure

element) atoms alternating in the lattice. Covalent bonds in gallium arsenide are formed with equal numbers of gallium and arsenic atoms, each surrounded by four atoms of the other. The compound crystal has the same electrical properties as elemental semiconductors, with a bandgap of about 1.4 eV.

Considering the crystalline silicon again, at room temperature, enough energy is available in the form of thermally induced crystal-lattice vibrations to cause some of the covalent bonds to be broken. Electrons acquiring this energy are released from the valence band and move up to the conduction band.[2] These freed electrons can now contribute to current when an external electric field is applied, much like in a metal.

The vacancy created by the absence of an electron in a broken covalent bond is called a *hole.* The significance of a hole is that it contributes to current much like an electron. As shown in Figure 2.4, an electron freed from a covalent bond and moved into the conduction band leaves behind a positive charge in the void created. Thus, a hole has a positive charge equal in magnitude to the charge of an electron. If another electron is dislodged from a neighboring covalent bond, as when an external electric field is applied, it can move into the vacancy available, so that the hole then appears at the newly broken bond. This is equivalent to the movement of a positively charged particle in the opposite direction of the electron movement. Note that the newly dislodged valence electron does not acquire sufficient energy to become a conduction electron. Hence the hole movement contributes to positive current *in addition to* the current due to the movement of freed conduction electrons.

Although a hole (or a valence electron) is not a classical particle, in terms of current conduction it may be thought of as a positive *charge carrier,* similar to an electron, which is a negative charge carrier. Since the hole movement occurs because of the broken bonds (but not freed electrons), it is slower than

---

[2] Energy can also be supplied in the form of light by irradiating the semiconductor with a light source. This method of freeing charge carriers is used in photodetector applications (Section 2.9).

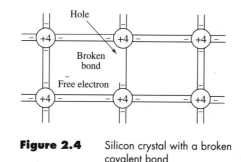

Figure 2.4   Silicon crystal with a broken covalent bond

conduction electrons. (Note that the movement of holes is due to the movement of valence electrons, which are tightly bound; hence, for the same applied electric field, holes appear to have more mass causing them to accelerate less than electrons.) Therefore, the mobility of holes is less than that of electrons. Both electron and hole mobilities, however, depend on the material. In addition, mobility is also dependent on temperature, the concentration of the impurity, and the applied electric field. Table 2.4 lists the mobilities in some of the common semiconductors.

Table 2.4   Electron and hole mobilities in semiconductors at 300 K

Material	Electron Mobility (m²/V-s)	Hole Mobility (m²/V-s)
Silicon	0.145	0.045
Germanium	0.39	0.19
Gallium arsenide	0.85	0.04

With two independent charge carriers—holes in the valence band and electrons in the conduction band—having opposite charges and moving in opposite directions, we would expect higher conductivity for a semiconductor; at room temperature, however, conductivity of a pure (intrinsic) semiconductor is low. We will now examine why.

2.2.2 CONDUCTIVITY OF SEMICONDUCTORS

If μ_n and μ_p are the mobilities of electrons and holes, respectively, the conductivity of a semiconductor is, from Equation (2.5),

$$\sigma = (n\mu_n + p\mu_p)e \qquad \textbf{(2.6)}$$

and the drift current density is

$$J = \sigma E = (n\mu_n + p\mu_p)eE \tag{2.7}$$

where n and p are the number of free electrons and holes available, or the concentrations of charge carriers per unit volume. Drift current (by holes or electrons) due to an applied electric field arises in field-effect transistors.

In an intrinsic semiconductor, an equal number of positive and negative charge carriers—holes and free electrons—are generated at any ordinary termperature above absolute zero. The *generation rate* of free electron-hole pairs of carriers in a pure semiconductor depends on the temperature and the bandgap energy of the material. Just as electrons receiving energy are freed and moved from valence band to conduction band, there are other electrons that are losing energy and falling back to the valence band. The reverse process of generation, in which a free electron falls from the conduction band and occupies a hole, is known as *recombination*. The repairing of broken bonds during recombination releases energy and annihilates both carriers. Clearly the recombination rate is proportional to the concentration of free electron-hole pairs available. At a constant temperature the two processes of generation and recombination are balanced and the semiconductor is in the state of *thermal equilibrium.*

The distribution of electrons with allowed energies (that is, energies outside the forbidden energy gap) in thermal equilibrium is governed by the *Fermi-Dirac probability distribution function.* This function, denoted by $f(E)$, represents the probability that an *allowable energy state E* will be occupied by the electron at absolute temperature T as

$$f(E) = \cfrac{1}{1 + \exp\!\left(\cfrac{E - E_f}{kT}\right)} \tag{2.8}$$

The function $f(E)$ is known as the *Fermi function,* in which E_f is a reference energy and k is the Boltzmann constant ($k = 8.62 \times 10^{-5}$ eV/K). E_f is called the *Fermi level* or *Fermi energy;* it is a characteristic constant of the material. Note that the probability of an empty energy state E, or the probability of finding a hole at state E, is $1 - f(E)$. With an equal number of valence band holes and conduction band electrons at ordinary temperatures, the probability that the state at energy $E = E_f$, the Fermi level, is filled by an electron (or hole) is 0.5 (Figure 2.5a). At absolute zero temperature ($T = 0$), $f(E) = 1$ for $E < E_f$, and $f(E) = 0$ for $E > E_f$. This shows that at $T = 0$ all allowed energy states below E_f are occupied and no electron is available at energies above E_f. At T above zero, there is some probability, $f(E)$, that an electron is at state $E > E_f$, and, correspondingly, there is the probability $1 - f(E)$ that no electrons are available with energies below E_f. As T increases, $f(E)$ also increases for $E > E_f$; hence, more electrons are available with higher energies. Combining the distribution function with the energy bands for intrinsic semiconductors (Figure 2.5b), it is clear that E_f lies in the forbidden band with $E_v < E_f < E_c$.

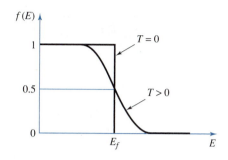

(a) Fermi Function

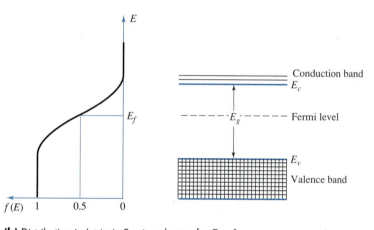

(b) Distribution in Intrinsic Semiconductors for $T > 0$.

Figure 2.5   Fermi distribution function and energy bands

Figure 2.5b shows small values for the probability $f(E)$ of finding an electron above the valence band for $T > 0$. Since energies in the range $E_v < E < E_c$ are not allowed, however, no electron can be found in the forbidden band. Note that the probability of a hole in the valence band, $1 - f(E)$, is small.

We calculate the number of electrons in the conduction band using the energy states at the bottom of the conduction band and the Fermi function. Denoting the number of effective energy states at the bottom of the conduction band by N_c, the intrinsic concentration of electrons n_i at temperature T is given by

$$n_i = N_c f(E_c) = \frac{N_c}{1 + \exp\left(\dfrac{E_c - E_f}{kT}\right)}$$

where the subscript i denotes intrinsic semiconductor. With $kT = 8.62 \times 10^{-5} \times 300 \approx 0.026 \, \text{eV}$ at 300 K and noting that $E_c > E_f$ by several kT, we can approximate the above equation as

$$n_i \approx N_c \exp\left[\frac{-(E_c - E_f)}{kT}\right] \tag{2.9a}$$

Using the expression $N_c = 2(2\pi m_n^* kT/h^2)^{3/2}$ for the number of effective energy states [Reference 6], the intrinsic concentration of electrons is given by

$$n_i = 2(2\pi m_n^* kT/h^2)^{3/2} \exp\left[\frac{-(E_c - E_f)}{kT}\right] \tag{2.9b}$$

where m_n^* is the effective mass of electrons and h is the Planck's constant (6.626×10^{-34} J-s, or 4.136×10^{-15} eV-s). Equation (2.9) shows the strong dependence of n_i on temperature. The effective mass of an electron in the conduction band differs from the free-electron mass due to the influence of crystal forces.

In a similar manner, the intrinsic concentration of holes in the valence band is given by

$$p_i = N_v \exp\left[\frac{-(E_f - E_v)}{kT}\right] \tag{2.10a}$$

or

$$p_i = 2(2\pi m_p^* kT/h^2)^{3/2} \exp\left[\frac{-(E_f - E_v)}{kT}\right] \tag{2.10b}$$

where $N_v = 2(2\pi m_p^* kT/h^2)^{3/2}$ and m_p^* is the effective mass of a hole.

Since $n_i = p_i$, from Equations (2.9) and (2.10), we can find the product of intrinsic concentrations by

$$n_i p_i = n_i^2 = p_i^2 = N_c N_v \exp\left[\frac{-(E_c - E_v)}{kT}\right]$$

or

$$n_i p_i = n_i^2 = p_i^2 = N_c N_v \exp\left(\frac{-E_g}{kT}\right) \tag{2.11a}$$

where $E_g = E_c - E_v$ is the bandgap energy, as shown in Figure 2.1. Hence,

$$n_i = p_i = \sqrt{N_c N_v} \exp\left(\frac{-E_g}{2kT}\right) \tag{2.11b}$$

The above equation, which describes the generation of an equal number of free electrons and holes—as well as the recombination of these carriers—in equilibrium is known as the *mass action law*. The product $n_i p_i$ is a constant for a given material in equilibrium at a given temperature. For silicon at 300 K using

$m_n^* = 1.1\, m_0$, $m_p^* = 0.56\, m_0$ (where $m_0 = 9.1 \times 10^{-31}$ kg is the electron rest mass), and $E_g = 1.12$ eV, the intrinsic concentration becomes

$$n_i = p_i \approx 4.8 \times 10^{21} T^{(3/2)} \exp\!\left(\frac{-5797 E_g}{T}\right) \Big/ m^3 \qquad \textbf{(2.11c)}$$

At 300 K (27°C) this equation gives $n_i = p_i \approx 1.5 \times 10^{16}/m^3$. Since silicon has about 5×10^{28} atoms/m^3, we note that an electron-hole pair is available for every 3.3×10^{12} atoms/m^3 by thermal generation. Because of this relatively low number of free carriers pure silicon is a poor conductor with a conductivity of 4.3×10^{-4} S/m at room temperature. Conductors such as copper, on the other hand, have conductivities that are 11 orders of magnitude higher than silicon.

Nevertheless, from Equation (2.11) it is clear that the number of carriers, and hence the conductivity, of a semiconductor can be increased manifold by raising its temperature. Conductivity of intrinsic silicon at 60°C, for example, increases to 29.3×10^{-4} S/m. This is due to increased concentration of the thermally generated electron-hole pairs from $1.5 \times 10^{16}/m^3$ at 27°C to $10^{17}/m^3$ at 60°C. This property of a semiconductor, which shows an exponential rise in conductivity (of approximately 6 per cent per degree centigrade of temperature rise for silicon), is used to advantage in temperature-sensitive devices known as *thermistors*.

2.3 DOPED OR EXTRINSIC SEMICONDUCTORS

A marked increase in the conductivity of an intrinsic semiconductor can be obtained by the addition of a small, carefully controlled number of atoms of a second material. The *impurity atoms*, or *dopants*, may replace 1 in 10^7 silicon atoms (with a dopant concentration on the order of 10^{21} atoms/m^3) to 1 in 100 silicon atoms (with a dopant concentration of 10^{26} atoms/m^3). The resulting semiconductor is the *doped*, or *extrinsic semiconductor*, which is the basic material used in microelectronic devices. By the choice of doping element, the number of free electrons or holes available as mobile charge carriers can be increased. The addition of such a relatively small number of impurity atoms compared with the number of atoms in intrinsic silicon, however, means that the material retains all the chemical properties of the otherwise pure silicon. As a result of doping, an extrinsic semiconductor has equilibrium concentrations of electrons and holes different from their intrinsic levels.

2.3.1 DONOR AND n-TYPE DOPING

Adjacent to silicon, a tetravalent element in the periodic tables (Table 2.3), are the elements phosphorus, arsenic, and antimony, all of which have five valence electrons. When an atom from such a pentavalent element replaces a silicon

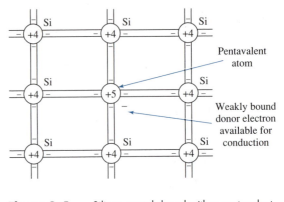

Figure 2.6   Silicon crystal doped with a pentavalent element: n-type doping

atom, the crystal structure shown in Figure 2.6 results. Four of the five valence electrons of the dopant atom form covalent bonds with neighboring silicon atoms. The fifth electron is so weakly bound to the impurity element that it requires very little energy—in the range of 0.039 eV for antimony to 0.049 eV for arsenic at room temperature—to become a free electron available for conduction. Since the doping element (arsenic, antimony, or phosphorus) donates an electron to intrinsic silicon, it is called a *donor impurity,* or simply *donor.* With the low binding energy, all of the donor electrons lie practically in the conduction band. The availability of these negative charge carriers, or free electrons, due to donor atoms in an otherwise neutral silicon crystal makes the doped semiconductor an *n-type semiconductor.*

When the fifth electron from a donor atom moves into the conduction band, the donor atom is locked in position surrounded by silicon atoms with covalent bonds. Hence, the positively charged donor ion does not contribute to current. Additionally, n-type doping decreases the number of holes from their intrinsic level. The decrease in hole concentration occurs because of the availability of a large number of free electrons some of which recombine with the intrinsic (thermally generated) holes. The net result is that n-type doping increases the free electron concentration above the intrinsic level while decreasing the number of holes. Thus, electrons are the dominant or *majority carriers* and holes are the *minority carriers* in an n-type semiconductor.

2.3.2   ACCEPTOR AND p-TYPE DOPING

If a trivalent element such as boron, gallium, or indium (Table 2.3) is added to an intrinsic silicon, three of the four covalent bonds in the surrounding silicon atoms are completed by the three valence electrons of the dopant as shown in Figure 2.7. The fourth unfilled bond leaves a hole. Since a hole contributes to current by accepting an electron from a neighboring silicon atom, the impurity element is called an *acceptor.* As with the n-type doping, the negative charge created by the dopant is immobile. It is located just above the edge of the valence

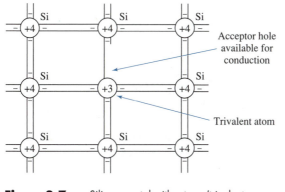

Figure 2.7   Silicon crystal with p-type (trivalent element) doping

band so that very little energy (0.045 eV for boron, for example) is required to ionize the acceptors. With net positive charge of free holes, which are the majority carriers, the semiconductor is called a *p-type* semiconductor.

2.3.3   CONDUCTIVITY OF A DOPED SEMICONDUCTOR

A semiconductor, either intrinsic or extrinsic, is electrically neutral. This charge neutrality means that the concentration of positive and negative charges must be the same. If N_D and N_A are the number of donor and acceptor atoms added, they represent the concentrations of immobile positive and negative ions, respectively. Now, if p and n represent the concentrations of holes and free electrons in the semiconductor (arising from doping and thermal generation combined), we have, to maintain charge neutrality, the fundamental equation

$$N_D + p = N_A + n \qquad \textbf{(2.12)}$$

In an n-type material, $N_A = 0$. Since the thermal generation of electron-hole pairs is independent of doping (and $N_D > 0$), there are now increased numbers of free electrons available; hence, more holes are lost by recombination than in the intrinsic case. As a result, we expect $n > n_i$ and $p < p_i$ in an n-type material. Similarly, for a p-type material with $N_D = 0$, we expect $n < n_i$ and $p > p_i$.

A second fundamental equation relates the concentrations of p and n under equilibrium conditions. With less than 1 percent of dopant atoms in the doped semiconductor, the thermal generation rate remains virtually the same as in the instrinsic material. Therefore, the product of the number of free electrons and holes is the same as n_i^2; that is,

$$pn = n_i^2 \qquad \textbf{(2.13)}$$

regardless of doping. This generalization of the mass action law from Equation (2.11) is significant when we are determining the concentration of one of the carriers given the other in a doped semiconductor.

From Equations (2.12) and (2.13), for a material doped with N_D and N_A, we now can write

$$N_D + \frac{n_i^2}{n} = N_A + n$$

or

$$n = \frac{(N_D - N_A)}{2} + \sqrt{\left(\frac{N_D - N_A}{2}\right)^2 + n_i^2} \qquad \textbf{(2.14)}$$

(Note that the positive sign is used for the second term so that for an undoped material, $n = n_i$.) Hence, for an n-type material, or one in which $N_A \ll n_i$ and $N_D \gg n_i$, we find the concentration of electrons to be

$$n \approx N_D$$

or

$$n_n \approx N_D \qquad \textbf{(2.15)}$$

where the subscript denotes n-type doping. Similarly, the concentration of holes in a p-type semiconductor is found to be $p \approx N_A$ or, more specifically,

$$p_p \approx N_A \qquad \textbf{(2.16)}$$

From Equation (2.15) or Equation (2.16), the density of holes (electrons) in an n-doped (p-doped) semiconductor is calculated using the generalized mass action law, Equation (2.13). Remember that Equations (2.15) and (2.16) neglect the thermal generation of electrons (holes) in n-doped (p-doped) semiconductors. This is justified in practical devices with doping densities on the order of $10^{20}/m^3$ or more while silicon has an intrinsic density of $n_i \approx 1.5 \times 10^{16}/m^3$ at room temperature.

In addition to changing the carrier concentrations, the presence of a large number of carriers in a doped semiconductor causes decreased mobility. Accordingly, the conductivity of a doped semiconductor increases nonlinearly with doping level, temperature, and applied electric field. Decrease in mobility (of about 30 percent in silicon at 300 K) occurs only at doping levels in excess of 10^{21} per m^3. At lower doping levels the mobilities are essentially constant.

Example 2.1

A silicon bar is doped with phosphorus atoms at a concentration level of $10^{20}/m^3$. What is the conductivity of the doped semiconductor at 300 K? (Remember that silicon has an intrinsic density of $n_i \approx 1.5 \times 10^{16}/m^3$ at this temperature.) Calculate the conductivity for the same number of indium atoms instead of phosphorus. If the doping level is increased to $10^{25}/m^3$ of phosphorus, which changes the mobility μ_n to 0.012 $m^2/V\text{-s}$, what is the conductivity?

Solution

With phosphorus donor atoms at $N_D = 10^{20}/m^3$, we find the concentration of electrons to be

$$n \approx N_D = 10^{20}/m^3$$

From the mass action law, Equation (2.13), we find for the concentration of holes,

$$p = \frac{(1.5 \times 10^{16})^2}{10^{20}} = 2.25 \times 10^{12}/\text{m}^3$$

Clearly, with eight orders of reduction in the hole concentration relative to the electrons, conductivity is primarily determined by the electrons; thus, using Equation (2.5),

$$\sigma = \mu_n en = 0.145 \times 1.6 \times 10^{-19} \times 10^{20} = 2.32\,\text{S/m}$$

Compared with the conductivity of intrinsic silicon at 4.3×10^{-4} S/m, we see that a light n-type doping increases σ by four orders.

Indium, with its three valence electrons, is an acceptor dopant, so that $p \approx N_A = 10^{20}/\text{m}^3$. Hence

$$n = 2.25 \times 10^{12}/\text{m}^3$$

and

$$\sigma = \mu_p ep = 0.045 \times 1.6 \times 10^{-19} \times 10^{20} = 0.72\,\text{S/m}$$

Again, a light doping with acceptor atoms raises the conductivity σ markedly. Due to the low mobility of holes, however, the conductivity of p-type silicon is lower than that of n-type silicon for the same level of concentration.

Recall that the density of silicon is approximately 5×10^{28} atoms/m^3. Thus, a doping concentration of $10^{20}/\text{m}^3$ corresponds to replacing 1 out of every 100 million silicon atoms by the dopant atom. This low level of impurity concentration alters electrical conductivity significantly, while retaining the chemical properties of pure silicon.

At $n \approx N_D = 10^{25}/\text{m}^3$, μ_n changes to 0.012 m^2/V-s. Therefore, the new conductivity is

$$\sigma = 0.012 \times 1.6 \times 10^{-19} \times 10^{25} = 1.92 \times 10^4\,\text{S/m}$$

This example illustrates the attainment of a high value of conductivity for a doped semiconductor material—comparable to that of metallic conductors—with heavy doping. Therefore, semiconductor devices use heavy donor doping (on the order of $10^{25}/\text{m}^3$), denoted by n$^+$, to make ohmic contacts (i.e., contacts offering negligible resistance to current flow); the internal, "active" regions (where device actions take place) have light to medium doped (10^{21} to $10^{24}/\text{m}^3$) semiconductor material.

2.3.4   DIFFUSION CURRENT

In addition to drift current due to an applied electric field, semiconductor devices exhibit another important component of current: *diffusion current*. This current exists when there is a spatial variation in the number of available free carriers.

Current in junction diodes and transistors is primarily due to the diffusion of minority carriers. As is the case with particle diffusion, such as a whiff of perfume in a room or a drop of dye in a glass of water, diffusion current arises due to both the random motion of electrons (or holes) and their concentration gradient. Figure 2.8(a) and (b) represents a semiconductor in which the hole (free electron) density is nonuniform along the x axis; that is, the carriers are more at $x = 0$, and they decrease with increasing x. Since the positively charged holes move (diffuse) from a region of high concentration ($x = 0$) to a region of low concentration ($x > 0$), positive current exists in the direction of $dn_p/dx < 0$. Electrons with lower density in the positive x direction diffuse toward $x > 0$ and hence positive current direction is the same as the direction of $dn_n/dx > 0$. Since diffusion is a natural result of random thermal motion and scattering from crystal lattice to produce a uniform distribution of carriers, a greater concentration gradient causes a larger current. Therefore, diffusion current densities for electrons and holes are proportional to the concentration gradient as given by

$$J_p = -eD_p \frac{dn_p}{dx} \qquad\qquad\text{(2.17a)}$$

$$J_n = eD_n \frac{dn_n}{dx} \qquad\qquad\text{(2.17b)}$$

for the one-dimensional case. The constants of proportionality D_p and D_n, in m²/s, are referred to as the hole and the electron *diffusion coefficients,* or *diffusivities.* Although we considered charge density variation only in the x direction, in general, for spatial variation of density the slope is replaced by the gradient ∇_p or ∇_n.

The diffusion current given by Equation (2.17) continues as long as a nonzero concentration gradient is present. Observe that the diffusion of charge

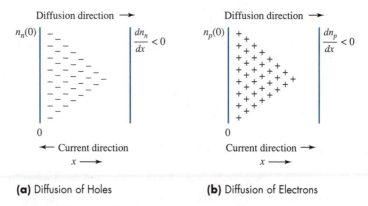

(a) Diffusion of Holes   **(b)** Diffusion of Electrons

Figure 2.8   Diffusion current

carriers is not due to a force of repulsion between like charges; it is caused by the statistical phenomenon of thermal motion like mobility. Mobility and diffusivity are related by the Einstein equation:

$$\frac{D_p}{\mu_p} = \frac{D_n}{\mu_n} = \frac{kT}{e} \qquad (2.18)$$

where k is the Boltzmann constant.

2.3.5 TOTAL CURRENT DENSITY

When an electric field E is applied to a semiconductor device with a positive charge (hole) gradient of ∇_p, both drift and diffusion currents are present. The total hole current density, therefore, is

$$J_p = e(p\mu_p E - D_p \nabla_p) \qquad (2.19)$$

For electrons, the corresponding total current density is

$$J_n = e(n\mu_n E + D_n \nabla_n) \qquad (2.20)$$

Equations (2.19) and (2.20) are known as the *transport equations* for holes and electrons, respectively.

If both electrons and holes are present in the device, the total current density is

$$J = J_p + J_n \qquad (2.21)$$

In most devices one of the two terms, with either holes or electrons, dominates. In junction diodes (Section 2.4) and npn transistors (Chapter 3), for example, the diffusion component due to externally injected electrons contributes to almost all of the current. In both n-channel and gallium arsenide field-effect devices, on the other hand, currents are due entirely to the drift of electrons caused by an externally applied electric field.

2.4 pn JUNCTION DIODE

If both donor and acceptor impurities are present in two physically separate regions of a wafer of a single-crystal semiconductor, we have a *pn junction*. This junction, which is located in the vicinity of the boundary between the p-type and the n-type regions (the metallurgical junction), is the basic building block in nearly all microelectronic devices.

A semiconductor device that is formed by a single pn junction is called a *junction diode,* or simply *diode*. The diode has useful electrical properties and many applications. The two terminals of the diode, which make ohmic contacts with the p region and the n region, are referred to as the *anode* and the *cathode,* respectively. Figure 2.9 shows the structure and the circuit symbol of a pn junction diode. The arrowhead symbol, which goes from the p region (anode) to the n region, shows the direction of positive current, as we shall see later.

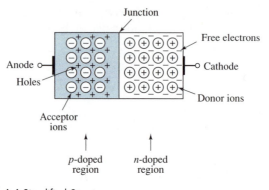

(a) Simplified Structure

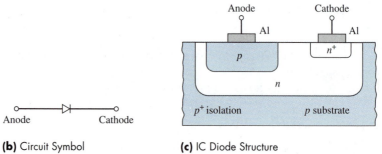

(b) Circuit Symbol   **(c)** IC Diode Structure

Figure 2.9   pn Junction diode

   Junction diodes are fabricated using the process of alloying, growing, or diffusion, during which the n and the p doping levels and the geometry are precisely controlled. A discrete diode is formed by introducing a controlled amount of p-type impurities within a single crystal of an n-type semiconductor material. Diodes and other devices in integrated circuits have a common body, or *substrate,* of relatively heavily doped p- (or n-) type silicon, on which a lightly doped n-type region (cathode) is grown. (Recall that heavy doping is used to make low resistance contacts.) The p-type region (anode) is formed on this n region by diffusion. To isolate the diode from other devices on the same substrate, the active n (cathode) and the p (anode) regions are surrounded by heavily doped p regions. Contacts to the anode and the cathode are made via aluminum bonding pads.

   The n-type region of the diode has a large concentration of mobile electrons (the majority carriers) freed from the donor atoms, and a small amount of thermally generated holes (the minority carriers). Likewise, in the p-region there is a large concentration of mobile holes (majority carriers) from the acceptor atoms and a small amount of thermally generated free electrons (minority carriers). Although practical diodes exhibit a somewhat gradual variation in the impurity level near the metallurgical junction, for simplified analysis we use the ideal diode, and assume an abrupt transition in the concentration level from p to n. In addition, we consider the flow of charge carriers in a diode as one-dimensional between the p and n regions. With two types of carriers, the terminal current, in general, is given by the current density equations, Equations (2.19), (2.20), and (2.21). We now consider characteristics of the diode with the anode and the cathode left open and with a voltage applied between them.

2.4.1   OPEN-CIRCUIT BEHAVIOR OF pn JUNCTION DIODE

When a pn junction is formed, a large concentration gradient exists at the boundary between the n-type and the p-type regions; that is, the boundary is assumed abrupt. This causes diffusion of the mobile carriers across the junction and subsequent recombination with the free carriers on the other side. Consequently, uncovered bound negative charges (immobile ions) are left on the p-region close to the junction; similarly, uncovered bound positive charges are left on the n-region close to the junction. Figure 2.10a illustrates the bound positive and negative charges in the n and the p regions. In this figure, the impurity ions (represented by encircled + and −) and carriers are shown in each region for the case with more donor atoms than acceptors (i.e., the density of donors is greater than the density of acceptors, $N_D > N_A$). The region close to the metallurgical junction which is depleted of free charge carriers is called the *carrier depletion region* or *space charge region*. Since the depletion region has bound positive charges on the n-side and bound negative charges on the p-side, an electric field is established near the boundary. This field is created instantaneously when the pn junction is fabricated. The potential difference across the depletion region, which acts as a barrier for further diffusion, is known variously as *barrier potential, diffusion potential, contact potential, built-in voltage,* or *equilibrium voltage.* Barrier potential, denoted by V_0, is about 1 V for most pn junctions. In thermal equilibrium, the diffusion of free carriers overcoming the barrier potential and the drift (due to the electric field in the depletion region) of thermally generated carriers are balanced exactly, and hence no external current exists, as must be the case with open terminals. In addition, note that the entire semiconductor is electrically neutral. It is important to note further that an attempt to measure the contact potential by placing voltmeter terminals across the junction device causes new contact potentials to form at each meter terminal and cancel V_0; that is, as an equilibrium quantity, no current flow can result from V_0. The electrostatic potential V_0, however, must be overcome by an external source in a circuit to cause current flow.

   We calculate the electric field in the depletion region and the contact potential based on our assumption of one-dimensional charge flow. The basic equations are the definition of electric field E, given by

$$E = -dV/dx \qquad \textbf{(2.22)}$$

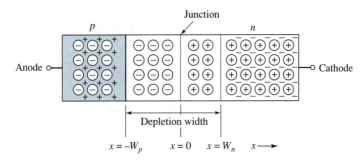

(a) Depletion Region and Uncovered Charges

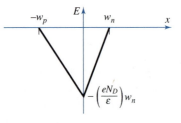

(b) Electric Field

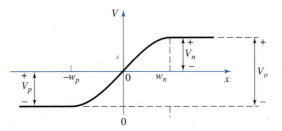

(c) Potential across the Junction

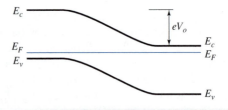

(d) Energy Levels in Equilibrium

Figure 2.10   Equilibrium of diode under open-circuit condition

and Poisson's equation, which relates the potential to the charge enclosed as

$$\frac{d^2V}{dx^2} = \frac{dE}{dx} = -\frac{\rho}{\epsilon}$$

(2.23)

where ρ is the volume charge density in C/m^3 at x and $\epsilon = \epsilon_r \epsilon_0$ is the permittivity, in farads per meter (F/m), of the region enclosing the charge. ϵ_r is the dielectric constant (relative permittivity) of the region and ϵ_0 is the permittivity of free space. Although the charge density in a semiconductor arises from both mobile and immobile electrons and holes, in the depletion region only immobile ions due to impurity atoms are significant in number. Hence, we find that the charge density is given by

$$\rho = -N_A e \quad \text{in the region} \quad -w_p < x < 0$$

and

$$\rho = N_D e \quad \text{in the region} \quad 0 < x < w_n$$

where N_A is the concentration (doping) level (per m^3) of holes in the p region and N_D that of electrons in the n region; w_p and w_n are the widths of the depletion region in the p and the n sides (see Figure 2.10a). Note that outside of the depletion region, that is for $x > w_n$ and $x < -w_p$, the field is zero because of charge neutrality.

For the p side, integrating $dE/dx = -N_A e/\epsilon$,

$$E = -\left(\frac{N_A e}{\epsilon}\right)(x + w_p) \quad \text{for} \quad -w_p < x < 0$$

(2.24)

and, for the n side, from $dE/dx = N_D e/\epsilon$,

$$E = \left(\frac{N_D e}{\epsilon}\right)(x - w_n) \quad \text{for} \quad 0 < x < w_n$$

(2.25)

The resulting electric field for $-w_p < x < w_n$ is shown in Figure 2.10b.

It is clear from Equations (2.24) and (2.25) that the maximum field intensity E_{max} occurs precisely at the metallurgical junction, $x = 0$. E_{max} is given by

$$E_{max} = -\left(\frac{eN_D}{\epsilon}\right)w_n = -\left(\frac{eN_A}{\epsilon}\right)w_p$$

(2.26)

For continuity of the electric field at $x = 0$, then,

$$N_D w_n = N_A w_p$$

(2.27)

Equation (2.27) shows that the depletion region is longer on the side where the doping is lighter; that is, for $N_A < N_D$, $w_p > w_n$. This is evidently true since the entire semiconductor is electrically neutral, so that the magnitude of space charge on either side of $x = 0$ must be the same. It follows that, with higher levels of free electrons ($N_D > N_A$) diffusing from the n region, they must travel farther in the lightly doped p region before recombination.

Because of the linearly varying electric field, with increasing magnitude from $x = -w_p$ to $x = 0$, thermally generated electrons (minority carriers) from the p side drift toward the n side. This drift current is exactly balanced by the diffusion current due to electrons (majority carriers) from the n side to the p side. In a similar manner the drift and diffusion currents due to holes are canceled. Therefore, equilibrium is reached with zero terminal current at any given temperature. The widths, w_p and w_n, of the depletion region in equilibrium are such that the total charge in the region is zero. Note, however, that there is movement of carriers within the depletion region due to drift (caused by the equilibrium voltage) and diffusion (due to concentration gradient), each canceling the other.

We calculate the equilibrium contact potential by equating the two current densities. If n_0 is the equilibrium concentration of electrons in the depletion region and dn_0/dx is the gradient, the drift and the diffusion current densities of electrons are given by

$$J_{n,\text{drift}} = -en_0\mu_n E$$

$$J_{n,\text{diff}} = eD_n\frac{dn_0}{dx}$$

Equating the two densities and using the Einstein's relation,

$$\frac{D_n}{\mu_n} = \frac{kT}{e}$$

we arrive at

$$-Edx = \left(\frac{kT}{e}\right)\left(\frac{dn_0}{n_0}\right) \tag{2.28}$$

We obtain the contact potential V_0, which is the potential difference between the n and the p regions, by integrating the above equation from $x = -w_p$ on the p side to $x = w_n$ on the n side. Note that the electron concentration is at the thermal equilibrium level of n_{p0} at $x = -w_p$, and it is n_{n0} at $x = w_n$. Hence, for the contact potential,

$$V_0 = -\int_{-w_p}^{w_n} E\,dx = \int_{n_{p0}}^{n_{n0}}\left(\frac{kT}{e}\right)\left(\frac{d_{n0}}{n_0}\right)$$

or

$$V_0 = \left(\frac{kT}{e}\right)\ln\left(\frac{n_{n0}}{n_{p0}}\right) \tag{2.29}$$

Since $n_{n0} = N_D$ and $n_{p0} = n_i^2/N_A$,

$$V_0 = \left(\frac{kT}{e}\right)\ln\left(\frac{N_D N_A}{n_i^2}\right) \tag{2.30}$$

Equation (2.30) shows that the contact potential depends on the doping levels and the intrinsic concentration, which varies with temperature. Figure 2.10c shows the potential variation in the depletion region. Contact potential, for normal levels of doping, is below 1 V at room temperature.

From the equilibrium transfer rate of electrons from one region to the other it can be shown (Problem 3) that the Fermi levels are constant and continuous throughout the two regions, as Figure 2.10d illustrates. (The continuity of the equilibrium Fermi level also applies to junctions between a metal and a semiconductor and between two dissimilar semiconductors.) With valence band holes on the p side and conduction band (free) electrons on the n side, the conduction band energy level, relative to Fermi level, is higher on the p side than on the n side. The contact potential separates the two levels in energy as eV_0.

Because of the barrier potential at equilibrium, few carriers possess energy to overcome V_0. Hence, no conduction occurs even when the diode terminals are shorted together.

From Equation (2.24) and (2.25) we can determine the potential difference between the metallurgical junction and the edges of the depletion region as

$$V_p = -\int_{-w_p}^{0} E \, dx = \left(\frac{N_A e}{\epsilon}\right) \int_{-w_p}^{0} (x + w_p) \, dx$$

or

$$V_p = \left(\frac{N_A e}{\epsilon}\right)\left(\frac{w_p^2}{2}\right)$$

and, similarly,

$$V_n = \left(\frac{N_D e}{\epsilon}\right)\left(\frac{w_n^2}{2}\right)$$

Hence we arrive at the equation for the total barrier potential,

$$V_0 = V_n + V_p = \left(\frac{e}{2\epsilon}\right)(N_A w_p^2 + N_D w_n^2) \tag{2.31}$$

Using Equations (2.27) and (2.31) we can now calculate the widths of the depletion regions, w_p and w_n, as

$$w_p = \sqrt{\frac{(2\epsilon/e)V_0}{(N_A^2/N_D) + N_A}} \tag{2.32a}$$

$$w_n = \sqrt{\frac{(2\epsilon/e)V_0}{(N_D^2/N_A) + N_D}} \tag{2.32b}$$

and the total width of the depletion region is

$$w = \sqrt{\left(\frac{2\epsilon}{e}\right)V_0\left(\frac{1}{N_A} + \frac{1}{N_D}\right)} \tag{2.33}$$

Equation (2.32) shows that the depletion region has the same width on the p and the n sides for the same levels of donor and acceptor doping. In practical diodes, however, $N_D \neq N_A$. In such cases the width depends strongly on the lighter doping, as we saw from Equation (2.27).

Example 2.2

A silicon pn junction diode has a concentration of $10^{22}/m^3$ of donor atoms and $10^{24}/m^3$ of acceptor atoms. Calculate the contact potential and the widths of the depletion regions at 300 K. Use $\epsilon_r = 11.9$ for silicon and $\epsilon_0 = 8.854 \times 10^{-12}$ F/m.

Solution

At 300 K, the intrinsic concentration of silicon is $n_i = 1.5 \times 10^{16}/m^3$, and $kT/e = 0.0258$ V.

Hence, from Equation (2.30),

$$V_0 = 0.0258 \ln\left[\frac{10^{24} \times 10^{22}}{(1.5 \times 10^{16})^2}\right] = 0.812 \text{ V}$$

Using the relative permittivity of silicon $\epsilon_r = 11.9$, and $\epsilon_0 = 8.854 \times 10^{-12}$ F/m in Equation (2.32),

$$w_p = 32.5 \times 10^{-10} \text{ m} \quad \text{and} \quad w_n = 32.5 \times 10^{-8} \text{ m}$$

2.4.2 FORWARD-BIASED DIODE

Biasing the junction of a semiconductor device refers to the application of an external dc voltage between the terminals of the n and the p regions. External bias voltage establishes dc, or quiescent, operating conditions, on which time-varying conditions may be superimposed. When a positive voltage V is applied to the p-side (anode) relative to the n-side (cathode) of a diode, as shown in Figure 2.11a, the junction is said to be *forward-biased,* since it causes forward (positive) current flow from the p side to the n side internally. With negligible resistance of the bulk of the semiconductor and at the contacts with external terminals, all the applied voltage appears across the depletion region. The height of the potential barrier is, therefore, reduced by the applied forward voltage to $V_0 - V$ (Figure 2.11c). Correspondingly, the Fermi levels in the p and the n sides (E_{Fp} and E_{Fn}) are separated in energy by eV. Because of the lowering of the potential barrier from V to $V_0 - V$, majority carriers (holes from the p side and electrons from the n side) cross the junction and become injected minority carriers on the other side. Since the majority holes from the p side diffuse across the junction, there is a high concentration of holes near the boundary of the depletion region on the n side. As these holes travel farther on the n side, they encounter the majority electrons and recombination takes place. The process of injection and recombination of holes continues with the external voltage source supplying the energy to overcome the contact potential. In a similar manner, majority electrons from the n side are injected into the p side where, as they diffuse farther from the junction, they are lost in recombination. Since the electrons diffusing from the n side to the p side constitutes positive current in the same direction as the holes crossing in the reverse direction, positive current flows from the p side to the n side. This current, which is sustained by the applied forward bias, consists of the diffusion components of both holes and electrons.

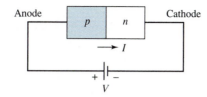

(a) Circuit with Bias

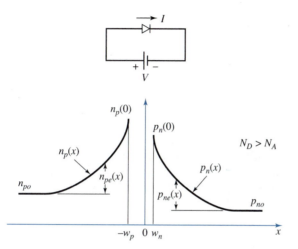

(b) Minority Carrier Concentration Profile

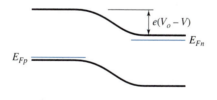

(c) Energy Levels

Figure 2.11 Forward biasing of diode

With the electric field confined to the narrow depletion region, no carriers are available except in the neutral regions; hence, drift current due to applied voltage is negligible.

Observe that the minority carriers on each side are in excess of their thermal equilibrium level at the edge of the depletion region. Farther into the junction, they drop off exponentially by recombination with the majority carriers. Therefore, the concentration of minority carriers in each region away from the junction is maintained at its equilibrium level. In addition, the injected minority carriers are still low compared to the majority carrier (doping) levels. This situation, in which the injected carriers in each region due to external bias voltage is negligible compared to the thermal equilibrium levels of majority carriers, is referred to as the *low-level injection*. Under low-level injection, minority carrier concentration in excess of its thermal equilibrium value is much smaller than the doping (majority carrier) concentration. As a result, the majority carrier levels remain virtually unchanged.

With our assumption of negligible drift current and low-level injection, we calculate the diode forward current under steady-state conditions using the diffusion equation and a simplified continuity equation. If $n_p(x)$ is the minority electron density in the p region at $x < 0$, then the one-dimensional electron diffusion current density, from Equation (2.17b), is

$$J_n(x) = eD_n \left[\frac{dn_{pe}(x)}{dx} \right] \tag{2.34}$$

where

$$n_{pe}(x) = n_p(x) - n_{p0} \tag{2.35}$$

is the electron concentration in excess of the thermal equilibrium level. From Equation (2.29) with the barrier height given by $V_0 - V$ and with $n_n(0) = n_n(w_n)$, the concentration of electrons at the boundary of the depletion region, $x = -w_p$ (Figure 2.11b), is given by

$$n_p(0) = n_p(-w_p) = n_n(w_n) \exp\left[\frac{-(V_0 - V)}{(kT/e)} \right] \tag{2.36}$$

The thermal equilibrium level (without applied bias) is

$$n_{p0} = n_{n0} \exp\left[\frac{-V_0}{(kT/e)} \right] \tag{2.37}$$

Because of low-level injection, $n_n(w_n) = n_{n0} = N_D$. Hence, with the substitution of Equation (2.37) in Equation (2.36),

$$n_p(-w_p) = n_{p0} \exp\left[\frac{V}{(kT/e)} \right] \tag{2.38}$$

Equation (2.38), which relates the minority electron concentration near the junction to the applied voltage, is known as the *law of the junction*. A similar equation relates the minority hole concentration to the applied voltage.

The excess electron concentration at the boundary $x = -w_p$ is

$$n_{pe}(-w_p) = n_{p0} \left\{ \exp\left[\frac{V}{(kT/e)} \right] - 1 \right\} \tag{2.39}$$

These excess electrons recombine with the majority holes as they move farther in the p region. Hence,

$$n_{pe}(-\infty) = 0 \quad \text{or} \quad n_p(-\infty) = n_{p0} \tag{2.40}$$

Equations (2.39) and (2.40) are the boundary conditions for the steady-state *continuity equation* for electrons in the p region. The continuity equation states that at any instant the time rate of increase of particles (carriers) within an elemental volume must be equal to the increase in net flow (current) minus the rate at which they are lost (by recombination) within the volume. That is,

$$\left(\frac{\delta n_{pe}}{\delta t}\right)A\,\Delta x = \frac{[I_n(x) - I_n(x + dx)]}{e} - A\Delta x \text{ (rate of recombination)}$$

where A is the area of the plane vertical to the one-dimensional flow of carriers in the x direction. $I_n(x)$ and $I_n(x + dx)$ represent currents entering and leaving an elemental volume, $A\Delta x$. In the limit as $\Delta x \to dx$, the above equation reduces to

$$\frac{\delta n_{pe}}{\delta t} = \frac{dJ_n/dx}{e} - \frac{n_{pe}(x)}{\tau_n}$$ **(2.41)**

where τ_n is the average lifetime of excess electrons in the p region before they are lost in recombination. τ_n is called the *minority carrier lifetime* of electrons in the p region. Under steady-state conditions, that is, $\delta n_{pe}/\delta t = 0$, with diffusion current given by Equation (2.34), Equation (2.41) becomes

$$D_n\left[\frac{d^2 n_{pe}(x)}{dx^2}\right] = \frac{n_{pe}(x)}{\tau_n}$$ **(2.42)**

The solution of Equation (2.42) with the boundary conditions given by Equations (2.39) and (2.40) is

$$n_{pe}(x) = n_{p0}\left\{\exp\left[\frac{V}{(kT/e)}\right] - 1\right\}\exp\left[\frac{(w_p + x)}{\sqrt{D_n \tau_n}}\right]$$ **(2.43)**

Equation (2.43) describes the excess (injected) electron concentration (above the thermal equilibrium level) in the p region in terms of the equilibrium level and the applied bias voltage.

A similar equation for the excess holes in the n region can be derived. Figure 2.11b shows the minority carrier profiles in the p and the n regions.

Since the electron current is the same throughout the region, at $x = -w_p$, from Equation (2.34),

$$J_n(-w_p) = \left[\frac{eD_n}{\sqrt{D_n \tau_n}}\right]n_{p0}\left\{\exp\left[\frac{V}{(kT/e)}\right] - 1\right\}$$

Denoting by $L_n = \sqrt{D_n \tau_n}$, the electron current is given by

$$I_n = -\left(\frac{Ae\,D_n}{L_n}\right)n_{p0}\left[\exp\left(\frac{eV}{kT}\right) - 1\right]$$ **(2.44)**

The hole current is similarly given by

$$I_p = \left(\frac{Ae\,D_p}{L_p}\right)p_{n0}\left[\exp\left(\frac{eV}{kT}\right) - 1\right]$$ **(2.45)**

where $L_p = \sqrt{D_p \tau_p}$. L_n and L_p are known as the minority carrier *diffusion lengths*. L_n and L_p are the mean distances the carriers travel in the p and the n regions respectively, before recombining with the opposite carrier type. With exponential decay, the excess carrier densities drop to $1/e$ of their peak values at the depletion boundaries. L_n and L_p in silicon are typically in the range of 1 to 100 μm.

Adding the two diffusion current components above, the total current I in the diode due to forward-bias voltage V is given by

$$I = Ae\left[\left(\frac{D_p}{L_p}\right)p_{n0} + \left(\frac{D_n}{L_n}\right)n_{p0}\right]\left[\exp\left(\frac{eV}{kT}\right) - 1\right] \qquad \textbf{(2.46)}$$

or

$$I = I_s\left[\exp\left(\frac{eV}{kT}\right) - 1\right] \qquad \textbf{(2.47)}$$

where

$$I_s = Ae\left[\left(\frac{D_p}{L_p}\right)P_{n0} + \left(\frac{D_n}{L_n}\right)n_{p0}\right] \qquad \textbf{(2.48a)}$$

or, using the carrier lifetimes,

$$I_s = Ae\left[\left(\frac{L_p}{\tau_p}\right)p_{n0} + \left(\frac{L_n}{\tau_n}\right)n_{p0}\right] \qquad \textbf{(2.48b)}$$

Equation (2.47) is the *ideal diode equation*. For negative values of V such that $|V| \gg kT/e$ in Equation (2.47), I is constant at $I \approx -I_s$. For this reason I_s is called the *reverse saturation current*, or simply the *saturation current*, of the diode. As we can see from Equation (2.48), the saturation current I_s depends on the area, the minority carrier lifetimes, and the minority carrier levels. For a material with given τ_p and τ_n, the lightly doped region, as we saw earlier, has more minority carriers; hence, this doping has a higher influence on I_s.

The ideal diode equation we derived is for the step, or abrupt, junction, in which the doping level changes from n to p abruptly at $x = 0$. In practice, diodes may have linear, or graded, doping levels from the p to the n regions. Also, the condition of low-level injection may not be satisfied. In addition, the calculation of the current throughout the semiconductor, which is assumed to arise from diffusion in the neutral region alone, does not take into account secondary effects such as the generation and recombination in the depletion region.

Practical diodes account for these cases of grading, high-level injection, and secondary effects by using an empirical constant n, known as the *ideality factor*, or *emission coefficient*, in the diode equation as

$$I = I_s\left[\exp\left(\frac{eV}{nkT}\right) - 1\right] \qquad \textbf{(2.49)}$$

The value of n is between 1 and 2. For graded junctions, in which the doping from p to n varies gradually, for example, $n \approx 2$ gives a better fit for the currents arrived at by Equation (2.47) with measured values. In general, discrete two-terminal diodes have $n \approx 2$ while for integrated circuit diodes, $n \approx 1$ is used. In addition, n may vary from approximately 2 at low forward bias to 1 at high bias.

If the forward-bias voltage V is increased in an attempt to reduce the barrier potential to zero, a large current will result in the diode according to Equation (2.49). At large currents, however, the bulk resistance of the semiconductor and the contact resistances at the terminals cause significant drops in voltage. As a result, the diode forward currents are limited by these resistances as V approaches V_0. For this reason, at high forward voltages one must consider the diode resistances in using Equation (2.49).

As we have seen, with a constant forward-bias voltage V there is a steady (exponential) gradient of minority carriers in the n and the p regions. This charge gradient causes diffusion and forward current flows from cathode to anode via the external circuit. The steady current is maintained by the external voltage source that replenishes the electron supply. If the forward-bias voltage is changed, the concentration levels of minority carriers in both the regions also must change to new values. This change in stored charges with forward-bias voltage exhibits capacitive effect resulting in *diffusion capacitance,* which we consider in Section 2.4.6.

2.4.3 REVERSE-BIASED DIODE

If a positive voltage is applied to the cathode relative to the anode as shown in Figure 2.12a, the diode is said to be reverse-biased. Because of the external bias in the same direction as the barrier potential, the field across the junction is increased above the equilibrium level. Therefore, holes in the p region and electrons in the n region are moved away from the junction. The depletion region is thus widened and no forward current flow occurs. The increased barrier potential across the junction reduces the minority carrier concentrations near the junction in accordance with the law of the junction, Equation (2.38). Farther away from the junction the carriers reach their thermal equilibrium levels. Figure 2.12b shows the density levels of minority carriers under reverse bias. The increased electrostatic potential across the depletion region increases the separation of energy levels, as Figure 2.12c illustrates.

Because of the concentration gradient in the thermally generated minority carriers, there is diffusion of carriers across the junction: minority electrons from the p region diffuse to the depletion region and are swept to the n region by the field across the junction, while minority holes in the n region move in the opposite direction to the p region. This results in a small positive current from cathode to anode, known as the *reverse current,* opposite to forward current direction. The thermal generation component, to a great degree, is given by the reverse saturation current I_s in the ideal diode equation. The ideal diode equation, therefore, is useful in describing the current–voltage characteristics for both forward- and reverse-bias conditions.

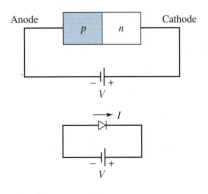

(a) Circuit

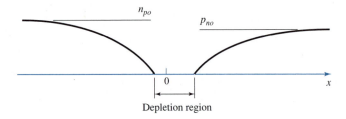

(b) Minority Carrier Concentration Profile

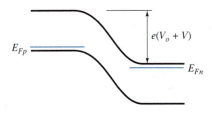

(c) Energy Levels

Figure 2.12   Reverse biasing of diode

In practical diodes, reverse current increases above the saturation current I_s for increasing reverse bias. With increased width of the depletion region, the electron-hole pairs generated within the depletion region become significant, unlike in a forward-biased diode. At room temperature, the reverse current is on the order of hundreds of pA to a few μA, depending on the material, doping levels, and dimensions. Temperature dependence of I_s arises from the diffusion constants and the equilibrium minority concentration levels. Experimental results show that I_s doubles approximately for every 10°C rise in temperature.

As with forward bias, a change in reverse-bias voltage causes a change in immobile charge density by widening the depletion region. This capacitive effect results in the *depletion* or *transition capacitance* (see Section 2.4.6).

2.4.4.   IDEAL DIODE CHARACTERISTICS

The ideal diode equation given in Equation (2.49) relates the current I through and the voltage V across terminals of the diode, up to the breakdown region. (In section 2.7 we examine the Zener diode, which is a junction diode designed for operation in the breakdown region.) We can rewrite this equation as

$$I = I_s \left[\exp\left(\frac{V}{nV_T}\right) - 1 \right] \qquad (2.50)$$

where

$$V_T = kT/e \qquad (2.51)$$

is known as the *thermal* voltage since it is the volt-equivalent of temperature. Using

k = Boltzmann's constant, 1.38×10^{-23} J/K,
e = magnitude of electronic charge, 1.62×10^{-19} C, and
T = absolute temperature in Kelvin $\approx 273 +$ temperature in degrees Celsius,

at a room temperature of $21°C \approx 294$ K, we get $V_T \approx 25$ mV.

In Equation (2.50) current I is measured positive if it flows from anode to cathode internal to the diode, and voltage V is measured positive if the anode is positive relative to the cathode. As we saw earlier, Equation (2.50) is a close approximation to the actual behavior of a junction diode in both the forward- and the reverse-biased conditions.

For forward-bias voltages with $V \gg nV_T$, Equation (2.50) can be approximated as

$$I \approx I_s \exp\left(\frac{V}{nV_T}\right) \qquad (2.52)$$

This is a good approximation covering several decades of forward current in the diode with very little change in forward voltage. Since V_T is only a few 10s of mV at ordinary temperature, Equation (2.52) is applicable for V above a few hundred mV.

When a reverse-bias voltage of V such that $|V| \gg nV_T$ is applied, the current is approximated by the constant $-I_s$. Based on the two ranges of $V \gg nV_T$ and $-V \ll nV_T$, the ideal current-voltage characteristics given by Equation (2.50) are shown graphically in Figure 2.13 for $n = 1$.

2.4.5   TEMPERATURE DEPENDENCE

Variation of diode current-voltage characteristics with temperature arises from the thermal voltage V_T and the reverse saturation current I_s. Using the mass-action law (Equations (2.13) and (2.11a)) in Equation (2.48), it can be verified

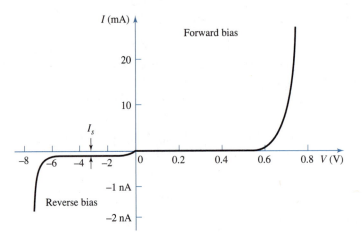

Figure 2.13   Ideal diode current-voltage characteristics

that I_s depends on the square of the intrinsic concentration, n_i^2. Therefore, from Equation (2.11c), it is clear that I_s is strongly dependent on temperature. In addition, the diffusion constants vary linearly with T, as seen from Equation (2.18). Since variation of the intrinsic concentration n_i is dominant at approximately 6 percent per °C, theoretically the overall variation in I_s is estimated at 15 percent per °C. Experimental results show that I_s doubles for every 8- to 10-°C rise in temperature; that is, it increases at the rate of 10 percent to 12 percent/°C.

We evaluate the overall variation of forward current with temperature at a fixed forward bias as follows. From Equation (2.52), for $n = 1$,

$$dI/dT \approx I\left[\left(\frac{dI_s}{dT}\right)\left(\frac{1}{I_s}\right) - \left(\frac{V}{V_T^2}\right)\left(\frac{dV_T}{dT}\right)\right]$$

Since $V_T = \dfrac{kT}{e}$,

$$\frac{dV_T}{dT} = \frac{k}{e} = \frac{V_T}{T}$$

Therefore,

$$\left(\frac{1}{I}\right)\left(\frac{dI}{dT}\right) \approx \left(\frac{dI_s}{dT}\right)\left(\frac{1}{I_s}\right) - \left(\frac{V}{V_T}\right)\left(\frac{1}{T}\right) \qquad \textbf{(2.53)}$$

Example 2.3 | If the forward bias of a diode operating at 25°C is maintained at 700 mV, what is the percentage change in forward current with temperature? The reverse saturation current of the diode varies 12 percent per °C.

If the operating temperature of the diode rises by 10°C, how much does I change from its initial value of 5 mA at room temperature?

Solution

At 25°C, $V_T = kT/e = 25.7$ mV
From Equation (2.53), we have

$$\left(\frac{dI}{dT}\right)\left(\frac{1}{I}\right) = 0.12 - \left(\frac{700}{25.7}\right)\Big/298 = 0.0286 \, /°C, \quad \text{or} \quad 2.86 \text{ percent per °C}$$

For a 10°C rise, I increases to $(1.0286)^{10} = 1.326$ of its value at 25°C, or by 32.6 percent. Hence, the current at 35°C is 6.63 mA.

We evaluate the variation of the forward voltage with temperature in a similar manner for a fixed forward current. From Equation (2.52), again with $n = 1$,

$$\ln I = \ln I_s + \frac{V}{V_T}$$

Taking the derivative and noting that I is constant,

$$\frac{dV}{dT} = \frac{V}{T} - \left(\frac{V_T}{I_s}\right)\left(\frac{dI_s}{dT}\right) \tag{2.54}$$

With the second term on the right of the above equation more dominant, the forward voltage is found to decrease with temperature at a rate of 2 to 2.5 mV/°C in practical diodes. Note that for the same forward current the diode V-I characteristic shifts to the left as shown in Figure 2.14 for increasing temperature.

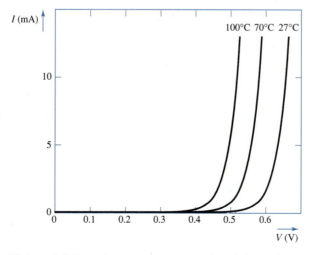

Figure 2.14   Temperature variation of diode forward characteristic

Since the general shape of the diode characteristic curve is unchanged, this property is used in conjunction with a Zener diode (Section 2.7) or the emitter-base junction of a bipolar junction transistor (Section 5.3) to obtain a temperature-compensated reference voltage.

Example 2.4	Calculate the temperature coefficient for the forward voltage at 25°C for the previous example.

Solution

From Equation (2.54),

$$dV/dT = 700/298 - 25.7 \times 0.12 = -0.735 \, \text{mV/°C}$$

Examples 2.3 and 2.4 point out the need to operate semiconductor devices at constant temperatures to maintain the operating conditions.

2.4.6   DIODE CAPACITANCES

Diffusion Capacitance   As we have seen in Figure 2.11, the region next to the junction in a forward-biased diode has excess minority carriers contributing to diffusion (forward) current. Concentration levels of these carriers vary with the bias voltage, giving rise to an incremental capacitance. This capacitance, known as the *diffusion capacitance,* is derived based on the rate of change of excess carriers with voltage. Let the diode have a much higher level of acceptor doping than donors (i.e., $N_A \gg N_D$), so that the forward current in Equation (2.46) is due to holes only. The concentration of excess minority holes $p_{ne}(x)$, on the neutral n side, similar to $n_{pe}(x)$ in Equation (2.43), is given by

$$p_{ne}(x) = p_{n0}\left[\exp\left(\frac{eV}{kT}\right) - 1\right]\exp\left(\frac{W_n - x}{L_p}\right)$$

Integrating the above equation for the excess of injected positive charge Q_p on the n side, we have

$$Q_p = \int_{W_n}^{\infty} Aep_{ne}(x) \, dx = AeL_p p_{n0}\left[\exp\left(\frac{eV}{kT}\right) - 1\right] \tag{2.55}$$

From the definition of $C = (dQ/dV)$, the diffusion capacitance is given by

$$C_D = AeL_p p_{n0} \exp\left(\frac{eV}{kT}\right)\left(\frac{e}{kT}\right)$$

Using Equation (2.52) and the thermal voltage $V_T = kT/e$, we write the above equation in terms of the forward current as

$$C_D = \left(\frac{I}{V_T}\right)\tau_p = \left(\frac{I}{V_T}\right)\left(\frac{L_p^2}{D_p}\right) \qquad \textbf{(2.56a)}$$

where $\tau_p = L_p^2/D_p$ is the minority carrier (hole) mean lifetime in the n region. For general emission coefficient n, the diffusion capacitance is given by

$$C_D = \left(\frac{I}{nV_T}\right)\tau_p = \left(\frac{I}{nV_T}\right)\left(\frac{L_p^2}{D_p}\right) \qquad \textbf{(2.56b)}$$

Recall that the excess minority holes in the n region are lost in recombination with an average lifetime of τ_p. (Note that L_p is the average distance that the holes diffuse before recombination.) Hence, the external biasing source must replenish the excess positive charge given by Equation (2.55) every τ_p seconds to sustain the forward current I. This leads to the steady forward current expression,

$$I = Q_p/\tau_p = \left(\frac{AeL_p p_{n0}}{\tau_p}\right)\left[\exp(eV/kT) - 1\right]$$

which, using $L_p = \sqrt{D_p \tau_p}$, is the same as Equation (2.45). This derivation shows that the diode forward current is proportional to the stored excess charge due to minority carriers. Such a linear relationship between current I and stored charge Q is referred to as the *charge-control description* of the device. Note that the excess charge, given by the charge-control equation, Equation (2.55), is negative under reverse bias; hence, the current direction for $V < 0$ is opposite to the forward current direction.

If the forward current is due to both types of carriers as given by Equation (2.46), the total diffusion capacitance can be obtained as the sum of the contributions by the two types of carriers. Equation (2.56) shows that the diffusion capacitance varies linearly with the forward current. Therefore, under a reverse-bias condition, which results in a negligible minority carrier level and, hence, forward current, C_D is negligible. The diode capacitance is then due to the stored immobile charge in the depletion region. Typically, the value of C_D is on the order of a nF, depending on the minority carrier lifetime and the forward current.

If the voltage is varied during forward bias, at low-level injection the profile of the excess minority carriers also varies. As we have seen, the variation for both the profile and the total charge taking part in the diffusion current is exponential with the forward voltage. This is akin to the charging of a capacitor; for this reason, the diffusion capacitance is commonly known as the *storage capacitance*. Clearly, as the forward voltage varies, there is delay in charging or discharging C_D to the new profile. This delay is particularly significant when the bias voltage changes from forward to zero or reverse, causing a storage delay in pn junction switching circuits (Section 2.6).

Junction Capacitance   For a reverse-biased diode there is immobile charge stored in the depletion region. Note that the depletion region, which has no carriers, behaves like a dielectric. The charge stored in the depletion region varies with the applied reverse voltage, much like in a parallel-plate capacitor. The associated capacitance, which relates the change in charge to the applied bias voltage, is known as the *depletion, junction, transition,* or *barrier capacitance.* Similar to the capacitance of a parallel-plate capacitor, the junction capacitance is given by

$$C_T = \frac{A\epsilon}{w}$$

where A is the cross-sectional area and ϵ is the dielectric permittivity of the depletion region (silicon) with width w. Since the depletion width w varies with applied bias voltage V, if we replace V_0 in Equation (2.33) with the net potential across the junction, we have

$$C_T = A \sqrt{\frac{e\epsilon \, N_A N_D}{2(N_A + N_D)(V_0 - V)}} \tag{2.57}$$

In Equation (2.57), V is negative for reverse bias (i.e., with the anode at lower potential relative to the cathode). We rewrite this equation, derived for the abrupt (step-graded) junction, to show explicitly the bias voltage dependency on C_T as

$$C_T = \frac{C_0}{\sqrt{\left(1 - \dfrac{V}{V_0}\right)}} \tag{2.58a}$$

where

$$C_0 = A \sqrt{\frac{e\epsilon \, N_A N_D}{2(N_A + N_D)V_0}}$$

is the *equilibrium* or *zero-bias capacitance.*
In general, for any grading,

$$C_T = \frac{C_0}{\left(1 - \dfrac{V}{V_0}\right)^m} \tag{2.58b}$$

where m is the *grading coefficient.* The value of m is between 1/3 and 1/4; for abrupt junctions, $m = 1/2$.

The variation of the capacitance of the reverse-biased pn junction with applied bias voltage is used in tuning high-frequency communication circuits. In such applications as television and radio receivers, for instance, the diodes are called *varactors.* Varactors are also used in high-speed bipolar switching circuits, which we will examine in Section 4.4.9.

The zero-bias junction capacitance is typically in the range of 10 to 100 pF. Reverse voltage in the range of 0 to -50 V causes C_T to decrease by a factor of nearly 10.

From the above discussion, we note that the value of the diffusion capacitance C_D dominates over the junction capacitance C_T for the forward-biased junction, while C_D is negligible for reverse bias. Both capacitances are, however, nonlinear elements represented by linear approximations.

2.5 DIODE MODELS

While the ideal diode equation (Equation (2.47) closely models the *V-I* characteristics, for most digital circuit applications a simplified model would generally suffice. In this section we first develop simplified diode models for quick hand calculations. For more accurate calculation using computer simulation, we then look at the model used in MicroSim™ PSpice.®

2.5.1 SIMPLIFIED MODELS FOR HAND CALCULATIONS

Consider, for example, the diode circuit shown in Figure 2.15. Voltage V_D across the diode and the current I in the circuit can be solved for by using Kirchhoff's voltage law and the ideal diode equation as

$$V_S = V_D + IR \tag{2.59}$$

$$I = I_s\left[\exp\left(\frac{V_D}{V_T}\right) - 1\right] \tag{2.50}$$

With the diode forward-biased, we can obtain the simultaneous solution of these two equations graphically, as Figure 2.16 shows for $R = 1$ kΩ, $V_S = 10$ V, and $I_s = 1$ pA. The straight line representing Equation (2.59) in Figure 2.16 is the *load line,* as it describes the current-voltage characteristic of the load R in the circuit. The load line passes through $V_D = 0$, $I = V_S/R$, and $V_D = V_S$, $I = 0$, with a slope of $-1/R$. The intersection of the diode V-I characteristic (experimental or manufacturer's graph) and the load line gives the unique solution to V_D and I.

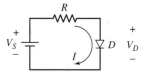

Figure 2.15   A diode circuit

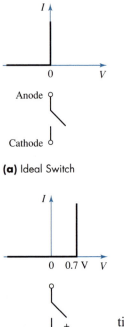

(a) Ideal Switch

(b) Switch with Voltage Drop

(c) Piecewise Linear

Figure 2.17
Diode models

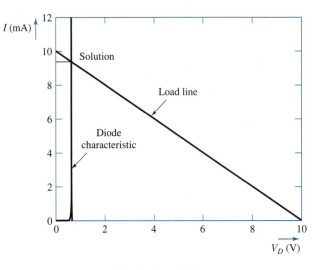

Figure 2.16   Graphical solution for the circuit in Figure 2.15

The analytical solution can be obtained by solving the transcendental equation resulting from Equations (2.50) and (2.59),

$$V_D = V_S - RI_s\left[\exp\left(\frac{V_D}{V_T}\right) - 1\right] \tag{2.60}$$

The iterative solution of Equation (2.60) requires a computer or a calculator for high accuracy. The solution for $R = 1\,k\Omega$, $V_S = 10\,V$, and $I_s = 1\,pA$ in Figure 2.15 is $V_D = 0.5742\,V$ and $I = 9.4362\,mA$.

For most applications in digital circuits, an approximate solution may be obtained rapidly by using a piecewise linear model of the diode.

Ideal Switch Model   The simplest model represents the diode as an ideal switch, that is, one with zero voltage drop when it is conducting (forward-biased), and open circuit when it is off (reverse-biased). (See Figure 2.17a.) Using this model for the circuit in Figure 2.15, we see that $I = 10\,mA$ and $V_D = 0$. Although this solution ignores the diode forward voltage drop, it is reasonably accurate if other circuit voltages are much higher than the nominal diode forward drop of about 0.7 V. Hence, for a quick estimation of diode current, the ideal switch model may be used.

Switch with Offset Voltage   Since the diode voltage V_D changes very little over a large range of forward current, this model represents the diode by a constant voltage drop of about 0.7 V. From the diode I-V characteristic this voltage drop is observed to be the nominal voltage for any diode carrying a forward current in excess of about a tenth of its rated maximum current. When the diode is reverse-biased, or when the forward voltage is below 0.7 V, very little current flows and therefore an open circuit may represent the diode. Using this model, shown in Figure 2.17b, the solution for the circuit in Figure 2.15 is $I = 9.3$ mA and $V_D = 0.7$ V. Observe that this solution is obtained by assuming that the diode is conducting. This is a reasonable assumption with a 10 V source in the circuit.

Piecewise Linear Approximation   To account for the finite resistance of a conducting, forward-biased diode, the ideal diode forward characteristic is replaced by a piecewise linear approximation as shown in Figure 2.17c. In this model, the nonlinear exponential characteristic is approximated by two linear relationships, namely, zero current for $V < V_{\text{cut-in}}$, and by a straight line with a finite slope for $V > V_{\text{cut-in}}$. The cut-in voltage, which separates the two linear regions, is the voltage at the onset of conduction, and depends on the current rating of the device. At room temperature it may, for most applications, be approximated by 0.6 V. This is justified by the fact that if a diode is carrying its intended or operational current I_{on} at $V = 0.7$ V, then at currents I_{off}, which are about 1 percent of I_{on}, it may be considered off even if $V > 0$. From Equation (2.50),

$$I_{\text{on}} = I_s\left[\exp\left(\frac{V_{\text{on}}}{V_T}\right) - 1\right] \approx I_s \exp\left(\frac{V_{\text{on}}}{V_T}\right)$$

and

$$I_{\text{off}} = I_s\left[\exp\left(\frac{V_{\text{off}}}{V_T}\right) - 1\right] \approx I_s \exp\left(\frac{V_{\text{off}}}{V_T}\right),$$

where $V_{\text{on}}, V_{\text{off}} \gg V_T$. Hence,

$$\frac{I_{\text{on}}}{I_{\text{off}}} = 100 = \exp\left[\frac{(V_{\text{on}} - V_{\text{off}})}{V_T}\right]$$

Solving for the difference in voltage,

$$V_{\text{on}} - V_{\text{off}} = V_T \ln 100 \approx 115 \text{ mV}$$

Therefore, the cut-in voltage is $V_{\text{cut-in}} = 0.7 - 0.115 \approx 0.6$ V.

The slope of the model in the I-V characteristic for $V > V_{\text{cut-in}}$ is the reciprocal of the diode on-, or forward-resistance, $1/R_F$. With R_F and $V_{\text{cut-in}}$ included, the model appears as shown in Figure 2.17c. To use this model the diode on-resistance must be estimated from the I-V characteristics. If the diode in Figure 2.15 has a forward resistance of 10 Ω, (that is, the slope ≈ 0.1 in the neighborhood of 10 mA) for example, $I = (10 - 0.6)/1.001 = 9.307$ mA and $V_D = 0.693$ V.

Clearly, all three models shown in Figure 2.17 yield results close to the graphical solution. Depending on the accuracy needed, therefore, any of the three models may be used. This approach is particularly valid for digital circuits in which voltages and resistances are much higher than those of the forward-biased diode. For circuits with low voltages (below $1\,V$) and low resistances (below $100\,\Omega$), in general, the piecewise linear model with $V_{\text{cut-in}}$ and R_F must be employed for better approximation.

Note that in all of the above models the diode is considered open when reverse-biased. This is again a valid approximation when the circuit resistances are much less than the reverse resistance R_R, given by the ratio of reverse-bias voltage to reverse current. R_R, as seen from the diode characteristic in Figure 2.13, is in the range of a few hundreds of $k\Omega$ to several $M\Omega$. Both R_F and R_R are ratios of steady diode voltage to current under constant forward- and reverse-bias conditions, respectively. Small variations of R_F and R_R around the bias points are ignored. For this reason the model resistances are referred to as large signal values; the models are, therefore, known as *large-signal,* or low-frequency, models. Circuits such as static logic gates, which operate at one of several bias voltages, are analyzed using large-signal models.

Example 2.5

Determine the diode currents and the output voltage V_o in the circuit shown in Figure 2.18.

Solution

Since both V_A and V_B are higher than V_S, it appears that both the diodes may be forward-biased. But a little thought shows that if diode D_A conducts, it will bring the voltage V_o to no more than $6 - 0.7 = 5.3\,V$. This voltage will heavily forward-bias diode D_B and raise V_o above $5.3\,V$. Therefore, it is reasonable to assume that D_B conducts and D_A is off.

Modeling the conducting diode D_B using the ideal switch with a drop of $0.7\,V$, the current I_B is given by

$$I_B = \frac{(15 - 0.7 - 4)}{(1 + 2)} = 3.43\,\text{mA}$$

Hence, the voltage V_o is

$$V_o = 4 + 3.43 \times 2 = 10.87\,\text{V}$$

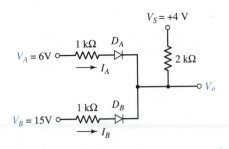

Figure 2.18   Circuit for Example 2.5

Since $V_o > V_A$, the diode D_A is indeed reverse-biased as assumed. Hence $I_A \approx 0$ except for the reverse saturation current.

Note that we used the ideal switch and the switch with voltage drop models in the above example: the ideal switch model to quickly determine which diode is likely to conduct, and then the refined model to calculate the currents. We are justified in neglecting the diode forward resistance, since the circuit resistances are higher than $R_F = 0.7/I_B \approx 204\,\Omega$. The solution is close to the one obtained using the more accurate MicroSim™ PSpice® model (Section 2.5.2).

We shall use the above procedure of assuming that certain diodes conduct, based on known voltages and the ideal switch model; then proceed with calculations for all node voltages and currents; and, finally, ensure that the results justify our initial assumptions.

When the bias voltage of a forward-biased diode varies incrementally over a *quiescent,* or operating bias point, one must consider *variation* of the anode-cathode resistance in analyzing the change in the diode current. The *incremental, dynamic,* or *ac resistance* offered by the still forward-biased diode, denoted by r_F, is obtained from Equation (2.50) as a ratio of change in forward voltage to change in forward current about the operating point.

Since the forward current is approximated as

$$I \approx I_s \exp(V/V_T)$$

we have

$$r_F = \frac{dV}{dI} \approx \frac{V_T}{I_s \exp(V/V_T)}$$

or

$$r_F = \frac{V_T}{I} \qquad\qquad \text{(2.61a)}$$

The general equation, for any grading of the junction, is

$$r_F = \frac{nV_T}{I} \qquad\qquad \text{(2.61b)}$$

where n is the ideality factor.

Comparing r_F to the static, or dc resistance $R_F = V/I$ at the operating point, it is clear that the effect of r_F is significant only in the small neighborhood of the bias point. If the operating point moves farther from the bias point, one must use a different value of r_F, corresponding to the new current I. R_F, on the other hand, gives a gross model for the forward-biased diode as the slope of the straight line approximating the I-V characteristic at $V = V_{\text{cut-in}}$.

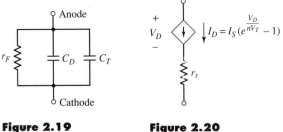

Figure 2.19
Dynamic model of a
forward-biased diode

Figure 2.20
MicroSim PSpice dc model
of a diode

Using the ac resistance given in the above equation, the diffusion capacitance given by Equation (2.56) may be written as

$$C_D = \left(\frac{I}{V_T}\right)\tau_p$$

$$= \tau_p/r_F \tag{2.62}$$

Note that the product $C_D r_F$ is the *diffusion time constant,* which relates the time taken for the excess minority carrier distribution to change with changing forward bias. Internally, the process is characterized by the minority carrier lifetime τ_p.

Analysis of the dynamic operation of a diode must include the capacitance and the dynamic resistance. At high-frequency operation, for example, the effect of junction and diffusion capacitances and variation of the anode-cathode resistance are needed in order to study the transient behavior of the circuit. The small-signal dynamic, or ac, model, shown in Figure 2.19, includes r_F, C_T, and C_D. As seen earlier, C_D is negligible under reverse-biased conditions, while both capacitances are present in the forward-biased junction. When biased to carry normal operating current, however, C_D is significantly higher than C_T.

2.5.2   MICROSIM™ PSPICE® MODEL

We now proceed to examine the computer modeling for circuit analysis that requires a high degree of accuracy. Figure 2.20 shows the model used for computer simulation of the pn junction diode in MicroSim PSpice. Other simulation programs use similar models. Some of the MicroSim PSpice model parameters that are frequently used and their symbols are given in Table 2.5.

The model uses the ideal diode equation (Equation (2.50)) with a general value for the emission coefficient. The resistance r_s models the ohmic, or contact

Table 2.5   MicroSim PSpice diode model parameters

Symbol	MicroSim PSpice Symbol	Parameter	Default	Units
I_s	IS	Saturation current	1.0E-14	A
r_s	RS	Contact resistance	0	Ω
n	N	Emission coefficient	1	
C_0	CJO	Zero-bias junction capacitance	0	F
V_0	VJ	Built-in potential	1	V
m	M	Grading coefficient	0.5	
τ_t	TT	Transit time	0	s

resistance, between the semiconductor and the external terminals. The capacitance C_0 takes into account the junction (depletion) capacitance, contact potential, and the grading coefficient, as given by Equation (2.58b). Diffusion capacitance is modeled by the transit time. Note, from Equation (2.56), that C_D depends on the minority carrier lifetime. Since the MicroSim PSpice model uses an empirical equation, the transit time TT in the model is approximately equal to the minority carrier lifetime τ_p.

In MicroSim PSpice, the effect of temperature on the diode behavior is simulated by specifying the temperature (or a range of temperature) in the .TEMP statement. Default temperature is 27°C. The .TEMP statement produces calculation of temperature-dependent quantities such as I_s and V_T at specified temperatures using specified (or default) temperature coefficients. For details the reader is referred to the MicroSim PSpice manual [7].

Simulate the circuit shown in Figure 2.18 using MicroSim PSpice with default diode parameters and $I_s = 1\,\text{pA}$. Assume the contact and bulk resistance of the diode to be $10\,\Omega$. | **Example 2.6**

Solution

The circuit shown in Figure 2.18 is redrawn in Figure 2.21a with the nodes labeled for the PSpice input file, Figure 2.21b. Simulation output is shown in Figure 2.21c. Observe that the off diode current is the saturation current I_s. Large-signal resistance of the conducting diode is $R_F = (V_6 - V_3)/I_B = 0.603/3.466 = 174\,\Omega$. It can be verified that this is close to the value that would result if we used the piecewise linear model in Example 2.5.

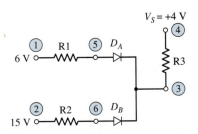

(a) Circuit with Nodes Labeled

```
* DIODE CIRCUIT SIMULATION
R115 1K
R226 1K
R334 2K
DA53 SWITCH
DB63 SWITCH
VA10 DC 6V
VB20 DC 15V
VS40 DC 4V
.MODEL SWITCH D (IS=1E-13 RS=10)
.OPTIONS NOPAGE
.END
```

(b) MicroSim PSpice Input File

```
**** 9/10/95 11:08:26 ******** Evaluation PSpice (January 1993) **********

 * DIODE CIRCUIT SIMULATION

****      CIRCUIT DESCRIPTION

*****************************************************************************

R115 1K
R226 1K
R334 2K
DA53 SWITCH
DB63 SWITCH
VA10 DC 6V
VB20 DC 15V
VS40 DC 4V
.MODEL SWITCH D (IS=1E-13 RS=10)
.END
```

(c) Simulation Output

Figure 2.21   MicroSim PSpice simulation

```
**** 9/10/95 11:08:26 ******** Evaluation PSpice (January 1993)

 * DIODE CIRCUIT SIMULATION

****        Diode MODEL PARAMETERS



            SWITCH
      IS   100.000000E-15
      RS    10




****       SMALL SIGNAL BIAS SOLUTION      TEMPERATURE =  27.000 DEG C

* * * * * * * * * * * * * * * * * * * * * * * * * * * * * * * * * * * * * * * * * * * * * * * * * * * * * * * * * * * * * * * * * *

   NODE    VOLTAGE      NODE    VOLTAGE      NODE    VOLTAGE      NODE    VOLTAGE

 (    1)    6.0000  (     2)   15.0000  (     3)   10.8920  (     4)    4.0000

 (    5)    6.0000  (     6)   11.5540




       VOLTAGE SOURCE CURRENTS
       NAME          CURRENT

       VA          4.992E-12
       VB         -3.446E-03
       VS          3.446E-03

       TOTAL POWER DISSIPATION    3.79E-02   WATTS


          JOB CONCLUDED

          TOTAL JOB TIME                .87
```

(c) Simulation Output (*concluded*)

2.6   DIODE SWITCHING RESPONSE

When the bias voltage of a forward-biased diode is switched abruptly to a negative voltage, the diode current does not reverse instantaneously. We analyze this response of the diode for the circuit shown in Figure 2.22a, in which the applied voltage V_i is switched from positive (V_F) to negative (V_R) at $t = 0$.

For $t < 0$, the constant forward bias causes the minority carrier levels in the p and n regions as shown in Figure 2.11b, and a steady forward current $I_F = (V_F - V_D)/R$ results in the circuit. With a steady reverse voltage applied for $t > 0$, we expect the minority carrier profile to change to that shown in Figure 2.12b. This changeover requires the following sequence of events to take place: the diffusion capacitance, which is dominant during forward bias ($t < 0$), is fully discharged; the depletion region is widened; and the depletion capacitance is charged in the opposite direction. The two capacitances involved in the sequence cause the time delays t_s and t_t as shown in Figure 2.22b.

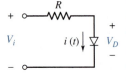

(a) Circuit

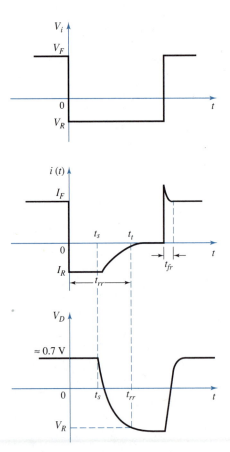

(b) Waveforms

Figure 2.22   Diode switching response

```
* DIODE SWITCHING RESPONSE
VI      1    0   PULSE 5 -3 10N 0.05N 0.05N 30N 50N
DI      2    0   SWITCH
RS      1    2   1K
.MODEL SWITCH D (IS=1E-12 RS=10 CJO=5P TT=10N)
.TRAN 1N 50N
.PROBE
.END
```

(c) MicroSim PSpice Input File

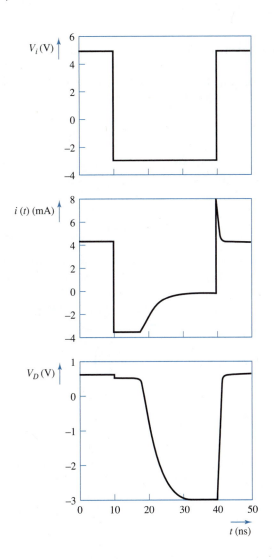

Figure 2.22   (concluded)

The *storage time* t_s is the time taken by the diffusion capacitance to discharge, that is, for the excess minority carriers to drop down to zero. During this time $0 < t < t_s$, as the excess carriers are drawn back across the junction, external voltage supports the diode current as $I_R = (V_R - V_D)/R$. Note that this current is in the opposite direction to I_F. The diode voltage V_D drops slightly from its forward-bias value due to the large number of carriers near the junction causing lower resistance.

Storage time delay may be estimated using the charge-control model of the diode. For $t \le 0$, the diode is carrying a steady forward current $i(t) = I_F$ with stored excess hole charge Q_p in the n region (see Equation (2.55)) given by

$$Q_p = I_F \tau_p \tag{2.63}$$

where τ_p is the minority hole lifetime in the n region. (Here again we assume that the p region is heavily doped relative to the n region.) For $t > 0$, the excess charge may be assumed to discharge at a constant rate of I_R. Hence, we find the approximate storage delay by

$$t_s \approx \frac{I_F \tau_p}{I_R}$$

A better approximation for t_s is obtained using a time-dependent charge-control equation. For $t > 0$, the current $i(t)$ is the sum of two components as given by

$$i(t) = \frac{Q_p(t)}{\tau_p} + \frac{dQ_p}{dt} \tag{2.64}$$

The first term in the above first-order differential equation corresponds to steady current due to the recombination and injection of holes every τ_p seconds. The second term accounts for the variation of excess hole charge because of time-varying bias. Using the initial condition $Q_p(0^+) = I_F \tau_p$ and $i(t) = I_R$ for $t > 0$, we can readily see that the excess charge decay is given by

$$Q_p(t) = t_p \left[(I_F - I_R) \exp\left(\frac{-t}{\tau_p}\right) + I_R \right] \quad \text{for} \quad t \ge 0$$

At $t = t_s$, the excess charge decays to zero. Hence, for $Q_p = 0$,

$$t_s = \tau_p \ln\left(\frac{I_F - I_R}{-I_R}\right) \tag{2.65}$$

The above estimate is based on the assumption that the minority carriers follow exponential distribution at every instant during the transition from forward to reverse bias.

The *transition time* t_t is the time taken by the depletion capacitance to charge to the applied reverse voltage. During t_t the depletion region widens from its zero-bias width as the minority carriers in each side drop from their equilibrium levels. It is nominally the time required for the reverse current to reach

10 percent of its initial value I_R, when the reverse current is close to the saturation current I_s.

The sum of t_s and t_t is the *reverse recovery time, t_{rr}*. Manufacturers specify t_{rr}, which is the turn-off delay of the diode, for typical operating conditions. It is clear from Equation (2.63) that at higher forward current, more charge is stored. Hence, longer storage time t_s is needed to discharge Q_p, as we see from Equation (2.65). If the reverse voltage V_R is increased (within the breakdown limit), the discharge current I_R becomes large and the storage time is reduced. To speed up the recovery process for given I_F and I_R, however, the minority carrier lifetime must be reduced. For example, silicon doped with gold, which can be an acceptor or a donor, has reduced τ_p. Additionally, lifetime in the compound semiconductor gallium arsenide is several orders smaller than in silicon. Another way of obtaining a faster switching diode is by a metal-semiconductor junction known as the Schottky diode, which we discuss in Section 2.8.

The *forward recovery time* t_{fr} is the time delay of the diode in switching from *off* state to *on* state. Since t_{fr} is the time taken to charge the diffusion capacitance, at low-level injection it is on the order of the minority carrier lifetime τ_p. The value of t_{fr}, therefore, is much smaller than that of t_{rr}; hence, the speed limitation of a diode comes from switching the device from *on* to *off* state. In practice, t_{fr} is reduced further by high-level injection which occurs during the transition of the applied voltage V_i from V_R to V_F. Because of high-level injection, the diode voltage shows an overshoot before settling down. Figures 2.22c and d show a PSpice simulation for the diode switching response.

2.7 ZENER DIODE

When the applied reverse bias in a diode is increased to a certain voltage (the knee point in the I-V characteristic shown in Figure 2.13), the reverse current increases abruptly and the diode enters one of two breakdown regions. In a heavily doped diode, a reverse voltage of about 6 V produces a large electric field in the depletion region. This high field pulls free (valence) electrons off of the silicon atoms in the depletion region by rupturing the covalent bonds. These freed mobile electrons contribute to a large reverse current, which is limited by external circuitry. The diode voltage is nearly constant, independent of the reverse current under this condition. The diode is then said to be operating in the *Zener breakdown* region.

The second mechanism of breakdown, called the *avalanche breakdown,* occurs when the reverse voltage is above 6 V. Due to the high reverse voltage applied, thermally generated electrons and holes acquire sufficient energy to knock off carriers from the silicon atoms in the depletion region. The newly freed carriers in turn create additional carriers in a cumulative process, or avalanche multiplication. Again the magnitude of the reverse current is deter-

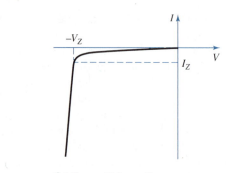

Anode   Cathode

$-$ V_Z $+$

(a) Circuit Symbol

(b) Current-Voltage Characteristics in the Breakdown Region

Figure 2.23   Zener diode

mined by the external circuitry and the diode voltage is relatively constant over a range of currents.

Diodes operating in either type of breakdown mechanism are commonly known as *Zener diodes*. The circuit symbol and the current-voltage characteristics of a Zener diode are shown in Figure 2.23. Observe the negligible change in the diode reverse voltage for a large change in reverse current. Because of this low incremental resistance $r_z = \Delta V/\Delta I$, Zener diodes are used in digital circuits to obtain constant reference voltages (V_z in Figure 2.23) under varying currents.

Although a Zener diode is primarily intended for operation in the reverse bias, it can be biased to operate in the forward bias as well. Note further that the breakdown process of a Zener diode is reversible if the diode power dissipation is within its rated capacity. Zener diodes are used in voltage regulators for obtaining fairly constant voltage supplies under varying source voltages and load currents. Reference voltages in the range of 2 to 200 V with power dissipation of up to 100 W are commonly available.

Although relatively independent of reverse current, the breakdown voltage of a Zener diode does vary with temperature. The temperature coefficient of the breakdown voltage ranges from about -4 mV/°C for a 3 V reference diode to about 4 mV/°C for a 15 V reference diode. It is possible to produce a reference voltage that is almost independent of temperature in the vicinity of 5 V where a combination of Zener and avalanche breakdowns occur. Additionally, by cascading devices with an appropriate negative temperature coefficient (such as a junction diode operating in the forward bias mode, Section 2.4.5) the effect of the positive temperature coefficient of the Zener diode may be minimized, as shown in Figure 2.24, to obtain stable breakdown voltages above 6 V.

The Zener diode is modeled in PSpice in the same manner as a junction diode, with specified knee voltage (V_z) and knee current (I_z) where reverse breakdown begins to occur (see Figure 2.23b).

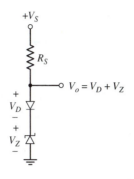

Figure 2.24
A temperature-compensated
Zener diode voltage regulator

2.8   SCHOTTKY DIODE

A Schottky diode (Figure 2.25a) is formed by making contact between a metal and an n-type semiconductor. Aluminum, platinum, or tungsten, for example, forms a rectifying contact, or diode, when combined with a lightly doped[3] n-type $(N_D \leq 10^{23}/m^3)$ material. A metal-semiconductor contact potential arises because of the difference in carrier concentrations in the two materials. In metals, electrons require energy $(e\Phi_B$, in electron volts), known as *work function*, to be freed from their Fermi level to a position outside the material, that is, to the vacuum level. An electron at the vacuum level is free from the influence of the given material and has zero kinetic energy. The work function of a metal used for a Schottky diode (about 5 eV) is slightly higher than that of the semiconductor. Hence, at the metal-semiconductor contact there exists a field. This field is supported by the diffusion of electrons from the semiconductor to the metal, and a positive charge layer is formed on the semiconductor. Note that the diffusion of electrons from the semiconductor creates a space-charge region with immobile positive charge in the semiconductor and an equal amount of negative surface-charge (image) in the metal (Figure 2.25b). At thermal equilibrium, the Fermi level in the two materials is maintained at the same level. The energy difference between the edge of the conduction band and the Fermi level is the *Schottky barrier* $e\Phi_B$, which is independent of doping. The Schottky barrier height varies between 0.6 eV for titanium to 0.9 eV for platinum in n-type silicon. The metal-semiconductor equilibrium contact potential V_0 is the difference between the work functions of the metal and the semiconductor. It is also the potential difference between the barrier height Φ_B and $(E_c - E_F)$ for the

---

[3] Recall that at heavy doping levels $(N_D \geq 10^{25}/m^3)$ the contact resistance is negligibly small, independent of the direction of current flow, as needed for ohmic contacts.

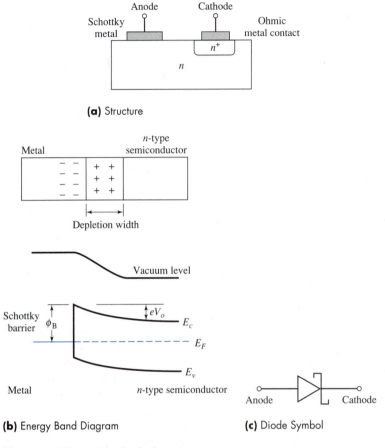

(a) Structure

(b) Energy Band Diagram

(c) Diode Symbol

Figure 2.25   Schottky diode

semiconductor (see Figure 2.25b). This potential barrier, known as the *Schottky barrier potential,* is about 0.5 V. V_0 is calculated using the depletion width and the donor density in the semiconductor. As in a pn junction, the presence of the barrier potential prevents bidirectional flow of current. The rectifying property of metal-semiconductor contact, which was known long before the development of pn junction devices, makes it useful in switching devices. When the metal side, which is the anode, is biased positive relative to the n-type material, the barrier is reduced and forward current flows by the drift of electrons from the semiconductor to the metal. The electrons that reach the metal during conduction are indistinguishable from the large supply of free electrons present in the metal. Hence, there are no minority carriers present in the neighborhood of the junction that must take part in recombination during reverse bias. Accordingly, when the bias is reversed, the electrons that crossed the junction from the semiconductor need not return from the metal. This virtually eliminates the storage time delay. Schottky diodes typically have approximately 50 ps of reverse recovery time. Note further that the fabrication of the unipolar, or majority carrier, Schottky diode does not require p diffusion, as in a pn junction. The junction must, however, be formed on a microscopically clean surface, which makes it more expensive to produce.

The forward voltage drop of a Schottky diode, whose circuit symbol is shown in Figure 2.25c, is usually lower than that of pn junction diodes. This is because of the increased reverse saturation currents of several orders in the Schottky diodes compared to those in pn junction diodes. Forward voltage varies from 0.3 V to 0.5 V for silicon, depending on its making contact with aluminum, tungsten, or platinum. The diode equation, Equation (2.50), is still applicable with the increased value of I_s. The saturation current I_s is a function of the barrier height and temperature. The low forward voltage of the Schottky diode is used to advantage in low-voltage digital circuits (such as integrated injection logic families, Chapter 5). Note that the diffusion capacitance, which arises from minority carrier storage effect, is zero. Junction capacitance C_T associated with reverse bias, however, is present, similar to that given for a pn junction. Due to the low value of the barrier potential, C_T is slightly higher for Schottky diodes than for pn junction diodes.

Schottky diodes formed using n-type gallium arsenide in direct contact with a metal behave similarly to Si-metal junctions, with a slightly higher forward voltage. GaAs Schottky diodes are used extensively in the GaAs technology of digital circuits (see MESFETs in Chapters 7 and 8). Schottky diodes are also formed using a p-type semiconductor and a metal. Because of their low barrier heights (lower than those using an n-type semiconductor), however, they are of limited interest.

The MicroSim PSpice model of a Schottky diode for computer simulation is identical to that of a pn junction diode with zero transit time, lower barrier height, and higher saturation current.

Metal-Semiconductor Ohmic Contact   As we saw in Section 2.3, contact between a metal and a semiconductor can be made ohmic; that is, current conduction can be made bidirectional with low potential barrier and low contact resistance. Ohmic contact between n-type silicon and aluminum (a low work function material) for external terminal connection with an IC, for example, is achieved by heavy donor doping at the interface; with a large enough number of free electrons, the n+ region is indistinguishable from the metal. Electrons easily overcome the low barrier and tunnel through an extremely narrow deple-tion region in either direction. In the case of contact with a p-type region, wire bonding with aluminum or gold (which acts as p-type impurity) is used with p$^+$ doping.

2.9   LIGHT-EMITTING DIODE

A light-emitting diode (LED) is a pn junction diode, which, when sufficiently forward-biased, emits spontaneous radiation in ultraviolet, visible, or infrared regions. Electron-hole recombination of injected carriers in a forward-biased silicon junction diode releases energy in the form of heat. In a compound semiconductor such as gallium arsenide or gallium phosphide, when conduction electrons drop back to the valence band and lose their excess energy, photons are emitted, one for each recombination. These photons have wavelength λ that depends on the semiconductor and the dopant. The energy ΔE released during recombination is slightly above the bandgap energy E_g. ΔE is related to quanta of discrete energy as $\Delta E = h\nu = hc/\lambda$, where h is the Planck constant, ν is the frequency of radiative emission, and c is the speed of light in vacuum. For visible light with maximum sensitivity of the eye, wavelength must be between 0.4 μm and 0.7 μm, which corresponds to a bandgap energy of between 1.8 eV and 3 eV. LEDs for the visible spectrum are made of compounds that can emit red, orange, yellow, or green light, depending on the composition. An LED made of gallium phosphide ($E_g \approx 2.2$ eV) doped with nitrogen, for example, emits green light ($\lambda \approx 0.56\,\mu$m). A gallium-arsenide-phosphide LED ($E_g \approx 1.8$ eV) provides visible red light ($\lambda \approx 0.69\,\mu$m). More recently, the compound AlInGaP has been developed for producing high luminous efficiency yellow and orange light. Blue light with high efficiency is produced using GaN.

The intensity of emitted light, which is concentrated near the junction, is proportional to the number of carriers participating in recombination; hence, higher forward bias results in higher current and brighter light. Since the recombination energy ΔE is higher than the bandgap energy E_g, emission has a spectral distribution. To obtain monochromatic light, a colored plastic lens is used; this lens serves as an optical filter and enhances contrast. Visible LEDs are used as indicator lamps and self-luminous display devices. The forward drop of LEDs is typically 1.5 V to 1.7 V at currents of 10 to 20 mA. The current-voltage characteristic of an LED is identical to that of a silicon junction diode with increased forward drop and lowered reverse breakdown voltage (<10V). Figure 2.26a shows the circuit symbol for an LED.

LEDs emitting light in the infrared are used as optoisolators (Figure 2.26b), where an input signal is isolated from the output. Gallium arsenide and gallium-indium-arsenide-phosphide semiconductors, for example, emit light in the infrared range. The photodiode, which is a light sensor, is a pn junction diode operating under reverse bias. When light impinges on the photodiode, photons interact with the valence electrons in the depletion region and generate free electron-hole pairs by photoionization. Because of the field created in the depletion region, these carriers move in the opposite direction and cause reverse current (in addition to the small thermal generation component). The reverse current of the photodiode is thus proportional to the incident light. To increase the efficiency of the light detection, an intrinsic semiconductor material is sandwiched between the layers of p and n regions. The resulting p-i-n diode has a wider depletion region so that more of the incident light takes part in electron-hole pair generation. The photodiode (detector) and the LED are enclosed in the same opaque package so that light proportional to the input signal directly illuminates the detector. The electrical isolation of input and output is extremely useful in transmitting digital signals at very high speed and in highly noisy environments.

Anode

Cathode

(a) Circuit Symbol

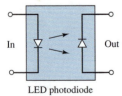

In   Out

LED photodiode

(b) Opto-isolator

Figure 2.26
Light-emitting diode

Infrared LEDs are used in the transmission of digital signals at optical frequencies. Light intensity is modulated by varying the diode current in accordance with the signal, and transmitted via fiber optic cables. A photodetector at the receiving end converts the optical signal back to electrical signal. Laser diodes are increasingly used in fiber optic transmission; these are similar in structure to LEDs but emit highly directional coherent radiation.

2.10   DIODES IN DIGITAL CIRCUITS

Diodes find many applications in electronic circuits. They are used, for example, in power supplies as rectifiers, in communication circuits as modulators and demodulators, and in signal processing circuits as waveform shapers and voltage limiters. In this section we consider the digital logic applications of diodes.

2.10.1   DIODE LOGIC CIRCUITS

Diodes can be used to perform OR and AND logic functions as shown in Figure 2.27. Let us assign, arbitrarily, the following voltages for logic low and logic high:

$$0 < V_L < 0.7 \, \text{V} \quad \text{and} \quad 4 \, \text{V} < V_H < 5 \, \text{V}$$

Then, in Figure 2.27a we see that output Y performs the OR function of the two inputs. For a logic high input of 5 V, the output is 4.3 V (assuming a 0.7 V drop for the on diode), and it is 0 V when both inputs are at 0 V. Output Y in the AND circuit of Figure 2.27b has a logic high voltage of 5 V and a logic low voltage of 0.7 V.

A problem with these simple logic circuits is the diode forward voltage drop of 0.7 V. If, for example, the output of the OR gate in Figure 2.27a is connected to the input of another identical OR gate, the second gate output will be no

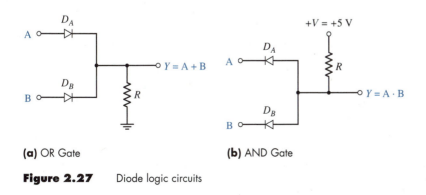

(a) OR Gate   **(b)** AND Gate

Figure 2.27   Diode logic circuits

higher than 3.6 V, a value unacceptable for our assigned logic high state. In the case of two cascaded AND gates with one input at 0V at the first gate, output of the second gate will be at 1.4 V; again this is not in the acceptable range of logic values.

Although we can compensate for the diode drop by adding a buffer diode (see Figure P2.10), there still remains the limitation due to loading. The logic high output of Figure 2.27b, for example, drops when it is connected to the input of an OR gate as shown in Figure 2.27a. Modification of diode logic circuits by the addition of one or more active devices (buffer amplifiers) to drive loads results in saturation logic circuits (Chapter 4). Active devices are also needed to implement logic inverter circuits.

2.10.2   TRIGGER GENERATOR

A simple diode circuit that is frequently used for generating a sharp trigger of a given polarity from a pulse input is shown in Figure 2.28. To obtain a narrow trigger pulse, the time constant of this circuit must be small compared to the pulse interval (see problem 2.11).

Other applications of diodes in logic circuits include voltage reference generators for ECL circuits, voltage clamps for TTL circuits, and logic level shifters in GaAs circuits.

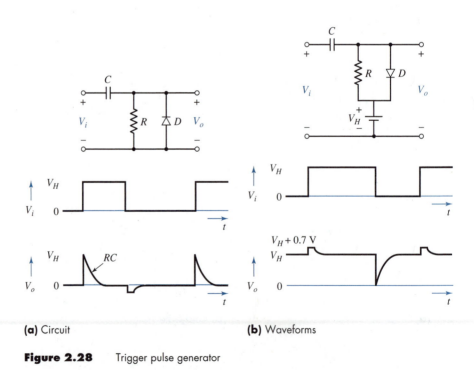

(a) Circuit   **(b)** Waveforms

Figure 2.28   Trigger pulse generator

SUMMARY

This chapter introduced the basic current conduction mechanisms in semiconductors.

Silicon is the most widely used single-element semiconductor. Other commonly used semiconductors include germanium, gallium arsenide, and indium phosphide.

Conductivity of a pure, or intrinsic, semiconductor is increased by doping it with certain impure elements. Extrinsic, or doped, semiconductors can be n-type, with a majority of electrons available for conduction, or p-type, with positively charged holes available as charge carriers.

Both drift current due to applied electric field and diffusion current due to concentration gradient of charge carriers are present in semiconductors.

The pn junction diode, a basic microelectronic device, is formed by doping a semiconductor with both n- and p-type impurities. Due to diffusion and recombination of holes and electrons, there is a carrier-depleted, space-charge region close to the junction of the p and the n regions. External forward-bias voltage is required to overcome the barrier potential arising from the space-charge.

With bias voltage, a diode operates as a unidirectional switch, offering negligible resistance when forward-biased and a large resistance when reverse-biased. Forward current varies exponentially with voltage while the reverse current is almost independent of bias. Reverse saturation current depends on doping level, area, and temperature.

Because of immobile charge storage in the depletion region and carriers near the edges of the depletion region, a diode has transition and diffusion capacitances. These capacitances contribute to diode switching delays.

A Zener diode is used to obtain a constant voltage, equal to the diode breakdown voltage, under varying currents.

A Schottky diode, formed by a metal-semiconductor junction, is a majority carrier, or unipolar device, with negligible storage delay.

Compound semiconductors are used to form pn junction light-emitting diodes, which emit radiation in the visible or infrared spectrum during electron-hole recombination. Photodiodes, which convert incident light to current, are used with LEDs in optoisolator applications.

Diode circuits can realize AND and OR logic functions; forward voltage drop and loading of diodes, however, limit their applications.

REFERENCES

1. Millman, J., and C. C. Halkias. *Integrated Electronics: Analog and Digital Circuits and Systems.* New York: McGraw-Hill, 1972.

2. Pulfrey, D. L., and N. G. Tarr. *Introduction to Microelectronic Devices.* Englewood Cliffs, N. J.: Prentice Hall, 1989.

3. Muller, R. S., and T. I. Kamins. *Device Electronics for Integrated Circuits.* New York: Wiley, 1986.

4. Sze, S. M. *Semiconductor Devices: Physics and Technology.* New York: Wiley, 1985.

5. Hodges, D. A., and H. G. Jackson. *Analysis and Design of Digital Integrated Circuits.* New York: McGraw-Hill, 1988.

6. Streetman, B. G. *Solid State Electronic Devices.* Englewood Cliffs, N. J.: Prentice Hall, 1990.

7. *PSpice Circuit Analysis User's Guide.* Irvine, Calif: MicroSim Corporation, 1991.

8. Werner, K. "Higher Visibility for LEDs." *IEEE Spectrum,* July 1994, pp. 30–39.

REVIEW QUESTIONS

1. What is bandgap of a semiconductor?

2. How can the conductivity of an intrinsic semiconductor be increased?

3. How is drift current related to applied field?

4. Why are single crystal semiconductors used in devices?

5. How does conduction in a doped semiconductor differ from that in a metal?

6. What causes diffusion current in a semiconductor?

7. How does doping a pure silicon bar with phosphorus atoms increase its conductivity?

8. What are n- and p-type semiconductors?

9. What causes the built-in potential in a diode? What is its effect?

10. What factors govern the reverse saturation current of a diode?

11. Under what biasing condition does diffusion capacitance dominate over junction capacitance?

12. If the circuit resistances in a diode logic circuit are in the range of 100 Ω to 500 Ω and the source voltages are in the 1 V to 2 V range, what diode model is more appropriate?

13. What is the normal bias for the operation of a Zener diode?

14. How does a Schottky diode differ in construction and performance from a pn junction diode?

15. How does ohmic contact in a metal differ from rectifying contact?

16. How does the function of an LED differ from that of a pn junction diode?

17. What is the normal biasing of a photodiode? How does it increase current?

18. What is the source of power for a diode OR gate?

19. If a series of three diode OR gates are cascaded together, what is the logic high voltage at the third gate output?

20. In terms of logic implementation, what is the limitation of diode gates, if loading and diode drop can be compensated for?

21. To obtain a sharp trigger in a diode trigger circuit, how must you choose the time constant?

PROBLEMS

1. Calculate the conductivities of intrinsic gallium arsenide at 300 K and at 370 K if the corresponding concentrations are $2.25 \times 10^{12}/m^3$ and $10^{15}/m^3$. What is the percentage rate of change of conductivity per degree increase in temperature?

2. If the donor and the acceptor doping levels in a silicon diode are $1.5 \times 10^{21}/m^3$ and $10^{23}/m^3$, calculate the contact potential at 300 K. What are the densities of minority carriers? What are the depletion widths on the n and the p sides? What is the maximum value of the electric field? What is the contact potential at 100°C?

3. Under equilibrium conditions with no current, show that the Fermi level in the n- and p-type regions in a junction diode is constant throughout. Note that the rate of carrier transport from one region to the other is proportional to the number of filled states in one region and the corresponding number of empty states in the other at energy E.

4. If the minority carrier lifetime in the n region is 100 ns for a diode (with $N_A \geq N_D$), calculate the diffusion capacitance at 300 K while carrying a current of 2 mA.

5. For the circuit shown in Figure P2.5, calculate the diode voltage and current using simplified models and compare with the iterative solution using Equation (2.60).

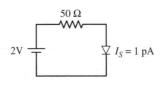

Figure P2.5

*6. A forward-biased diode with $I_s = 1$ pA carries a current of 5 mA at 300 K. Calculate the static and dynamic resistances. If the temperature rises to 350 K, what are the new resistances at the same forward current? Assume I_s changes by 1 percent/K.

*7. For the circuit shown in Figure P2.7, calculate the indicated currents and the voltage V_o. Verify your results by running a MicroSim PSpice simulation.

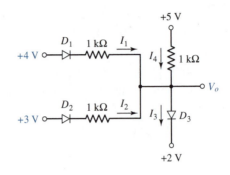

Figure P2.7

8. For the circuit shown in Figure 2.24, with $R_s = 400\,\Omega$ $V_z = 5.3\,\text{V}$, determine the output voltage V_o if V_s varies from 10 V to 12 V. Assume that the diode has a forward resistance of 50 Ω and the Zener diode has a resistance of 20 Ω. If diode has a temperature coefficient of $-2\,\text{mV/°C}$, and the Zener diode voltage varies by 3 mV/°C, how much does V_o change over a 50°C rise?

9. For the diode circuit shown in Figure P2.9, calculate the voltage V_o. Assume a drop of 1.5 V for the LED. What is the current in the LED?

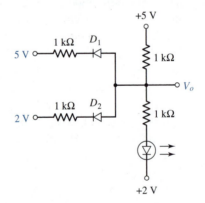

Figure P2.9

10. Verify the logic operation of the buffered diode gates shown in Figure P2.10. Calculate V_o when a 2 kΩ resistor is connected from each output to ground.

*11. A trigger circuit output is connected to the input of a logic inverter as shown in Figure P2.11. If the input logic low and high threshold voltages of the inverter are 1.2 V and 1.8 V respectively, sketch the waveform at the output of the inverter.

   Perform a MicroSim PSpice transient simulation for the output of the trigger circuit and verify your result.

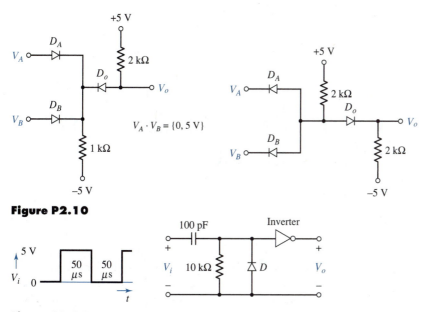

Figure P2.10

Figure P2.11

*12. If the pulsewidth of the inverter output must be no more than 500 ns in the previous problem, determine R and C.

13. *Experiment: Diode parameter measurement.* In this experiment, determine the diode parameters I_s, n, and t_p for the 1N914 diode.

a. To estimate I_s, apply a reverse bias to the diode as shown in Figure P2.13 using the negative supply voltage V_S. Measure the reverse current for $-1\,V \geq V_D \geq -5\,V$. Note that the diode must be sufficiently reverse-biased so that the current I_D (<0) is relatively constant, but not operating in the breakdown region. Since I_s is extremely small, use large resistance for R (in the MΩ range) so that there is a measurable drop in voltage across R. Estimate I_s.

b. Apply forward bias to the diode and measure the forward current I_D and forward voltage V_D for $I = 500\,\mu A$, $1\,mA$, $5\,mA$, and $10\,mA$. Use a variable resistor R in the range of $1\,k\Omega$ to $10\,k\Omega$ to limit the diode current.

c. Using the estimated value of I_s in (a) and the data from (b), plot in (I/I_s) as a function of forward voltage V. From the slope of the straight-line graph, estimate the value of nV_T. Using the equation $\ln(I_2/I_1) \approx (V_2 - V_1)/nV_T$ for $I_1 = 1\,mA$ and $I_2 = 2\,mA$, recalculate nV_T; from this value of nV_T, reestimate I_s. Compare the two sets of I_s and n.

d. In the circuit shown in Figure 2.22a, apply a square wave voltage V_i with $V_F = 5\,V$ and $V_R = -5\,V$. Study the effect of R on the switching delays using $R = 1\,k\Omega$ and $R = 10\,k\Omega$. From the storage delay measurement, estimate the minority carrier lifetime using Equation (2.65).

3

INTRODUCTION TO BIPOLAR JUNCTION TRANSISTORS

INTRODUCTION

In this chapter we study the basic structure and the terminal currents of the *bipolar junction transistor* (BJT), commonly referred to as the *transistor*. A transistor has two pn junctions formed on a single piece of silicon, germanium, or the compound GaAs. Transistors are used extensively in electronic circuits as voltage and current amplifiers and switching devices.

We use the basic pn junction theory developed in Chapter 2 to study the operation and modeling of the transistor. We derive the analytical expressions relating voltages and currents using the large-signal Ebers-Moll model of a BJT. For small-signal linear amplification, we describe the hybrid-π model. Using these models, we study different modes of operation and different circuit configurations of the transistor. For analyzing the operation of the BJT in any mode, we examine the unified charge-control model and its relationship to the other two models. Following this, we consider the model of the BJT for computer simulation using MicroSim PSpice and discuss some of the commonly used model parameters. The chapter ends with a brief discussion of power dissipation and thermal effects.

3.1 SIMPLIFIED STRUCTURE OF A BJT

A BJT is a three-terminal device with two pn junctions formed in series opposition. Figure 3.1a shows the simplified structure of an npn BJT. (We use this structure for purposes of analysis only; practical BJT structures are illustrated in Figure 3.3.) As Fig. 3.1a shows, the two junctions are formed between two n-type regions and a p-type region. These regions are identified as the *emitter,* the *base,* and the *collector,* and the leads connected to the three regions are labeled accordingly. The complement of the npn transistor is the pnp transistor (Figure 3.2a) which has an n-type base formed between a p-type emitter and a p-type collector. The symbols for both types of transistors are shown in Figures 3.1b and 3.2b; the arrow heads in the emitter leads point to the direction of positive external current flow under normal operating modes.

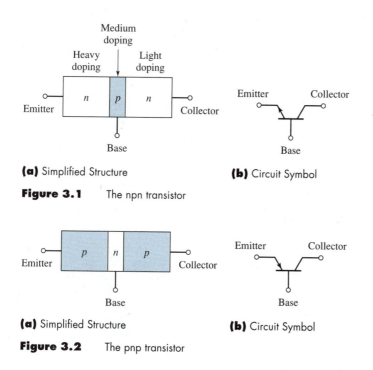

Medium doping

Heavy doping   Light doping

| Emitter | n | p | n | Collector |

Base

(a) Simplified Structure

Emitter   Collector

Base

(b) Circuit Symbol

Figure 3.1   The npn transistor

| Emitter | p | n | p | Collector |

Base

(a) Simplified Structure

Emitter   Collector

Base

(b) Circuit Symbol

Figure 3.2   The pnp transistor

The structure of the discrete npn transistor depicted in Figure 3.3a shows a heavily doped n^+ substrate on which a relatively large but thin, lightly doped film of n-type region is grown for the collector. After oxidizing the collector surface, a small window is opened by etching, and the base is formed by diffusing p-type impurity in the window. This base region is reoxidized, another window is opened, and heavy n-type impurity is diffused to form the emitter region.[1] The emitter is designed to be heavily doped so as to inject majority carrier electrons into the base. The base, on the other hand, is a thin region with medium doping. The collector region is wide, with the smallest doping. Typical doping levels are: $10^{26}/m^3$ for the emitter, $10^{24}/m^3$ for the base, and $10^{21}/m^3$ for the collector. Aluminum leads are attached to the three regions in a discrete transistor. Note that the narrow emitter region is surrounded by the wide but thin base region, so that carriers leaving the emitter must pass through the base. The asymmetric doping and geometries for the two junctions result in different currents in the device when the emitter and collector terminals are interchanged. The main purposes of the heavily doped n^+ substrate between the collector region and the terminal are to provide physical strength and to serve as an ohmic contact. Therefore, it will not be included in the analysis of voltages and currents. Multiple transistors in integrated circuits have a similar structure, as Figure 3.3b

---

[1] As an alternative to the vertical structure, emitter, base, and collector can be arranged laterally. Lateral pnp transistors, which are more difficult to fabricate, are used in I^2L circuits (Chapter 5).

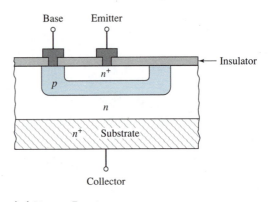

(a) Discrete Transistor

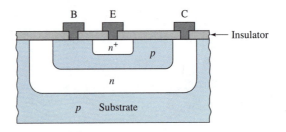

(b) IC Transistor

Figure 3.3   Structures of practical npn transistors

shows, with each npn transistor isolated from others by heavily doped p regions or by large oxide regions.

The two pn junctions are identified as the *emitter-base junction* (EBJ) and the *collector-base junction* (CBJ). The path from emitter to collector is via the two series pn junctions connected in series opposition. Hence, the BJT is a normally off device. When the junctions are biased, current across one junction is controlled by the current (or voltage) associated with the other junction; that is, the BJT operates as a dependent current source. Both majority and minority carriers participate in the current conduction process. Since the operation of the device depends on both negative and positive charge carriers (i.e., electrons and holes), the device is called a *bipolar* transistor.

Although both npn and pnp transistors are commercially available as discrete devices, we shall consider the npn transistor in our analysis for the following reasons. The analysis for the npn transistor can be readily modified for the pnp transistor with appropriate voltage polarities and current directions. As we saw in Chapter 2, the mobility of electrons, which are the dominant carriers of current in npn transistors, is higher than that of the holes in semiconductors at room temperature. Therefore, npn transistors operate at faster switching speeds than comparable pnp transistors, and as a result, integrated circuits for switching applications are primarily made using npn transistors.

3.2   CURRENTS IN A BJT

Currents in a BJT can be described using the *forward active, linear,* or *normal mode* of operation shown in Figure 3.4a. Although the forward-active mode of operation is encountered primarily in linear circuit applications, its analysis provides us with insight into the current components that we can extend to other modes of operation.

As shown in Figure 3.4a, the EBJ is forward-biased and the CBJ is reverse-biased; that is, the base is more positive relative to the emitter, and the collector is more positive relative to the base. The forward-bias V_{BE} (>0) lowers the potential barrier at the EBJ from the built-in potential of V_{0E} to $V_{0E} - V_{BE}$, while the reverse-bias V_{BC} (<0) raises the barrier height at the CBJ from V_{0C} to $V_{0C} - V_{BC}$, as shown in the energy level diagram of Figure 3.4b. The lowering of forward bias at the EBJ causes electrons to be injected or *emitted* from the emitter region into the base, and holes to be injected from the base into the emitter. Due to the heavy doping of the emitter, however, far more electrons reach the base from the emitter than do holes reach the emitter from the base. Thus there is a steady flow of minority carriers (electrons) into the base for sufficient forward bias of EBJ by steady voltage V_{BE}. Since the base width is designed to be very small—less than one diffusion length of electrons—very few of the injected electrons are lost in recombination in the base region. The remaining large number of minority electrons diffuse toward the collector. Near the CBJ there is a wide depletion region created by the reverse bias voltage V_{BC}. The diffusing electrons reaching near the CBJ are, therefore, swept across by

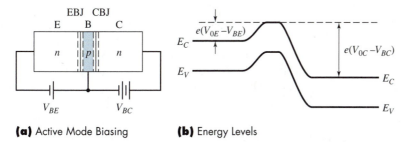

(a) Active Mode Biasing   **(b)** Energy Levels

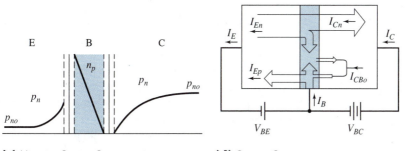

(c) Minority Carrier Concentration   **(d)** Current Components

Figure 3.4   The npn transistor in forward active mode

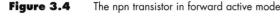

the electric field in the depletion region. The collector, thus, *collects* most of the electrons injected by the emitter. Note that the electron injection into the base depends exponentially on the base-emitter junction forward voltage V_{BE} [Equation (2.55)]. Hence, the number of electrons reaching the collector junction increases exponentially as V_{BE} increases. The minority carrier concentration in each region is shown in Figure 3.4c. The straight line profile in the base neglects the small amount of electrons recombining in the thin base region.

Holes that are lost in recombination in the base, and those injected from the base into the emitter, constitute the base current. Both of these components of base current are proportional to the foward bias at the EBJ.

The collector current is primarily due to the diffusion of the injected electrons from the emitter that reach the collector. Since the CBJ is reverse-biased, a small reverse saturation current caused by the thermal generation of holes also contributes to the collector current. These holes diffuse across the base and reach the emitter.

From the foregoing discussion it is clear that if the forward bias across the EBJ is increased, more electrons leave from the emitter and, consequently, more current results in the base and the collector. Increase in the reverse bias voltage V_{CB}, on the other hand, has little effect on any of the currents in the first-order model we examine. In addition, current in the base region is mainly due to diffusion of minority carriers. These carriers are accelerated by the field across the CBJ depletion region and reach the collector; hence, the collector current arises primarily by the drift of minority carriers from the base region.

Since the sum of the terminal currents must be zero (Figure 3.4d), we have the first basic relation for BJT currents as

$$I_C + I_B = I_E \qquad\qquad \textbf{(3.1)}$$

Note that the collector and the base currents, I_c and I_B, are into the transistor while the emitter current I_E is out of the transitor. The emitter current I_E is the sum of two components: I_{En} due to electrons injected into the base, and I_{Ep} due to holes injected from the base to the emitter (Figure 3.4d). The ratio, $\gamma = I_{En}/I_E$, is the *emitter injection efficiency*. Since only I_{En} contributes to collector current, γ must be as close to unity as possible for $I_C \approx I_E$. High emitter efficiency is achieved by much heavier doping of the emitter than the base, as mentioned. Because of recombination in the base, only a fraction of the electron current I_{En} reaches the collector. The ratio of the electron current at the collector, I_{Cn}, to the electron current of the emitter, I_{En}, is the *base transport factor*, $B = I_{Cn}/I_{En}$. To achieve I_C close to I_E, we make the base transport factor high by several methods. We reduce the loss of electrons due to recombination by making the base width smaller than a diffusion length, and by the low-level doping of the base. Additionally, we make the geometry such that the collector region surrounds the base, so that the diffusing electrons are readily swept by the electric field in the depletion region of the CBJ.

Denoting the fraction of the total emitter current reaching the collector by $\alpha = \gamma B$, the collector current is given by

$$I_C = \alpha I_E + I_{CBO} \qquad\qquad \textbf{(3.2)}$$

where I_{CBO} stands for the current from collector to base with emitter open, that is, the CBJ saturation current. α is called the *common-base current gain*. This terminology arises from the circuit configuration in which the base terminal is the reference, or common node, to both the emitter and the collector (Figure 3.4a).

From Equations (3.1) and (3.2), we can now write

$$I_E = \frac{(I_B + I_{CBO})}{(1 - \alpha)} \tag{3.3}$$

Defining β as

$$\beta = \alpha/(1 - \alpha) \tag{3.4}$$

the emitter current in terms of the base current is now given by

$$I_E = (\beta + 1)I_B + (\beta + 1)I_{CBO} \tag{3.5}$$

The collector current in terms of the base current is, therefore, given by

$$I_C = \beta I_B + (\beta + 1)I_{CBO} \tag{3.6}$$

The quantity β, defined in Equation (3.4), is called the *common-emitter current gain*. β is measured experimentally for small signal variations using Equation (3.6) as

$$\beta = \frac{\Delta I_C}{\Delta I_B} \tag{3.7}$$

where the saturation current I_{CBO} is constant at a given temperature. This value of β is called the *small-signal current gain, or β_{ac}*.

For large signal variations, as in switching applications, we have, from Equation (3.6),

$$\beta_{dc} = \frac{[I_C - I_{CBO}]}{[I_B + I_{CBO}]} \tag{3.8}$$

If we define the cutoff mode of a transistor (more about this mode later) to correspond to $I_C = I_{CBO}$, $I_E = 0$, and $I_B = -I_{CBO}$, then the *large-signal β*, or β_{dc}, is given by

$$\beta_{dc} = \frac{\text{Change in collector current from cutoff to active mode}}{\text{Change in base current from cutoff to active mode}}$$

(Note that if the emitter is open ($I_E = 0$), I_{CBO} flows out of the base.)

Since the reverse saturation current I_{CBO} is on the order of $10^{-15}\,\mu$A, and I_C is in mA, we may neglect I_{CBO} and write

$$\beta_{dc} \approx \frac{I_C}{I_B} \tag{3.9}$$

for most of the active range. A typical value for β is in the range of 50 to 200, with specially made high-gain transistors having a value of up to 1000. Both β_{ac} and β_{dc} vary with the dc operating point and temperature.

3.3   EBERS-MOLL MODEL OF A BJT

To study the terminal currents and voltages under different biasing conditions of the two junctions, we can model the npn transistor (Figure 3.5a) as two pn junction diodes connected back to back, with a common p region as shown in Figure 3.5b. Terminal currents and voltages are shown in Figure 3.5c. The Ebers-Moll model, shown in Figure 3.5d, is applicable at different biasing voltages of two junctions (that is, the modes of operation of the BJT) under static conditions. The two diodes at the EBJ and the CBJ are represented by D_E and D_C, each of which accounts for the current crossing the junction due to the voltage across it. The second component of the terminal current I_C (I_E) arises from the injected carriers from the emitter (collector) diffusing across the base and reaching the collector (emitter). This feedback, or coupling of current from each junction to the other, distinguishes the operation of a transistor from that of two noninteracting series-connected diodes like those shown in Figure 3.5b.

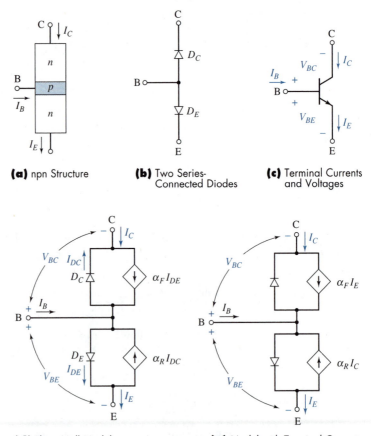

(a) npn Structure

(b) Two Series-Connected Diodes

(c) Terminal Currents and Voltages

(d) Ebers-Moll Model

(e) Model with Terminal Currents

Figure 3.5   Development of the Ebers-Moll model

Depending on the biasing of the junctions, one or more of the components of the terminal currents may be negligible.

Recall that in the forward active (normal) mode, α is the ratio of the collector current due to electrons reaching the collector to the total emitter current. The *forward,* or *normal active current gain,* denoted by αF (or α_N), in Figure 3.5 is the fraction of the injected minority carriers from the emitter into the base that reaches the collector. I_{DE}, which is the diode current that crosses the emitter junction, is a function of the EBJ bias voltage. Similarly, I_{DC} represents the diode current crossing the CBJ. When the CBJ is forward-biased and the EBJ is reverse-biased, the roles of emitter and collector are reversed, so that electrons are injected from the collector and a portion of these are collected at the emitter. This mode of operation is called the *inverse,* or *reverse active mode.* The fraction of the collector-injected electrons that reach the emitter in this mode is the *inverse,* or *reverse active current gain* and is denoted by α_R (or α_I). (The inverse active mode of operation, although it does not result in identical terminal behavior as does the forward active mode, is used in the transistor-transistor-logic (TTL) family of switching circuits, for example.) A typical value for α_R is in the range of 0.02 to 0.5 for integrated circuit transistors and 0.4 to 0.8 for discrete transistors. α_F, on the other hand, is designed to be close to unity.

The terminal currents are related to the diode currents (Figure 3.5d) as

$$I_C = \alpha_F I_{DE} - I_{DC} \qquad \textbf{(3.10a)}$$

$$I_E = I_{DE} - \alpha_R I_{DC} \qquad \textbf{(3.10b)}$$

Using the ideal diode equation [Equation (2.50)] with the ideality factor $n = 1$, the diode currents are written as

$$I_{DC} = I_{CS}\left[\exp\left(\frac{V_{BC}}{V_T}\right) - 1\right] \qquad \textbf{(3.11a)}$$

$$I_{DE} = I_{ES}\left[\exp\left(\frac{V_{BE}}{V_T}\right) - 1\right] \qquad \textbf{(3.11b)}$$

where I_{CS} and I_{ES} represent the reverse saturation currents for the CBJ and the EBJ respectively. I_{CS} is larger than I_{ES} by a factor of 2 to 50 because of the much larger area of the CBJ compared to the EBJ. Both currents are on the order of 10^{-15} A and are dependent on temperature.

Combining Equation (3.10) and (3.11), the terminal currents of an npn transistor in the Ebers-Moll model are given by

$$I_C = \alpha_F I_{ES}\left[\exp\left(\frac{V_{BE}}{V_T}\right) - 1\right] - I_{CS}\left[\exp\left(\frac{V_{BC}}{V_T}\right) - 1\right] \qquad \textbf{(3.12a)}$$

$$I_E = I_{ES}\left[\exp\left(\frac{V_{BE}}{V_T}\right) - 1\right] - \alpha_R I_{CS}\left[\exp\left(\frac{V_{BC}}{V_T}\right) - 1\right] \qquad \textbf{(3.12b)}$$

The base current $I_B = I_E - I_C$ is

$$I_B = I_{ES}(1 - \alpha_F)\left[\exp\left(\frac{V_{BE}}{V_T}\right) - 1\right] + I_{CS}(1 - \alpha_R)\left[\exp\left(\frac{V_{BC}}{V_T}\right) - 1\right] \qquad \textbf{(3.12c)}$$

In addition to the terminal currents given above, the four diode currents are related by the reciprocity condition

$$\alpha_F I_{ES} = \alpha_R I_{CS} \tag{3.13}$$

Equations (3.12) and (3.13) are the basis for the analysis and design of circuits using the BJT, operating under any of the four modes of operation which we discuss in the following section. In these equations the diode saturation currents may be eliminated (Problem 3.1), so that the emitter and the collector currents become

$$I_C = \alpha_F I_E - I_{CBO}\left[\exp\left(\frac{V_{BC}}{V_T}\right) - 1\right] \tag{3.14a}$$

$$I_E = \alpha_R I_C + I_{EBO}\left[\exp\left(\frac{V_{BE}}{V_T}\right) - 1\right] \tag{3.14b}$$

where

$$I_{CBO} = I_{CS}(1 - \alpha_F \alpha_R) \tag{3.15a}$$

and

$$I_{EBO} = I_{ES}(1 - \alpha_F \alpha_R) \tag{3.15b}$$

are the collector and the emitter junction reverse saturation currents respectively, with the other terminal open. As the saturation currents I_{CBO} and I_{EBO} are obtained using the emitter or the collector open, these are known as the *open-circuit saturation currents,* while I_{ES} and I_{CS} are known as the *short-circuit saturation currents.* Figure 3.5e shows the Ebers-Moll model using the terminal currents. The four model parameters—α_F, α_R, I_{ES}, and I_{CS} (or I_{EBO} and I_{CBO}), three of which are independent—can be obtained experimentally. With these equations we are now ready to look at the different modes of operation.

3.4   MODES OF OPERATION

Each of the two pn junctions in a BJT may be biased in the forward or in the reverse direction. Table 3.1 lists the four modes of operation depending on the biasing of the two pn junctions.

Table 3.1   Modes of operation for a BJT

EBJ	CBJ	Mode
Forward	Reverse	Forward active
Reverse	Forward	Inverse active
Reverse	Reverse	Cutoff
Forward	Forward	Saturation

Forward/Normal Active Mode   As stated earlier, this is the normal, or linear, mode of operation in which the emitter-base junction is forward-biased ($V_{BE} > 0$) and the collector-base junction is reverse-biased ($V_{BC} < 0$). To determine the approximate currents in this mode, we assume that $V_{BE} \geq 4V_T$ and $V_{BC} \leq -4V_T$. Note that V_T is about 25 mV at room temperature, and that at junction forward voltages below $4V_T$ (≈ 100 mV), the currents are negligibly small. Hence, for linear operation, sufficient biasing requires $V_{BE} \geq 4V_T$ and $V_{BC} \leq -4V_T$.

The currents from Equations (3.12) are

$$I_C \approx \alpha_F I_{ES} \exp\left(\frac{V_{BE}}{V_T}\right) + I_{CS} \qquad \textbf{(3.16a)}$$

$$I_E \approx I_{ES} \exp\left(\frac{V_{BE}}{V_T}\right) + \alpha_R I_{CS} \qquad \textbf{(3.16b)}$$

Since the first term in each of the above equations is large for $V_{BE} \geq 4V_T$, we approximate the collector current using Equation (3.14) as

$$I_C \approx \alpha_F I_E + I_{CBO} \qquad \textbf{(3.17a)}$$

The base current is given by

$$I_B \approx I_{ES}(1 - \alpha_F)\exp\left(\frac{V_{BE}}{V_T}\right) - I_{CS}(1 - \alpha_R) \qquad \textbf{(3.17b)}$$

Comparing Equation (3.17a) to Equation (3.2), we see that the forward active current gain $\alpha_F \approx \alpha$.

Since $\alpha_F \approx 1$, Equation (3.17b) shows that the base current is an extremely small fraction of I_E. If we neglect the saturation junction currents, it is clear that the collector current is more than the base current by a large factor,

$$I_C/I_B \approx \alpha_F/(1 - \alpha_F) = \beta_F \qquad \textbf{(3.18)}$$

It is this large (base-to-collector) current gain that makes the forward active mode of the BJT useful in amplifier applications. Using the common-emitter current gain β_F, the terminal currents are given by Equations (3.5) and (3.6), which are repeated here.

$$I_E = (\beta + 1)I_B + (\beta + 1)I_{CBO} \qquad \textbf{(3.5)}$$

$$I_C = \beta I_B + (\beta + 1)I_{CBO} \qquad \textbf{(3.6)}$$

Note that when there is no confusion, α and β will be used in the forward active mode to identify α_F and β_F.

The simplified models corresponding to the above equations are shown in Figure 3.6.

Reverse/Inverse Active Mode   If the EBJ is reverse-biased by $V_{BE} \leq -4V_T$ and the CBJ is forward-biased by $V_{BC} \geq 4V_T$, that is, if the roles of the emitter and the collector are reversed from their forward active mode, the BJT

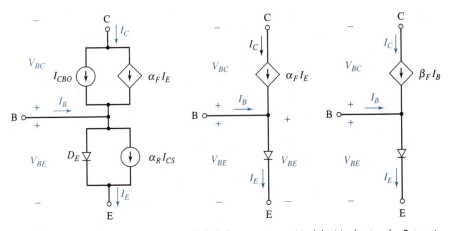

(a) Model Based on Equations (3.16) and (3.17)   **(b), (c)** Approximate Models, Neglecting the Saturation Currents

Figure 3.6   Simplified model of forward active BJT

is said to be operating in the *reverse,* or *inverse active mode.* The currents in this mode are given by

$$I_C \approx -\alpha_F I_{ES} - I_{CS} \exp\left(\frac{V_{BC}}{V_T}\right) \qquad \textbf{(3.19a)}$$

$$I_E \approx -I_{ES} - \alpha_R I_{CS} \exp\left(\frac{V_{BC}}{V_T}\right) \qquad \textbf{(3.19b)}$$

$$I_B \approx -I_{ES}(1 - \alpha_F) + I_{CS}(1 - \alpha_R)\exp(V_{BC}/V_T) \qquad \textbf{(3.19c)}$$

Clearly the base current is much higher than that in the forward active mode; the collector current, however, is only slightly above I_B, since

$$I_C \approx \frac{-I_B}{(1 - \alpha_R)} = -(1 + \beta_R)I_B \qquad \textbf{(3.20)}$$

and

$$\beta_R = \frac{\alpha_R}{(1 - \alpha_R)} < 1$$

by the nonsymmetrical design of doping and geometry. The inverse mode is, therefore, not suitable for current amplification. But the low value of β_R (from as low as 0.01 to a maximum of unity) is advantageous in the TTL family of switching circuits. Figure 3.7 shows the simplified model of a BJT operating in the reverse active mode.

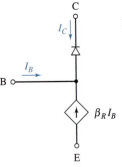

Figure 3.7
Simplified model of BJT operating in reverse active mode

Cutoff Mode   In this mode, both the junctions are reverse-biased by at least $4V_T$. The currents are given by

$$I_C \approx -\alpha_F I_{ES} + I_{CS} \tag{3.21a}$$

$$I_E \approx -I_{ES} + \alpha_R I_{CS} \tag{3.21b}$$

$$I_B \approx -I_{ES}(1 - \alpha_F) - I_{CS}(1 - \alpha_R) \tag{3.21c}$$

Using the reciprocity condition given in Equation (3.13), we may write the collector and the emitter currents as

$$I_C \approx I_{CS}(1 - \alpha_R) \tag{3.22a}$$

$$I_E \approx -I_{ES}(1 - \alpha_F) \tag{3.22b}$$

Since these currents are extremely small (even smaller than the reverse saturation currents of the junctions), the device may be considered virtually open-circuited, or *off,* and modeled as shown in Figure 3.8. This model of the BJT as an open switch is justified from the very large impedance between the collector and the emitter. Note that the currents given by Equation (3.22) are close to $I_C \approx I_{CBO}$ and $I_E \approx 0$, which we used to define the cutoff mode in Equation (3.8). It can be shown that the EBJ bias voltage must be below

$$V_{BE(\text{cutoff})} = V_T \ln(1 - \alpha_F)$$

(see Problem 3.2) to achieve these low cutoff currents.

We must note that in order to operate the transistor in cutoff mode with extremely small terminal currents, it is not necessary to reverse-bias the EBJ. When $V_{BE} = 0$, for example, $I_C \approx I_{CS}$ and $I_E \approx \alpha_R I_{CS} = \alpha_F I_{ES}$. These currents are equally negligible as those given in Equation (3.22), since the current gains α_F and α_R are close to zero near curoff for silicon devices. Thus, a silicon BJT can be driven into cutoff mode with zero bias for the EBJ. This is a distinct advantage when operating saturating logic circuits with a single positive power supply.

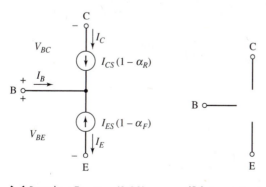

(a) Based on Equation (3.22)   **(b)** Approximate Model

Figure 3.8   BJT in cutoff mode

The cutoff currents discussed above and given in Equations (3.21) and (3.22) are based on the ideal diode equation for the two junctions. In practice, however, cutoff currents are higher (on the order of a few nA) due to surface leakage around the junctions and to thermal generation by collision in the depletion region. The open-circuit model is still valid because the useful range of currents is typically many orders of magnitude higher than the saturation currents.

If a transistor is in cutoff mode, what is the minimum voltage required to bring it into conduction? We can keep the CBJ reverse-biased but forward-bias the EBJ, so that the device can go into the forward active mode. This requires about 0.6 V for V_{BE} which, we recall from Section 2.5 and from the diode equation [Equation (3.16b)], is the (cut-in) voltage needed by the emitter-base pn junction to conduct. Note that we can activate the CBJ by raising V_{BC} to about 0.6 V while keeping the EBJ reverse-biased. This, of course, takes the transistor to the reverse active mode.

Saturation Mode   If both the emitter-base and the collector-base junctions are forward-biased with $V_{BE} \geq 4V_T$ and $V_{BC} \geq 4V_T$, the currents are given by

$$I_C \approx \alpha_F I_{ES} \exp\left(\frac{V_{BE}}{V_T}\right) - I_{CS} \exp\left(\frac{V_{BC}}{V_T}\right) \qquad \textbf{(3.23a)}$$

$$I_E \approx I_{ES} \exp\left(\frac{V_{BE}}{V_T}\right) - \alpha_R I_{CS} \exp\left(\frac{V_{BC}}{V_T}\right) \qquad \textbf{(3.23b)}$$

$$I_B \approx I_{ES}(1 - \alpha_F) \exp\left(\frac{V_{BE}}{V_T}\right) + I_{CS}(1 - \alpha_R) \exp\left(\frac{V_{BC}}{V_T}\right)$$

or,

$$I_B \approx (1 - \alpha_F)I_{DE} + (1 - \alpha_R)I_{DC} \qquad \textbf{(3.23c)}$$

The last equation indicates that the BJT can be in saturation for any positive base current I_B, since the right side of the equation is positive. The collector current is determined by external circuitry, that is, by the power supply and the collector circuit resistance.

To see the effect of saturation on the collector-emitter voltage, consider the circuit shown in Figure 3.9. When the voltage V_i is increased beyond the EBJ

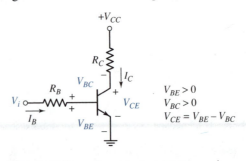

Figure 3.9   Transistor in saturation mode

cut-in voltage of about 0.6 V, the transistor stays in the forward active mode if $V_C > V_B$, or

$$V_{CC} - I_C R_C > V_B$$

In the forward active mode, the above equation may be written as

$$V_{CC} - \beta I_B R_C > 0.7 \, \text{V}$$

The CBJ remains reverse-biased until $V_C = V_B$, at which point

$$I_C = \frac{(V_{CC} - V_{CE})}{R_C} = \frac{(V_{CC} - V_{BE})}{R_C}$$

(since $V_{BC} = 0$), and

$$I_B = I_C/\beta$$

Any further increase in base current due to increase in V_i makes V_C lower than V_B, thereby forward-biasing the CBJ. Now the transistor is operating in the saturation mode. The collector-emitter voltage under this condition is

$$V_{CE}(\text{sat}) = V_{BE}(\text{sat}) - V_{BC}(\text{sat}) \qquad \textbf{(3.24)}$$

This voltage is approximately zero, since the forward-biased CBJ has almost the same voltage as the forward-biased EBJ. Analytical expression for the theoretical value of V_{CE} can be derived using the Ebers-Moll model equations given in Equations (3.23) (see Problem 3.4) as

$$V_{CE}(\text{sat}) = V_T \ln\left[\frac{1 + (I_C/I_B)(1 - \alpha_R)}{\alpha_R - (I_C/I_B)(\alpha_R/\alpha_F)(1 - \alpha_F)}\right] \qquad \textbf{(3.25)}$$

where I_C and I_B are the saturation currents.

In practice, because of the lower doping level and larger area of the collector, V_{BC} (≈ 0.7 V) is slightly less than V_{BE} (≈ 0.8 V), and hence V_{CE} is about a few tenths of a volt when the device is on the verge of going into saturation.

Note that the collector current, which is modeled as dependent on I_B, increases linearly with the base current in the active mode (Figure 3.6c) until the "edge-of-saturation" is reached. Maximum base current in the linear mode, which is the minimum current required to cause forward bias of the CBJ, is the *edge-of-saturation current* $I_B(EOS)$, and it is given by

$$I_B(EOS) = \frac{I_C(\text{sat})}{\beta} = \frac{V_{CC} - V_{CE}(\text{sat})}{\beta R_C} \qquad \textbf{(3.26)}$$

At this base current the collector current is the maximum for the circuit. If a base current I_B larger than the edge-of-saturation current $I_B(EOS)$ is applied, the "overdrive" base current cannot cause a corresponding increase in the collector current; instead, it flows through the emitter and increases the CBJ forward bias. This causes the saturation voltage $V_{CE}(\text{sat})$ to drop slightly. For switching-circuit applications, the assumption of $V_{CE}(\text{sat}) = V_{BE}(\text{sat}) - V_{BC} \approx 0.8 - 0.7 = 0.1$ V may be used. This voltage is assumed if the transistor is well in saturation, that is, if the ratio of saturation base current to minimum base current $I_B(EOS)$ is at least 2. The ratio, $I_B/I_B(EOS)$, is referred to as the *overdrive factor*. A design with large overdrive of base current is typical to allow for variations in the values of the common-emitter current gain β.

The emitter-base voltage in saturation is slightly higher than the typical 0.7 V under forward active mode, due to the ohmic drop in the base region caused by the large base current. For hand calculations, we may assume $V_{BE}(sat)$ to be constant at 0.8 V. Based on this discussion, Figure 3.10a shows the model of the BJT as it operates in the saturation mode. For a simplified approximation, the two junctions may be replaced by short circuits, as shown in Figure 3.10b, so that the terminal currents are determined by external circuit elements. It must be emphasized that a BJT in saturation carries the same (maximum) collector current regardless of the amount of base current above $I_B(EOS)$, and that the collector-emitter voltage is essentially constant at approximately 0.1 V. ($V_{CE}(sat)$ decreases slightly with overdrive; see Problem 3.4.) Note, further, that it is the collector circuit that saturates with the maximum possible current, which depends on the supply voltage V_{CC} and the resistance R_C.

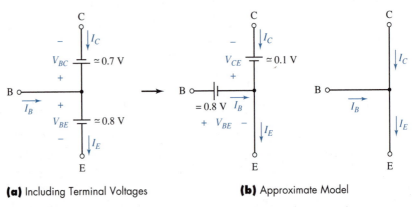

(a) Including Terminal Voltages   **(b)** Approximate Model

Figure 3.10   Model of a BJT in saturation

Example 3.1

Determine the base and collector currents and the junction voltage of the transistor in the circuit shown in Figure 3.11 for (a) $V_i = 0.2$ V, (b) $V_i = 2$ V, and (c) $V_i = 4$ V. Use $\beta = 50$. What is the mode of operation for the transistor in each case?

Solution

(a) Since the EBJ requires a minimum forward bias of 0.6 V, there is negligible base current at $V_i = 0.2$ V. Therefore, $V_{BC} < 0, I_C \approx 0$, and the transistor mode is cutoff. This results in

$$V_{BE} \approx 0.2 \text{ V} \quad \text{and} \quad V_{CE} \approx 5 \text{ V}$$

where the EBJ and the CBJ reverse saturation currents are neglected.

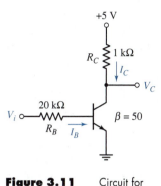

Figure 3.11   Circuit for Example 3.1

(b) For $V_i = 2$ V, which is above the minimum forward bias of 0.6 V, we may assume the BJT to be operating in either the forward active or the saturation mode. Assuming active mode, the base voltage is

$$V_{BE} \approx 0.7 \text{ V},$$

and the base current is

$$I_B = \frac{V_i - V_{BE}}{R_B} = \left[\frac{2 - 0.7}{20}\right] \text{ mA} = 65 \ \mu\text{A}$$

The active-mode collector current is, therefore,

$$I_C = \beta I_B = 3.25 \text{ mA}$$

Hence, $V_C = 5 - 3.25 \times 1 = 1.75$ V

At this voltage the CBJ is reverse-biased, while the EBJ is forward-biased, as assumed. Therefore, the BJT is operating in the active mode.

(c) With $V_i = 4$ V, it is likely that the EBJ is forward-biased high enough and the transistor is in saturation. Using $V_{BE}(\text{sat}) = 0.8$ V, then,

$$I_B = \left[\frac{4 - 0.8}{20}\right] = 160 \ \mu\text{A}$$

If $V_{CE}(\text{sat}) = 0.1$ V, $I_C = \left[\frac{5 - 0.1}{1}\right]$ mA $= 4.9$ mA

For $\beta = 50$, a minimum base current of $I_B(EOS) = 4.9/50$ mA $= 98 \ \mu$A is needed for saturation. Since $I_B > I_B(EOS)$, the transistor is in saturation, as assumed.

Note that if we had assumed active mode, with $V_{BE} = 0.7$ V,

$$I_B = \left[\frac{4 - 0.7}{20}\right] \text{ mA} = 165 \ \mu\text{A}$$

This base current demands a collector current of $I_C = \beta I_B = 8.25$ mA in active mode. Clearly, this is more than the maximum possible current in the collector circuitry of $I_{C(\text{sat})} = 4.9$ mA, so that active mode is not possible.

Note that a BJT is operating in saturation mode if it carries the maximum collector current possible for the given circuit parameters. Clearly, the saturation, or the maximum collector current, is constant in a given circuit. The minimum base current $I_B(EOS)$ to cause saturation, however, depends on the common-emitter current gain β as given by Equation (3.26). Therefore, variation of β with temperature and forward currents, and between devices, must be considered when designing circuits to operate in saturation.

Example 3.2 | If the transistor in Example 3.1, Part (b), is replaced with another having a lower value for β, what is the minimum value of β required for operation in saturation mode at $V_i = 4\,\text{V}$?

Solution

For saturation at any given base current, $\beta I_B \geq I_C$. Therefore,

$$\beta \geq \frac{I_C}{I_B} = \frac{4.9\ \text{mA}}{160\,\mu\text{A}} = 30.6$$

The Ebers-Moll model provides an analytic procedure for determining the operating mode of a transistor. Since this model is based on the operation of the coupled diodes under steady bias conditions, it is primarily used in large-signal analysis and in computer simulation. To understand higher-order effects not covered by the model, such as the narrowing of the base width under reverse bias conditions, and to analyze small-signal operation, we now study the BJT current-voltage characteristics.

3.5 BJT CHARACTERISTICS

With three terminals available, a BJT is operated as a two-port network with one of the terminals common to both input and output circuits. In this section we consider the BJT characteristics for the three cases of keeping (1) the emitter, (2) the collector, and (3) the base as the common terminal. Appropriate current-voltage characteristics of output (input) with either voltage or current at input (output) as a controlling variable are necessary for the understanding of the circuit behavior.

3.5.1 COMMON-EMITTER CHARACTERISTICS

In this configuration the emitter is common to both the input and the output nodes in a circuit. Voltage or current input is applied at the base circuit and the output is taken at the collector circuit, as Figure 3.12 shows. Because of the high current gain between the base and the collector as $I_C = \beta I_B$ in the forward active mode of operation (see Figure 3.6c), the common-emitter circuit configuration

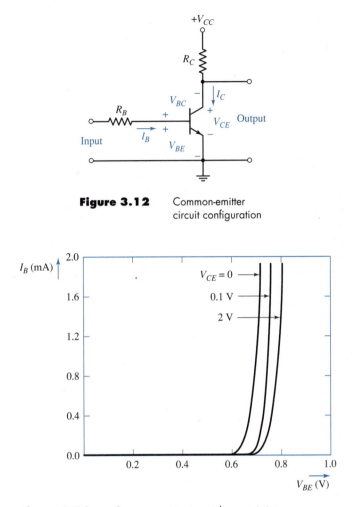

Figure 3.12   Common-emitter
circuit configuration

Figure 3.13   Common-emitter input characteristics

is the most widely used configuration in both analog and digital circuit applications.

From Equation (3.17b), using $\alpha = \beta/(\beta + 1)$, the base current is related to the input base-emitter voltage as

$$I_B \approx \left(\frac{I_{ES}}{\beta_F + 1}\right)\exp\left(\frac{V_{BE}}{V_T}\right) - I_{CS}(1 - \alpha_R) \qquad \textbf{(3.27)}$$

which is similar to the junction diode behavior. The exponential variation of I_B with V_{BE} is shown in Figure 3.13. We will examine the effect of V_{CE} on the I_B–V_{BE} behavior later in this section.

From Equation (3.27) we determine the *dynamic,* or *incremental, base resistance* r_π at the operating point (I_B, V_{BE}), as

$$r_\pi = \frac{dV_{BE}}{dI_B}\bigg|(I_B, V_{BE}) = \frac{(\beta_F + 1)V_T}{I_{ES}\exp\left(\dfrac{V_{BE}}{V_T}\right)}$$

or, using Equations (3.16b) and (3.5) and neglecting the junction saturation currents,

$$r_\pi \approx \frac{V_T}{I_B} \tag{3.28}$$

Similar to the dynamic resistance r_F of a diode (Equation (2.61)), r_π represents the incremental resistance offered by the emitter-base junction for small variations of the junction voltage around the operating point. Depending on the base current level, r_π may range from a few tens of ohms to a few thousand ohms.

In addition to r_π, higher-order analysis would include consideration of the resistance of the metal contact between the active base region and the base terminal, referred to as the *base-spreading resistance*. This resistance, denoted by r_b, is typically from a few ohms to tens of ohms.

Note that the base (input) current controls the collector (output) current as

$$I_C = \beta I_B + (\beta + 1)I_{CBO} \tag{3.6}$$

where the collector current I_C is given by

$$I_C = \alpha_F I_{ES}\exp\left(\frac{V_{BE}}{V_T}\right) + I_{CS} \tag{3.16a}$$

for the case where V_{BE} is the input variable.

Equation (3.16a) describes the output current–input voltage transfer characteristic; it shows that the output (collector) current varies exponentially with the input (base-emitter) voltage. Clearly, if the EBJ voltage V_{BE} is below about 0.6 V, there are negligible base and collector currents, as given by Equations (3.27) and (3.16a), and the BJT is off.

With the collector current dependent on the base current I_B, a family of graphs showing variation of I_C with V_{CE} for different base currents describes the output behavior. Figure 3.14 depicts the common-emitter output (collector) characteristics for a typical switching transistor at different input (base) currents.

Equations (3.27) and (3.16a) indicate that both I_B and I_C are independent of V_{CB}. Reverse-biasing of the CBJ, however, widens the depletion region near the collector, and hence, narrows the effective base width. This effect is known as *base-width modulation* or the *Early effect*. At $V_{CE} = 0$, that is, with the collector shorted to the emitter, the EBJ behaves like a forward-biased diode (Figure 3.13). For $V_{CE} = V_{BE} - V_{BC} > 0$, the collector-base junction is reverse-biased. Hence, the depletion region at the CBJ widens and the effective base width is reduced. Because of the narrower base width, fewer of the injected carriers from the emitter are lost in recombination in the base. As a result, base

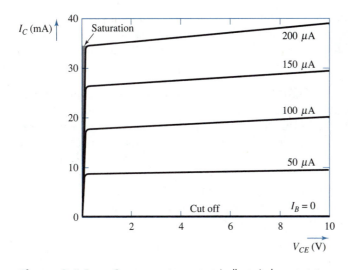

Figure 3.14   Common-emitter output (collector) characteristics

current is reduced at higher V_{CE} as shown in Figure 3.13. Additionally, at constant base current, base-width modulation causes the collector current, and hence α_F, to increase slightly with increasing V_{CE}. Figure 3.14 shows the finite slope in the I_C–V_{CE} graph due to base-width modulation. At very high V_{CE}, the base depletion region at the collector-base junction may extend to the emitter base junction and bring the effective base width to zero. This condition, known as *punch-through,* causes voltage breakdown of the BJT resulting in excessive emitter current. Therefore, for safe operation, the transistor must be operated below the punch-through voltage of the CBJ.

From the I_C–V_{CE} characteristics, we can identify the forward active mode of operation as the region where $I_B \gg I_{CBO}$ and V_{CE} is above about 0.5 V. Note that for $V_{CE} \geq 0.5$ V, the CBJ is not adequately forward-biased. It is clear that I_C varies linearly with I_B in this region. Considering the graphs for $I_B = 50\,\mu$A and $I_B = 100\,\mu$A, for example, we see that I_C changes by 9 mA for a fixed V_{CE} of 2 V. Hence, from Equation (3.6), neglecting I_{CBO}, small-signal current gain $\beta = \Delta I_C/\Delta I_B = 180$, which is the ac β. Current gain β_{ac} is slightly higher at higher V_{CE}. At $V_{CE} = 8$ V, for example, $\beta_{ac} \approx 200$. This increase in β, arising from an increase in the collector current at higher V_{CE}, is due to the Early effect of narrowed base width at higher reverse-bias of the CBJ. The dc (large signal) β, which is the ratio I_C/I_B, also varies from about 180 at $V_{CE} = 2$ V, to about 190 at $V_{CE} = 8$ V, both measured at $I_B = 50\,\mu$A.

In addition, β_{dc} also increases slightly at higher base current. At $I_B = 200\,\mu$A, for example, β_{dc} is about 178 for $V_{CE} = 2$ V, and it is about 192 at $V_{CE} = 8$ V. Careful design must take this variation of β into account for guaranteed operation of the BJT in any mode.

Operation of the transistor in the cutoff mode, in which the terminal currents are extremely small, is identified as the region corresponding to $I_B = 0$ in the collector characteristics. At zero base current, the collector current is

$$I_C = (\beta + 1)I_{CBO} = I_{CEO}$$

from Equation (3.6). I_{CEO} is the collector-to-emitter current with the base open, similar to the open-circuit saturation currents I_{CBO} and I_{EBO}. This current is on the order of a few nA for transistors designed for low current operation (below 1 A). At this low collector current of I_{EBO}, the (output) resistance between collector and emitter is very high at cutoff (conventionally, infinite); hence, the approximate model of open circuit shown in Figure 3.8b is applicable at $I_B = 0$.

I_{CBO}, and hence I_{CEO}, both vary with the CBJ reverse-bias voltage due to thermal generation in the depletion region. In addition, all three saturation currents vary with temperature in a manner similar to that of a junction diode (see Section. 2.4.5).

Example 3.3

The transistor shown in Figure 3.15 is in cutoff, with $I_C = -I_B = I_{CBO}$ at $V_i = 0.2$ V, as in Example 3.1. As the temperature rises, however, I_{CBO} increases and causes an increase in V_{BE}. If $I_{CBO} = 50$ nA at 25°C and it rises at 10 percent/°C, calculate the change in base voltage at 90°C.

Solution

At 25°C,

$$V_{BE} \approx V_i = 0.2 \text{ V} \quad \text{and} \quad V_{CE} \approx 5 \text{ V}$$

At 90°C,

$$I_{CBO} = 50(1.1)^{65} \text{ nA} = 24.5 \, \mu\text{A}$$

Hence,

$$V_{BE} = V_i + I_{CBO} R_B = 0.2 + 0.49 = 0.69 \text{ V}$$

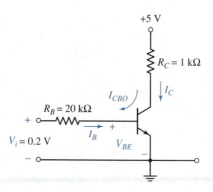

Figure 3.15   Transistor in cutoff mode at 25°C

Clearly, at this base voltage the EBJ is sufficiently forward-biased so that there will be significant collector current—much more than I_{CBO}—as given by Equation (3.16a). Hence the transistor will no longer be in cutoff.

The saturation region in the collector characteristics of Figure 3.14 corresponds to V_{CE} below about 0.5 V. In this range of V_{CE}, the CBJ is forward-biased by a few hundred millivolts and the collector current is limited by external circuitry. The forward-biased CBJ causes injection of electrons from the collector into the base, in addition to the emitter injection. The total current across the base, however, is limited by the collector circuit resistance. Hence, in saturation the p-type base region has excess minority electrons that do not contribute to current flow. Charge in the base due to these mobile carriers simply accumulates in the neutral base region as the base current is increased beyond the minimum required, $I_B(EOS)$ [Equation (3.28)]. For $I_B \geq I_B(EOS)$, the collector current saturates at

$$I_C = I_C(\text{max}) = I_C(\text{sat}) = \frac{V_{CC} - V_{CE}(\text{sat})}{R_C}$$

From this it is clear that the saturation collector current and hence the minimum base current for saturation are dependent on the collector circuit resistance and the supply voltage V_{CC}. Figure 3.16 shows load lines superimposed on the collector characteristics of the common-emitter circuit of Figure 3.12 for three cases of R_C and V_{CC}. These load lines describe the current–voltage variation in R_C according to

$$I_C = \frac{V_{CC} - V_{CE}}{R_C}$$

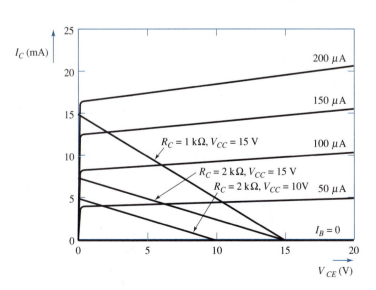

Figure 3.16 Common-emitter circuit saturation characteristics

For $V_{CC} = 15\,\text{V}$ and $R_C = 1\,\text{k}\Omega$, for example, the maximum current available through the collector circuit resistance R_C is 15 mA; collector circuit current, however, saturates at 14.9 mA due to $V_{CE}(\text{sat})$ of approximately 0.1 V. For $V_{CC} = 10\,\text{V}$ and $R_C = 2\,\text{k}\Omega$, the circuit saturation current is 5 mA. Hence, it is actually the collector circuit current that saturates rather than the transistor. Still, it is commonly referred to as saturation of the BJT. In saturation, the resistance from collector to emitter ($ = V_{CE}(\text{sat})/I_C(\text{sat})$) is extremely small; it is about $10\,\Omega$, for example, at $I_B = 150\ \mu\text{A}$ in Figure 3.16.

Note that the minimum base current required for saturation depends on the saturated collector current. For $V_{CC} = 15\,\text{V}$ and $R_C = 1\,\text{k}\Omega$ (Figure 3.16), $I_B(EOS)$ is slightly below $200\ \mu\text{A}$, while for $V_{CC} = 10\,\text{V}$ it is a little over $50\ \mu\text{A}$. At $I_B \geq 100\ \mu\text{A}$, the transistor is well into saturation for $V_{CC} = 10\,\text{V}$ and $R_C = 1\,\text{k}\Omega$. This causes a large number of excess carriers in the base region. When the transistor is to be brought from this level of saturation to cutoff, the excess carriers must be removed from the base. The removal of excess carriers causes delay in switching from saturation to cutoff.

Many BJT logic circuit families operate by switching the transistors between saturation and cutoff modes. Chapter 4 considers these saturation logic families. To reduce the delay in switching between logic states because of the presence of excess carriers in saturation, active mode of operation is used in the emitter-coupled logic (ECL) families, which we study in Chapter 5. As we have seen, the common-emitter circuit with the BJT in active mode has high current gain with low base (input) currents. For this reason, common-emitter circuits are commonly employed as amplifiers in linear, or analog, circuit applications. In these applications the BJT is biased to operate in the active mode in the absence of any signal to be amplified. As with the diodes, the no-signal bias, or operating, point is known as the *quiescent,* or *Q-point*. When the input base current (or voltage) changes from its Q-point, the collector current varies by a large factor (β), causing current amplification. Because of the exponential variation of I_C with V_{BE} [Equation (3.16a)], variation of input from the Q-point must be limited to a small range to maintain linear operation. If the input signal causes a large change in the base drive, the transistor is driven to cutoff or saturation. These points are illustrated in the following example.

Example 3.4

Figure 3.17 shows a common-emitter amplifier circuit. If the BJT has the characteristics shown in Figure 3.18, determine the dc operating points at the input and the output. What is the change in the output (collector) voltage for 1 V peak sinusoid at the signal source v_s? If the input signal amplitude is increased, what is the maximum sinusoidal amplitude available at the collector? What is the corresponding input amplitude?[2]

---

[2] The dc quantities are denoted by upper case variables and upper case subscripts (example: V_{BE}, I_B) while the time-varying quantities use lower case variables and subscripts (example: v_{be}, i_b). The total instantaneous (dc + time-varying) quantities are denoted by lower case variables and upper case subscripts (example: v_{BE}, i_B).

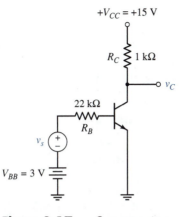

Figure 3.17   Common-emitter
amplifier circuit

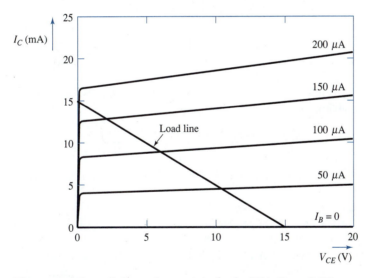

Figure 3.18   Collector characteristics for the BJT in Figure 3.17

Solution

Setting the signal source voltage v_s to 0, we determine the quiescent point by first assuming the active mode of operation for the BJT. The dc base current is given by

$$I_B = \frac{V_{BB} - V_{BE}}{R_B} = \frac{3 - 0.7}{22} \text{ mA} \approx 100 \ \mu A$$

At the collector side, the load line equation is

$$I_C = \frac{V_{CC} - V_{CE}}{R_C} \quad \text{or} \quad I_C = 15 - V_{CE}$$

At the intersection of the load line and the characteristic for $I_B = 100\,\mu\text{A}$ (Figure 3.18), we have

$$I_C \approx 9\,\text{mA} \quad \text{and} \quad V_{CE} \approx 6\,\text{V} = V_C$$

As v_s varies from 0 to 1 V peak, the base current goes from 100 μA to $(4 - 0.7)/22 = 150\,\mu\text{A}$. This raises the collector current to $i_C \approx 13.5\,\text{mA}$ and the collector voltage changes to $v_C = v_{CE} \approx 2.5\,\text{V}$ (Figure 3.18).

As v_s decreases to -1 V, i_B drops to $(2 - 0.7)/22\,\text{mA} \approx 50\,\mu\text{A}$. At this base current, $I_C \approx 5\,\text{mA}$ and $v_C = v_{CE} \approx 10.5\,\text{V}$ (Figure 3.18).

Thus, for a change of 2 V peak-to-peak at input v_s, the collector voltage changes by $2.5 - 10.5 = -8\,\text{V}$, giving a *voltage gain A_v* of

$$A_v = \frac{\text{Change in collector voltage}}{\text{Change in signal (ac) voltage}} = -8/2 = -4$$

Note that for increasing input signal voltage v_s, the total collector current i_C increases, causing the total instantaneous collector voltage to decrease. Hence there is a phase reversal of 180° from input (base voltage) to output (collector voltage).

If R_C is a load that responds to change in current, the circuit has a *current gain A_i* given by

$$A_i = \frac{\text{Change in collector current}}{\text{Change in signal (ac) current}} = \frac{(13.5 - 5)}{(0.15 - 0.05)} = 85$$

Clearly, A_i for the given circuit is the same as β_{ac} for this example. (In general, $A_i \neq \beta_{ac}$.)

For the given value of R_C, we see from the load line in Figure 3.18 that the transistor goes to saturation with $i_C \approx 15\,\text{mA}$ and $v_C \approx 0.1\,\text{V}$ at a base current of approximately 180 μA; at $i_B = 0$, it is cutoff with $i_C \approx 0$ and $v_C \approx 15\,\text{V}$. Therefore, for symmetrical sinusoidal output, the collector voltage can swing a maximum of $\pm 6\,\text{V}$ from its quiescent value of 6 V. This gives a peak-to-peak amplitude of 12 V. Note that the voltage swing in the given circuit is limited by saturation. (If the Q-point is moved to 7.5 V by using lower base bias current or higher collector circuit resistance R_C, for example, V_C can have a maximum swing of ± 7.5 V before the BJT goes to saturation or cutoff.) The corresponding input signal swing is obtained from the base circuit. The BJT is in the active mode at

$$v_C = v_C \,|\, \text{at } Q + 6 = 12\,\text{V} \quad \text{with} \quad i_B \approx 35\,\mu\text{A}$$

Hence,

$$v_s = V_{BE}(\text{active}) + i_B R_B - V_{BB} \approx 0.7 + 0.035 \times 22 - 3 = -1.53\,\text{V}$$

At the other extreme, with

$$v_C = v_C \mid \text{at } Q - 6 \approx 0\,\text{V}$$

and with the BJT entering saturation at $i_B \approx 180\,\mu\text{A}$,

$$v_s = 0.7 + 0.180 \times 22 - 3 = 1.66\,\text{V}$$

Hence, the input signal swing is approximately $\pm 1.5\,\text{V}$.

Note that the voltage gain at the maximum output is -4. Therefore, the common-emitter circuit is performing linear amplification over the range of $-1.5\,\text{V} < v_s < 1.5\,\text{V}$.

Low-Frequency ac Equivalent Circuit–Hybrid-π Model To speed up the small-signal, or ac, analysis performed in the above example, it is useful to model the active mode BJT as shown in Figure 3.19a. Because of its shape, this model is known as the *hybrid-π model*. It simplifies the analysis of the circuit by ignoring the junction capacitances. As we saw earlier, the resistance r_π in the model represents the incremental base resistance arising from the forward-biased EBJ. r_b is the resistance between the interior base node b' to the external base terminal b. This resistance is commonly referred to as the *base spreading resistance*. The resistance r_μ is the incremental resistance of the reverse-biased CBJ, arising from the Early effect. The dependent current source, which links the two junctions, depends on the ac base current and uses incremental β. The dependence of the ac collector current on the CBJ reverse-bias, or the Early effect, accounts for the collector-emitter resistance r_o. Note that an increase in

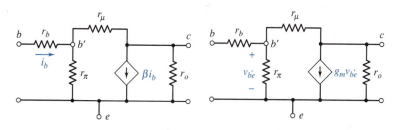

(a) With i_b as Independent Variable   **(b)** With $v_{b'e}$ as Independent Variable

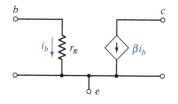

(c) Simplified Model

Figure 3.19   Low-frequency hybrid-π model

the CBJ reverse-bias increases the collector-emitter voltage. Hence, r_o represents the incremental slope, dv_{ce}/di_c, in the collector characteristics shown in Figure 3.14. Similarly, r_μ represents dv_{cb}/di_c in the common-base collector characteristics (see Section 3.5.3). An alternative form of the hybrid-π model that uses the internal base voltage is shown in Figure 3.19b. For small variations of the active base-to-emitter voltage from its dc value V_{BE} to $V_{BE} + v_{be}$, the total collector current, from Equation (3.16a), is

$$I_C + i_c = \alpha_F I_{ES} \exp\left(\frac{V_{BE} + v_{be}}{V_T}\right)$$

from which the small-signal ac component of current is given by

$$i_c = \alpha_F I_{ES} \exp\left(\frac{V_{BE}}{V_T}\right)\left[\exp\left(\frac{v_{be}}{V_T}\right) - 1\right] \approx \alpha_F I_{ES} \exp\left(\frac{V_{BE}}{V_T}\right)\frac{v_{be}}{V_T}$$

or,

$$i_c \approx \left(\frac{I_C}{V_T}\right)v_{be} \tag{3.29}$$

Equation (3.29) is valid for $v_{be} \ll V_T \approx 26\,\text{mV}$. (Why?) Hence, the small-signal collector current is related to the small-signal EBJ voltage by the linearized approximation

$$i_c \approx g_m v_{be} \tag{3.30}$$

where g_m is the *transconductance* in A/V. The transconductance g_m defines the ratio of incremental variation in output (collector) current to incremental variation in input (base) voltage in the common-emitter configuration, as

$$g_m = \frac{di_c}{dv_{be}} = \frac{I_C}{V_T} \tag{3.31}$$

It is readily seen that

$$g_m = \frac{I_C}{V_T} = \frac{\beta I_B}{V_T} = \frac{\beta}{r_\pi} \tag{3.32}$$

Clearly, the low-frequency hybrid-π model is an approximation for the active mode of operation at a given bias point. The equation for g_m above, for example, neglects the effect of resistances r_μ and r_o. This approximate model is useful under conditons of small voltage variations across the EBJ in the forward active mode.

The low-frequency hybrid-π model shown in Figure 3.19 does not include junction capacitances, whose effects (see Section 3.6) are considered negligible at the frequency range of interest. The model can be simplified further for rapid analysis as shown in Figure 3.19c depending on the values of the circuit resistances relative to model resistances. The series resistance r_b, for example, is negligible at tens of ohms when circuit resistances are in the range of a few thousands of ohms or higher. Similarly, the effects of the reverse-biased CBJ resistance r_μ and the output resistance r_o at hundreds of kΩ are insignificant if the circuit resistances connected in parallel with these are only a few kΩ. The hybrid-π model is most widely used for simplified small-signal ac analysis. Example 3.5 illustrates the application of the model.

Estimate the parameters of the simplified hybrid-π model for the BJT in the circuit of Figure 3.17 using the characteristics shown in Figure 3.18. Neglect the base spreading resistance r_b. Determine the voltage and current gains of the amplifier using the model.

Example 3.5

Solution

From the load line and the collector characteristics shown in Figure 3.18, at the operating point of $V_{CE} = 6$ V, $I_C = 9$ mA, and $I_B = 100\,\mu$A,

$$\beta_{dc} \approx \frac{9\,\text{mA}}{100\,\mu\text{A}} = 90$$

and

$$\beta_{ac} \approx \frac{(13 - 4.5)\,\text{mA}}{(150 - 50)\,\mu\text{A}} = 85$$

$$r_\pi = \frac{V_T}{I_B} \approx \frac{25\,\text{mV}}{100\,\mu\text{A}} = 250\,\Omega$$

Hence, from the equivalent circuit shown in Figure 3.20, voltage gain A_v is given by

$$A_v = \frac{v_c}{v_s} = \frac{-\beta_{ac} i_b R_C}{v_s}$$

where

$$i_b = \frac{v_s}{(R_B + r_\pi)}$$

From this we have

$$A_v = v_c/v_s = \frac{-\beta R_C}{(R_B + r_\pi)} = \frac{-85 \times 1}{(22 + 0.25)} \approx 3.82$$

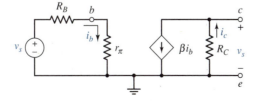

Figure 3.20   Equivalent circuit for the amplifier of Figure 3.17

Current gain A_i is given by

$$A_i = i_c/i_b = \beta_{ac} = 85$$

Clearly, these gains are close to the graphically estimated values shown in Example 3.4.

In addition to having a high current gain and a reasonable voltage gain, the common-emitter configuration has moderate input and output resistances. Because of the voltage and current gains, common-emitter configuration is extensively used in linear and digital circuit applications. When a design requires higher input and lower output resistances for driving high current loads (at lower voltage gain), we employ the common-collector configuration.

3.5.2 COMMON-COLLECTOR CHARACTERISTICS

The common-collector configuration of a BJT circuit, shown in Figure 3.21, keeps the collector at zero signal voltage; input (base) and output (emitter) signal voltages are referenced to ground or collector voltage. With $I_E \approx I_C$, the static characteristics at the input and the output are similar to those for the common-emitter configuration. For small-signal ac conditions, the low-frequency hybrid-π model is as shown in Figure 3.22. From the equivalent circuit it is readily seen that the ac voltage gain is given by

$$A_v = \frac{v_o}{v_s} = \frac{(\beta + 1)R_E}{[r_b + r_\pi + R_B + (\beta + 1)R_E]} \tag{3.33}$$

The above equation shows us that voltage gain A_v is less than, but very close to, unity for any reasonably large value of β. Hence, the emitter (output) voltage follows the base (input) voltage. For this reason the common-collector configuration is known as the *emitter follower*.

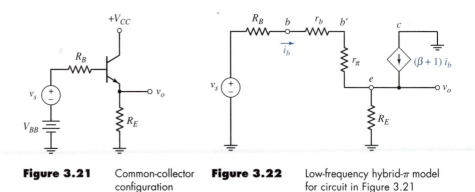

Figure 3.21   Common-collector configuration

Figure 3.22   Low-frequency hybrid-π model for circuit in Figure 3.21

Although the voltage gain is lower than unity, the common-collector circuit configuration provides higher input resistance and lower output resistance than those of the common-emitter configuration. These resistances are given by Problem 3.9.

$$R_i = \frac{v_s}{i_b} = R_B + r_b + r_\pi + (\beta + 1)R_E \tag{3.34}$$

$$R_o = \left[\frac{(R_B + r_b + r_\pi)}{(\beta + 1)}\right] \parallel R_E \tag{3.35}$$

With typical β in the range of 100 to 1,000, the above equations show large R_i and small R_o. Hence the common-collector circuit is an excellent buffer for use in interfacing a high-impedance source to a low-impedance load. Common-collector output drivers, for example, are used in emitter-coupled logic (ECL) circuits (Chapter 5).

3.5.3 COMMON-BASE CHARACTERISTICS

In a common-base circuit, the input is applied at the emitter and the output is taken from the collector as shown in Figure 3.23. Input characteristics (Figure 3.24) describe the EBJ behavior as a function of the CBJ bias voltage. As with the common-emitter configuration, input conduction is negligible until the junction is sufficiently forward-biased. Therefore, Figure 3.24 shows a cut-in voltage of approximately 0.55 V for a typical switching transistor, similar to diode cut-in voltage. Observe that with the collector terminal open, effective base width is essentially the same as the metallurgical width; hence the recombination rate in the base increases and higher input bias is needed for the same input current as with $V_{CB} > 0$ (cf. Figure 3.13).

The output (collector) characteristics in Figure 3.25 show almost constant collector current independent of V_{CB} as long as the CBJ is reverse-biased. This is because in the active mode the collector current, given by Equation (3.14a) as

$$I_C = \alpha_F I_E - I_{CBO}\left[\exp\left(\frac{V_{BC}}{V_T}\right) - 1\right] \tag{3.14a}$$

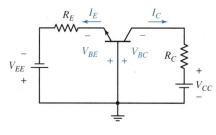

Figure 3.23　Common-base circuit configuration

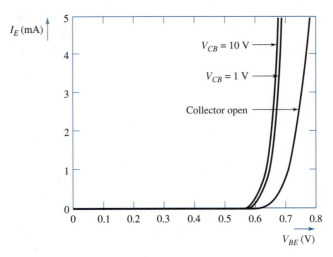

Figure 3.24   Common-base input characteristics

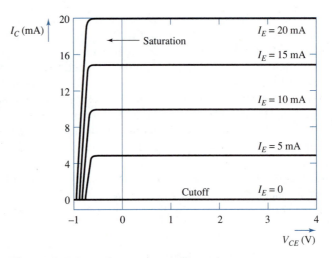

Figure 3.25   Common-base collector characteristics

depends largely on the emitter current. For $I_E = 0$, $I_C \approx I_{CBO}$, which is essentially constant for $V_{BC} < -4 \, V_T$. For $I_E > 0$, I_C is almost the same as I_E in the active mode, since $\alpha_F \approx 1$. As with the common-emitter configuration, the Early effect is evident in the output characteristics. At higher reverse-bias voltage V_{CB}, the effective base width is reduced, causing less recombination in the base; thus I_C increases slightly with V_{CB}. The nonzero slope of the collector current characteristic contributes to r_μ in the hybrid-π model (Figure 3.19). Voltage breakdown by punch-through occurs at high V_{CB}. In addition, high reverse-bias voltage at the CBJ may cause avalanche breakdown of the collector-base junction.

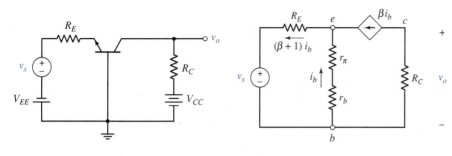

Figure 3.26   Common-base amplifier

For $V_{CB} < 0$ and with $I_E > 0$, both the junctions are forward-biased and the transistor is in saturation. Cutoff occurs for $I_E = 0$.

For voltage amplification, ac signal may be superimposed with dc bias as shown in Figure 3.26a. From the simplified low-frequency equivalent circuit shown in Figure 3.26b, it is clear that the common-base configuration has a current gain of approximately $\alpha = \beta/(\beta + 1)$, which is less than unity. Voltage gain is given by

$$A_v = v_o/v_s \approx \frac{\beta R_C}{r_b + r_\pi + (\beta + 1)R_E} \qquad \textbf{(3.36a)}$$

which may be further approximated as $A_V \approx R_C/R_E$ for any reasonable value of β and $R_E > r_b + r_\pi$.

Input resistance is small, as given by

$$R_i = (r_b + r_\pi) \| R_E \approx r_b + r_\pi \qquad \textbf{(3.36b)}$$

The low value of R_i arises from the forward-biased EBJ. The reverse-biased CBJ gives a large output resistance, as

$$R_o = r_\mu \| R_C \approx R_C$$

Because of the low input resistance, common-base amplifiers are useful for amplifying signals from transducers, which typically have high output resistances.

3.6 BJT Capacitances and the Hybrid-π Model

The hybrid-π model we used in the previous sections simplifies the analysis of BJT circuits by neglecting the junction capacitances. As with junction diodes, the EBJ and the CBJ capacitances limit the frequency range of operation and cause delays in switching between different operating modes. The model that is valid at high frequencies for the common-emitter configuration is shown in Figure 3.27.

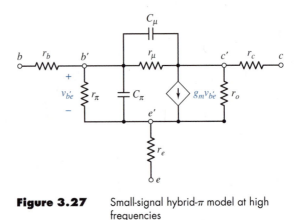

Figure 3.27   Small-signal hybrid-π model at high frequencies

Similar to the internal base node represented by b' in the model, c' and e' represent the active collector and emitter nodes. We denote the resistances between the external terminals and the active regions by r_c and r_e. As seen in Section 3.5.2, r_μ and r_o arise from the finite incremental resistance of the reverse-biased CBJ and the Early effect. C_π represents essentially the diffusion capacitance of the forward-biased EBJ, which has a resistance of r_π. For the reverse-biased CBJ, C_μ is the junction (depletion) capacitance between active collector and base regions. C_π is on the order of a few hundred pF, while C_μ is typically below 10 pF. Although C_μ is much smaller than C_π, its effect becomes significant at high voltage gain between output (collector) and input (base).[3] In addition to the above parameters, analysis of circuits using IC transistors includes the capacitance between collector and substrate (which is at ground potential).

For large-signal variation, as in switching circuit applications, equivalent junction capacitances may be defined [Reference 5]. The equivalent capacitances for a voltage transition from V_1 to V_2 are derived from the small-signal voltage-dependent capacitances [Equation (2.58)] and expressed as functions of V_1 and V_2. Alternatively, the charge-control model, which we discuss in the following section, considers charge storage in the base region as a means of analyzing the static and dynamic operations of the BJT.

3.7   CHARGE-CONTROL MODEL [Ref. 1]

In the preceding sections we discussed the Ebers-Moll model and the low- and high-frequency small-signal hybrid-π models. These models are useful in analyzing the BJT operating in one mode or at one Q-point at a time. However,

---

[3] This is known as the Miller effect, in which a feedback element between input and output may be replaced by its equivalent shunt-connected elements at input and output. Here, the equivalent shunt capacitance in the common-emitter configuration, for example, is C_μ times the voltage gain between the base and the collector.

parameters of the hybrid-π model, for example, must be recalculated for accurate analysis as the operating point changes. For dynamic analysis of the transistor we now consider the *charge-control model*, which describes the operation of the BJT based on the charge stored in the base region. This model, which is applicable to all modes of operation, relates terminal currents to the excess charge stored in the base region.

3.7.1   FORWARD ACTIVE MODE

The starting point in the charge-control model is the distribution of excess minority carriers in the base region (compare diode charge-control, Section 2.4.6). Using the npn transistor as our example, in the forward active mode, the neutral base (p) region has a high concentration of injected electrons from the emitter near the EBJ, as shown in Figure 3.28. As most of these electrons diffuse across the base and reach the collector, their density drops to zero near the CBJ. With uniform doping density and a small base width, there is negligible recombination in the base region. Therefore, with injection occurring at the emitter end of the base, distribution of minority electrons drops almost linearly to zero at the collector end. This injection is related to the thermal equilibrium concentration of electrons n_{bpo} in the p-type base and the EBJ forward-bias voltage. Hence, at the emitter end ($x = 0$), the excess minority carrier density above the thermal equilibrium level of $n_{bpo} \approx n_i^2/N_A$, is given by

$$n_{be}(0) = n_{bp0}\left[\exp\left(\frac{V_{BE}}{V_T}\right) - 1\right] \qquad \text{(3.37)}$$

(Recall that n_i is the intrinsic concentration and N_A is the doping concentration in the p-type base region.)

At the collector end, $n_{be}(W)$, where W is the width of the base region, is slightly below zero because of the reverse-bias $V_{CB} > 0$. For simplicity, however,

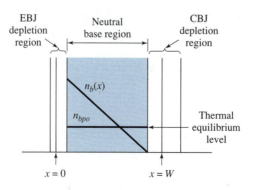

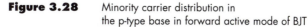

Figure 3.28   Minority carrier distribution in the p-type base in forward active mode of BJT

$n_{be}(W)$ is usually assumed to be zero. Also, the depletion widths at both junctions are neglected so that the effective base width is W. Additionally, as mentioned previously, recombination within the base is ignored. With these assumptions the distribution of the excess carriers in the base is linear as given by

$$n_{be}(x) = n_{bp0}\left[\exp\left(\frac{V_{BE}}{V_T}\right) - 1\right]\left(1 - \frac{x}{W}\right) \qquad \textbf{(3.38)}$$

Defining by Q_F the excess minority carrier charge,

$$Q_F = \int_O^W eAn_{be}(x)\,dx$$

where e is the charge of an electron and A is the cross-sectional area at the EBJ. Since the distribution is linear, from Equation (3.38),

$$Q_F = \left(\frac{1}{2}\right)eWAn_{be}(0)$$

$$= \left(\frac{1}{2}\right)eWAn_{bp0}\left[\exp\left(\frac{V_{BE}}{V_T}\right) - 1\right] \qquad \textbf{(3.39)}$$

where the thermal equilibrium level

$$n_{bp0} \approx \frac{n_i^2}{N_A} \qquad \textbf{(3.40)}$$

for the npn transistor.

Now, if we assume the collector current is entirely due to diffusion of the excess carriers in the base, for the linear distribution, the steady-state collector current I_C [see Equation (2.13)] is given, from Equation (3.38), by

$$I_C = -eAD_b\left(\frac{dn_{be}}{dx}\right)\bigg|_{x=0} \qquad \textbf{(3.41)}$$

$$= (eAD_b/W)n_{bp0}\left[\exp\left(\frac{V_{BE}}{V_T}\right) - 1\right]$$

where D_b is the *electron diffusion coefficient* in the base region.

Using Equation (3.39), we rewrite I_C as

$$I_C = Q_F\left(\frac{2D_b}{W^2}\right) \qquad \textbf{(3.42)}$$

The quantity

$$\tau_F = \frac{W^2}{2D_b} \qquad \textbf{(3.43)}$$

is defined as the *mean forward transit time,* which is the average time taken by the minority electrons to reach the collector in the forward active mode. The collector current is, therefore, given by

$$I_C = \frac{Q_F}{\tau_F} \qquad \textbf{(3.44)}$$

The excess minority charge in the neutral base region is expressed as a function of the EBJ bias voltage as

$$Q_F = Q_{F0}\left[\exp\left(\frac{V_{BE}}{V_T}\right) - 1 \right] \qquad \textbf{(3.45)}$$

where $Q_{F0} = (1/2)eWAn_{bp0}$ is a physical parameter dependent on the device geometry and the dopant profile.

The steady-state base current is proportional to the rate of recombination of minority electrons in the base and the rate of injection of holes from the base to the emitter. Since both these rates are directly proportional to the excess minority charge Q_F, base current is obtained as

$$I_B = \frac{Q_F}{\tau_{BF}} \qquad \textbf{(3.46)}$$

where τ_{BF} is the effective *minority carrier lifetime,* that is, the average lifetime of the excess minority carriers in the p-type base region before recombination.

Since the collector current and the base current in the steady-state (forward active) mode are related by $I_C = \beta_F I_B$, then from Equations (3.44) and (3.46), we have

$$\frac{I_C}{I_B} = \beta_F = \frac{\tau_{BF}}{\tau_F} \qquad \textbf{(3.47)}$$

Equation (3.47) expresses the common-emitter dc current gain β_F in terms of the device time-constants τ_{BF} and τ_F. Clearly, for large β_F, the ratio of minority carrier lifetime to transit time must be made large. A narrow base region (small W) reduces transit time τ_F; heavier doping of the emitter, on the other hand, increases emitter efficiency and raises τ_{BF}. Higher lifetime, however, increases the delay in switching a BJT from cutoff to saturation (see Section 4.1). For faster switching, therefore, τ_{BF} must be small. Certain impurities such as gold and copper in the base region provide lower lifetimes.

Equations (3.43) and (3.45) express the static terminal currents in terms of the excess charge stored in the base region; that is, the BJT is described as a charge-controlled device. The steady-state current equation, Equation (3.6), for example, describes the BJT as a current-controlled device.

We describe the dynamic behavior of the transistor by the instantaneous currents, which are obtained from the time variations of the stored charge. We assume that the terminal voltage and current variations are small enough such that the instantaneous distribution of excess minority carriers in the base is still approximately linear. In addition, we assume low-level injection, so that the majority carriers in the neutral p-type base region are always holes. Denoting the total excess base charge stored due to the two junction voltages by $Q_F(t)$, the collector and the base currents are given by

$$i_B = \frac{Q_F(t)}{\tau_{BF}} + \frac{dQ_F(t)}{dt} \qquad \textbf{(3.48)}$$

$$i_C = \frac{Q_F(t)}{\tau_F} \qquad \textbf{(3.49)}$$

Note that the total excess charge in the above equations is given by the area of the excess charge distribution in the base region. Recall from the continuity equation given in Equation (2.41) that the excess base charge $Q_F(t)$ is obtained by integrating Equation (2.41) for the length of the base. The first term on the right-hand side for the terminal current i_B in Equation (3.48) accounts for the steady recombination of minority carriers in the base and the injection of majority carriers in the base and the injection of majority carriers from the base into the emitter. The second term describes the instantaneous change of excess charge with time.

Although the junction voltage v_{BE} is varying with time, it is assumed, as mentioned earlier, that the time rate is small enough for the excess charge distribution ($n_b(x)$ in Figure 3.28) to change linearly in succession; as a result of this quasi-static distribution, $Q_F(t)$ varies incrementally with v_{BE}. This total excess charge $Q_F(t)$ reaches the collector with mean forward transit time τ_F; hence, the collector terminal current is given by $Q_F(t)/\tau_F$ in Equation (3.49).

Note that the quasi-static distribution of $Q_F(t)$ implies that the emitter and the collector currents, which are proportional to the slopes of the excess base charge distribution $n_b(x)$, are approximately the same. Since the instantaneous base current in most circuit applications is small compared to the collector current, the assumption of linear variation of $n_b(x)$ in succession is justified.

Combining the base and the collector currents, the emitter current is given by

$$i_E = \frac{Q_F(t)}{\tau_F} + \frac{Q_F(t)}{\tau_{BF}} + \frac{dQ_F(t)}{dt} \tag{3.50}$$

Note that the "steady" part of the emitter current in Equation (3.50) is

$$I_E = Q_F(t)\left[\frac{1}{\tau_F} + \frac{1}{\tau_{BF}}\right]$$

Using Equation (3.45), the excess charge is given by

$$Q_F(t) = Q_{F0}\left[\exp\left(\frac{v_{BE}}{V_T}\right) - 1\right] \tag{3.51a}$$

and the quasi-steady emitter current becomes

$$I_E = Q_{F0}\left(\frac{1}{\tau_F} + \frac{1}{\tau_{BF}}\right)\left[\exp\left(\frac{v_{BE}}{V_T}\right) - 1\right] \tag{3.51b}$$

where v_{BE} is the total instantaneous EBJ voltage.

From the exponential dependence (diode form) of I_E on v_{BE}, it is clear that the diode has a saturation current of

$$I_{ES} = Q_{F0}\left(\frac{1}{\tau_F} + \frac{1}{\tau_{BF}}\right) \tag{3.52}$$

Figure 3.29 shows the charge-control model described by Equations (3.48) to (3.50). The diode in this figure represents the path for the steady emitter current. The time-varying current component $dQ_F(t)/dt$ arising from the time-

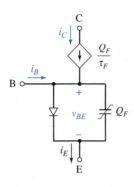

Figure 3.29   Charge-control model for forward (emitter-to-base) injection

dependent excess charge in the base is represented by the nonlinear capacitor. This charge, unlike $Q = VC$ in a linear capacitor, is given by

$$Q_F(t) = Q_{F0}\left[\exp\left(\frac{v_{BE}}{V_T}\right) - 1\right] \qquad \textbf{(3.53)}$$

The nonlinear, voltage-dependent charge storage is indicated by the line across the capacitor in Figure 3.29. The dependent current source at the collector terminal is controlled by the charge $Q_F(t)$ on the nonlinear capacitor.

3.7.2   REVERSE ACTIVE MODE

In the reverse active mode of an npn transistor, in which the CBJ is forward-biased and the EBJ is reverse-biased, majority carrier (electron) injection occurs from collector to base. Some of these carriers are collected at the emitter due to the field across the EBJ depletion region. As seen in Section 3.4, with the roles of the emitter and the collector reversed, the terminal currents are different from their corresponding active mode values as a result of nonsymmetric geometry and doping. These currents are derived based on the excess charge stored in the neutral base region, similar to those in the forward active mode.

Denoting the reverse transit time and the minority carrier lifetime in the base by τ_R and τ_{BR} respectively, the total instantaneous currents in the reverse active mode are given by

$$i_{ER} = \frac{-Q_R(t)}{\tau_R} \qquad \textbf{(3.54)}$$

$$i_{BR} = \frac{Q_R(t)}{\tau_{BR}} + \frac{dQ_R(t)}{dt} \qquad \textbf{(3.55)}$$

$$i_{CR} = -\frac{Q_R(t)}{\tau_R} - \frac{Q_R(t)}{\tau_{BR}} - \frac{dQ_R(t)}{dt} \qquad \textbf{(3.56)}$$

where

$$Q_R(t) = Q_{R0}\left[\exp\left(\frac{V_{BC}}{V_T}\right) - 1\right] \qquad \textbf{(3.57)}$$

is the total charge stored in the base with

$$Q_{R0} = \left(\frac{1}{2}\right)eWAn_{bp0} \qquad \textbf{(3.58)}$$

(Compare Equations (3.48) to (3.50) with Equations (3.54) to (3.56).) Since injection from the collector is smaller than that from the emitter, Q_R is less than Q_F.

3.7.3   COMPLETE CHARGE-CONTROL MODEL

In addition to the neutral base region, there is also charge stored in the depletion regions of the two junctions. Since this charge varies with the junction voltage, there is current associated with it as $i = dq/dt$. Recall that in the case of a junction diode, the charge variation in the depletion region is accounted for by the voltage-dependent junction (depletion) capacitance. For the BJT, the two charging currents for the two nonlinear capacitances are denoted by dQ_{VE}/dt and $-dQ_{VC}/dt$, where the charges are functions of the voltages, V_{BE} and V_{BC}, denoted by V_E and V_C at the EBJ and the CBJ. For the CBJ, the depletion region charge current is in the same direction as the reverse active mode component $-dQ_R/dt$, that is, they both increase in the opposite direction of the normal forward collector current. Combining the forward and reverse injection components with the depletion region charging currents, we have

$$i_C = \frac{Q_F}{\tau_F} - Q_R\left[\frac{1}{\tau_R} + \frac{1}{\tau_{BR}}\right] - \frac{dQ_R}{dt} - \frac{dQ_{VC}}{dt} \qquad \textbf{(3.59)}$$

$$i_B = \frac{Q_F}{\tau_{BF}} + \frac{Q_R}{\tau_{BR}} + \frac{dQ_F}{dt} + \frac{dQ_R}{dt} + \frac{dQ_{VC}}{dt} + \frac{dQ_{VE}}{dt} \qquad \textbf{(3.60)}$$

$$i_E = -\frac{Q_R}{\tau_R} + Q_F\left[\frac{1}{\tau_F} + \frac{1}{\tau_{BF}}\right] + \frac{dQ_F}{dt} + \frac{dQ_{VE}}{dt} \qquad \textbf{(3.61)}$$

Figure 3.30 shows the complete charge-control model for an npn transistor. Note the three sets of terms arising from (1) forward injection, (2) reverse injection, and (3) charging of depletion region, in Equations (3.59) to (3.61) and in Figure 3.30. With Q_F and Q_R carrying the exponential variations with the junction voltages, the two diodes have saturation currents given by

$$I_{CS} = Q_{R0}\left[\frac{1}{\tau_R} + \frac{1}{\tau_{BR}}\right] \qquad \textbf{(3.62)}$$

for the collector diode, and, for the emitter diode, repeating Equation (3.52),

$$I_{ES} = Q_{F0}\left[\frac{1}{\tau_F} + \frac{1}{\tau_{BF}}\right] \qquad \textbf{(3.63)}$$

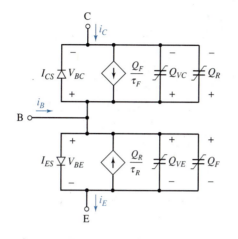

Figure 3.30 Complete charge-control model for an npn transistor

3.7.4 COMPARISON OF CHARGE-CONTROL MODEL WITH EBERS-MOLL AND HYBRID-π MODELS

Ebers-Moll Model The large-signal, static Ebers-Moll model can be derived from the complete charge-control model. To establish the equivalence between the two models, we set the time-dependent terms in Equations (3.59)–(3.61) to zero. This results in

$$I_C = \frac{Q_F}{\tau_F} - Q_R\left[\frac{1}{\tau_R} + \frac{1}{\tau_{BR}}\right]$$

and

$$I_E = -\frac{Q_R}{\tau_R} + Q_F\left[\frac{1}{\tau_F} + \frac{1}{\tau_{BF}}\right]$$

Replacing Q_F and Q_R with their diode-like variations, as in Equation (3.45), gives

$$I_C = \left(\frac{Q_{F0}}{\tau_F}\right)\left[\exp\left(\frac{V_{BE}}{V_T}\right) - 1\right]$$

$$- \left[Q_{R0}\left(\frac{1}{\tau_R} + \frac{1}{\tau_{BR}}\right)\right]\left[\exp\left(\frac{V_{BC}}{V_T}\right) - 1\right] \tag{3.64}$$

and

$$I_E = Q_{F0}\left[\frac{1}{\tau_F} + \frac{1}{\tau_{BF}}\right]\left[\exp\left(\frac{V_{BE}}{V_T}\right) - 1\right]$$

$$- Q_{R0}\left(\frac{1}{\tau_R}\right)\left[\exp\left(\frac{V_{BC}}{V_T}\right) - 1\right] \tag{3.65}$$

Comparing the above with Equation (3.12), we have the diode saturation currents as given in Equations (3.62) and (3.63):

$$I_{CS} = Q_{R0}\left(\frac{1}{\tau_R} + \frac{1}{\tau_{BR}}\right) \tag{3.62}$$

$$I_{ES} = Q_{F0}\left[\frac{1}{\tau_F} + \frac{1}{\tau_{BF}}\right] \tag{3.63}$$

and α_F and α_R are given by

$$\alpha_F = \frac{\tau_{BF}}{\tau_F + \tau_{BF}} \tag{3.66}$$

and

$$\alpha_R = \frac{\tau_{BR}}{\tau_R + \tau_{BR}} \tag{3.67}$$

In addition, the reciprocity relation given in Equation (3.13) becomes

$$Q_{R0}\tau_F = Q_{F0}\tau_R \tag{3.68}$$

and the zero-bias charges are related to the device parameters as

$$Q_{F0} = \alpha_F I_{ES}\tau_F \tag{3.69}$$

and

$$Q_{R0} = \alpha_R I_{CS}\tau_R \tag{3.70}$$

Hybrid-π Model   We obtain the parameters of the linearized, small-signal hybrid-π model from the charge-control model by considering the forward active mode of operation. Since the CBJ is reverse-biased with $|V_{BC}| \ll V_T$, reverse injection charge is $Q_R \approx -Q_{R0}$, independent of time, for small signal variations. In addition, using Equation (3.70)

$$Q_R(t)\left[\frac{1}{\tau_r} + \frac{1}{\tau_{BR}}\right] \approx -\alpha_R I_{CS}\left(1 + \frac{\tau_r}{\tau_{BR}}\right)$$

Since I_{CS} is extremely small (in the pA or lower range), and $\tau_R/\tau_{BR} = \beta_R < 1$, the above term is negligibly small in the collector current given by Equation (3.59). Hence, the small-signal collector current is given by

$$i_c = \frac{Q_F(t)}{\tau_F} - \frac{dQ_{VC}}{dt} \tag{3.71}$$

and the base current by

$$i_b = \frac{Q_F(t)}{\tau_{BF}} + \frac{dQ_F(t)}{dt} + \frac{dQ_{VC}}{dt} + \frac{dQ_{VE}}{dt} \tag{3.72}$$

Corresponding currents for the hybrid-π model (Figure 3.27) are obtained by approximating the terminal-to-active region resistances to zero and the early effect resistances to infinity. With $V_{b'e} \approx V_{be}$, then

$$i_c = g_m v_{be} + C_\mu \frac{dv_{cb}}{dt} \tag{3.73}$$

and

$$i_b = \frac{v_{be}}{r_\pi} + C_\pi \frac{dv_{be}}{dt} - C_\mu \frac{dv_{cb}}{dt} \tag{3.74}$$

Comparing the above equations with Equations (3.71) and (3.72), we can see that the forward active current gain is given by

$$g_m r_\pi = \frac{\tau_{BF}}{\tau_F} = \beta_F \tag{3.75}$$

where $g_m = \dfrac{I_C}{V_T}$. In addition, it can be shown (Problem 3.10) that the steady charge Q_{F0} is related to dc bias current I_B as

$$Q_{F0} = I_B \tau_{BF} = I_C \tau_F \tag{3.76}$$

The term $C_\pi dv_{be}/dt$ in the base current of Equation (3.74) arises from the variation of stored base charge with EBJ voltage. Hence, neglecting the depletion component of collector current, from Equations (3.71) and (3.72),

$$\frac{dQ_F(t)}{dv_{be}} = C_\pi = \frac{d(i_C \tau_F)}{dv_{be}} = g_m \tau_F$$

Other higher-order terms in the hybird-π model can be similarly related to the charge-control model [see Reference 1].

A BJT has the following parameters: $\tau_F = 2\,\text{ns}$, $\tau_{BF} = 200\,\text{ns}$, $\tau_R = 30\,\text{ns}$, and $\tau_{BR} = 15\,\text{ns}$. If it is carrying a steady base current of $0.5\,\text{mA}$ and a collector current of $5\,\text{mA}$, determine the forward and reverse charge and the total charge stored in the base. What is the mode of operation? | **Example 3.6**

Solution

The forward and reverse current gains are

$$\beta_F = \frac{\tau_{BF}}{\tau_F} = 100 \quad \text{and} \quad \beta_R = \frac{\tau_{BR}}{\tau_R} = 0.5$$

From Equations (3.57) and (3.58), for steady currents, we have

$$\left(\frac{Q_F}{2}\right) - Q_R\left(\frac{1}{30} + \frac{1}{15}\right) = 5$$

$$\left(\frac{Q_F}{200}\right) + \left(\frac{Q_R}{15}\right) = 0.05$$

Solving the above equations, we get

$$Q_F = 11.3\,\text{pC} \quad \text{and} \quad Q_R = 6.65\,\text{pC}$$

Hence the total charge stored in the base is $Q_F + Q_R = 17.95\,\text{pC}$. Since $I_c > \beta_F I_B$, the device is in saturation.

The charge-control model equations for the pnp transistor are derived in a similar manner to those for the npn transistor. With the two zero-bias charges and the four time-constants that are device-dependent, the model is used in the analysis of dynamic performance of BJT switching circuits.

3.8   MICROSIM PSPICE MODEL FOR BJTS

While the charge-control model is useful for accurate analysis of BJT circuits in any mode of operation, the MicroSim PSpice model uses a graduated (multilevel model) approach. The simplest is the dc model, based on the Ebers-Moll model of Figure 3.5, as shown in Figure 3.31 for npn transistor. This model is governed by the following terminal current equations.

$$I_C = I_S\left[\exp\left(\frac{V_{BE}}{V_T}\right) - \exp\left(\frac{V_{BC}}{V_T}\right)\right] - \left(\frac{I_S}{\beta_R}\right)\left[\exp\left(\frac{V_{BC}}{V_T}\right) - 1\right] \quad \textbf{(3.77a)}$$

$$I_E = I_S\left[\exp\left(\frac{V_{BE}}{V_T}\right) - \exp\left(\frac{V_{BC}}{V_T}\right)\right] + \left(\frac{I_S}{\beta_F}\right)\left[\exp\left(\frac{V_{BE}}{V_T}\right) - 1\right] \quad \textbf{(3.77b)}$$

These are the Ebers-Moll model equations given by Equation (3.12) with

$$I_S = \alpha_F I_{ES} = \alpha_R I_{CS}, \quad \beta_R = \frac{\alpha_R}{1 - \alpha_R} \quad \text{and} \quad \beta_F = \frac{\alpha_F}{1 - \alpha_F}$$

The default values for this model are $\beta_F = 100$ and $\beta_R = 1$, and I_S is set to provide an EBJ voltage of approximately 0.6 V.

Junction capacitances are considered at the next (higher) level by specifying zero-bias capacitances and grading factors for the two junctions. These capaci-

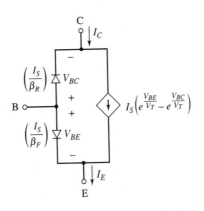

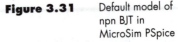

Figure 3.31   Default model of npn BJT in MicroSim PSpice

tances (C_{BE} and C_{BC}) are determined as

$$C_{BE} = \left(\frac{T_F I_S}{V_T}\right)\exp\left(\frac{V_{BE}}{V_T}\right) + \frac{C_{JE}}{[1 - V_{BE}/V_{JE})]M_{JE}} \tag{3.78}$$

$$C_{BC} = \left(\frac{T_R I_S}{V_T}\right)\exp\left(\frac{V_{BC}}{V_T}\right) + \frac{C_{JC}}{[1 - V_{BC}/V_{JC})]M_{JC}} \tag{3.79}$$

where T_F and T_R are the forward and the reverse transit times, C_{JE} and C_{JC} are the zero-bias depletion capacitances, V_{JE} and V_{JC} are the built-in potentials, and M_{JE} and M_{JC} are the grading factors (see Section 2.4.6). In addition, this level includes the parasitic resistances, labeled RE, RC, and RB, from external terminals to internal active regions.

At the final (highest) level, the MicroSim PSpice model incorporates higher-order effects to obtain simulation results that are closer to experimentally observed results in practical transistors. The Early effect and the resulting base width modulation are modeled by including the term $[1+(V_{CB}/V_{AF})]$ in the collector current, where V_{AF} is the Early voltage. The value for the early voltage is obtained by extrapolating the forward active mode collector characteristic to $I_C = 0$ in the common-emitter mode. Figure 3.32 shows the extension of the linear portion of the I_C–V_{CE} graph shown in Figure 3.14. The voltage at which the converging lines intersect is the Early voltage, $-V_{AF}$. V_{AF} is typically between 50 and 100 V. The slope of the lines emanating from $-V_{AF}$ corresponds to the output conductance,

$$g_o = \frac{1}{r_o} = \frac{dI_C}{dV_{CE}}$$

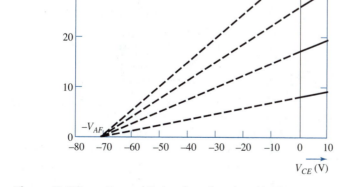

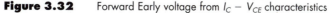

Figure 3.32   Forward Early voltage from I_C – V_{CE} characteristics

Since V_{AF} is usually much larger than the operating voltage V_{CB}, g_o is approximated as

$$g_o \approx \frac{dI_C}{dV_{CB}} = \frac{I_C}{V_{AF}} \tag{3.80}$$

Including the Early Voltage V_{AF} in the MicroSim PSpice model modifies the collector current in the forward active mode as

$$I_C = I_S \exp\left(\frac{V_{BE}}{V_T}\right)\left[1 + \left(\frac{V_{CE}}{V_{AF}}\right)\right] \tag{3.81}$$

which causes nonzero slope in the simulated I_C–V_{CE} characteristics. A similar voltage V_{AR} is defined for the BJT operating in the reverse active mode.

The MicroSim PSpice model includes other higher-order effects, such as the depletion region leakage and recombination effect, and high-level injection, by specifying appropriate parameters [see References 6, 8]. For IC transistors, the MicroSim PSpice model contains substrate junction parameters. Additional parameters in the model include temperature coefficients for parasitic resistances and β.

Table 3.2 lists some of the commonly used MicroSim PSpice model parameters for reasonable accuracy in simulation results. The complete list of parameters, with a total of 40, are given in Reference 8.

The following example illustrates BJT circuit analysis using MicroSim PSpice.

Table 3.2   MicroSim PSpice Model Parameters for BJT

Symbol	PSpice Symbol	Parameter	Default Value	Unit
I_S	IS	Saturation current	1.0E-16	A
β_F	BF	Ideal maximum forward β	100	
β_R	BR	Ideal maximum reverse β	1	
V_{AF}	VAF	Forward Early voltage	∞	V
V_{AR}	VAR	Reverse Early voltage	∞	V
r_b	RB	Zero-bias (maximum) base resistance	0	Ω
r_e	RE	Emitter ohmic resistance	0	Ω
r_c	RC	Collector ohmic resistance	0	Ω
C_{je}	CJE	Base-emitter zero-bias capacitance	0	F
V_{0e}	VJE	Base-emitter built-in potential	0.75	V
C_{jc}	CJC	Base-collector zero-bias capacitance	0	F
V_{0c}	VJC	Base-collector built-in potential	0.75	V
τ_F	TF	Ideal forward transit time	0	s
τ_R	TR	Ideal reverse transit time	0	s
m_e	ME	Base-emitter grading factor	0.33	
m_c	MC	Base-collector grading factor	0.33	

Example 3.7

Simulate the circuit of Figure 3.17 for 1 V peak sinusoidal input at 1 kHz applied at v_s. Use the following parameters: $\beta = 85$, RC = 1 Ω, RB = 10 Ω, CJC = 10 pF, CJE = 20 pF, T_R = 50 ns, and T_F = 400 ps. (Note that β is obtained from the characteristics shown in Figure 3.18; other parameters are typical for a general-purpose transistor.)

Solution

Figure 3.33a shows Figure 3.17 redrawn with node numbers included. Figure 3.33b shows the MicroSim PSpice input file. From the MicroSim PSpice

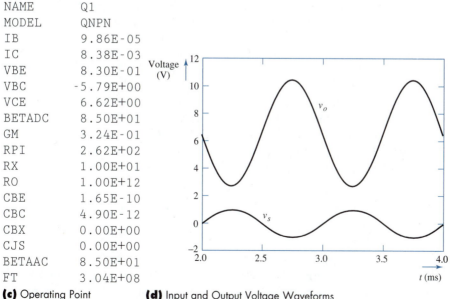

```
VBB      1    0    3V
VS       2    1    SIN (0 1V 1000Hz)
RBB      2    3    22K
Q1       4    3    0   QNPN
RC       5    4    1K
VCC      5    0    15V
.MODEL QNPN NPN(BF=85 RC=1 RB=10
+ CJC=10p CJE=20p TR=50n TF=400p)
.OP
.TRAN 10u 4m 2m 10u
.OPTIONS LIMPTS=4500
.PROBE
.END
```

(a) Circuit with Nodes labeled   **(b)** MicroSim PSpice Input file

```
NAME       Q1
MODEL      QNPN
IB         9.86E-05
IC         8.38E-03
VBE        8.30E-01
VBC       -5.79E+00
VCE        6.62E+00
BETADC     8.50E+01
GM         3.24E-01
RPI        2.62E+02
RX         1.00E+01
RO         1.00E+12
CBE        1.65E-10
CBC        4.90E-12
CBX        0.00E+00
CJS        0.00E+00
BETAAC     8.50E+01
FT         3.04E+08
```

(c) Operating Point   **(d)** Input and Output Voltage Waveforms

Figure 3.33   MicroSim PSpice simulation for the circuit of Figure 3.16

output, the dc operating point (Figure 3.33c) is at $I_B = 98.6\,\mu A$, $I_C = 8.34\,mA$, $V_{BE} = 0.83\,V$, and $V_{CE} = 6.62\,V$. These values are close to the hand-calculated values at the bias point obtained in Example 3.4. As seen in Figure 3.33d, output sinusoid has a peak-to-peak amplitude of approximately $10.5 - 2.75 = 7.75\,V$ for 2 V peak-to-peak input. Hence, the voltage gain is -7.75.

3.9   POWER DISSIPATION AND THERMAL EFFECTS

Power dissipation in a BJT must be considered for both safe operation of the device and stability of the operating point. Dissipation in the emitter-base junction is negligible due to the low base current and voltage; hence, the device dissipation is usually given by

$$P_D = V_{CB}I_C \approx V_{CE}I_C \qquad (3.82)$$

Since dissipation raises operating temperature, P_D is specified as a function of temperature for discrete transistors. As it does with diodes, temperature affects junction voltages and saturation currents (see Example 3.3). In addition, thermal generation increases the minority carrier lifetime τ_{BF}. The transit time τ_F also increases due to increase in mobility with temperature, but at a lower rate. The net result is an increase in current gain β. Also, the reverse saturation currents of the two junctions increase with temperature at the rate of 5 to 15 percent/°C. This increase in saturation current coupled with an increase in β causes higher dissipation, which, in turn, raises temperature further. If this process continues, overheating and thermal runaway can occur, causing permanent damage to the device. For safe operation, manufacturers specify an operating temperature range of $-65°C$ to $200°C$. Clearly, from Equation (3.82), a BJT operating in the linear mode (as in an amplifier) dissipates large power and raises temperature; hence, the circuit design must provide for adequate heat dissipation to prevent thermal runaway. Static operation of BJTs in either cutoff or saturation, on the other hand, takes negligible power (in cutoff because $I_C = 0$ and in saturation because $V_{CE} \approx 0$). Logic circuits operating in these modes (TTL circuits, for example), are, therefore, preferable to those operating in the active mode (ECL circuits) in terms of power dissipation and heating.

SUMMARY

This chapter introduced the basic principles of bipolar junction transistors. A transistor has three terminals: emitter, base, and collector. These terminals in an npn device are connected to an n-type, a p-type, and another n-type region, thereby forming two pn junctions. In a pnp transistor an n-type region is formed between two p-type regions. In terms of doping levels, the emitter has the highest density and the collector has the lowest. Dominant carriers for current conduction in npn devices are electrons, and in pnp devices, holes. Both npn and pnp discrete transistors are commercially available. Because holes have lower mobility than electrons, switching circuits and ICs employ predominantly npn transistors.

A transistor is a current amplifying device; a small change in base current causes a large change in emitter current. With current gain β of 100 or more, collector and emitter currents are approximately equal.

A transistor operates in one of four modes, depending on the biasing of the two junctions. In the forward active mode, the emitter-base junction is forward-biased while the collector-base junction is reverse-biased. This is the mode of the transistor in linear amplifier applications. When both junctions are reverse-biased, the transistor carries negligible currents and is in the cutoff mode. Cutoff mode current arises from thermal generation at the junctions. When both the junctions are forward-biased, a BJT is operating in the saturation mode. The device in saturation may be replaced by short-circuits. Reverse-biasing the emitter-base junction and forward-biasing the collector-base junction causes the device to operate in the inverse (reverse) active mode. Because of the difference in the doping levels of emitter and collector, the currents in the inverse active mode are not the same as in the forward active mode.

The Ebers-Moll model describes the BJT as two junction diodes connected back-to-back with feedback between the junctions. This model is suitable for analyzing the device currents and voltages under static, large-signal conditions.

Analysis in the neighborhood of a dc operating point is carried out using the hybrid-π model. This model includes resistances and capacitances whose values depend on the bias conditions.

Based on which of the three terminals is at signal (ac) ground, that is, common to input and output, a BJT circuit is called common-emitter, common-collector, or common-base configuration. With large voltage and current gains, the common-emitter configuration is widely used in analog and digital circuits. The common-collector circuit, also known as the emitter follower, has approximately unity voltage gain, high current gain, high input impedance, and low output impedance.

The charge-control model represents a BJT in terms of charge stored in the base region. Minority carrier lifetimes and transit times in the base are used to determine static terminal currents. Dynamic current due to time-varying input is obtained as the rate of charge flow. In addition to the time-constants, the model parameters include physical parameters describing zero-bias base charges at the two junctions.

Power dissipation in a BJT is determined by the product of the collector current and the collector-emitter voltage. A raise in dissipation increases the operating temperature, which increases the reverse saturation current and changes bias conditions. Dissipation must be limited to specified maximum values to avoid a cumulative rise in temperature and the resulting thermal runaway.

REFERENCES

1. Gray, P. E., and C. L. Searle. *Electronic Principles: Physics, Models, and Circuits.* New York: Wiley, 1969.

2. Muller, R. S., and T. I. Kamins. *Device Electronics for Integrated Circuits.* New York: Wiley, 1986.

3. Millman, J., and A. Grabel. *Microelectronics.* New York: McGraw-Hill, 1987.

4. Burns, S. G., and P. R. Bond. *Principles of Electronic Circuits.* St. Paul, Minn.: West Publishing, 1987.

5. Hodges, D. A., and H. G. Jackson. *Analysis and Design of Digital Integrated Circuits.* New York: McGraw-Hill, 1988.

6. Pulfrey, D. L., and N. G. Tarr. *Introduction to Microelectronic Devices.* Englewood Cliffs, N.J.: Prentice Hall, 1989.

7. Streetman, B. G. *Solid State Electronic Devices.* Englewood Cliffs, N.J.: Prentice Hall, 1990.

8. Tuinenga, P. W. *SPICE: A Guide to Circuit Simulation and Analysis Using PSpice.* Englewood Cliffs, N.J.: Prentice Hall, 1995.

REVIEW QUESTIONS

1. If an ohmmeter is connected between the emitter and the collector terminals of an npn transistor, what will it read? Will the reading be different if the meter terminals are reversed?

2. Why must the base region be thin and wide?

3. What causes the difference in currents when the emitter and the collector terminals are exchanged?

4. State the biasing requirements for each mode of operation.

5. What mode of operation is suitable for linear amplification? Why?

6. What is the voltage across the EBJ of an npn transistor at cutoff?

7. In a common-emitter configuration, what is the current in the collector circuit at cutoff?

8. If the BJT, a common-emitter circuit with $V_{BE} = 0$, is cutoff at 25°C, can it stay in cutoff at 100°C? Explain.

9. Can a BJT be used as an amplifier with emitter and collector exchanged? Explain.

10. What is the Early effect? How does it manifest in the output characteristics of common-emitter configuration?

11. Compare the three configurations—common-emitter, common-collector, and common-base—in terms of small-signal voltage gain, input resistance, and output resistance.

12. For linear operation of a common-emitter amplifier, how large can the base-emitter junction voltage change?

13. Define transconductance.

14. Name the two significant parameters in the low-frequency hybird-π model. How are they determined in a given circuit?

15. What accounts for r_o and r_μ in the hybird-π model? How significant are they?

16. If a BJT with $\alpha = 0.5$ is used in the common-base configuration, what is the ratio of collector current to emitter current?

17. What accounts for C_π and C_μ in the hybird-π model?

18. Name the four time-constants in the charge-control model. How are they related to the current gains?

19. How is power dissipation in a BJT defined?

20. What is thermal runaway? What causes it?

21. Compare the power dissipation levels of two BJTs, one operating in the active mode and the other switching between cutoff and saturation at a low rate.

PROBLEMS

1. Using Equations (3.10)–(3.13) and (3.15), verify Equation (3.14).

2. Defining cutoff by $I_E = 0$ and $I_C = I_{CBO}$, show that the cutoff base-emitter voltage is given by

$$V_{BE(\text{cutoff})} = V_T \ln(1 - \alpha_F)$$

(Hint: Show, using reciprocity, that $\alpha_F I_{EBO} = \alpha_R I_{CBO}$).
Calculate $V_{BE(\text{cutoff})}$ at room temperature for $\alpha_F = 0.98$.

3. Using Equation (3.12) and the reciprocity condition, show that the base voltage corresponding to $I_B = 0$ and large reverse-bias of the CBJ is given by

$$V_{BE} \doteq V_T \ln\left[\frac{1 + (\alpha_F/\alpha_R)(1 - \alpha_R)}{(1 - \alpha_F)}\right]$$

Calculate this voltage for $\alpha_F = 0.995$ and $\alpha_R = 0.5$.

* 4. Using the Ebers-Moll model equations, the reciprocity condition, and Equation (3.24), show that in saturation

$$V_{CE}(\text{sat}) = V_T \ln \frac{[1/\alpha_R + \beta_S/\beta_R]}{[1 - \beta_S/\beta_F]}$$

where $\beta_S = I_C(\text{sat})/I_B(\text{sat})$.
Calculate $V_{CE}(\text{sat})$ for $\beta_R = 0.5$, $\beta_F = 100$, and $\beta_S = 1, 5,$ and 10.

5. Determine the collector and base currents and voltages in the circuit shown in Figure P3.5 for (a) $V_i = 0.2$ V, (b) $V_i = 1$ V, and (c) $V_i = 2$ V. State the mode of operation in each case.

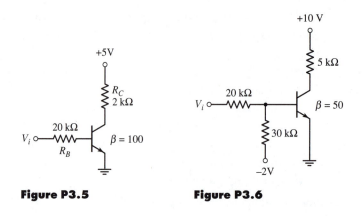

Figure P3.5   **Figure P3.6**

If variation of β among devices is ±20 percent, what is the range of R_B for keeping the BJT in saturation at $V_i = 3V$?

6. Determine the collector and the base currents and voltages in the circuit shown in Figure P3.6 for (a) $V_i = 3$ V, (b) $V_i = 1$ V, and (c) $V_i = 4$ V. State the mode of operation in each case.

7. Determine the collector and the base currents and voltages in the circuits shown in Figure P3.7a, b, and c.

If R_c can be varied, what is the smallest value required to operate the BJT in saturation in each circuit?

8. Derive the input and output resistances of the circuit shown in Figure P3.8. Assume that the capacitors C_C and C_B are so large in values that their effect is negligible at the frequency range of interest.

9. Using the equivalent circuit shown in Figure 3.21 verify Equations (3.34) and (3.35).

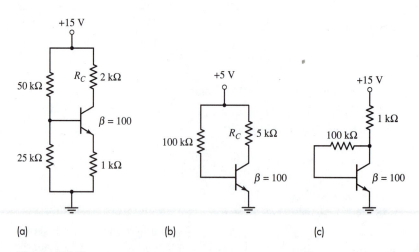

(a)   (b)   (c)

Figure P3.7

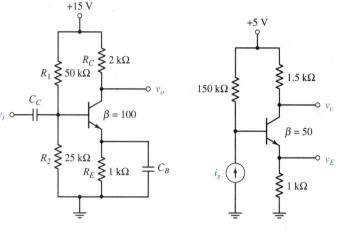

Figure P3.8   **Figure P3.12**

10. Assuming $V_{BE} \gg V_T$ and $v_{be} \ll V_T$ and using Equations (3.71), (3.73), (3.75), and (3.32), verify Equation (3.76).

11. A BJT operating in the forward active mode carries a steady current of $I_c = 5\,\text{mA}$ and $I_B = 50\,\mu\text{A}$. It has the following parameters: $\tau_F = 2\,\text{ns}$, $\tau_R = 30\,\text{ns}$, and $\beta_R = 0.5$, base width $W = 1\,\mu\text{m}$, EBJ area $A = 10^{-8}\,\text{m}^2$, and the base doping concentration is $N_A = 10^2/\text{m}^3$. Determine zero-bias base charge Q_{FO} and the forward injection base charge Q_F. What is the EBJ voltage?

*12. Figure P3.12 shows a phase splitter circuit. If the current source supplies $10\,\mu\text{A}$ peak sinusoidal current, determine and sketch the voltage waveforms at the collector and emitter. Verify your results using MicroSim PSpice. You may use default parameters for the BJT, or the following values, which are available for Q2N2222A on the disk in the student (evaluation) version of MicroSim PSpice.

.model Q2N2222A NPN (Is = 14.34f Xti = 3 Eg = 1.11 Vaf =
+ 74.03 Bf = 255.9 Ne = 1.307 Ise = 14.34f Ikf = .2847
+ Xtb = 1.5
+ Br = 6.092 Nc = 2 Isc = 0 Ikr = 0 Rc = 1 Cjc = 7.306p
+ Mjc = .3416
+ Vjc = .75 Fc = .5 Cje = 22.01p Mje = .377 Vje = .75
+ Tr = 46.91n Tf = 411.1p
+ Itf = .6Vtf = 1.7 Xtf = 3 Rb = 10)

13. Calculate the voltage gain v_c/v_s, where the ac signal v_s is applied to the base via a large capacitor in the circuit shown in Figure p3.7c.

*14. Determine the output voltage V_C for $V_i = 0$ to 5 V, in steps of 0.5 V for the circuit shown in Figure P3.5. State the mode of operation at each input voltage. Sketch the V_C–V_i transfer characteristic from your results. What is the range of input that is suitable for linear amplification? Run a MicroSim PSpice simulation for the transfer characteristic.

*15. Figure P3.15 shows a Darlington circuit. Determine the ratio of currents, I_{E2}/I_{B1}. Using the simplified hybrid-π model, calculate the ac voltage gain v_o/v_i, output resistance R_o, and input resistance R_i. If R_B is varied, can both Q_1 and Q_1 saturate? Explain.

16. Determine and sketch the voltage at V_o in the circuit shown in Figure P3.16 for the input waveform applied at V_i. Assume a β of at least 200. State the modes of operation for the transistor during $V_i = 0$ and $V_i = 4$ V. Sketch V_o if V_i switches between 0 and 5 V.

*17. Perform a MicroSim PSpice simulation for Problem 3.16 using the 2N2907A transistor, and compare results with calculated values.

18. *Experiment:* Most commonly used BJT parameters are determined in this experiment for a commercial switching transistor such as 2N2222A. In the circuit shown in Figure P3.18,

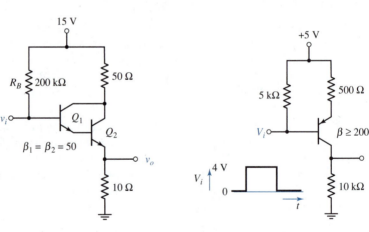

Figure P3.15   **Figure P3.16**

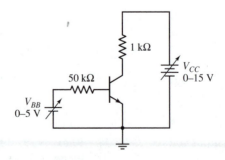

Figure P3.18

a. Vary the supplied V_{BB} and V_{CC} so that the BJT is well in the active mode. Measure the dc currents, I_B and I_c, and voltages, V_{BE} and V_{BC}.

b. Now vary V_{BB} slightly so that I_B changes by ΔI_B (about 10 percent). Measure the corresponding collector current.

   Determine β_{dc} and β_{ac} from these two pairs of dc current measurements. From one set of junction voltages and terminal currents, and using Equation (3.72b) for $I_E = I_B + I_C$, determine I_s. Using $I_s = \alpha_F I_{ES}$ and $\alpha_F = \beta_F/(1 + \beta_F)$, where $\beta_F = \beta_{dc}$, calculate I_{ES}.

c. Interchange the emitter terminal with the collector and adjust V_{BB} and V_{cc} so that the BJT operates in the (inverse) active mode. Measure the dc currents, I_B and I_E. Determine $\beta_R = I_E/I_B$. Calculate $\alpha_R = \beta_R/(1 + \beta_R)$ and $I_{CS} = I_s/\alpha_R$.

d. Calculate the dynamic base resistance, $r_\pi = V_T/I_B$, and the transconductance, $g_m = I_c/V_T$ at the forward active mode operating point (Step a).

e. Obtain data for the I_C–V_{CE} characteristics at $I_B = 20\,\mu\text{A}$, $30\,\mu\text{A}$, and $40\,\mu\text{A}$. From the I_C–V_{CE} graph, determine the Early voltage V_{AF} as shown in Figure 3.32, and β_{dc} and β_{ac}.

f. Compare the values of β obtained in Steps (b) and (e).

chapter 4

BIPOLAR JUNCTION TRANSISTOR SATURATION LOGIC FAMILIES

INTRODUCTION

In this chapter we begin the study of digital circuits using bipolar junction transistors. The earliest logic device implemented using BJTs was the basic inverter. Although the discrete BJT inverter does not belong to any particular family, or series, of digital circuits in the IC technologies, it is the precursor of all microelectronic digital logic families. Implementation of the BJT inverter as an integrated circuit in the early 1960s ushered in the technology that resulted in all of the different digital IC families.

Following the first use of the BJT inverter in the 1960s came the *resistor-transistor-logic* (RTL), *the diode-transistor-logic* (DTL), and the currently popular *transistor-transistor-logic* (TTL) series of integrated logic circuits.

A characteristic feature of RTL, DTL, and TTL families is that the BJTs operate by switching between saturation (on) and cutoff (off) modes of operation. For this reason these three families are generally called the *saturation logic families.* Two other bipolar logic families, the *emitter-coupled logic* (ECL) and the *integrated-injection logic* (I²L) families, operate by switching a constant current between two parts of a circuit. These families are referred to as the *current mode logic* (CML) families. We study the CML families in Chapter 5.

For the purpose of comparison between the saturation families and the other logic families, this chapter begins with the study of voltage transfer characteristics (VTC), noise margins, switching response, and fan-out of a BJT inverter. Next, we discuss the characteristics and the limitations of RTL and DTL families. Following this, we analyze the TTL family of logic circuits and its subseries of Schottky TTL circuits. The chapter concludes with a discussion of the modified circuit configurations and available gate circuits in the TTL series.

4.1 THE BJT INVERTER

A BJT logic inverter circuit is shown in Figure 4.1. Qualitatively, we can see that for an input voltage V_i below the cut-in voltage, the transistor is off and the output voltage V_o is close to the supply voltage V_{CC}. For a sufficiently high input voltage

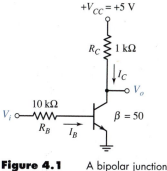

Figure 4.1 A bipolar junction transistor inverter

(and with the appropriate choice of the two resistors), the BJT is driven to saturation so that the output is nearly at zero voltage. Thus for the input increasing from zero voltage to V_{CC}, the output decreases from about V_{CC} to $V_{CE(sat)}$, causing the inverter action.

4.1.1 VOLTAGE TRANSFER CHARACTERISTICS (VTC)

To study the noise margins and the fan-out capability of the BJT inverter, let us consider the static behavior of the circuit when the input V_i is increased from zero to V_{CC}.

With the piecewise linear parameters of the transistor assumed as given in Table 4.1, it is clear that for input voltage V_i below 0.6 V—the base-emitter cut-in voltage—collector current is essentially zero, and the transistor is in the cutoff mode. Output voltage V_o is very nearly equal to the supply voltage V_{CC} when no load is connected. The input-output voltage transfer characteristic (VTC), therefore, shows a constant output voltage V_o of 5 V for V_i going from zero to 0.6 V. As V_i is increased above the first breakpoint at $V_i = V_{BE}$(cut-in) $= 0.6$ V and $V_o \approx V_{CC}$, the transistor enters forward active mode. The collector current ($I_C = \beta I_B$) causes a voltage drop in the collector resistor and the collector voltage $V_c = V_o = V_{CC} - I_c R_c$ falls. The active mode of oper-

Table 4.1	Piecewise linear parameters for the diode and the BJT	
Diode:	Cut-in voltage	0.6 V
	Forward voltage	0.7 V
BJT:	$V_{BE(\text{cut-in})}$	0.6 V
	$V_{BE(\text{active})}$	0.7 V
	$V_{BE(\text{sat})}$	0.8 V
	$V_{CE(\text{sat})}$	0.1 V

ation continues for increasing V_i until the collector voltage drops to about 0.1 V. Now the collector-base junction (CBJ) is forward-biased and the mode changes to saturation. Input voltage higher than that required to forward bias the CBJ drives the transistor heavily into saturation.

The input voltage at the second breakpoint is calculated using the condition that the transistor is on the verge of going from active to saturation mode.

Example 4.1

Calculate the input voltage V_i at which the transistor in Figure 4.1 goes from active to saturation mode of operation. Use $\beta = 50$.

Solution

At the edge of saturation (EOS), the CBJ is forward-biased by 0.7 V while the EBJ has about 0.8 V. Hence $V_{CE(EOS)} = 0.1$ V and the collector current is given by

$$I_C = I_{C(sat)} = \frac{V_{CC} - V_{CE(EOS)}}{R_C} = \frac{5 - 0.1}{1} = 4.9 \text{ mA}$$

The corresponding base current required in forward active mode is

$$I_{B(EOS)} = \frac{I_C}{\beta} = \frac{4.9}{50} \text{ mA} = 98 \text{ } \mu\text{A}$$

Since

$$I_{B(EOS)} = \frac{V_{i(EOS)} - V_{BE(sat)}}{R_B},$$

by rearranging, we can now find the input voltage at the edge of saturation as

$$V_{i(EOS)} = I_{B(EOS)} R_B + V_{BE(sat)} = 0.098 \times 10 + 0.8 = 1.78 \text{ V}$$

Hence, at $V_i = 1.78$ V, the transistor is on the verge of switching from active to saturation mode of operation.

From the above example, the second breakpoint for the BJT inverter of Figure 4.1 is at $V_i = 1.78$ V, and $V_o \approx 0.1$ V. For $V_i \geq 1.78$ V, the transistor is driven into saturation and the output voltage stays at $V_{CE(sat)} \approx 0.1$ V. Figure 4.2

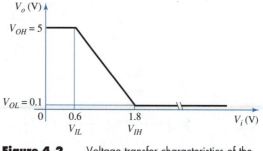

Figure 4.2   Voltage transfer characteristics of the BJT inverter

shows the piecewise voltage transfer characteristic, which has been obtained by neglecting the higher-order variations in the output voltage.

4.1.2 LOGIC LEVELS AND NOISE MARGIN

Because of the piecewise linear variation of V_o with V_i, we determine the logic levels for the BJT inverter based on unambiguous regeneration of output levels. From the VTC in Figure 4.2 it is clear that the logic levels are as follows:

Logic low: Input: $0 \leq V_i \leq 0.6$ V or $V_{IL} = 0.6$ V

Output: $V_{OL} = 0.1$ V

Logic high: Input: 1.78 V $\leq V_i \leq 5$ V or $V_{IH} \approx 1.8$ V

Output: $V_{OH} = 5$ V (No-load)

Observe that for input in the range of $0 \leq V_i \leq V_{IL}$, output of the inverter in Figure 4.1 is guaranteed at the logic high level of 5 V. Since the logic low voltage at output is contained within the range for low-level input, regeneration of low logic level in a cascade of inverters is guaranteed. Clearly, regeneration is also guaranteed for logic high, with $V_{IH} \leq V_i \leq 5$ V and $V_{OH} = 5$ V.

From the about voltages, the logic swing of the BJT inverter is given by

$$V_{ls} = V_{OH} - V_{OL} = 5 - 0.1 = 4.9 \text{ V}$$

The noise margins are given by

$$NM_L = V_{IL} - V_{OL} = 0.6 - 0.1 = 0.5 \text{ V}$$

$$NM_H = V_{OH} - V_{IH} = 5 - 1.8 = 3.2 \text{ V}$$

The width of transition, or indeterminate region, is $V_{tw} = V_{IH} - V_{IL} = 1.2$ V.

Note that the above voltages are applicable when no load is connected at the output of the inverter. When a load, such as a LED or another logic inverter, is connected to the collector of the BJT, the logic levels may change because of the load requirements. In addition, the voltage transfer characteristic is also dependent on temperature. Since the junction saturation currents and voltages vary with temperature (see Example 3.3), the VTC and hence the noise margins are affected by the operating temperature. At high temperatures, for example, the logic low and logic high input voltages are decreased due to lowered EBJ voltage.

4.1.3 FAN-OUT

Fan-out of the inverter is the number of identical circuits that the inverter can drive before V_o enters the transition region.

When the output of a BJT inverter (the driver) is at logic low, it can drive practically any number of identical inverters (loads). Since each load transistor is operating in the cutoff mode, no current is drawn via the collector resistor of the driver, and the output voltage remains at logic low. The situation is different,

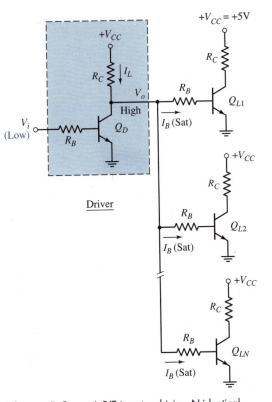

Figure 4.3   A BJT inverter driving N identical inverters at logic high output

however, for the logic high output of the driver. Figure 4.3 shows an inverter at logic high output driving N identical inverters. Since the saturation base current of each load inverter is supplied by the driver, output voltage V_o drops from its open-circuit voltage of V_{CC}. The gate fan-out, therefore, is limited by the number of load inverters that can be driven into saturation while maintaining a satisfactory logic high level of $V_o \geq V_{IH}$. This number depends on the current gain β of the driver transistor, and the base overdrive required for the load transistors.

Referring to Figure 4.3, the sum of the base currents of the load gates, which is the current through the collector resistor of the driver, is given by

$$I_L = N\, I_{B(\text{sat})} = N\, k\, I_{B(EOS)} \tag{4.1}$$

where k is the overdrive factor defined by

$$k = \frac{I_{B(\text{sat})}}{I_{B(EOS)}} \tag{4.2}$$

and

$$I_{B(EOS)} = \frac{I_{C(EOS)}}{\beta} = \frac{V_{CC} - V_{CE(\text{sat})}}{\beta R_C} \tag{4.3}$$

Therefore, the output voltage V_o is given by

$$V_o = V_{CC} - NkI_{B(EOS)} R_C$$

$$= V_{CC} - \frac{Nk[V_{CC} - V_{CE(sat)}]}{\beta} \qquad \textbf{(4.4)}$$

For regeneration of logic levels at the load gates, this voltage V_o must satisfy the input logic high level. Hence

$$V_{CC} - \frac{Nk[V_{CC} - V_{CE(sat)}]}{\beta} \geq V_{IH}$$

or

$$N \leq \frac{\beta[V_{CC} - V_{IH}]}{k[V_{CC} - V_{CE(sat)}]} \qquad \textbf{(4.5)}$$

Determine the number of identical inverters that the circuit shown in Figure 4.1 can drive at logic high output. It is required that the output must not drop below $V_{IH} = 1.8$ V with all load inverters connected, so that they can be driven to saturation. Assume that the load transistors are operating at minimum base current required for saturation. Use $\beta = 50$ for all transistors.

Example 4.2

To account for variations in β, if a base current of 25 percent more than the minimum is needed to drive each load into saturation, what is the fan-out?

Solution

For the given circuit parameters, the saturation (maximum) collector current of each load is

$$I_{C(sat)} = \frac{V_{CC} - V_{CE(sat)}}{R_C} = \frac{5 - 0.1}{1} = 4.9 \text{ mA}$$

Hence, the minimum base current required for saturation is

$$I_{B(EOS)} = \frac{I_{C(sat)}}{\beta} = \frac{4.9}{50} \text{ mA} = 98 \text{ } \mu\text{A}$$

With N loads, collector current supplied by the driver at logic high level is

$$I_C = 98N \text{ } \mu\text{A}$$

Hence the driver output voltage falls to

$$V_o = V_{CC} - I_C R_C$$

For $V_o \geq V_{IH}$ (Equation (4.4)), then,

$$5 - 98N \times 10^{-6} \times 1 \times 10^3 \geq 1.8 \text{ V}$$

which gives $N \leq 32.65$. Hence, the logic high fan-out with minimum base drive for the loads is 32. Note that we cannot round up the fan-out: the higher fan-out would cause the driver output voltage to go below the minimum logic high voltage of V_{IH}.

Note that in the cutoff (logic high) state, the circuit has an output resistance of R_C. With an input resistance of approximately R_B, then, N loads connected at the collector of the driver drops V_C to

$$V_C = V_o \approx \frac{R_C V_{CC}}{(R_C + R_B/N)}$$

With 25 percent base overdrive, each load requires

$$I_{B(min)} = 1.25 \times 98 = 122.5 \ \mu A$$

Hence, for base overdrive factor $k = 1.25$,

$$N \leq \frac{5 - 1.8}{0.1225} = 26.12$$

which gives a fan-out of 26.

Observe that the logic high fan-out is the gate fan-out for the saturated BJT inverter. The value of N obtained using Equation (4.5) has come at a cost: since we have allowed the logic high voltage to drop to V_{IH}, the logic high noise margin NM_H is now zero. That is, when all of the N load gates are connected, the driver output is just enough to be considered logic high by the loads. Any noise voltage at the output of the driver can cause all the load inputs to go into the transition, or indeterminate, region. To prevent this, a tradeoff between noise margin and fan-out is generally made in inverter and other gate designs. (see problem 4.2). Note that even with fewer than the maximum number of loads connected, the logic high noise margin decreases from its no-load noise margin, which is an undesirable characteristic of the BJT inverter.

4.1.4   TRANSIENT RESPONSE

The speed at which the BJT inverter can change its logic states is limited by the delays of the transistor in switching between saturation and cutoff modes of operation. We can estimate these delays from the charge stored in the base and the depletion regions of the collector-base and the emitter-base junctions as given by the charge-control model. We obtain reasonable approximations of the delays by modeling the charge storage in terms of diffusion and transition capacitances. Figure 4.4 shows the waveforms resulting from the dynamic behavior of the inverter to a switching signal.

When the input voltage V_i (Figure 4.4a) is at $V_L(\leq V_{IL})$, the transistor is operating in the cutoff mode with zero base current, and the emitter-base junction (EBJ) is zero-, or reverse-biased. As V_i changes to V_H at $t = t_0$, the base current I_B (Figure 4.4b) jumps to approximately V_H/R_B. The collector current I_C (Figure 4.4c), however, does not attain its value of $I_{C(sat)}$ instantaneously due to two factors. First, the EBJ transition capacitance must be charged to forward-bias (cut-in) voltage. As the EBJ becomes forward-biased, the junction

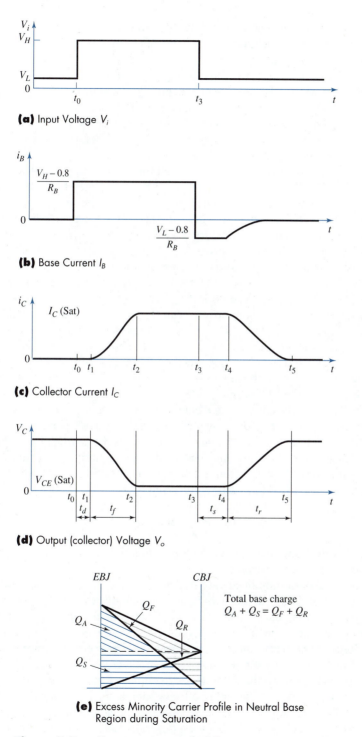

(a) Input Voltage V_i

(b) Base Current I_B

(c) Collector Current I_C

(d) Output (collector) Voltage V_o

(e) Excess Minority Carrier Profile in Neutral Base Region during Saturation

Figure 4.4   Transient response of a BJT inverter

capacitance becomes predominantly diffusion capacitance, which controls the collector current. The first factor contributes to the *delay time* $t_d = t_1 - t_0$ shown in Figure 4.4d. At t_1, the transistor is brought from cutoff to active mode, and the collector current begins to increase. Now the forward-bias diffusion capacitance at the EBJ charges exponentially from cut-in voltage while I_C rises as βI_B. At $t = t_2$, I_C reaches its maximum of $I_{C(\text{sat})}$. Now the EBJ voltage is $V_{BE(\text{sat})}$ (≈ 0.8 V) and the collector (output) voltage is $V_{CE(\text{sat})}$ (≈ 0.1V). Base current I_B ($= [V_H - V_{BE(\text{sat})}]/R_B$) in excess of $I_{C(\text{sat})}/\beta$ contributes to excess charge in the neutral base region. The *rise time* of the collector current is the delay $t_r = t_2 - t_1$. Note that this is also the fall time of the output (collector) voltage (Figure 4.4d). The sum of the delay and the rise times is the *turn-on delay*. Figure 4.4e shows the concentration profile of the excess minority carriers in the base region. The rectangular region in this figure corresponds to the excess charge proportional to the overdrive base current, $I_B - I_{C(\text{sat})}/\beta$.

The delay in the turn-off process of the transistor is caused primarily by the removel of the excess minority carriers in the base region. When the input voltage switches from $V_H(\geq V_{IH})$ to $V_L(\leq V_{IL})$, the base current switches to $I_B = [V_L - V_{BE(\text{sat})}]/R_B$. The collector current continues at its saturation level until the excess charge is removed from the base region. Now the EBJ capacitance discharges toward V_L. At $t = t_4$, the transistor enters the active mode. The *storage,* or *saturation delay,* $t_s = t_4 - t_3$, is the interval during which I_C remains at $I_{C(\text{sat})}$ after V_i has switched to V_L. As the EBJ capacitance continues to discharge in the active mode, I_C begins to drop. At $t = t_5$, the predominantly depletion capacitance of the EBJ is charged to V_L, and the collector current is turned off. The *fall time* of the collector current (or the rise time of the collector voltage) is given by $t_f = t_5 - t_4$, during which the transistor goes from active to cutoff mode. The sum of the storage and the fall times is the *turn-off delay*. Figure 4.4 also shows the high-to-low and low-to-high propagation delay times, t_{PHL} and t_{PLH}, which are the delays from input transitions to 50 percent points at output voltage waveforms. These delays are given by

$$t_{PHL} \approx t_d + \frac{t_f}{2} \quad \text{and} \quad t_{PLH} \approx t_s + \frac{t_r}{2}$$

Note that the above definitions of propagation delays assume abrupt transitions at the input voltage waveform.

Because of the nonlinear variation of emitter-base capacitance with voltage, an exact calculation of the switching delays requires complex modeling and analysis by a computer. Exact analysis must also include the nonlinear collector-base capacitance, which is typically smaller than the emitter-base capacitance. In addition, one must consider the capacitance between collector and ground, which includes the substrate capacitance in integrated circuits. In the following analysis, we estimate the delay times assuming average currents charging the capacitances at the emitter and collector junctions. The charge-control model enables the estimation of net charge stored in the neutral and the depletion base regions as the device moves between different modes of operation. Externally, the base and collector currents are used to determine the same charge.

From Chapter 3, the complete charge-control model equations are given by

$$i_C = \frac{Q_F}{\tau_F} - Q_R\left[\frac{1}{\tau_R} + \frac{1}{\tau_{BR}}\right] - \frac{dQ_R}{dt} - \frac{dQ_{VC}}{dt} \qquad \textbf{(3.59)}$$

$$i_B = \frac{Q_F}{\tau_{BF}} + \frac{Q_R}{\tau_{BR}} + \frac{dQ_F}{dt} + \frac{dQ_R}{dt} + \frac{dQ_{VC}}{dt} + \frac{dQ_{VE}}{dt} \qquad \textbf{(3.60)}$$

$$i_E = -\frac{Q_R}{\tau_R} + Q_F\left[\frac{1}{\tau_F} + \frac{1}{\tau_{BF}}\right] + \frac{dQ_F}{dt} + \frac{dQ_{VE}}{dt} \qquad \textbf{(3.61)}$$

where

$$Q_F(t) = Q_{F0}\left[\exp\left(\frac{V_{BE}}{V_T}\right) - 1\right] \qquad \textbf{(3.53)}$$

$$Q_R(t) = Q_{R0}\left[\exp\left(\frac{V_{BC}}{V_T}\right) - 1\right] \qquad \textbf{(3.57)}$$

$$Q_{F0} = \left(\frac{1}{2}\right)eWAn_{bp0} \qquad \textbf{(3.45)}$$

$$Q_{R0} = \left(\frac{1}{2}\right)eWAn_{bp0} \qquad \textbf{(3.58)}$$

and V_E and V_C are the voltages at the EBJ and the CBJ. The terms τ_F and τ_R are the average forward and reverse transit times of minority carriers in the base region, and τ_{BF} and τ_{BR} are the minority carrier lifetimes in the base region during forward and reverse active modes.

Turn-On Delay   Although the transistor changes modes during $t_0 < t < t_1$—from cutoff to saturation via cut-in and active modes—we simplify the analysis by considering only two transitions: off to active mode, and active to saturation mode.

Transition from Off to Active Mode—Delay Time   During cutoff in the interval $t < t_0$, with reverse- (or zero-) biased EBJ and CBJ, the forward and reverse charges, $Q_F(t)$ and $Q_R(t)$, are negligible. Hence, we may approximate the base and the collector currents in the above equations as

$$i_B = dQ_{VC}/dt + dQ_{VE}/dt \qquad \textbf{(4.6)}$$

$$i_C = -dQ_{VC}/dt \qquad \textbf{(4.7)}$$

These currents arise from the charge stored in the depletion regions of the two junctions.

As the EBJ voltage changes from V_L at $t = t_0$ to V_{BE}(active) $= 0.7$ V at $t = t_1$, the change in stored charge in the depletion regions is given by

$$\int_{t_0}^{t_1} i_B \, dt = [Q_{VC}(t_1) - Q_{VC}(t_0)] + [Q_{VE}(t_1) - Q_{VE}(t_0)] \qquad \textbf{(4.8)}$$

The left side of Equation 4.8 may be approximated by using the average base current assuming linear variation from t_0 to t_1; the right side represents charge stored in the junction (depletion) capacitances.

To calculate the base current and the charge we assume that the voltage across the EBJ changes from $V_E = V_L$ at $t = t_0$ to $V_E = 0.7$ V at $t = t_1$, where the transition from cut-in to active mode is considered rapid. In the same interval, voltage V_C at the CBJ changes from $V_{BC}(t_0) = V_L - V_{CC}$ to $V_{BC}(t_1) = 0.7 - V_{CC}$. If C_{E1} and C_{C1} are the junction capacitances, which are dependent on the voltage changes ΔV_E and ΔV_C, the total stored charge changes by

$$\Delta Q = \Delta V_E \, C_{E1} + \Delta V_C \, C_{C1} \tag{4.9}$$

Base currents at t_0 and t_1 are given by

$$i_B(t_0) = \frac{V_i(t_0) - V_{BE}(t_0)}{R_B} = \frac{V_H - V_L}{R_B} \tag{4.10}$$

$$i_B(t_1) = \frac{V_i(t_1) - V_{BE}(t_1)}{R_B} = \frac{V_H - 0.7}{R_B} \tag{4.11}$$

We combine Equations (4.8)–(4.11) to find the *delay time* as

$$t_1 - t_0 = \frac{\Delta Q}{i_{B\,(\text{avg})}} = \frac{2[\Delta V_E \, C_{E1} + \Delta V_C \, C_{C1}]}{[i_B(t_0) + i_B(t_1)]} \tag{4.12}$$

Transition from Active Mode to Saturation—Rise Time   In the forward active mode, the charge-control model gives the base and collector currents as

$$i_B = \frac{Q_F}{\tau_{BF}} + \frac{dQ_F}{dt} + \frac{dQ_{VC}}{dt} + \frac{dQ_{VE}}{dt} \tag{4.13}$$

$$i_C = \frac{Q_F}{\tau_F} \tag{4.14}$$

where we ignore the dQ_{VC}/dt term in i_C in light of the large QF/t_F term. Again, we use the average base current on the left side of Equation (4.13) to determine the change in total charge and hence estimate the rise time. In addition, we assume that the steady base current during $V_i = V_H$ is sufficient to drive the transistor into saturation.

Integrating Equation (4.13),

$$(t_2 - t_1)I_{B\,(\text{avg})} = \int_{t_1}^{t_2} \left(\frac{Q_F}{\tau_{BF}}\right) dt + \Delta Q_F + \Delta Q_{VC} + \Delta Q_{VE} \tag{4.15}$$

where $t_r = t_2 - t_1$ is the *rise time* of the collector current waveform (which, we recall, is also the *fall time* of the collector voltage waveform).

Since $V_{BE}(t_1) = 0.7$ V and $V_{BE}(t_2) = 0.8$ V, average base current is given by

$$I_{B(\text{avg})} = \frac{\dfrac{[V_H - 0.7]}{R_B} + \dfrac{[V_H - 0.8]}{R_B}}{2} \tag{4.16}$$

The first term on the right side of Equation (4.15), which represents the recombination charge ΔQ_{Fr}, is approximated by the average charge using Equation (4.14) as

$$\Delta Q_{Fr} = [Q_F(t_1) + Q_F(t_2)]\frac{(t_2 - t_1)}{2\tau_{BF}}$$

$$= [i_C(t_1) + i_C(t_2)]\frac{(t_2 - t_1)\tau_F}{2\tau_{BF}}$$

or

$$\Delta Q_{Fr} \approx i_{C(sat)}\frac{(t_2 - t_1)\tau_F}{2\tau_{BF}} = i_{C(sat)}\frac{(t_2 - t_1)}{2\beta} \qquad \textbf{(4.17)}$$

if the collector current at the onset of the active mode is neglected. Clearly, the contribution to base charge from recombination becomes significant if the lifetime in the base region is small.

The other terms are given by

$$\Delta Q_F = Q_F(t_2) - Q_F(t_1) = \tau_F[i_C(t_2) - i_C(t_1)] \approx \tau_F I_{C(sat)} \qquad \textbf{(4.18)}$$

$$\Delta Q_{VE} = \Delta V_E\, C_{E2} \qquad \textbf{(4.19)}$$

$$\Delta Q_{VC} = \Delta V_C\, C_{C2} \qquad \textbf{(4.20)}$$

where C_{E2} and C_{C2} are the capacitances at EBJ and CBJ. The junction voltage changes are

$$\Delta V_E = 0.8 - 0.7 = 0.1 \text{ V} \quad \text{and} \quad \Delta V_C = 0.7 - (0.7 - V_{CC}) = V_{CC} \qquad \textbf{(4.21)}$$

Using Equations (4.15)–(4.20) we can estimate the rise time. But note that t_r calculated using Equations (4.15)–(4.20) is approximate; it corresponds to collector current going from 0 to $I_{C(EOS)}$, rather than from $0.1 I_{C(sat)}$ to $0.9\, I_{C(sat)}$.

Determine the approximate delay and rise times for the inverter shown in Figure 4.1 with $\beta_F = 50$, $\beta_R = 0.7$, $\tau_F = 0.2$ ns, and $\tau_R = 20$ ns. The input voltage levels are $V_L = -1$ V and $V_H = 5$ V. As a first-order approximation to the junction capacitances, use the following values, assumed to be constant throughout the operating regions. Off to active mode capacitances are $C_{E1} = 0.4$ pF and $C_{C1} = 0.1$ pF; active to saturation mode capacitances are $C_{E2} = 0.8$ pF and $C_{C2} = 0.1$ pF. (More exact capacitances may be calculated using zero-bias values, C_{je0} and C_{jc0}, and the change in junction voltages, ΔV_E and ΔV_C.) | **Example 4.3**

Solution

In going from cutoff to active mode during t_d, the emitter junction voltage changes by

$$\Delta V_E = V_{BE}(t_1) - V_{BE}(t_0) = 0.7 - (-1) = 1.7 \text{ V}$$

The collector junction voltage also changes by the same amount,

$$\Delta V_C = 1.7 \text{ V}$$

From Equation (4.9), the change in base charge is

$$\Delta Q = 1.7 \times 0.4 + 1.7 \times 0.1 = 0.85 \text{ pC}$$

This charge arises from the change in base current from

$$i_B(t_0) = \frac{5 - (-1)}{10} = 0.6 \text{ mA}$$

to

$$i_B(t_1) = \frac{5 - 0.7}{10} = 0.43 \text{ mA}$$

Hence, using the average base current of

$$I_{B\,(\text{avg})} = \frac{0.6 + 0.43}{2} = 0.515 \text{ mA},$$

the delay time is given by

$$t_d(0.515 \text{ mA}) = 0.85 \text{ pC}$$

or

$$t_d = \frac{0.85 \text{ pC}}{0.515 \text{ mA}} = 1.65 \text{ ns}$$

In the interval $t_1 < t < t_2$, the base current changes from $i_B(t_1) = 0.43$ mA to

$$i_B(t_2) = \frac{5 - 0.8}{10} = 0.42 \text{ mA}$$

Hence, during the rise in collector current (or fall in collector voltage),

$$I_B(\text{avg}) = 0.425 \text{ mA}$$

In $t_r = t_2 - t_1$, the changes in junction voltages are

$$\Delta V_E = 0.1 \quad \text{and} \quad \Delta V_C = V_{CC} = 5 \text{ V}$$

and the collector current changes from 0 to

$$I_{C(\text{sat})} = \frac{5 - 0.1}{1} = 4.9 \text{ mA}$$

Thus, the charge changes are given by

$$\Delta Q_{VE} = 0.1 \text{ V} \times 0.8 \text{ pF} = 0.08 \text{ pC}$$

$$\Delta Q_{VC} = 5 \text{ V} \times 0.1 \text{ pF} = 0.5 \text{ pC}$$

$$\Delta Q_F = 4.9 \text{ mA} \times 0.2 \text{ ns} = 0.98 \text{ pC}$$

and the recombination charge, from Eqation (4.17), is

$$\Delta Q_{Fr} = 4.9 \text{ mA} \times \frac{t_r}{100} = 0.049 t_r \text{ mC}$$

Then, from Equation (4.15),

$$0.425 \text{ mA} \times t_r = 0.08 \text{ pC} + 0.5 \text{ pC} + 0.98 \text{ pC} + 0.049 t_r \text{ mC}$$

or

$$t_r = \frac{1.56 \text{ pC}}{[0.425 - 0.049] \text{ mA}} = 4.15 \text{ ns}$$

Without the recombination charge ΔQ_{Fr}, the collector current rise time is

$$t_r = \frac{1.56 \text{ pC}}{0.425} = 3.67 \text{ ns}$$

Clearly, the effect of recombination in the base increases t_r, with fewer carriers available for conduction.

The total forward recovery time for the transistor to go from off to saturation as the input voltage changes from -1 V to 5 V is

$$t_{fr} = t_f + t_r = 1.65 + 3.67 = 5.32 \text{ ns}$$

The above delay, t_{fr}, is the total turn-on delay.

The high-to-low propagation delay of the output (collector) voltage is

$$t_{PHL} = t_f + \frac{t_r}{2} = 3.49 \text{ ns}$$

Turn-Off Delay   Proceeding in the same manner as for turn-on delay, we calculate the charge stored in the base region and use the average current to determine the storage delay and fall times.

Transition from Saturation to Active Mode—Storage Delay   When the transistor is in saturation during $t_2 > t > t_3$, the base current I_B is in excess of the edge-of-saturation value given by $I_{B(EOS)} = I_{C(sat)}/\beta$. Because both junctions are forward-biased, this overdrive base current builds up forward and reverse charge (minority carrier) storage in the base as $Q_F + Q_R$. This excess charge is governed by the base current, given by

$$i_B = \frac{Q_F}{\tau_{BF}} + \frac{Q_R}{\tau_{BR}} + \frac{dQ_F}{dt} + \frac{dQ_R}{dt} \tag{4.22}$$

This equation neglects the change in charge in the depletion region, $dQ_{VC}/dt + dQ_{VE}/dt$, since V_{BC} and V_{BE} are constant.

The collector current in saturation is

$$i_C = \frac{Q_F}{\tau_F} - Q_R \left[\frac{1}{\tau_R} + \frac{1}{\tau_{BR}} \right] \tag{4.23}$$

The total charge QB in the base region (Figure 4.4e) may be split into two components:

$$Q_B = Q_F + Q_R = Q_A + Q_S \tag{4.24}$$

where Q_A arises from the active mode base current $I_{B(EOS)}$, and Q_S is due to the overdrive base current $I_{B(OD)}$. Q_F and Q_R represent charges due to forward-biased EBJ and CBJ, respectively. Note that Q_A represents the minimum forward charge required to bring the transistor to the edge of saturation (with no reverse component), while Q_S arises from both forward and reverse components. The overdrive base charge Q_S does not contribute to collector current. The total (static) base current is given by

$$I_B = I_{B(EOS)} + I_{B(OD)} \tag{4.25}$$

where the components are related to charges as

$$Q_A = I_{B(EOS)} \tau_{BF} \tag{4.26}$$

and

$$Q_S = I_{B(OD)} \tau_S \tag{4.27}$$

In Equation (4.27), τ_s represents the *saturation time constant,* similar to forward and reverse transit time constants. τ_S determines the rate at which the overdrive base charge Q_S decays to zero when turned off. It can be shown that τ_S is related to τ_F and τ_R as

$$\tau_S = \frac{\alpha_F(\tau_F + \alpha_R\tau_R)}{(1 - \alpha_F\alpha_R)} \tag{4.28}$$

When turned off, the instantaneous base current $i_B(t)$ becomes

$$i_B(t) = \frac{Q_A}{\tau_{BF}} + \frac{Q_S}{\tau_S} + \frac{dQ_S}{dt} \tag{4.29a}$$

where Q_A and Q_S decay at different rates. Note also that the constant overdrive base charge must be removed first, so that $dQ_A/dt = 0$ until $Q_S = 0$.

Eliminating Q_A using Equation (4.26), the instantaneous base current becomes

$$i_B = I_{B(EOS)} + \frac{Q_S}{\tau_S} + \frac{dQ_S}{dt} \tag{4.29b}$$

or

$$i_B = \frac{I_{C(EOS)}}{\beta_F} + \frac{Q_S}{\tau_S} + \frac{dQ_S}{dt} \tag{4.29c}$$

This equation relates the overdrive base charge Q_S to the instantaneous base current after turn-off. Q_S is removed during *storage delay* $t_s = t_4 - t_3$, where t_s is determined using Equation (4.29). The initial condition is that at $t = t_3$, the overdrive base charge Q_S from Equation (4.27) is given by

$$Q_S(t_3) = \tau_S I_{B(OD)}\big|t < t_3 = \left[I_B(t_{3-}) - \frac{I_{C(sat)}}{\beta_F}\right]\tau_S \tag{4.30}$$

where the forward base current in saturation is given by

$$I_B(t_{3-}) = \frac{V_H - V_{BE(\text{sat})}}{R_B} \qquad \textbf{(4.31)}$$

Solving the first-order differential equation in Equation (4.29) using the initial condition in Equation (4.30), it can be shown that

$$Q_S(t) = \tau_S \left[\left\{ I_B(t_{3+}) - \frac{I_{C(\text{sat})}}{\beta_F} \right\} + \{ I_B(t_{3-}) - I_B(t_{3+}) \} \exp\left(\frac{-t}{\tau_S} \right) \right] \qquad \textbf{(4.32)}$$

where the base current is assumed to change instantaneously from $I_B(t_{3-})$ to

$$I_B(t_{3+}) = \frac{V_L - V_{BE(\text{sat})}}{R_B} \qquad \textbf{(4.33)}$$

At $t = t_s$, Q_S becomes zero. Hence, we find the storage delay as

$$t_s = \tau_S \ln \left\{ \frac{[I_B(t_{3-}) - I_B(t_{3+})]}{[I_{C(\text{sat})}/\beta_F - I_B(t_{3+})]} \right\} \qquad \textbf{(4.34)}$$

Transition from Active to Cutoff—Fall Time   With $Q_S = 0$ at $t = t_4$, the base now discharges Q_A and the collector current falls off exponentially from $I_{C(\text{sat})}$. The *fall time* for I_C is determined similar to the rise time in the active mode. Using Equation (4.13) for the interval $t_5 < t < t_4$, Equation (4.15) becomes

$$(t_5 - t_4)i_{B(\text{avg})} = \int_{t_4}^{t_5} \left(\frac{Q_F}{\tau_{BF}} \right) dt + \Delta Q_F + \Delta Q_{VC} + \Delta Q_{VE} \qquad \textbf{(4.35)}$$

Here again the average base current in $t_5 < t < t_4$ (from edge of saturation to active mode) is used for the left side of Equation (4.35). The right side can be seen as the same as in Equation (4.15), corresponding to off-to-active transition, but changing in the opposite direction. Note further that in the absence of reverse base drive, that is, if $V_L = 0$, $i_{B(\text{avg})}$ would be small and the base charge decay would be primarily due to recombination. As with the rise time calculation, we can approximate the recombination term by the average charge in linear discharge using Equation (4.14) as

$$\Delta Q_{Fr} \approx [I_C(t_5) + I_C(t_4)](t_5 - t_4)\frac{\tau_F}{2\tau_{BF}} = I_C(t_4)\frac{\tau_f}{2\beta}$$

Similar to the rise time t_r, the fall time t_f estimates the time for the collector current to drop from $I_{C(EOS)}$ to 0.

In addition to the fall time, *recovery time* is needed to charge the depletion capacitances to steady reverse voltages. Since the transistor is assumed off for $t > t_4$, only the depletion charge contributes to the average base current given by

$$I_{B(\text{avg})} = \frac{V_L - V_{BE(\text{active})}}{2R_B}$$

Hence, we determine the approximate recovery time using the charge calculated for the delay time estimation.

The following example illustrates the reverse recovery time calculations.

Example 4.4	Determine the approximate storage delay and fall time for the inverter shown in Figure 4.1. What are the turn-off and propagation delays? Use the data given in Example 4.3.

Solution

First, we calculate the saturation time constant:

Since $\alpha_F = 50/51 = 0.98$ and $\alpha_R = 0.7/1.7 = 0.4$,

$$t_S = 0.98 \left(\frac{0.2 + 0.4 \times 20}{1 - 0.98 \times 0.4} \right) = 13.2 \text{ ns}$$

The base currents at t_3 are

$$I_B(t_{3-}) = \frac{5 - 0.8}{10} = 0.42 \text{ mA}$$

and

$$I_B(t_{3+}) = \frac{-1 - 0.8}{10} = 0.18 \text{ mA}$$

The collector current, from Example 4.3, is

$$I_{C(\text{sat})} = 4.9 \text{ mA}$$

and the base current at the edge of saturation, at $t = t_{4-}$, is

$$I_{B(\text{EOS})} = \frac{I_{C(\text{sat})}}{\beta_F} = \frac{4.9}{50} \text{ mA} = 98 \ \mu\text{A}$$

Hence the stored excess base charge Q_S decays as (Equation (4.32))

$$Q_S(t) = 13.2 \left[-0.18 - 0.098 + (0.42 + 0.18) \exp \left(\frac{-t}{13.2} \right) \right.$$

$$= 13.2 \left[0.6 - 0.278 \exp \left(\frac{-t}{13.2} \right) \right]$$

Solving for $Q_S(t) = 0$ at $t = t_s$, the storage delay is given by

$$t_s = 13.2 \ln \left[\frac{0.6}{0.278} \right] = 10.2 \text{ ns}$$

Average base current during the fall time of I_C is

$$I_{B(\text{avg})} = \frac{[I_B(t_{4+}) + I_B(t_s)]/2}{[(-1 - 0.8)/10 + (-1 - 0.7)/10]/2}$$

$$= -0.175 \text{ mA}$$

Using the values for charges from Example 4.3, Equation (4.35) becomes

$$t_f I_{B(\text{avg})} = \Delta Q_F + \Delta Q_{VC} + \Delta Q_{VE} + \Delta Q_{Fr}$$

or $(-0.175 \text{ mA}) t_f = -0.98 \text{ pC} - 0.5 \text{ pC} - 0.08 \text{ pC} + 0.049 t_f \text{ mC}$

Solving,

$$t_f = \frac{-1.56}{-(0.175 + 0.049)}$$

$$= 7 \text{ ns}$$

Without the recombination charge term $0.049t_f$ mC, the fall time increases to

$$t_f = \frac{-1.56}{-0.175} = 8.9 \text{ ns}$$

The decrease in average charge due to recombination clearly aids in reducing the fall time.

During the recovery interval, base charge current is

$$I_{B(\text{avg})} = \frac{I_B(t_5)}{2} = \frac{-0.17}{2} = -0.085 \text{ mA}$$

The junction transition capacitances, C_{E1} and C_{C1}, are charged to $V_L = -1$ V and $V_{BC} = V_L - V_{CC} = -4$ V, respectively, during the recovery time. Hence,

$$I_{B(\text{avg})} \times t_{\text{rec}} = \Delta Q_{VE} + \Delta Q_{VC} = \Delta V_E C_{E1} + \Delta V_C C_{C1}$$

or, substituting values,

$$-0.085 \text{ mA} \times t_{\text{rec}} = -1.7 \times 0.4 - 1.7 \times 0.1 = -0.85 \text{ pC}$$

Hence,

$$t_{\text{rec}} = 10 \text{ ns}$$

Total delay in switching the transistor from saturation to cutoff as the input switches from 5 V to -1 V is

$$t_{rr} = t_s + t_f + t_{\text{rec}} = 10.2 + 7 + 10 = 27.2 \text{ ns}$$

This is the turn-off delay or *reverse recovery time*.

Low-to-high propagation delay for the output (collector) voltage is

$$t_{PLH} = t_s + \frac{t_f}{2} = 13.7 \text{ ns}$$

The above analysis simplified the calculations by using fixed average capacitances at the two junctions as the transistor moved from one mode at $t = t_1$ to another at $t = t_2$. Since the junction capacitances depend on the bias voltages, their values in each mode of operation may be determined as in the MicroSim PSpice model (Section 3.8) given by

$$C_{BE} = \left(\frac{T_F I_S}{V_T}\right) \exp\left(\frac{V_{BE}}{V_T}\right) + \frac{C_{JE}}{\left(1 - \frac{V_{BE}}{V_{JE}}\right)^{M_{JE}}} \qquad \textbf{(3.78)}$$

$$C_{BC} = \left(\frac{T_R I_S}{V_T}\right) \exp\left(\frac{V_{BC}}{V_T}\right) + \frac{C_{JC}}{\left(1 - \frac{V_{BC}}{V_{JC}}\right)^{M_{JC}}} \qquad \textbf{(3.79)}$$

where both the diffusion and the transition components are included.

Alternatively, one may use large-signal equivalent capacitances (Section 3.6) as [Reference 4]

$$C_{BE} = -C_{je0}\left[\frac{V_{0e}}{\Delta V_{BE}(1 - m_e)}\right]\left\{\left[1 - \left(\frac{V_{BE}(t_1)}{V_{0e}}\right)\right]^{(1-m_e)}\right.$$
$$\left. - \left[1 - \left(\frac{V_{BE}(t_2)}{V_{0e}}\right)\right]^{(1-m_e)}\right\}$$

(4.36)

$$C_{BC} = -C_{jc0}\left[\frac{V_{0c}}{\Delta V_{BC}(1-m_c)}\right]\left\{\left[1 - \left(\frac{V_{BC}(t_1)}{V_{0c}}\right)\right]^{(1-m_c)}\right.$$
$$\left. - \left[1 - \left(\frac{V_{BC}(t_2)}{V_{0c}}\right)\right]^{(1-m_c)}\right\}$$

(4.37)

where $\Delta V_{BE} = V_{BE}(t_1) - V_{BE}(t_2)$
and $\Delta V_{BC} = V_{BC}(t_1) - V_{BC}(t_2)$

For quick hand calculations, we may use an estimate of the junction capacitances based on their zero-bias values and the average bias voltage in each interval.

From the foregoing discussion we can see that the turn-off time, which is the predominant delay in switching the transistor from saturation to cutoff state, is dependent on the base overdrive and the low-level voltage of input. A large negative voltage V_L (or a negative power supply in the base circuit) will increase the magnitude of the reverse base current and speed up the removal of the excess charge from the base. While this fast removal of the base charge reduces the storage delay t_s, it is now necessary to charge the EBJ capacitance from a large negative voltage to $V_{BE(sat)}$, which results in the increase of the delay time t_d. Also, for this method the input signal must be bipolar, requiring the use of two supply voltages. Instead of resorting to this, BJT circuits operating with single voltage sources employ nonlinear pull-down circuitry ("active pull-down") (see Section 4.4.7) or use speed-up capacitors (see Section 4.1.5) for faster discharge of excess base charge.

We estimated the above values for the delays based solely on the emitter junction capacitance. If the feedback collector junction capacitance cannot be neglected, its effect must be included at both input and output terminals. In general, however, parasitic load capacitance is higher than junction capacitances. As a result, the delays due to exponential charging and discharging of load capacitance via output resistance (see Section 1.6.2) may become more significant than the intrinsic delays of the transistor.

MicroSim PSpice simulation of the BJT inverter switching delays may be obtained as shown in Figure 4.5. For this figure, typical zero-bias junction capacitance values were used. These zero-bias capacitances yield approximately the same average values at the bias conditions as we used in Examples 4.3

```
* BJT INVERTER SWITCHING RESPONSE
Q        3   2   0   NPNT
RB       1   2   10K
RC       4   3   1K
VI       1   0   PULSE(-1V 5V 10n 0.05n 0.05n 20n 50n)
VCC      4   0   DC 5.0
.MODEL NPNT NPN (IS=2E-16 BF=50 BR=0.7 TR=20E-9
+ TF=0.2E-9 RB=50 RC=20 CJC=0.15p CJE=0.3p)
.TRAN 0.5n 50n
.PROBE
.END
```

(a) Input File

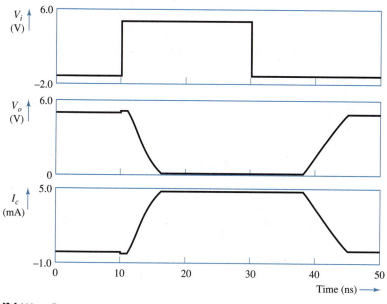

(b) Wave Forms

Figure 4.5   MicroSim PSpice simulation of BJT inverter switching response

and 4.4. (The average values we used in the examples can be obtained using Equations (4.36) and (4.37) with a junction grading coefficient of 0.33 and a built-in potential of 0.8 V for both junctions.) In addition, a junction saturation current of $I_S = 2 \times 10^{-16}$ is used based on the saturation collector current of

$$I_{C(EOS)} = 4.9 \text{ mA at } V_{BE(EOS)} = 0.8 \text{ V} \quad \text{and} \quad \beta_F = 50$$

and using the approximation $I_C \approx \alpha_F I_S \exp^{(VBE/VT)}$. Other parameters for more accurate modeling can be determined from the terminal behavior of the circuit [see Reference 4].

From the MicroSim PSpice simulation we get the following values for the delays: $t_d = 1$ ns, $t_r = 5$ ns, $t_s = 8$ ns, and $t_f = 7$ ns. Compare these with the estimated values of $t_d = 1.7$ ns, $t_r = 4.2$ ns, $t_s = 10.2$ ns, and $t_f = 7$ ns from Examples 4.3 and 4.4. Clearly, there is a good agreement between the values we estimated using approximate linear charging models and the simulated results.

In general, switching delays for a BJT inverter using a general-purpose switching transistor are typically given by $t_d = 10-40$ ns, $t_r = 10-40$ ns, $t_s = 20-200$ ns, and $t_f = 10-60$ ns, depending on base drive. Based on these values, the average propagation delay of a BJT inverter without any load is approximately 100 ns. Since each load gate contributes to capacitance at the output of the driver, the average delay of the inverter with fan-out is greater than 100 ns.

4.1.5   SWITCHING SPEED IMPROVEMENT

The switching response of the BJT inverter can be improved if large impulses of currents are used to charge and discharge the EBJ capacitance during input transitions. A small capacitor C_S, connected across the base resistor R_B as shown in Figure 4.6, can accomplish this.

The speed-up capacitor provides a low-impedance path for the base to discharge the excess charge during turn-off. As a result, the reverse base current is much higher when the input voltage V_i changes from V_H to V_L and the turn-off is faster.

During $V_i = V_H$, the capacitor C_S is charged to Q_{SC} given by

$$Q_{SC} = I_{BF} R_B C_S = [V_H - V_{BE(\text{sat})}] C_S \qquad \textbf{(4.38)}$$

The initial impulse of charging current due to the voltage step V_H-V_L injects carriers into the base. Consequently, the collector current rises abruptly.

As V_i is maintained at V_H, the base has stored charge (refer to Figure 4.4e) given by

$$Q_B = \text{Total excess base charge} = Q_A + Q_S$$
$$= \tau_{BF} I_{B(EOS)} + [I_{BF} - I_{B(EOS)}]\tau_S \qquad \textbf{(4.39)}$$

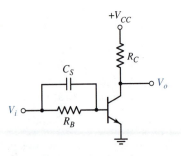

Figure 4.6   BJT inverter with speed-up capacitor

When V_i changes from V_H to V_L, the capacitor discharges and removes the excess base charge abruptly. Adjusting the capacitor value so that $Q_{SC} = Q_B$ eliminates the delay due to the saturation time constant τ_s; hence, the storage delay is reduced significantly. In addition, the charging current impulse as V_i switches from V_L to V_H reduces the turn-on delay. Note that the capacitor provides a low-impedence path for both charging and discharging the base.

For the circuit shown in Figure 4.6 with the parameters as given in Examples 4.3 and 4.4, calculate the value of the speed-up capacitor to reduce switching delays. | **Example 4.5**

Solution

From the parameters for the circuit and the transistor, the base charge is given by (Equation (4.39)):

$$Q_A = 10 \text{ (ns)} \left(\frac{4.9}{50} \text{ mA} \right) = 0.98 \text{ pC}$$

$$Q_S = \left(\frac{4.2}{10} - \frac{4.9}{50} \text{ mA} \right) (13.2 \text{ ns}) = 4.2 \text{ pC}$$

Hence the total excess base charge is

$$Q_B = 0.98 + 4.2 = 5.18 \text{ pC}$$

To pull this charge out of the base, the capacitor must be charged to Q_B or above by the voltage $(V_H - V_{BE(\text{sat})})$ during saturation of the transistor with base current, $I_{BF} = 0.42$ mA.
Hence,

$$C_S = \frac{Q_B}{V_H - V_{BE(\text{sat})}} = \frac{5.18 \text{ pC}}{4.2 \text{ V}} = 1.2 \text{ pF}$$

Figure 4.7 shows the results of MicroSim PSpice simulation of the inverter of Example 4.4 with a speed-up capacitor of 1.2 pF. Notice that the turn-on and turn-off delays are significantly reduced using the calculated value of the capacitor. Overshoot and undershoot of the collector voltage waveform arise from the coupling of input voltage to the collector via the collector-base junction capacitance.

Improving the switching speed of the inverter using a capacitor across the base resistor is a practical solution for discrete circuits. For ICs, however, capacitors above tens of picofarad are not a desirable option because of the large silicon areas required to fabricate them. As with diodes, storage delay may be reduced by reducing the minority carrier lifetime τ_{BF} using impurities (referred to as *lifetime killers*) such as copper and gold. We will consider other techniques for reducing the switching delays later, in connection with the TTL gate circuits.

```
* BJT INVERTER SWITCHING RESPONSE WITH CS=1.2pF
Q         3   2   0   NPNT
RB        1   2   10K
CS        1   2   1.2E-12
RC        4   3   1K
VI        1   0   PULSE(-1V 5V 10n 0.5n 0.5n 20n 50n)
VCC       4   0   DC 5.0
.MODEL NPNT NPN (IS=2E-16 BF=50 BR=0.7 TR=20E-9
+ TF=0.2E-9 RB=50 RC=20 CJC=0.15p CJE=0.3p)
.TRAN 0.5n 50n
.PROBE
.END
```

(a) Input File

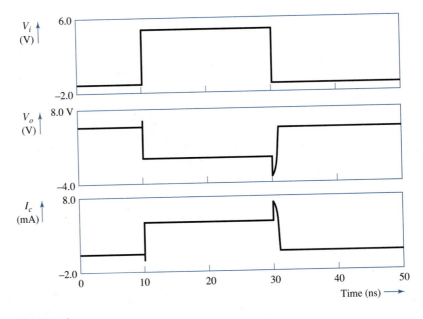

(b) Waveforms

Figure 4.7   MicroSim PSpice simulation of the switching response of the inverter in Figure 4.6 with $C_S = 1.2$ pF

4.2   RESISTOR-TRANSISTOR-LOGIC (RTL) FAMILY

Adding another transistor in parallel with the transistor of the BJT inverter results in a two-input NOR gate, as shown in Figure 4.8. This circuit was introduced as an integrated circuit in the early 1960s. Since it contains only

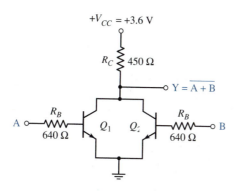

Figure 4.8   A two-input RTL NOR gate

resistors and transistors, it is known as the *resistor-transistor-logic*(RTL) gate. Note that from the simple NOR logic realization, all logic functions can be implemented in the RTL family.

It is easily verified that the circuit in Figure 4.8 implements the NOR function; if one of the inputs is at logic high, the transistor connected to that input goes into saturation and the output is at logic low. When both inputs are at logic low, both transistors are in cutoff, and the output is at logic high.

The voltage transfer characteristic of the NOR gate operating as an inverter with one of the inputs connected to logic low is identical to that of the BJT inverter shown in Figure 4.2. If both inputs are connected together, the logic high threshold voltage V_{IH} drops slightly, due to the current sharing of the two collectors (Problem 4.3). The noise margins (which depend on the power supply voltage and the number of load gates connected) and the fan-out are comparable to those of the BJT inverter.

4.2.1   WIRED-AND

Outputs of two RTL gates implementing the functions $Y_1 = \overline{(A + B)}$ and $Y_2 = \overline{(C + D)}$ may be tied together to obtain $Z = Y_1 \cdot Y_2 = \overline{(A + B)} \cdot \overline{(C + D)} = \overline{A} \cdot \overline{B} \cdot \overline{C} \cdot \overline{D} = \overline{A + B + C + D}$, without any additional circuitry. This implementation of the AND function is known as *implied AND,* or *wired-AND.* [1] While the wired-AND, in effect, increases the number of inputs, or *fan-in,* to the RTL gate, it also causes a large collector current in the saturated transistor. The increase in the collector current requires a larger input voltage or higher value of β to remain in saturation (see Problem 4.5). For this reason, wired-AND of RTL gates is generally not preferred.

| [1] Wired-AND is also possible with other logic families. See Section 5.4 for ECL wire-ANDing.

4.2.2   PROPAGATION DELAY

The propagation delay of an RTL gate is on the order of 10 ns. This delay is comparable to those of the other families. Average power dissipation of an RTL gate, however, is rather high. Note that the entire circuit dissipates a power of $P = V_{CC}I_{CC}$, where I_{CC} is the power supply current to the circuit. At $V_{CC} = 3.6$ V, $R_B = 640$ Ω and $R_C = 450$ Ω, the gate dissipation is about 15 mW (Problem 4.6). This results in a delay-power product of 150 pJ. The high power consumption and the decreasing noise margin with increasing fan-out has resulted in the demise of the RTL logic family over the years.

The RTL family of intergrated circuits included NOR gates with two, three, and four inputs, and flip-flops and four-bit shift registers.

4.3   DIODE TRANSISTOR-LOGIC (DTL) FAMILY

The *diode-transistor-logic* (DTL) family eliminates the problem of decreasing output voltage with increasing load. Figure 4.9 shows the circuit of a simplified version of a two-input DTL NAND gate.

The two diodes in this circuit, D_A and D_B, perform the logic AND operation, while the transistor Q inverts the results of A · B at the output Y. Diode D_1, which carries the base current for Q, raises the logic low voltage V_{IL}.

With either input at logic low ($V_i = 0.1$ V, for example), one of the two input diodes conducts and clamps the voltage at the junction of the diodes to $V_P = V_D \approx 0.8$ V. Since V_P is not sufficiently high to forward-bias both D_1 and Q, only D_1 conducts via R_2. (In addition to providing a conducting path for D_1 at low input, R_2 enables fast turn-off of Q, as we see in the next section.) With a base voltage V_B below $V_{BE(cut-in)}$, Q is off and the output is at logic high level. Hence $V_{OH} = V_{CC}$. As V_i increases, the current through R_1 is steered into the base of Q when V_P reaches 1.3 V. The logic low threshold voltage, therefore, is $V_{IL} =$

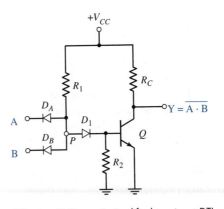

Figure 4.9   A simplified two-input DTL NAND circuit

$V_P - V_D = 0.6$ V. In the absence of D_1, however, both the input diode/s and the transistor will conduct for $V_i \leq 0.1$ V; V_{IL}, therefore, will be reduced to almost zero. Diode D_1 in the path of the base current of Q increases V_P and hence V_{IL}. As a result, logic low noise margin is increased to a reasonable value (see below).

When both the inputs are at logic high level, the input diodes are reverse-biased, and the transistor Q saturates with the output at logic low level. V_{OL} is therefore given by the saturation voltage of $V_{CE(sat)} \approx 0.1$ V, which occurs at $V_i \geq V_{IH} = 0.8$ V.

The logic low noise margin ($NM_L = V_{IL} - V_{OL}$) of this simplified DTL gate is only 0.5 V. This can be increased to a more useful value with the addition of a second diode D_2 in series with D_1, as is done in the early version of the discrete DTL gates.

4.3.1   DISCRETE DTL NAND GATE

Figure 4.10 shows the discrete version of the early DTL NAND gate. In addition to the second diode D_2, this circuit also uses a negative supply voltage, which aids in the turn-off of Q. The input labeled X (for expander) is used to increase the number of inputs (fan-in) when necessary.

To obtain the voltage transfer characteristics, consider the circuit with the expander input X left open and the two inputs A and B tied together. For $V_i(= V_A = V_B) = 0$, the two input diodes are forward-biased by the supply voltage and the point P is held at $V_D = 0.7$ V. For this voltage at P the diodes D_1 and D_2 conduct very little current via R_2, and the transistor Q is in cutoff. The output (collector) voltage V_o is therefore at $V_{OH} = V_{CC}$.

When V_P reaches two diode-drops above the cut-in voltage of the transistor, Q begins to conduct and the output voltage falls. Hence,

$$V_{IL} = V_{BE(cut-in)} + 2V_D - V_D = 1.3 \text{ V}$$

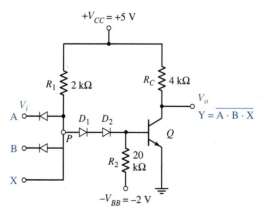

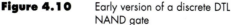

Figure 4.10   Early version of a discrete DTL NAND gate

As V_i is increased above 1.3 V, V_P rises and Q goes well into its active mode. When V_P reaches a voltage of $V_{BE(sat)} + 2\ V_D = 2.2$ V, Q enters saturation. Further increase in V_i clamps V_P at 2.2 V, which reverse-biases the input diodes, while transistor Q is kept in saturation with a constant base drive. Therefore,

$$V_{IH} = V_P - V_D = 1.5\ \text{V},$$

and

$$V_{OL} = V_{CE\,(sat)} = 0.1\ \text{V}$$

Figure 4.11 shows the voltage transfer characteristics for the discrete DTL inverter.

The noise margins for the discrete DTL gate are

$$NM_L = V_{IL} - V_{OL} = 1.3 - 0.1 = 1.2\ \text{V}$$

and

$$NM_H = V_{OH} - V_{IH} = 5 - 1.5 = 3.5\ \text{V}$$

Resistor R_2 and the negative power supply provide a path for removing the excess base charge when the transistor goes from saturation to cutoff. If $R_2 = \infty$, the base must discharge through the reverse-biased diodes D_1 and D_2, at the rate of the diode reverse saturation current. But with a finite value of R_2, the excess minority carriers are removed more rapidly, at an initial rate of $[V_{BE(sat)} + V_{BB}]/R_2$. Thus, the rate of base discharge can be increased with a smaller R_2, or larger V_{BB}. The problem with this is that a lower value for R_2 results in increased power drain, and the use of V_{BB} requires an additional power supply.

Expander input X is used for increasing the number of inputs, or fan-in. For each additional input above two, an external diode is connected at X similar to D_A and D_B. Expander input is useful when the DTL gate is fabricated as an integrated circuit on a single chip.

Since an open-circuited input does not draw current through the input diode, it is equivalent to a logic high value. However, unused inputs that are left floating act as antennas and pick up noise signals. To prevent false switching at

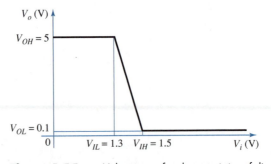

Figure 4.11   Voltage transfer characteristics of discrete DTL gate

the output, therefore, each unused input of a DTL gate must be disabled by connecting it to logic high voltage or supply voltage V_{CC}.

Fan-Out   A DTL gate at logic high output needs to supply only the reverse saturation currents of the diodes at the input of the load gates. Therefore, output high voltage V_{OH} and noise margin NM_H of the DTL gate are essentially constant with varying loads. This is an advantage over an RTL gate. At logic low output, the saturated transistor of the driver must sink the input diode current from each load gate as shown in Figure 4.12. The fan-out of a DTL gate is therefore limited by the amount of load current the collector can sink at logic low output and still be in saturation.

Referring to Figure 4.12, the collector current of the driver transistor, when it is in saturation, is given by

$$I_C = I_3 + NI_L = I_3 + N(I_1 - I_2)$$
$$= \frac{V_{CC} - V_{CE \text{ (sat)}}}{R_C} + N\left[\frac{V_{CC} - V_{PL}}{R_1} - \frac{V_{PL} - 2V_D + V_{BB}}{R_2}\right]$$

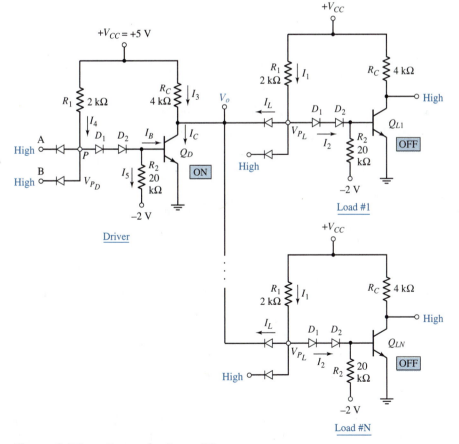

Figure 4.12   Fan-out of a discrete DTL gate

and the base current of the driver is given by

$$I_B = I_4 - I_5 = \frac{V_{CC} - V_{PD}}{R_1} - \frac{V_{BE\ (sat)} + V_{BB}}{R_2}$$

Fan-out is determined using $\beta I_B \geq I_C$ for Q_D in saturation.

Example 4.6

Determine the logic low fan-out of the DTL gate shown in Figure 4.12. Assume a β of 50 for all transistors.

Solution

At logic low output of the driver, each load gate sources a current of

$$I_L = I_1 - I_2$$

Since $V_{PL} = V_{CE\ (sat)} + V_D = 0.8$ V, and all load transistors are off,

$$I_1 = \frac{(5 - 0.8)}{2} = 2.1\ \text{mA},$$

and

$$I_2 = \frac{(0.8 - 2 \times 0.7 + 2)}{20} = 0.07\ \text{mA}$$

With $I_3 = \frac{(5 - 0.1)}{4} = 1.23$ mA, then,

$$I_C = I_3 + NI_L = 1.23 + 2.03N$$

Since the driver is in saturation, $V_{PD} = 2.2$ V. Therefore,

$$I_4 = \frac{(5 - 2.2)}{2} = 1.4\ \text{mA}$$

$$I_5 = \frac{(2 + 0.8)}{20} = 0.14\ \text{mA}$$

and

$$I_B = 1.26\ \text{mA}$$

With a β of 50, the driver transistor stays in saturation and logic low output of 0.1 V is maintained if

$$\beta I_B \geq I_C$$

or

$$50 \times 1.26 \geq 1.23 + 2.03N$$

This gives $N \leq 30$.

Hence, the logic low fan-out for the circuit of Figure 4.10, which is also the fan-out of the DTL family for the given parameters, is 30.

As with the BJT inverter, if all the 30 identical gates, as given in the above example, are connected to the output of the DTL gate in Figure 4.10, the driver goes to the edge of saturation at logic low output. With a 25 percent overdrive of base current to accommodate variations in current gain β, the fan-out reduces to $N = 24$. In either case, when fewer than the maximum number of loads are connected, the base is heavily overdriven; consequently, the turn-off delay time increases with decreasing number of load gates.

4.3.2 IC VERSION OF DTL GATE

The integrated circuit version of the DTL gate that became available in the 1960s is shown in Figure 4.13. This circuit operates with a single positive supply. Further, transistor Q_1 replaces one of the level-shifting diodes of the discrete DTL circuit. This change from diode to transistor increases the fan-out of the gate without significantly increasing fabrication cost or silicon area. Resistor R_2 connected between the collector and the base of Q_1 ensures that Q_1 does not saturate; any increase in the collector current results in the decrease of collector voltage and, consequently, a fall in the base current; hence, the collector current is reduced to its previous stable value and Q_1 stays in the active mode of operation. (Note that any current from power supply V_{CC} through R_2 causes the collector-base junction of Q_1 reverse-biased.) Note further that Q_1, and hence Q_2, are off if one of the inputs is at logic low.

The voltage transfer characteristics of the IC DTL gate are the same as those for the discrete version shown in Figure 4.11.

Since Q_1 is in the active mode when conducting, its current gain I_{E1}/I_{B1} is large. Therefore, a higher base drive is provided for the output transistor Q_2 than available in the discrete DTL gate. The high base drive enables Q_2 to saturate with a small value of β, thus giving the name β *saver* for Q_1. The high base drive, clearly, leads to an increase in the fan-out of the circuit.

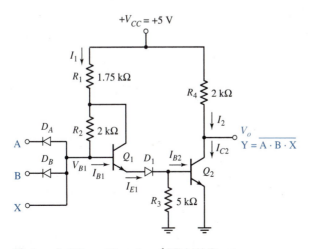

Figure 4.13   IC version of DTL NAND gate

Fan-Out As with the discrete version, fan-out of the IC DTL gate is determined by the logic low output driving capability. At logic low output, the base drive for Q_2 is determined from the emitter current I_{E1}. Referring to Figure 4.13, with Q_1 in active mode, and all the inputs at logic high, I_{E1} is given by

$$I_{E1} = I_1 = (\beta + 1) I_{B1} \qquad \textbf{(4.40)}$$

From the base-emitter circuit of Q_1, we have

$$V_{CC} - R_1 I_1 - R_2 I_{B1} - V_{B1} = 0 \qquad \textbf{(4.41)}$$

Solving for the base current of Q_1 using Equations (4.40) and (4.41),

$$I_{B1} = \frac{V_{CC} - V_{B1}}{R_1(\beta + 1) + R_2}$$

Hence

$$I_{E1} = (\beta + 1)I_{B1} = \frac{(\beta + 1)(V_{CC} - V_{B1})}{R_1(\beta + 1) + R_2} \qquad \textbf{(4.42)}$$

and the base current of the output transistor Q_2, which is in saturation, is given by

$$I_{B2} = I_{E1} - \frac{V_{BE2 \,(sat)}}{R_B} \qquad \textbf{(4.43)}$$

Fan-out is calculated from Equations (4.42) and (4.43) and the collector current of Q_2.

Example 4.7   Determine the fan-out of the IC DTL gate shown in Figure 4.13 using $\beta = 50$ for both transistors.

Solution

With Q_2 in saturation,

$$V_{B1} = V_{BE1} + V_{D1} + V_{BE2 \,(sat)} = 2.2 \text{ V}$$

From Equation (4.42), using $\beta = 50$, we find that $I_{E1} = 1.565$ mA. Therefore, from Equation (4.43), we obtain

$$I_{B2} = 1.565 - \frac{0.8}{5} = 1.405 \text{ mA}$$

Since the output is at logic low with $V_o = V_{CE(sat)} \approx 0.1$ V, both Q_1 and Q_2 are off in each of the N identical load gates. The input (sourcing) current I_{IL} of each load gate, then, is

$$I_{IL} = \frac{(V_{CC} - V_D - V_o)}{(R_1 + R_2)} = \frac{5 - 0.7 - 0.1}{1.75 + 2} = 1.12 \text{ mA}$$

In addition to the input low current from each load, Q_2 must sink I_2, the current through R_C, given by

$$I_2 = \frac{[V_{CC} - V_{CE \, (\text{sat})}]}{R_C} = \frac{5 - 0.1}{2} = 2.45 \text{ mA}$$

With a β of 50, Q_2 will remain in saturation with N load gates connected to its output if

$$NI_{IL} + I_2 \leq \beta I_{B2}$$

and solving for this case, we get $N = 60$.

Clearly, the IC DTL gate has a significantly higher fan-out than the discrete version.

4.3.3 PROPAGATION DELAY AND POWER DISSIPATION

The propagation delay of the IC version of DTL gates is typically 30 ns. This delay arises mainly from the turn-off delay of the output transistor and the charging time-constant of load capacitance to ground. Average power dissipation is approximately 15 mW, which gives a power-delay product of 450 pJ.

4.3.4 WIRED-AND OF DTL OUTPUTS

As with an RTL gate, the output impedance of a DTL gate is high (approximately equal to $R_4 = 2\text{k}\Omega$) at logic high output, and low (equal to the on-resistance of the output transistor) at logic low output. Therefore, it is possible to wire-AND two (or more) DTL gate outputs as shown in Figure 4.14. The

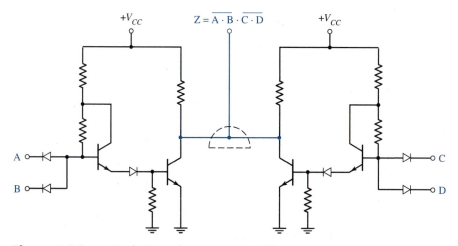

Figure 4.14   Wired-AND implementation using DTL gates

resulting output is the logic AND of the individual (untied) gate outputs; that is, the gate with the logic low output determines the final output.

An advantage of this wired-AND connection using DTL gates is that it can implement useful logic functions (unlike the RTL wired-AND connection, which only expands the number of inputs). An example of this is considered in Problem 4.11. Note that the wired-AND gate at logic high output has a lower output impedance than does a single gate, due to paralleling of the collector circuit resistances. Additionally, if the output transistor of only one gate is in saturation with all others turned off, the on transistor must now sink a high current—a situation similar to that in a wired-AND RTL gate. This increase in collector current must be taken into consideration when determining the fan-out of wired-AND gates.

4.3.5 HIGH-THRESHOLD DTL GATE

We can improve the noise margin of a DTL gate, as we saw earlier, by the addition of diodes in series with D_1. For operation in highly noisy environments, the noise margin can be increased significantly by using a Zener breakdown diode in place of a string of pn junction diodes. The resulting high-threshold DTL gate, shown in Figure 4.15a, operates at a higher supply voltage. The resistances are higher in this circuit to limit the increased power dissipation arising from the higher supply voltage. For $V_z = 6.9$ V, the circuit has noise margins of $NM_L = 7.4$ V and $NM_H = 7.3$ V. Figure 4.15b shows the VTC for the case of $V_z = 6.9$ V.

The higher noise margins have been achieved at the expense of increased propagation delays. The time constant with which the output transistor turns on or off is now large because of the large resistances. Typically, the propagation delay of a high-threshold gate is in the range of a few hundred nanoseconds.

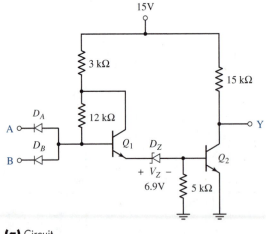

(a) Circuit

Figure 4.15   High-threshold DTL NAND gate

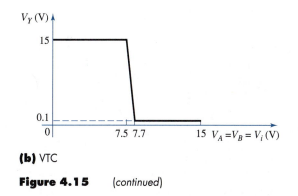

(b) VTC

Figure 4.15   (continued)

On the other hand, an added advantage arises from the use of the Zener diode. The positive temperature coefficient of the Zener diode (at voltages above 6 V) cancels the negative temperature coefficient of the base-emitter junction of Q_2. In addition, variation of input diode voltage with temperature is offset by the base-emitter voltage variation of Q_1, since the two vary in opposite directions. The net result is a negligible variation in the voltage characteristics of the gate with temperature.

The DTL family has better noise margin, increased fan-out, and faster response than the RTL family. The switching speed and fan-out of the DTL family are further improved in the transistor-transistor-logic family. The simple DTL circuit for logic formation, however, is used in modified forms in fast switching circuits such as Schottky and advanced TTL circuits.

4.4   TRANSISTOR-TRANSISTOR-LOGIC (TTL) FAMILY

The *transistor-transistor-logic* family has been one of the most widely used IC logic families since it evolved from the DTL family around 1965. Although the DTL and TTL families are input- and output-compatible with each other, the switching speed of a DTL circuit is limited by the large delay time in switching from logic low to logic high at the output. First, when the input switches from high to low level, the output transistor Q_2 in Figure 4.13 does not turn off until all excess charge stored in the base is removed to ground via the base resistor R_3. The initial rate of discharge of the base ($I_B \approx V_{BE \text{ (sat)}}/R_3$) is much smaller than the forward base current, $I_{B(\text{sat})}$. (Note that to prevent excessive power supply drain, R_3 cannot be made too small.) Thus, the output transistor turns off rather slowly. Secondly, the output voltage transition from low ($V_{CE \text{ (sat)}}$) to high (V_{CC}) occurs by exponentially charging the load capacitance via the collector resistor R_4. Therefore, for a given load capacitance, which arises from wiring and the reverse-biased diodes of load gates, the charging time depends on the large passive resistor R_4. Circuit design techniques that incorporate active base discharge and active collector pull-up improve the low-to-high switching speed in TTL circuits.

4.4.1 BASIC TTL INVERTER

To understand how the TTL circuit configuration alleviates the switching speed limitations of the DTL gate, consider the basic TTL inverter shown in Figure 4.16a. Comparing this primitive form of TTL circuit with the DTL inverter of Figure 4.16b (Figure 4.9 with only one input), observe that the transistor Q_1 retains the structure at the input: the EBJ of Q_1 replaces the input diode D_A while the CBJ replaces the level-shifting diode D_1. Q_1, therefore, performs the function of the diodes D_A and D_1 by steering current toward the input V_i or the base of Q_2, depending on the logic level at the input.

When the input is at logic high (above about 0.8 V), the emitter of Q_1 becomes more positive than the collector, which is held at or below $V_{BE2(sat)}$; hence Q_1 operates in the reverse active mode. The base current of Q_2 is then supplied by the collector of Q_1 acting as emitter. With appropriate choices of values for R_1 and R_C, Q_2 is driven to saturation with

$$I_{B2} = -I_{C1} = (1 + \beta_R)I_{B1}$$

where β_R is the reverse current gain of Q_1. When V_i switches to logic low (≈ 0.1 V), initially

$$V_{CE1} = V_{BE2\,(sat)} - V_i \approx 0.7 \text{ V}$$

and Q_1 operates in the forward active mode. Since the source of its collector current is the base of Q_2 ($I_{C1} = -I_{B2}$), excess charge stored in the base of the previously saturated Q_2 is now removed at the rate of $I_{C1} = \beta_F I_{B1}$. At this high rate ($\beta_F \gg 1$), the reverse base current I_{B2} drops to zero rapidly, and Q_2 is turned off. With Q_2 off, only the reverse current of the collector-base junction constitutes I_{C1}; hence Q_1 goes to (forward) saturation. Thus the turn-off of the output transistor Q_2 is speeded up by the large collector current drawn by the input transistor Q_1 operating in the active mode. The consequence of all this is that the

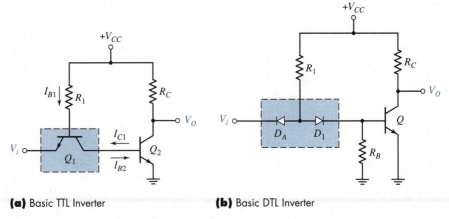

(a) Basic TTL Inverter   **(b)** Basic DTL Inverter

Figure 4.16   Evolution of TTL inverter from DTL inverter

storage time of the primitive TTL inverter is much less than that of the DTL inverter using a passive resistor to discharge the base.

The rise time of the output voltage is reduced by modifying the output circuit of the basic TTL inverter. We consider this modification in Section 4.4.3, after discussing the basic TTL gate and the multiemitter transistor.

4.4.2 MULTIEMITTER TRANSISTOR AND BASIC TTL NAND CIRCUIT

Replacing each input diode of the discrete DTL gate (Figure 4.9) with the emitter-base junction of a transistor, and having a common collector and a common base as shown in Figure 4.17, implements the discrete version of the basic TTL NAND gate. The operation of this circuit is readily verified. If input A (B) is at logic low (≈ 0), $Q_{1A}(Q_{1B})$ goes into saturation and the common base terminal P is held at 0.8 V above input. For input B (A) at logic high (>0.8 V), then, $Q_{1B}(Q_{1A})$ is cut off. With $I_{B1} = -I_{C1} = 0$, transistor Q_2 is cut off and the output is at logic high. When both A and B are at logic low, both the input transistors go into saturation and Q_2 is again cut off. For logic high inputs at A and B, both Q_{1A} and Q_{1B} operate in the reverse active mode, and the common collector supplies the base drive for Q_2 to go into saturation.

The logic high input current drawn by each emitter, assuming Q_{1A} and Q_{1B} are perfectly matched, is given by

$$-I_{EA} = -I_{EB} = \beta_R \cdot \frac{I_{B1}}{2}$$

In general, for an M-input NAND gate (M transistors paralleled at input) with all M inputs at logic high, the current drawn from each input source is given by

$$I_{IH} = -\beta_R \cdot \frac{I_{B1}}{M} \tag{4.44}$$

Equation (4.44) suggests the need for a low value of β_R. If an input is driven by the output of a similar gate (with its Q_2 in cutoff), current I_{IH} through the collector resistor of the driving gate causes a drop in the logic high level. To reduce this loading and increase the logic high fan-out of the driving gate, the reverse current gain β_R of the input transistors must be small (<1).

The paralleling of two transistors, such as Q_{1A} and Q_{1B} shown in Figure 4.17, with a common base and a common collector results, as far as the terminal behavior is concerned, in the single multiemitter transistor with the circuit symbol shown in Figure 4.18a. In the integrated circuit version, this multiemitter transistor is fabricated with two emitter regions and a collector region, all within the same base region, as shown in Figure 4.18b. The emitter regions are heavily doped relative to the collector, so that the reverse current gain β_R is below unity, as desired. Typical values of β_R in TTL circuits are in the range of 0.01 to 0.1. TTL gates with input transistors having 2 to 13 emitters

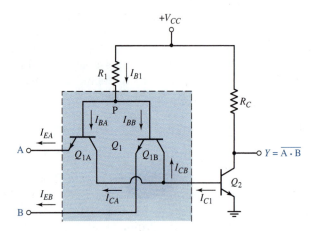

Figure 4.17   Basic two-input TTL NAND gate

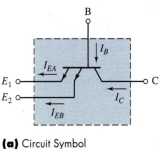

(a) Circuit Symbol

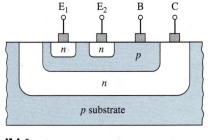

(b) Structure

Figure 4.18   Multiemitter transistor

(inputs) are available. The construction of multiemitter transistors is advantageous because it reduces silicon chip area and eliminates the parasitic capacitances that arise from individual devices.

4.4.3   STANDARD TTL NAND CIRCUIT

Figure 4.19 shows the circuit configuration of a standard (7400 series) two-input TTL NAND gate with typical resistor values. The output circuit consists of two transistors, Q_4 stacked on top of Q_3; this configuration is commonly called a *totem-pole output*. Transistor Q_2 serves to increase the logic low input voltage V_{IL}, similar to the diode D_2 in the DTL circuit of Figure 4.10. Observe that changes in voltages at the collector and the emitter of Q_2 are complementary, or out of phase. If V_{E2} increases due to an increase in the base current I_{B2}, then V_{C2} decreases, and vice versa. Q_2 thus acts as a *phase splitter* and ensures that only Q_4 or Q_3 (but not both) is turned on for any input. Now let us look at the operation of the circuit for logic low and high input conditions.

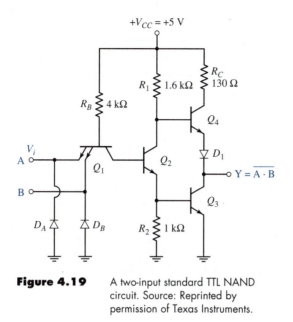

Figure 4.19   A two-input standard TTL NAND circuit. Source: Reprinted by permission of Texas Instruments.

When both the inputs, A and B, are at logic high level, Q_1 operates in the reverse active mode and supplies base current to Q_2. With sufficient base drive, Q_2 goes into saturation and, in turn, Q_3 is driven to saturation. The collector voltage of Q_2 is

$$V_{C2} = V_{CE2 \text{ (sat)}} + V_{BE3 \text{ (sat)}} = 0.9 \text{ V}$$

In the absence of the diode D_1 between the two transistors Q_3 and Q_4, this would give, for Q_4, a base-emitter voltage of

$$V_{BE4} = V_{B4} - V_{CE3(\text{sat})} = 0.8 \text{ V}$$

This voltage is sufficient to saturate Q_4. Consequently, there would be a steady power supply current of

$$\frac{[V_{CC} - V_{CE4 \text{ (sat)}} - V_{CE3 \text{ (sat)}}]}{R_C} = \frac{(5 - 0.1 - 0.1)}{130} = 36.9 \text{ mA}$$

Adding the diode D_1 between the emitter of Q_4 and the collector of Q_3 requires $V_{B4} = 0.6 + 0.7 + 0.1 = 1.4$ V to turn both D_1 and Q_4 on. With only 0.9 V for $V_{B4} (= V_{C2})$, both Q_4 and D_1 are turned off and the wasteful power supply current is eliminated.

Since Q_3 is saturated, lumped stray capacitance C_L present at the output (due to load and wiring) now discharges from its previous logic high level through the on-resistance of Q_3. With an on-resistance of typically 10 Ω, this discharge is completed quickly, and the output voltage falls to $V_{CE3 \text{ (sat)}}$ of about 0.1 V. Note that if the output is connected to another TTL (load) gate, the collector of Q_3 must, under static conditions, *sink* the input current of the load gate. Logic low fan-out of a TTL gate, therefore, is determined by the maximum input current Q_3 can sink in saturation.

Consider now the case when one of the inputs switches to logic low (≈ 0.1 V). With the EBJ forward-biased, the base voltage of Q_1 changes to 0.8 V. Since Q_2 and Q_3 are yet to be turned off, the collector of Q_1 is initially at 1.6 V. The CBJ of Q_1 is therefore reverse-biased and Q_1 enters forward active mode. The base current of Q_1 is given by

$$I_{B1} = \frac{V_{CC} - V_{B1}}{R_B} = \frac{5 - 0.8}{4} = 1.05 \text{ mA}$$

With a β of 50, this demands a collector current $I_{C1}(= -I_{B2})$ much in excess of the reverse current available from the base of Q_2. Therefore, Q_2 turns off quickly and Q_1 goes into saturation. Since $V_{E2} = V_{BE3} = 0$, Q_3 is now turned off. The output voltage, however, cannot change from its previous value of 0.1 V until the load capacitance C_L is charged to the logic high voltage. This charging takes place via the top transistor Q_4 of the totem-pole state in the following manner.

With Q_2 off, V_{C2} rises and Q_4 and D_1 are turned on. At this time V_{B4} is given by

$$V_{B4} = V_{BE4} + V_D + V_{CE3} = 0.7 + 0.7 + 0.1 = 1.5 \text{ V}$$

where active mode is assumed for Q_4. The base current of Q_4 is

$$I_{B4} = \frac{V_{CC} - V_{B4}}{R_1} = \frac{3.5}{1.6} = 2.19 \text{ mA}$$

If β is 50, this calls for a collector current in excess of 100 mA. Since the maximum (saturation) collector current of Q_4 when V_o is at 0.1 V is only

$$I_{C4(\text{sat})} = \frac{V_{CC} - V_{CE4(\text{sat})} - V_{D1} - V_{CE3(\text{sat})}}{R_C} = \frac{5 - 0.1 - 0.7 - 0.1}{0.13}$$

$$= 31.54 \text{ mA}, \tag{4.45}$$

Q_4 is in saturation.

A large emitter current of $I_{E4} = I_{B4} + I_{C4} = 33.73$ mA is now charging the load capacitance. The initial charging time constant is given by $C_L (R_C + R_D + R_{S4})$, where R_D is the diode forward resistance of less than 10 Ω, and R_{S4} is the saturation resistance of $Q_4(\approx 10 \ \Omega)$. As C_L charges with this time constant towards V_{CC}, the rising voltage V_o decreases I_{C4} and brings Q_4 out of saturation. Under static conditions with open-circuited output, therefore, no current is drawn from Q_4, and the transistor is at its cut-in point. The logic high (maximum) output voltage V_o in this steady state, therefore, is given by

$$V_o = V_{CC} - V_{BE \text{ (cut-in)}} - V_{D \text{ (cut-in)}} = 5 - 0.6 - 0.6 = 3.8 \text{ V}$$

neglecting the drop in R_1 due to leakage current. Although this voltage is considerably less than that of a DTL logic high voltage, it is reached significantly faster by charging the output capacitance via the saturated transistor Q_4.

It may seem reasonable to decrease the charging time constant further by eliminating the small resistor R_C altogether from the collector of Q_4. There is a need for R_C, however, and it arises as follows. Although only Q_4 or Q_3 is conducting under a steady output state, both transistors go into saturation during the brief transition when the output switches from low to high state. Consequently, the supply voltage V_{CC} would be short-circuited if R_C were not present. With a value of 130 Ω for R_C, the supply current in this interval is limited to about 32 mA as seen in Equation (4.45).

Large supply current transients during the low-to-high transitions at the output result in increased average power dissipation at high switching frequencies; they also cause large voltage transients on the V_{CC} supply line, which then distributes these "spikes" to all circuitry connected to it. To filter out these spikes, power supply bypass capacitors of 0.01 μF to 0.02 μF each are connected between the V_{CC} and the ground pins of each TTL IC package. Also, each IC package has diodes, shown as D_A and D_B in Figure 4.19, connected at the inputs (to each of the emitters of the multiemitter transistor) with their anodes grounded. These diodes are forward-biased when there are negative-going noise spikes arising at the inputs ("ringing" due to the lead inductances), and the negative amplitudes are thus clamped to about -0.7 V.

Since the output voltage is pulled up to the logic high level by the transistor Q_4 (an active device instead of the passive resistor as in the collector circuit of a DTL), the totem-pole stage is said to have an *active pull-up* circuit. In addition to causing a faster turn-off, the pull-up transistor Q_4 can supply (*source*) more current to loads and increase the logic high fan-out as we shall see in Section 4.4.5. Resistor R_2 provides a path to ground for the base of Q_3 to discharge when the input goes from logic high to low level. R_2 is therefore the *passive pull-down* resistor. The totem-pole stage thus has an active pull-up and a passive pull-down circuitry. The value for R_2 is chosen so that it provides a low impedance path to ground without excessive current drain from Q_2. The active pull-up and the passive pull-down circuitry of the totem-pole output stage therefore offers low impedance paths to charge and discharge load capacitances and thereby contributes to switching speed improvement over a DTL gate.

4.4.4   TRANSFER CHARACTERISTICS OF A STANDARD TTL NAND GATE

The piecewise voltage transfer characteristic of a standard TTL inverter (Figure 4.19 with a single emitter for Q_1) has three breakpoints as shown in Figure 4.20. For an input voltage V_i below V_{IL}, Q_1 is in saturation with the collector-base reverse leakage current of Q_2; that is, $I_{C1} = -I_{B2} \approx 0$. Q_2 and Q_3 are off. Q_4 is in active mode and the logic high output voltage V_o is 3.8 V, as we saw in the preceding section. The first breakpoint occurs when V_i reaches 0.5 V, for at this voltage, Q_1 is still in saturation and

$$V_{C1} = V_{BE2} = V_i + V_{CE\,(sat)} = 0.6 \text{ V}$$

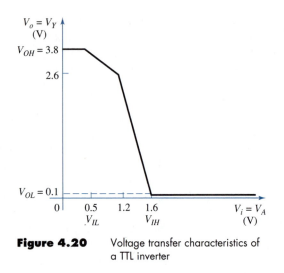

Figure 4.20   Voltage transfer characteristics of a TTL inverter

This voltage brings Q_2 out of its cutoff mode. Hence, $V_{OH} = 3.8$ V and $V_{IL} = 0.5$ V.

As V_i is increased above V_{IL}, Q_2 goes into active mode and the voltage drop across R_2 reaches the cut-in voltage for Q_3. At this point, V_i is given by

$$V_i = V_{BE3 \text{ (cut-in)}} + V_{BE2} - V_{CE1\text{(sat)}} = 1.2 \text{ V}$$

With $V_{R2} = V_{BE3 \text{ (cut-in)}} = 0.6$ V, $I_{E2} \approx 0.6/1 = 0.6$ mA, and, for a β of 50, $I_{C2} \approx 0.6$ mA.

Hence, the collector of Q_2 is at

$$V_{C2} = V_{CC} - I_{C2}R_2 = 4 - 0.6 \times 1.6 \approx 4 \text{ V}$$

With Q_4 conducting, output voltage is

$$V_o = V_{C2} - V_{BE4} - V_{D1} \approx 2.6 \text{ V}$$

Further increase in V_i diverts more of the base current of Q_1 to its CBJ and drives Q_2 and Q_3 to saturation. With the EBJ and CBJ of Q_1 at almost the same voltage, Q_1 is on the verge of entering saturation mode when

$$V_i \approx 2V_{BE \text{ (sat)}} = 1.6 \text{ V}$$

At this input, output $V_o = V_{CE3 \text{ (sat)}} = 0.1$ V. Hence, $V_{IH} = 1.6$ V and $V_{OL} = 0.1$ V.

For V_i above 1.6 V, Q_1 is in (reverse) saturation mode until V_i reaches about 2.2 V.

Since

$$V_{EC1} = V_i - V_{C1} = 2.2 - 1.6 \geq 0.6 \text{ V}$$

the EBJ of Q_1 is reverse-biased while the CBJ is forward-biased and Q_1 is in reverse active mode for $V_i \geq 2.2$ V.

Table 4.2   Modes of operation of transistors in a standard TTL circuit

Input	Q_1	Q_2	Q_3	Q_4
Low (≤ 0.5 V)	Saturation	Off	Off	Active
0.5 to 1.2 V	Saturation	Active	Off	Active
1.2 to 1.6 V	Reverse active	Saturation	Active	Active
High (≥ 1.6 V)	Reverse active	Saturation	Saturation	Off

Table 4.2 lists modes of operation for each transisor in Figure 4.19. We can determine the noise margins from the transfer characteristics as

$$NM_L = V_{IL} - V_{OL} = 0.5 - 0.1 = 0.4 \text{ V}$$

$$NM_H = V_{OH} - V_{IH} = 3.8 - 1.6 = 2.2 \text{ V}$$

The logic swing is

$$V_{ls} = V_{OH} - V_{OL} = 3.8 - 0.1 = 3.7 \text{ V}$$

The transition width is

$$V_{IH} - V_{IL} = 1.1 \text{ V}$$

4.4.5   FAN-OUT

Logic Low Fan-Out   Figure 4.21 shows N_L loads—identical TTL inverters— connected to the output of the driving gate, which is at logic low. At this output, Q_1 is in reverse active mode, Q_2 and Q_3 are in saturation, and Q_4 is in cutoff. Fan-out is determined by the amount of current Q_3 can sink and still be in saturation.

Example 4.8 illustrates the calculation of logic low fan-out.

Determine the logic low fan-out for the TTL circuit shown in Figure 4.19.   **Example 4.8**

Solution

From the circuit shown for the driver-load combination in Figure 4.21, input (sourcing) current I_{IL} from each load gate is given by

$$I_{IL} = \frac{(5 - 0.1 - 0.8)}{4} = 1.03 \text{ mA}$$

Hence, for N_L loads, the collector current of Q_3 is given by

$$I_{C3} = 1.03 N_L$$

This must be less than $\beta I_{B3}/\sigma$, where σ is the base overdrive factor.

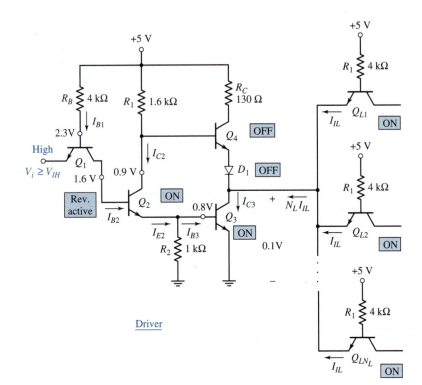

Figure 4.21   Logic low fan-out of a TTL gate

With the input at logic high, the voltages at the bases and collectors of Q_1, Q_2, and Q_3 are as given in Figure 4.21.

Base current of Q_1 operating in the reverse active mode is

$$I_{B1} = \frac{5 - 2.3}{4} = 0.68 \text{ mA}$$

If $\beta_R = 0.1$ for Q_1, then

$$I_{B2} = -I_{C1} = (1 + \beta_R) I_{B1} = 0.75 \text{ mA}$$

$$I_{C2} = \frac{5 - 0.9}{1.6} = 2.56 \text{ mA}$$

With a β_F of 50, these currents show that Q_2 is indeed in saturation.

The emitter current of the phase-splitter Q_2 is

$$I_{E2} = I_{C2} + I_{B2} = 3.31 \text{ mA}$$

Hence, the base current of the output transistor Q_3 is

$$I_{B3} = I_{E2} - \frac{V_{E2}}{R_2} = 2.5 \text{ mA}$$

If β_F for Q_3 is 50, then, for Q_3 to remain in saturation with N_L loads,

$$50 \times \frac{2.5}{\sigma} \geq 1.03 \ N_L$$

or $N_L \leq 121$ for $\sigma = 1$ (no overdrive), and $N_L \leq 97$ for $\sigma = 1.25$ (25 percent base overdrive).

Logic High Fan-Out   With no load connected, the logic high output of a standard TTL gate, as we saw earlier, is 3.8 V. Since Q_3 is in cutoff and Q_4 is in the active mode, any load (sink) current causes a drop in the logic high output voltage V_{OH} (see Figure 4.22). Assuming a constant voltage across the forward-biased EBJ of Q_4 and D_1, this decrease in V_{OH} occurs as a drop in the base resistor R_1 due to load current flow. Although V_{OH} can decrease to as low as $V_{IH}(=1.6 \text{ V})$, usually $V_{OH}(\text{min})$ is set between 3 V and 3.4 V to provide sufficient logic high

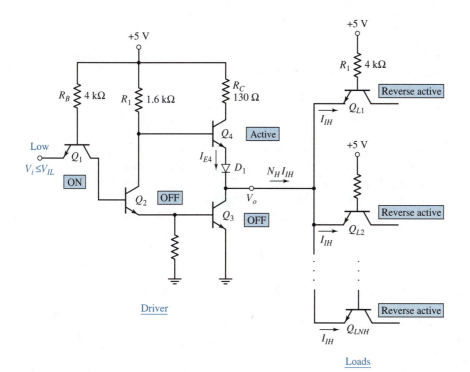

Figure 4.22   Logic high fan-out of a TTL gate

noise margin. Even for this small drop allowed for loading, the logic high fan-out of the gate is higher than that for logic low.

Example 4.9

Calculate the logic high fan-out for the standard TTL gate of Figure 4.19 using a drop of no more than 0.4 V at output when fully loaded.

Solution

Referring to Figure 4.19, the load (sinking) current I_{IH} from Example 4.8, drawn by each input at logic high level, is

$$I_{IH} = \beta_R I_{B1} = 0.068 \text{ mA}$$

With N_H loads, then, we have

$$I_{E4} = 0.068 N_H \quad \text{and} \quad I_{B4} = \frac{I_{E4}}{\beta_F + 1}$$

The above base current assumes that the output voltage with load does not drop significantly, so that Q_4 is still operating in the active mode.

For a maximum drop of 0.4 V in V_{OH}, the corresponding fan-out is given by

$$N_H I_{B4} R_1 = N_H R_1 \frac{I_{E4}}{(\beta_F + 1)} \leq 0.4 \quad \text{or} \quad N_H \leq 187$$

If a load gate has M inputs but only one is connected to the output of the driving gate, current I_{IH} drawn by the load gate is $I_{IH} = \beta_R I_{B1}/M$. Evidently this causes less loading on the driving gate than when all inputs are connected, and increases the fan-out N_H. Hence, any unused input of a TTL gate is usually tied to the power supply (if NAND logic), or ground (if NOR logic), instead of leaving them floating or connected to other inputs.

Heavy loading at logic high output will cause Q_4 to go into saturation (see Problem 4.14) and contribute to increased switching delays.

Since $N_H > N_L$, we determine the fan-out of the gate by the logic low fan-out, N_L. For the given β_R and β_F of our example, therefore, fan-out $N = N_L = 121$. Manufacturers typically specify a fan-out of 10 for guaranteed logic levels and propagation delay.

4.4.6   PROPAGATION DELAY AND POWER DISSIPATION

The propagation delay of a TTL gate is much smaller than that of a DTL gate because of the fast turn-off of the transistor Q_2 in Figure 4.19. The turn-off delay is primarily caused by the output transistor Q_3, which must discharge its excess base charge through resistor R_2. Average propagation delays of 10 ns are typical for standard TTL gates with a fan-out of 10. With increased fan-out, the delay time increases due to larger capacitive loads. Figure 4.23 shows a MicroSim

```
*  STANDARD  TTL  TRANSIENT  RESP.
Q11       4   3   1    MULTIE
Q12       4   3   2    MULTIE
Q2        6   4   5    NPNT
Q3        7   5   0    NPNT
Q4        9   6   8    NPNT
D1        8   7   SW
RB       10   3   4K
R1       10   6   1.6K
R2        5   0   1K
RC       10   9   130
RL       10  11   400; LOAD
CL        7   0   50PF
DL1      11   7   SW
DL2      11  12   SW
DL3      12  13   SW
DL4      13   0   SW
VI        2   0   PULSE (3V 0V 2N 0.1N 0.1N 30N 40N)
VHIGH     1   0   DC 5V
VCC      10   0   DC 5V
.MODEL MULTIE NPN (IS=2E-6 BF=100 BR=0.5
+ TF=1E-10 TR=1E-8 CJE=0.5PCJC=0.5P RC=5)
.MODEL NPNT NPN (TF=1E-10 TR=1E-8 RC=5 CJE=0.5P CJC=0.5P)
.MODEL SW D(IS=1E-14 TT=1E-10 RS=10 CJO=2P VJ=0.7)
.TRAN/OP 2N 45N
.PROBE
.OPTIONS NOPAGE
.END
```

(a) Simulation

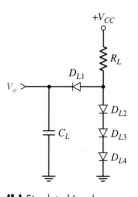

(b) Simulated Load

Figure 4.23   MicroSim PSpice simulation, switching characteristics of a standard TTL gate

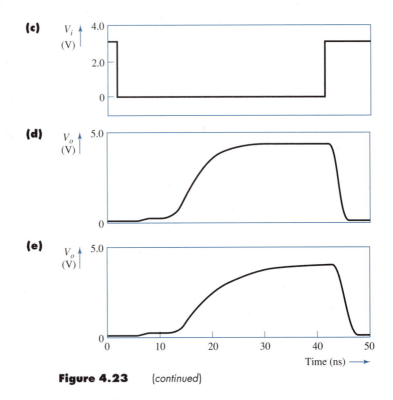

Figure 4.23   (continued)

PSpice simulation of the switching behavior of a standard TTL gate for capacitive loads of $C_L = 20$ pF and $C_L = 50$ pF. Note that the load, which is considered an identical TTL stage, is simulated by the diodes D_{L1} through D_{L4}, the input capacitance, and the reflected resistance of R_B at the input. From the waveforms it is clear that the standard TTL gate is not suitable for driving highly capacitive loads at fast switching speeds. (Observe that the storage time, which depends on Q_2, remains essentially the same for both capacitive loads while the rise time for V_o increases with C_L.)

Each standard TTL gate dissipates approximately 10 mW of power (see Problem 4.15), which gives a typical delay-power product of 100 pJ. Because of the relatively high power dissipation, cooling must be considered when a large number of the standard TTL gates are used in a system. Standard TTL ICs are available as the 74 series, which operate in the 0 to 70°C, and the 54 series for the −55°C to 125°C; both series have the same power dissipation and propagation delay. Manufacturers typically guarantee noise margins of $NM_H = 1.4$ V and $NM_L = 0.6$ V and a propagation delay of 10 ns for a maximum fan-out of 10 for standard 74 series TTL gates. For applications requiring low power dissipation, a subseries known as the low-power TTL, or 74L/54L series, is available in which the resistor values are scaled by nearly tenfold. The average propagation delay of the low-power gates with a dissipation of 1 mW is approximately 35 ns.

4.4.7   HIGH-SPEED TTL GATE

The turn-off speed of a standard TTL inverter can be improved by (1) sourcing a larger output current to charge the load capacitance, and (2) providing an active path for discharging the base of the lower transistor in the totem-pole. Figure 4.24 shows the circuit configuration and voltage transfer characteristic for a high-speed TTL series NAND gate which incorporates both of these features. The cascaded emitter-follower formed by the transistor pair Q_5 and Q_6

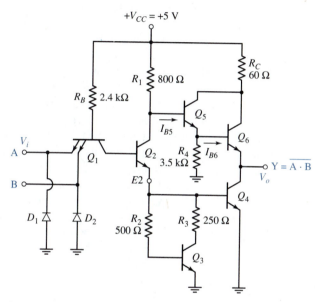

(a) Circuit

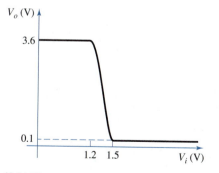

(b) VTC

Figure 4.24   A high-speed TTL gate
Source: Reprinted by permission
of Texas Instruments

is known as a *Darlington pair*. Since $I_{B6} \approx I_{E5} \approx \beta_F I_{B5}$, and $I_{E6} \approx \beta_F I_{B6}$, the load current is now increased to $I_L = I_{E6} \approx \beta_F{}^2 I_{B5}$ when Q_5 and Q_6 are in the active mode. Therefore, a high current gain and a large amount of sourcing current are available in the logic high output state. As a result, the load capacitance is charged quickly to the logic high voltage and the low-to-high delay time is reduced. Resistor R_4 ensures that I_{E5} is sufficiently high so that a reasonably large β is available in the forward active mode. (At low currents comparable to the saturation junction currents, β_F may be less than half its nominal value.) As the emitter current I_{E6} charges the load capacitance, the output voltage rises until $V_o \approx V_{CC} - 2V_{BE} = 3.5$ V. At this voltage Q_5 is driven to saturation. Q_6, however, cannot saturate, since the voltage level $V_{CB6} = V_{CE5(\text{sat})} \approx 0.1$ V reverse-biases the CBJ of Q_6. As a result, when the input switches from low to high, Q_6 is turned off quickly.

The function of the diode D_1 in the standard TTL circuit is now performed by Q_5. When the output is at logic low, Q_2 and Q_4 are in saturation, and hence, V_{B5} is clamped at 0.9 V. Q_6, therefore, cannot conduct with Q_4 in saturation.

The in-phase part of the phase-splitter circuit formed by Q_2 is now connected to an active pull-down circuitry instead of the passive resistor pull-down of the standard TTL circuit. The active pull-down circuitry provides a nonlinear resistance between the terminals E_2 (B_4) and ground. When V_{E2} is below about 0.7 V, Q_3 is off, and a\high resistance appears between B_4 and ground. As Q_3 begins to conduct at $V_{B4} = V_{E2} \approx 0.7$ V, the pull-down circuitry offers a resistance of slightly more than R_2. At low forward bias of the EBJ of Q_3, the total resistance to ground is approximately 600 Ω; this resistance drops to close to 400 Ω when V_{B4} increases to about 0.8 V. Figure 4.25 shows the nonlinear i-v characteristic of the pull-down circuit consisting of Q_3, R_2, and R_3. For comparison, the figure also shows the linear i-v characteristic of a 600 Ω resistor.

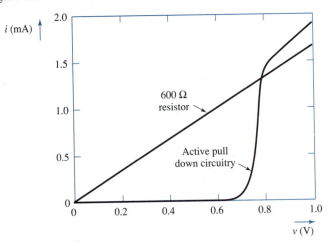

Figure 4.25   i-v characteristics of an active pull-down circuit and a passive (600 Ω) resistor

The lowering of resistance with voltage between the base of Q_4 and the ground terminal helps in reducing the switching delays of Q_4. When Q_2 begins to conduct, initially its emitter current flows almost entirely into the base of Q_4 until V_{B4} reaches about 0.7 V. This helps in faster turn-on of Q_4. As I_{E2} increases, Q_4 enters saturation and V_{B4} rises. Now the resistance offered by the pull-down circuitry decreases and part of I_{E2} is diverted to Q_3. Q_4 is therefore prevented from being excessively driven into saturation. As a result there is less base charge to be removed from Q_4, and the turn-off delay of Q_4 is reduced. A further reduction in the turn-off delay compared to a fixed resistor pull-down is provided by the active pull-down circuitry. During the turn-off of Q_4, the low resistance (400 Ω) of the pull-down circuitry due to $V_{B4} \approx 0.8$ V lasts until Q_4 comes out of saturation and V_{B4} drops below 0.7 V. Consequently, the base of Q_4 discharges a large current into the collector of Q_3, and this causes fast turn-off of Q_4.

Along with improving the switching speed, the active pull-down circuitry also sharpens the voltage transfer characteristics. At low input voltages at A and B in Figure 4.24, transistors Q_2, Q_3, and Q_4 are off because of the large resistance of the pull-down circuitry to ground. As we saw earlier, any small current from the emitter of Q_2 goes into the base of Q_4. When V_{BE4} reaches the cut-in voltage, resistance from B_4 to ground drops as Q_3 begins to conduct, and Q_2 goes well into the active mode. There results, then, the almost simultaneous conduction of Q_2, Q_3, and Q_4 when the input voltage is approximately $2V_{BE} - V_{CE(sat)} = 1.3$ V. At this input, the output voltage begins to drop from its logic high level. Further increase in input voltage to about 1.5 V drives Q_2, Q_3, and Q_4 into saturation, and the output drops to about 0.1 V. In the standard TTL circuit, however, the passive pull-down resistor (R_2 in Figure 4.19) enables the phase-splitter transistor Q_2 to conduct prior to the output totem-pole transistor. This results in the two breakpoints, the first occurring at an input of 0.5 V and the second at 1.2 V, as shown in Figure 4.20. With the active pull-down circuitry, output remains at logic high until all three transistors begin conduction at an input voltage of 1.3 V, and drops to logic low when input rises to 1.5 V. The transition in the output voltage from logic high to low, therefore, occurs over a much narrower range of input voltage: only 0.2 V, compared to 1.1 V for the standard TTL circuit. This sharp transfer characteristic, shown in Figure 4.26b, results in increased noise margins for the high-speed TTL circuit. Since the active pull-down circuit removes the knee portion of the transfer characteristic of the standard TTL circuit, it is known as a *squaring circuit*.

Resistor R_3 ensures that V_{BE4} does not fall to 0.1 V when Q_3 saturates. Resistor R_2 enables most of the emitter current I_{E2} to go into the base of Q_4 once Q_4 is turned on.

Typical propagation delays of the high-speed TTL gate circuits, available as the 74H series, are on the order of 6 to 8 ns. Figure 4.26 shows the PSpice simulation results of the switching behavior of the high-speed TTL circuit for a capacitive load of 20 pF. The improvement in the switching speed compared to the standard TTL circuit has come at a much higher power dissipation of 25 mW.

Figure 4.26

MicroSim PSpice simulation of the switching behavior of a high-speed TTL circuit

```
* HIGH-SPEED TTL INVERTER TRANSIENT RESP.
Q1       4    2    3    MULTIE
Q2       5    4    6    NPNT
Q3       10   7    0    NPNT
Q4       11   6    0    NPNT
Q5       8    5    9    NPNT
Q6       8    9    11   NPNT
DI       0    3         SW
RB       1    2         2.4K
R1       1    5         800
R2       6    7         500
R3       6    10        250
R4       9    0         3.5K
RC       1    8         60
RL       1    12        240
CL       11   0         20P
DL1      12   11        SW
DL2      12   13        SW
DL3      13   14        SW
DL4      14   0         SW
VI       3    0         PULSE (3V 0V 2N 0.05N 0.05N 15N 25N)
VCC      1    0         DC 5V
.MODEL MULTIE NPN (IS=2E-16 BF=100 BR=0.5 RC=5
+ CJC=0.5P CJE=0.5P TF=1E-10 TR=1E-8)
.MODEL NPNT NPN (RC=5 CJC=J0.5P CJE=0.5P TF=1E-10 TR=1E-8)
.MODEL SW D       (IS=1E-14 TT=1E-10 RS=10 CJO=2P VJ=0.7)
.TRAN/OP 2N 30N
.PROBE
.OPTIONS NOPAGE
.END
```

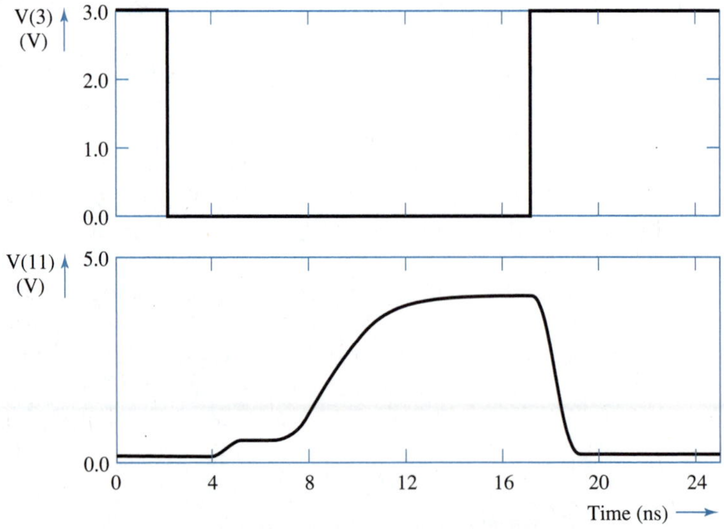

(a) Simulation

(b) Waveforms

4.4.8   SCHOTTKY TTL

A primary limitation in the switching speed of a TTL circuit arises from the storage delay of the saturated BJTs, as we have seen in the previous sections. We can therefore improve the switching speed considerably by preventing the transistors (all except Q_4 in Figure 4.19, for example) from saturating. Saturation can be controlled by one of two methods: (1) by clamping the collector-emitter voltage above saturation voltage, or (2) by limiting the base current to less than the edge-of-saturation value. Operating a BJT outside of saturation mode by keeping a low base current is the technique used to achieve high switching speed in the emitter-coupled logic circuits (Chapter 5). For TTL circuits, a low forward voltage diode such as a Schottky diode (Section 2.8) connected between the base and the collector of a BJT as shown in Figure 4.27a effectively takes the transistor out of saturation.[2] With its forward-bias voltage of typically 0.4 V, the Schottky diode conducts when the transistor is turned on and the collector voltage falls below the base voltage. The collector is then clamped at 0.4 V above the base instead of reaching the silicon forward junction voltage of 0.7 V. Figure 4.27b shows the circuit symbol for a Schottky-clamped (or Schottky) transistor, which has a Schottky barrier diode connected across the collector-base junction. A Schottky barrier diode is readily incorporated in a BJT by extending a Schottky metal (aluminum or platinum) over the base and collector as shown in Figure 4.27c. (Recall that a Schottky metal and a lightly doped n region form a rectifying junction.) Schottky-clamped transistors typically have a base-emitter voltage of 0.7 V at cut-in and 0.8 V in active mode.

When the diode conducts due to a large input current, the CBJ is forward-biased by only 0.4 V instead of 0.8 V. Hence, there is very little current in the CBJ. The collector-emitter voltage V_{CE} is now 0.4 V, which is more than the nominal saturation voltage of 0.1 V. Therefore, the BJT is in its active mode for the same amount of base current and for a base voltage of 0.8 V.

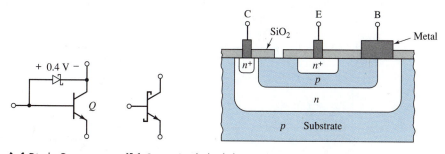

(a) Diode Connection   **(b)** Circuit Symbol   **(c)** Cross Section

Figure 4.27   A Schottky-clamped transistor

---

[2] A junction diode formed of germanium also has a low forward voltage of about 0.2 V; but the difficulty in fabricating both the silicon and the germanium junctions in the same chip has precluded its usage.

Adapting the high-speed TTL circuit of Figure 4.24 with transistors Q_1 to Q_5 replaced by Schottky transistors (note that Q_6 in Figure 4.24 cannot saturate) improves the switching speed significantly—(to 3 ns compared to 10 ns of the standard TTL 74/54 series)—and also increases the noise margins. Schottky TTL circuits are available as the 74S/54S series. As with the 74H series, the Schottky series TTL also has a high power dissipation of typically 20mW (see Problem 4.13). Figure 4.28 shows the switching response of a Schottky series TTL inverter.

With the squaring circuit in the pull-down stage of the Schottky TTL gate, the transition voltages are given by $V_{IL} = 2V_{BE(\text{cut-in})} - V_{CE1} = 1.1$ V, $V_{OH} = V_{CC} - 2V_{BE(\text{cut-in})} = 3.7$ V, $V_{IH} = 2V_{BE(\text{on})} - V_{CE1} = 1.2$ V, and $V_{OL} = V_{CE4(\text{on})} = 0.4$ V. Hence, the noise margins are: $NM_H = 2.5$ V and $NM_L = 0.7$ V. The logic swing is, therefore, $V_{ls} = 3.4$ V.

Since power dissipation and propagation delay depend on the current levels in the circuit, a design engineer may need to make a trade-off for applications requiring the same switching speed as a standard TTL gate but at a lower current drain from the supply voltage. Lower power dissipation is also required to limit total power dissipation in commonly used IC packages to approximately 100 mW.

Low-Power Schottky TTL Gate
A low-power version of the Schottky TTL NAND gate is available in the subfamily known as the 74LS/54LS series. Figure 4.29 shows the circuit, MicroSim PSpice input file, and VTC for this

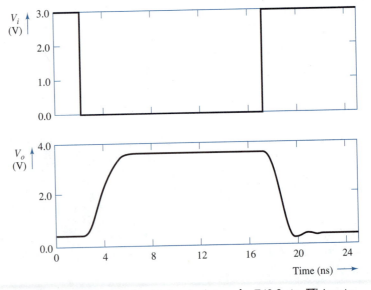

Figure 4.28   MicroSim PSpice simulation of a 74S Series TTL inverter switching response

low-power Schottky gate. Although this circuit is in the family of TTL gates, it is actually a DTL NAND gate in which Schottky diodes are used to perform the logic AND function of the inputs, and active pull-up and pull-down circuits are used at the output. Since the Schottky transistor Q_2 does not saturate, it can be turned off quickly without requiring a multiemitter input transistor as in a standard TTL circuit. Using advanced IC fabrication methods, the input Schottky diodes D_1 and D_2 are made with low parasitic capacitances so that the switching speed is not affected by the replacement of Q_1. Also, the use of input diodes and the large resistor R_{B1} increases the input resistance. In the absence of the CBJ of Q_1 at the input, however, the logic low threshold voltage, V_{IL}, and hence the low noise margin NM_L, are reduced slightly.

The large resistances used in the circuit lower the supply current drain. The accompanying speed reduction is compensated by the addition of Schottky diodes D_3 and D_4 between the Darlington and the phase-splitter stages. When the inputs are switched from logic low to high, Q_2 turns on and the collector voltage V_{C2} falls. With V_o at logic high level initially, both the diodes D_3 and D_4 are forward-biased. The base charge of Q_6 is therefore discharged via D_3 into the

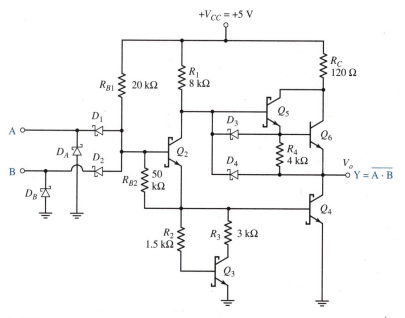

(a) Circuit

Figure 4.29   Low-power Schottky TTL NAND gate
Source: Reprinted by permission of Texas Instruments.

```
* VTC OF LOW-POWER SCHOTTKY TTL
D1    2   1 SBD
DA    0   1 SBD
Q2    4   25 SNPN
Q3    7   60 SNPN
Q4    10  50 SNPN
Q5    8   49 SNPN
Q6    8   910 NPNT
D3    9   4 SBD
D4    10  4 SBD
DQ2   2   4 SBD
DQ3   6   7 SBD
DQ4   5   10 SBD
DQ5   4   8 SBD
RB1   3   220K
RB2   2   550K
R1    3   48K
R2    5   61.5K
R3    5   73K
R4    9   104K
RC    3   8120
VI    1   0 ; INPUT
VCC   3   0 DC 5V
.MODEL SNPN NPN (IS=2E-16)
.MODEL NPNT NPN (IS=2E-14)
.MODEL SBD D(IS=1E-12)
.DC VI 0 5V 0.01
.PROBE
.OPTIONS NOPAGE
.END
```

(b) MicroSim PSpice Input File for VTC

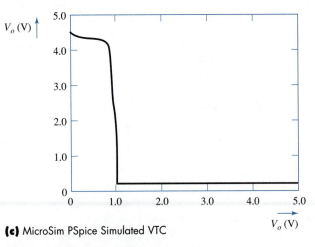

(c) MicroSim PSpice Simulated VTC

Figure 4.29 (continued)

collector of Q_2. Conduction of D_4 helps in discharging the load capacitance. Since the two diode currents form part of the collector current of Q_2, the base of Q_4 is initially supplied with a large current, which causes it to turn on and drop the output voltage faster. Thus the high-to-low transition at the output is speeded up by the bootstrap effect of D_3 and D_4. Resistor R_{B2}, connected between the input steering diode junction and the pull-down circuit, causes Q_4 to conduct even before Q_2 turns on.

When an input falls to logic low, Q_2 is turned off and Q_5 starts to conduct. Resistor R_4, which connects the emitter of Q_5 directly to the output, enables charging of the load capacitance even before Q_6 begins conduction. Hence the low-to-high transition delay at the output is reduced. As the charging current increases so that $V_{BE6} = V_{R4} = 0.7$ V, the emitter current of Q_5 is limited by R_4 to approximately 0.18 mA.

The propagation delay of a typical low-power Schottky TTL gate is 10 ns, which is the same as that for a standard TTL gate. Power dissipation, however, is only 2 mW, compared to the 10 mW dissipation of a standard TTL gate.

The MicroSim PSpice file in Figure 4.29b may be used to obtain the voltage transfer characteristics of the low-power Schottky TTL circuit in Figure 4.29c. As with the high-speed TTL circuit (Figure 4.24) the VTC has only two breakpoints arising from the active pull-down (squaring) circuit. The first breakpoint occurs when Q_4 and, almost simultaneously, Q_2 begin to conduct at $V_i = V_A = V_B = 2V_{BE(\text{cut-in})} - V_D = 0.9\,V = V_{IL}$. With Q_5 at the edge of conduction, the open-circuit logic high output voltage is $V_{OH} = 5 - 0.7 = 4.3$ V. As the input voltage increases slightly, Q_2 and Q_4 conduct heavily (without saturating), and the output voltage is brought down to $V_{OL} = V_{CE4} = 0.4$ V. The corresponding input voltage is $V_{IH} = 2V_{BE(\text{on})} - V_D = 1.1$ V. The noise margins and the logic swing are: $NM_H = 3.2$ V, $NM_L = 0.7$ V, and $V_{ls} = 3.9$ V.

4.4.9 ADVANCED SCHOTTKY TTL SUBFAMILIES

The 74F, 74AS, and 74ALS series of TTL gates, called the Advanced Schottky TTL subfamilies, began to appear in the 1980s. These subfamilies incorporate the recent developments in the BJT IC technology and improved circuit configurations. They achieve higher speed as well as lower power dissipation compared to all the earlier TTL versions.

FAST Series Schottky TTL   The 74F, or FAST (Fairchild Advanced Schottky TTL), series of gates have a circuit configuration at the input similar to the low-power Schottky TTL gates. Figure 4.30 shows the circuit of a two-input 74F series NAND gate. The input Schottky transistor Q_1 in the FAST series provides additional forward current gain to drive higher current loads, and also an additional junction voltage drop at the input. With the conventional silicon diodes performing the input logic function, the logic low voltage V_{IL} is increased to 1.4 V. The no-load logic high output voltage is $V_{OH} = V_{CC} - V_{BE5} = 4.3$ V, since at low output currents only Q_5 is conducting. (The drops in the base and

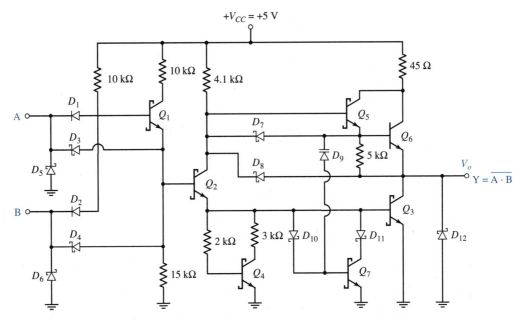

Figure 4.30   FAST Series Schottky TTL NAND gate
Source: Reprinted with perimission of National Semiconductor Corporation.

the emitter resistors of Q_5 are negligible.) The logic high input and the corresponding output voltages are $V_{IH} = 1.7$ V and $V_{OL} = 0.4$ V.

Switching speed improvement over the earlier Schottky TTL versions is achieved in several ways. Diodes D_7 and D_8 serve, as in the case of the low-power Schottky TTL circuit, to discharge Q_6 and the load capacitance respectively, during the high-to-low transition at the output.

Varactor diode D_9 helps in discharging the collector-base transition capacitance of Q_3 when V_o is rising. As Q_5 conducts and its emitter voltage rises, the reverse-bias capacitance of D_9 provides current to the base of Q_7 for a short interval. Q_7 therefore conducts and the CBJ capacitance of Q_3 is discharged via D_{11} and Q_7. In the absence of D_9 and Q_7, this discharge current would act as the base current for Q_3 which would, in turn, cause a current flow from supply voltage to ground via Q_6 and Q_3 until the capacitance is fully discharged. (Note that the nonlinear resistance provided by the squaring circuit of Q_4 is high since Q_2 is off.) D_7 and Q_2 provide a path for discharging the varactor diode D_9 during the low-to-high transition at output. Diodes D_3 and D_4 help to discharge the base of Q_2 rapidly when one of the inputs changes from high to low.

As in the case of the previous Schottky TTL circuits, the diodes D_5 and D_6 are used to clamp the negative signal transitions to -0.4 V and thus minimize reflections at the input. Similarly, the diode D_{12} serves to limit the negative output voltage to -0.4 V.

The FAST series of gates typically have a propagation delay of 2.5 ns and a power dissipation of 4 mW.

Advanced Schottky TTL (74AS Series) and Advanced Low-Power Schottky TTL (74ALS Series)   Figure 4.31 and 4.32 show the circuits of two-input NAND gates from the advanced Schottky (AS) and the advanced low-power Schottky (ALS) TTL subseries. These circuits feature the latest

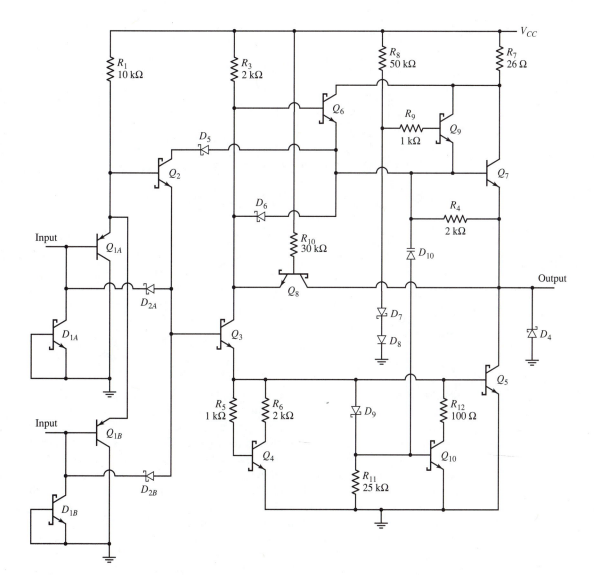

Figure 4.31   As series TTL NAND gate
Source: Reprinted by permission of Texas Instruments.

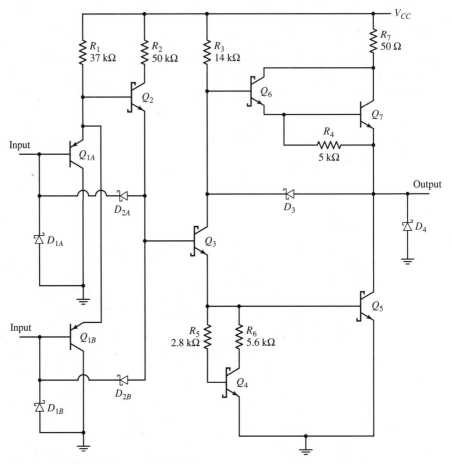

Figure 4.32   ALS series TTL NAND gate
Source: Reprinted by permission of Texas Instruments.

developments in bipolar fabrication technology and use complex subcircuits, similar to those in the FAST series, to achieve higher speeds.

Instead of diodes or multiemitter npn transistors, the pnp transistors Q_{1A} and Q_{1B} perform the logic AND function at the input in both of the circuits. Use of the pnp transistors reduces the logic low input current by a factor of $(\beta + 1)$. Also, their operation as emitter-followers causes faster high-to-low transition at the output, compared with the Schottky diodes of Figures 4.29 and 4.30. Input threshold voltage is still at $V_{IL} \approx 1.4$ V. Transistor Q_8 in the AS series and diode D_3 in the ALS series perform the same function as diode D_8 in the FAST series: discharging the load capacitance and aiding the turn off of the Darlington output pair.

The AS series gates achieve the fastest speed, at 1.5 ns, with a power dissipation of 8 mW. The ALS series has the lowest power dissipation at 1 mW and has an average propagation delay of 4 ns.

4.4.10 Performance Characteristics of TTL Subfamilies

Table 4.3 summarizes some of the principle performance characteristics of the TTL subfamilies of gates. These are the typical, or the average, values specified by the manufacturers. Manufacturers provide these conservative values as guaranteed minimum or maximum over a specified range of operating conditions.

Note that the improvement in switching speed from the standard (early) TTL to the high-speed TTL has resulted in more than twice the amount of power dissipation. The Schottky series brings down power dissipation and also improves speed over the high-speed series. Still, at 20 mw, the power per gate is higher than that for the standard version. Since there are many applications where power dissipation must be limited and the 3 ns speed is not required, the low-power Schottky series with its 2 mW power per gate and 10 ns speed has overtaken the standard TTL series in popularity. The advanced Schottky series gates offer the highest speed, surpassed only by the ECL family (Chapter 5), and the lowest power dissipation, rivaled by the CMOS family (Chapter 7).

4.4.11 Tristate TTL Output

With the increasing popularity of the TTL circuits, there are situations in which outputs of two or more circuits must be tied together to form a single output. A common example is a TTL memory unit (Chapter 10), where each output bit is connected with the corresponding *bits* of all the *words* in the memory. When a particular word is *addressed,* only the bits from that word activate the memory data output. Since the totem-pole output of a standard TTL circuit has a low impedance, it cannot be connected to any other outputs without causing a serious loading situation. Figure 4.33 illustrates the case in which the logic low output X of a gate and the logic high output Y of another gate are shown. When the two outputs are tied together, a large current results in the totem-pole stages of both gates that can cause damage to them.

Table 4.3 Performance characteristics of TTL subfamilies

Parameter	74	74H	74L	74S	74LS	74F	74AS	74ALS
max V_{IL}, V	0.8	0.8	0.8	0.8	0.8	0.8	0.8	0.8
min V_{IH}, V	2.0	2.0	2.0	2.0	2.0	2.0	2.0	2.0
max V_{OL}, V	0.4	0.4	0.4	0.5	0.5	0.5	0.5	0.5
min V_{OH}, V	2.4	2.4	2.4	2.7	2.7	2.7	2.7	2.7
$t_{pd}*$, ns	10	6	32	3	10	2.5	1.5	4
power, mW	10	25	1	20	2	5	8	1
d.p, pJ	100	150	32	60	20	13	12	4

* Varies with load capacitances

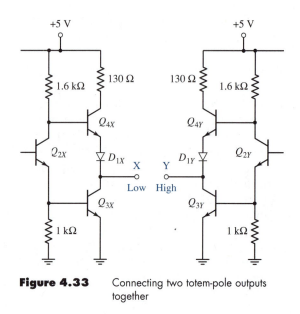

Figure 4.33   Connecting two totem-pole outputs together

A modified TTL circuit, called a *tristate TTL* circuit, eliminates the loading problem by disabling the output and putting it in a high impedance state, with the use of a control signal. The third state, other than the logic outputs of high and low, is called the high impedance (HiZ) state at the output. Tristate TTL outputs can be connected together to form a common output, or *bus*. Figure 4.34 shows symbols for different tristate circuits and an example of a bus connected output.

Figure 4.35 shows a basic TTL stage for the tristate NAND output. Note that this circuit is the same as the standard TTL NAND circuit of Figure 4.19, with an additional emitter in Q_1 and the diode D_2. When the control signal at G is logic high (≈ 5 V), diode D_2 and the EBJ of Q_1 associated with G are off. Hence the circuit functions as a standard TTL NAND gate. A logic low voltage at G (≈ 0.1 V from another TTL gate output, for example) turns Q_1 on to saturation independent of logic levels at A and B, and Q_2 is cut off. In a standard TTL circuit this will take the collector of Q_2 toward V_{CC}. The low voltage at G, however, forward-biases the diode D_2, clamping the collector voltage V_{C2} at approximately 0.8 V. At this voltage Q_3, Q_4, and D_1 are all off. Hence the output is an open circuit with only the pn junction leakage current flowing at Y.

Figure 4.36 shows the circuit of a typical TTL gate with tristate output. Here, transistors Q_2 and Q_4 perform the control operation similar to the extra emitter and the diode D_2 in Figure 4.35. When high, control signal G turns both Q_2 and Q_4 on and brings down the collector and the emitter voltages of Q_3. Both Q_5 and Q_6 are, therefore, turned off, and the output Y is left floating. For fast switching, the Schottky TTL series offers tristate output as shown in the inverting buffer circuit of Figure 4.37. As in Figure 4.36, low voltage at G enables the

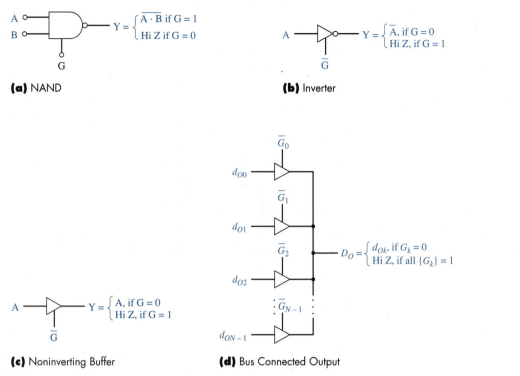

(a) NAND

(b) Inverter

(c) Noninverting Buffer

(d) Bus Connected Output

Figure 4.34   Examples of tristate gates

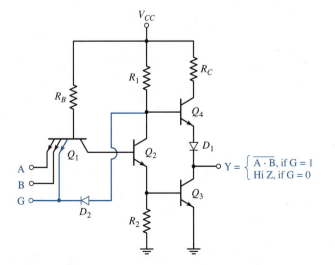

Figure 4.35   Basic tristate TTL NAND circuit

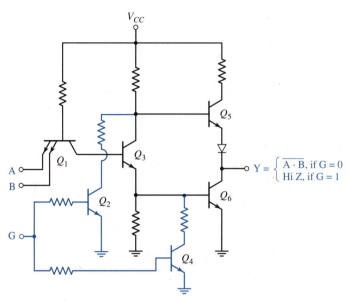

Figure 4.36   Typical tristate TTL gate

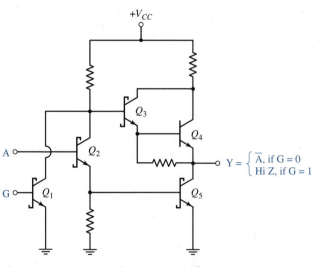

Figure 4.37   Schottky TTL tristate buffer (inverting)

buffer output. When G is high, Q_1 turns on and voltages at the collector and the emitter of Q_2 fall. Therefore, transistors Q_3, Q_4, and Q_5 are turned off, and the output goes to high impedance state.

Figure 4.38 shows the circuit of a 74126 TTL gate, which is a noninverting buffer with tristate output. This circuit functions the same way as the basic

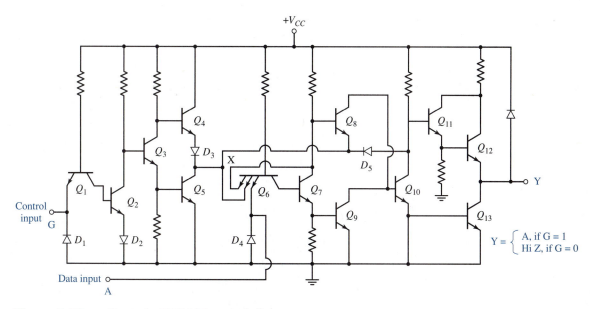

Figure 4.38   Circuit of a 74126 TTL tristate buffer

tristate circuit of Figure 4.35, with additional circuitry included for improved speed and load driving capability.

Here, control signal G is buffered by the input stage consisting of Q_1, Q_2, Q_3, Q_4, and Q_5 and is available as a totem-pole output $X(= G)$ at the collector of Q_5. The circuitry to the right of X is that of a high-speed logic AND gate giving $Y = A \cdot X$ when G is high. For a logic high control input G, the buffered signal X is high and hence the diode D_5 and the transistor Q_8 are turned off. The data input A is then transmitted to the output Y via the AND gate.

When G is at logic low, the emitter of Q_6 connected to X conducts and turns Q_7 off. Since the emitter of Q_8 is at a low voltage ($V_X \approx 0.1$ V), Q_8 is turned on, which brings the voltage at the collector of Q_7 to 0.8 V. Q_7 is therefore turned off, and, consequently, Q_9 goes off. With Q_8 on and Q_9 off, Q_{10} is also turned off. The conduction of the diode D_5 clamps the voltage at the base of Q_{11} to 0.8 V, which is not enough to turn Q_{12} on, and with Q_{10} off, Q_{13} is also off. Hence, Y is at the high-impedance state.

4.4.12   OPEN COLLECTOR AND EXPANDER GATES

As we saw in the previous section, totem-pole outputs of two or more TTL gates cannot be connected together to perform wired logic. To permit wire-ANDing, TTL gates called open-collector gates are available, in which the active pull-up transistor is replaced with an open circuit. An external resistor must be connected between the collector of the pull-down transistor and the supply voltage. Figure 4.39 shows the circuit of a two-input open-collector TTL NAND gate.

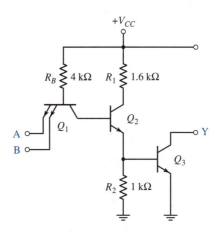

Figure 4.39
Open-collector TTL NAND gate
Source: Reprinted by permission of Texas Instruments.

The outputs of open-collector gates can be tied together with a single collector resistor to power supply voltage as shown in Figure 4.40. The symbol used for wire-ANDing is the dotted AND gate at the junction of all the outputs. The value of the resistor R_C depends on the number of gates connected and the allowable logic levels. Figures 4.41 and 4.42 show N open-collector gates wire-ANDed and driving L TTL gates. If all the N driving gates are at logic high as in Figure 4.41, the output voltage must not drop below the minimum acceptable logic high voltage. Hence,

$$V_o = V_{CC} - (N \cdot I_{OH} + L \cdot I_{IH})R_C \geq V_{OH} \qquad \textbf{(4.46)}$$

where the logic high input and output currents, I_{OH} and I_{IH}, are as indicated in the figure. Equation (4.46) gives the maximum value for R_C.

The minimum value for R_C is determined from the requirement that the sinking current of a single driver gate output, with all other drivers off (Figure 4.42), must not exceed the maximum it can withstand. Hence,

$$\frac{V_{CC} - V_{OL}}{R_C} \leq I_{OL} - L \cdot I_{IL} \qquad \textbf{(4.47a)}$$

To keep the output transistor in saturation, I_{OL} is limited by the base drive of Q_3 in Figure 4.39. That is,

$$I_{OL} = \beta_{F3} I_{\beta 3} \qquad \textbf{(4.47b)}$$

R_C may be chosen from the range of values given by Equations (4.46) and (4.47). Example 4.10 illustrates this calculation.

Example 4.10   Determined the range of R_C in Figure 4.40 if two open-collector gates are wire-ANDed and the output is used to drive three identical gates of open-collec-

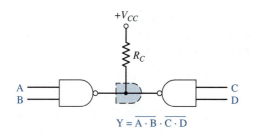

$$Y = \overline{A \cdot B \cdot C \cdot D}$$

Figure 4.40   Wire-ANDed output using
open-collector gates

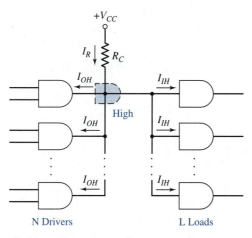

Figure 4.41   Wired-AND gates driving
loads—logic high output

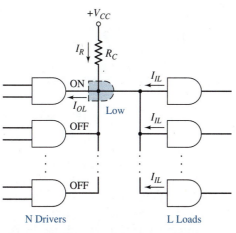

Figure 4.42   Wired-AND gates driving
loads—logic low output

tor family. It is required that the output must not drop below 4 V at logic high. Use $\beta_R = 0.4$ and $\beta_F = 50$ for all the transistors in the gates. Each gate has the circuit configuration as shown in Figure 4-39.

Solution

Here, $N = 2$ and $L = 3$.
Referring to Figure 4.41, at logic high output each of the three load gates is drawing a current of I_{IH} given by

$$I_{IH} = \beta_R I_{B1} = \frac{0.4(5 - 2.3)}{4} = 0.27 \text{ mA}$$

Since the drivers are off, $I_{OH} = 0$. Hence, from Equation (4.46),

$$V_{OH} = V_{CC} - L \cdot I_{IH} \cdot R_C \geq 4$$

This gives $R_C \leq 1235 \ \Omega$.
Referring to Figure 4.42, each load at logic low input of 0.1 V sources a current of

$$I_{IL} = \frac{5 - 0.9}{4} = 1.03 \text{ mA}$$

Thus, the worst-case maximum current sunk by a single driver gate is

$$I_T = \frac{V_{CC} - V_{OL}}{R_C} + L \cdot I_{IL} = \frac{4.9}{R_C} + 3.09$$

The base current of Q_3 at logic low output of a single driver gate is given by

$$I_{B3} = \frac{5 - 0.9}{1.6} + \frac{1.4(5 - 2.3)}{4} - \frac{0.8}{1} = 2.71 \text{ mA}$$

Hence, from Equation (4.47b), the output transistor Q_3 of the driver gate will remain in saturation if

$$\frac{4.9}{R_C} + 3.09 \leq 50 \times 2.71$$

This gives $R_C \geq 27 \ \Omega$.
From the maximum and minimum values we see that R_C can be in the range of

$$27 \ \Omega \leq R_C \leq 1235 \ \Omega$$

Open-collector gates provide the versatility of wire-ANDing with a large number of logic variables. Also, with appropriate pull-up resistors and a higher supply voltage for the output stage, a gate output can have large logic high levels. This is a particularly useful property for driving high voltage loads or for interfacing with logic families having higher logic levels. Problem 4.19 considers one such application. The disadvantage of on external pull-up resistor is that the switching delay time is increased, typically to 30 ns, due to passive (exponential) charging of parasitic capacitance. The circuit is also more sensitive to noise at the output.

Another way of realizing complex logic functions using outputs of different gates is by using *expander* and *expandable* TTL gates. Similar to the expander input in a DTL circuit (Figures 4.10, 4.13), where one or more diodes may be connected to increase fan-in, an expandable TTL gate has provision for connecting one or more multiemitter transistors. Figure 4.43 shows the structure of expander and expandable gates. As a basic TTL NAND gate, the expandable gate (Figure 4.43b) can be operated alone. When the outputs *xc* and *xe* of an expander (Figure 4.43a) are connected to the collector and the emitter of an expandable gate (Figure 4.43b), the resulting circuit realizes AND-OR-Invert function, as shown in Figure 4.43b. Observe that the complementary outputs at *xc* and *xe* of the expander are compatible with the totem-pole driver inputs.

4.4.13 LOGIC FUNCTIONS IN TTL FAMILY

Apart from the inverters and the NAND gates, the TTL family of logic circuits includes many different functions in the small-, medium-, and large-scale integrated circuits. These functions are implemented by modifying the basic TTL NAND gate circuit. As examples, Figures 4.44 and 4.45 show circuits for obtaining the NOR and the AND-OR-Inverter (AOI) functions in the standard TTL family. Note that both these circuits have configurations identical to the expander-expandable gate combination of Figure 4.43.

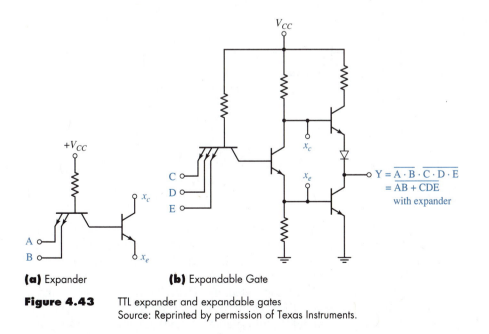

(a) Expander **(b)** Expandable Gate

Figure 4.43 TTL expander and expandable gates
Source: Reprinted by permission of Texas Instruments.

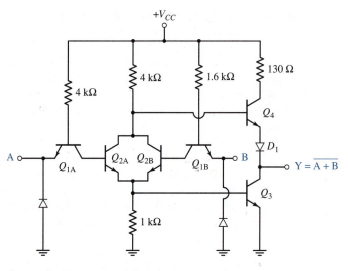

Figure 4.44   TTL NOR circuit (7402)

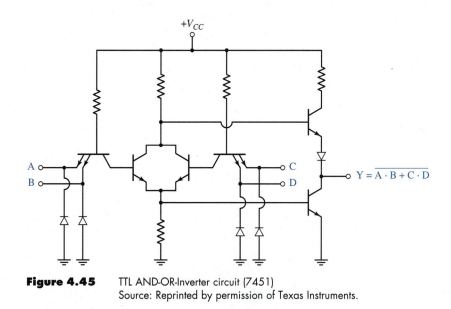

Figure 4.45   TTL AND-OR-Inverter circuit (7451)
Source: Reprinted by permission of Texas Instruments.

These circuits use the basic TTL configuration with more intermediate stages but without the improvements such as the active pull-down or the Darlington circuitry. Operation of these circuits is readily verified. In Figure 4.44, for example, the circuit functions as a standard inverter in the absence of Q_{1B} and Q_{2B}. With these two transistors connected, if A (B) is high, $Q_{1A}(Q_{1B})$ operates in the reverse-active mode and drives Q_{2A} (Q_{2B}) into saturation. As a result, Q_3 goes into saturation and the output Y is at logic low. Note that paralleling of Q_{2A} and

Q_{2B} clamps the common collector voltage at 0.9 V when either input is at logic high (similar to the operation in an RTL NOR gate). When both the inputs are at logic low, Q_{2A} and Q_{2B} are turned off, and the output goes to logic high. The totem-pole output provides load-driving capability. The AOI gate of Figure 4.45 functions in a similar manner, with an additional emitter provided in Q_{1A} and/or Q_{1B} for each additional input variable. Observe that the AOI circuit performs in a single chip the function of two open-collector NAND gates connected for wired-AND operation.

4.4.14   TTL CIRCUITS FOR LSI IMPLEMENTATION

Although the standard TTL circuit is popular for small-scale and medium-scale implementation of logic circuits, its large size and power dissipation make it unsuitable for fabricating highly complex systems on single chips. Circuit designers make modifications in the standard and the Schottky TTL circuits to increase circuit density and reduce power requirements. One way to reduce power dissipation is to design the circuit to operate at a low supply voltage. (Note that using large resistor values to reduce dissipation, as in low-power TTL circuits, for example, increases silicon area.) At low supply voltage, however, logic levels and noise margins are also reduced. Since noise generated within the same chip is low, this does not pose a problem; hence higher circuit density per silicon area can be obtained at low operating voltages.

Figure 4.46 shows a low-voltage and low-power TTL gate for use in a large-scale integrated circuit. This circuit, which realizes the NAND function, can operate at a supply voltage of 1 V to 3 V. Note that at logic low output, the base current for Q_2 is supplied directly from the supply voltage via R_1 and R_2, instead of from the multiemitter transistor. Outputs of different gates may be connected together with a single collector resistor, similar to open-collector output, to realize AND-OR-Invert function. At a typical supply voltage of 1.2 V (inside LSI circuits), each gate has a propagation delay of about 15 ns and dissipates an average of about 500 μW. Propagation delay is decreased at higher supply voltages at increased dissipation. Figure 4.47 shows two other versions of

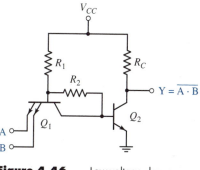

Figure 4.46   Low-voltage, low-power TTL circuit

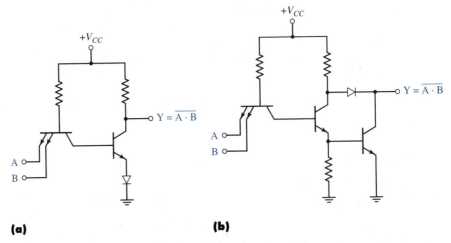

(a)   **(b)**

Figure 4.47   Examples of TTL circuits for LSI implementation

a TTL circuit for LSI implementation. Unlike the circuit of Figure 4.46, the multiemitter transistor operates in reverse active mode to supply base current to the output transistor at logic low output. Circuits in Figures 4.46 and 4.47 are readily interfaced with standard TTL circuits at input and output using appropriate supply voltage and resistances.

In high-speed LSI circuits, Schottky transistors are used in the basic DTL and TTL configurations. Figure 4.48 shows examples of lower-power, low-voltage Schottky TTL circuits for LSI implementation. These circuits have a typical delay of 5 to 10 ns and dissipate less than 1 mW of power at 2.5 V supply. A low-voltage LSI version is considered in Problem 4.27. Figure 4.49 shows examples of TTL circuits realizing logic functions in LSI systems.

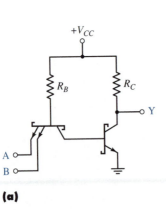

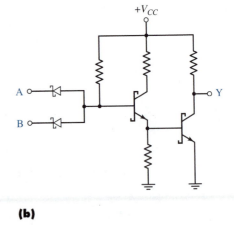

(a)   **(b)**

Figure 4.48   Schottky TTL circuits for LSI implementation

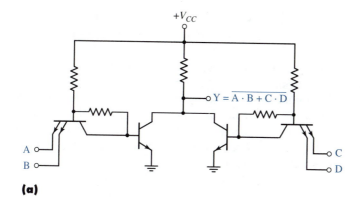

(a)

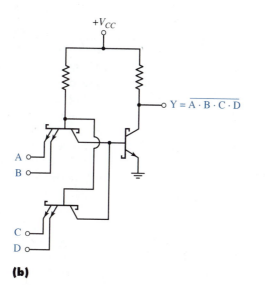

(b)

Figure 4.49   TTL circuits realizing logic functions in LSI systems

SUMMARY

This chapter considered the families of logic circuits that use the bipolar junction transistors. The transistors in all these families operate between saturation and cutoff mode; hence, these families are called the saturation logic families.

The output of a BJT inverter is the supply voltage in the logic high state, and the saturation collector-emitter voltage in the low state. A basic resistor-transistor-logic NOR function is realized by connecting one or more transistors in parallel with the inverter transistor. The inverter and the RTL family of IC logic gates have high noise margins but poor loading and switching characteristics.

The diode-transistor-logic family implements the NAND function in its basic gate. The circuit uses diodes to perform an AND function on input variables, and an output transistor complements the result. The DTL circuits have better noise margins and higher fan-out capability than the RTL circuits.

In the transistor-transistor-logic circuits, which are currently the most popular and versatile family of digital ICs, a multiemitter transistor replaces the input diodes of the DTL family. The multiemitter transistor, operating in the reverse active or saturation modes, drives a phase splitter circuit. In a standard TTL circuit, the in-phase and out-of-phase outputs of the phase splitter drive a two-transistor totem-pole output stage. The gate switching characteristics are greatly improved over the DTL family. Improved circuit designs, such as the active pull-down circuitry, result in the high-speed TTL family, which has increased switching speed and fan-out over the standard TTL family. Another version, called the low-power TTL family, uses higher resistances to reduce the power dissipation. Many logic functions are available in the standard, the high-speed, and the low-power TTL families. However, as the switching speed increases, the power dissipation also increases. For example, a low-power TTL gate dissipates 1 mW of power with a propagation delay of 32 ns, while a standard TTL gate dissipates 10 mW at a delay of 10 ns.

Further innovations in circuit design and in the bipolar device fabrication technology have led to several subseries of high-performance TTL circuits. These are the Schottky series, the low-power Schottky series, the FAST series, the advanced Schottky series, and the advanced low-power Schottky series of TTL circuits. Gates from these subfamilies of TTL circuits have Schottky barrier diodes fabricated between the base and the emitter of each saturating BJT. When the collector-base junction is forward-biased, the Schottky diode conducts and clamps the collector at 0.4 V below the base voltage. This effectively prevents the BJT from operating in the saturation mode. Hence, the turn-off delay of the BJT is greatly reduced. The advanced Schottky series TTL circuits have the highest switching speed with almost the lowest power dissipation, rivaled only by the ECL family in speed and the advanced CMOS family in power dissipation.

Logic circuits are available in the TTL family for connecting their outputs together in a bus, or for implementing wired-AND operations. Besides these and other simple logic functions, the TTL family has a variety of MSI and LSI circuits available. High density and low power dissipation are achieved in LSI systems using simpler versions of standard TTL circuitry.

REFERENCES

1. Millman, J., and H. Taub. *Pulse, Digital, and Switching Waveforms.* New York: McGraw-Hill, 1965.

2. Garret, L. S. "Integrated Circuit Digital Logic Families." Parts I, II, and III, *IEEE Spectrum,* October, November, December 1970.

3. Taub, H., and D. Schilling. *Digital Integrated Electronics.* New York: McGraw-Hill, 1977.

4. Hodges, D. A., and H. G. Jackson. *Analysis and Design of Digital Integrated Circuits.* New York: McGraw-Hill, 1988.

5. Muroga, S. *VLSI System Design, When and How to Design Very-Large-Scale Integrated Circuits.* New York: Wiley, 1982.

6. Elmasry, M. I. *Digital Bipolar Integrated Circuits.* New York: Wiley, 1983.

7. Barrett, J. C., et al. "Design Considerations for a High-Speed Bipolar Read-Only Memory," *IEEE JSSC* SC-5, no. 5 (October 1970), pp. 196–202.

8. Hamilton, D. J., and W. G. Howard. *Basic Integrated Circuit Engineering.* New York: McGraw-Hill, 1975.

9. *FAST Applications Handbook.* Santa Clara, CA: National Semiconductor Corp., 1987.

10. *ALS/AS Logic Data Book.* Dallas: Texas Instruments Inc., 1986.

Review Questions

1. What output logic state determines the fan-out of a BJT inverter? Why?

2. What causes the logic high output voltage of a BJT inverter to drop with load(s) connected?

3. Can a BJT inverter using a high β transistor have a high fan-out? Explain.

4. Which of the noise margins in a BJT inverter is affected by load gates? How?

5. What is the effect of base overdrive on the fan-out and propagation delays of a BJT inverter?

6. What is the basic logic function realized in the RTL family of gates (other than the inverter)?

7. What is fan-in of a logic gate? How can the fan-in of an RTL gate be increased?

8. The outputs of two two-input RTL NOR gates having inputs A, B, and C, D are tied together. What logic function is realized at the common output?

9. What is the advantage of wired-AND using RTL gates? What is the disadvantage?

10. What is the basic logic function realized in the DTL family of gates?

11. What is the function of the diode(s) in series with the base of the transistor in a discrete DTL gate?

12. How is the expander input X in a DTL gate used? Where is it connected when unused?

13. What output state determines the fan-out in a DTL gate?

14. What is the function of the transistor Q_1 in the IC version of a DTL gate?

15. If a DTL gate designed to drive 20 identical gates only has a load of 10 gates, how is the driver propagation delay affected?

16. How is the logic function $Y = \overline{A \cdot B \cdot C \cdot D}$ realized using a two-input DTL NAND gate with expander terminal?

17. The outputs of two two-input DTL NAND gates having inputs A, B, and C, D are tied together. What logic function is realized at the common output?

18. What causes the increase in noise margin in a high-threshold DTL gate?

19. What is the mode of operation of the multiemitter input transistor in a TTL gate when all inputs are at logic low? What is the mode when all inputs are at logic high?

20. What is the function of the diode in the path of the totem-pole output stage in a standard TTL gate?

21. Where should unused inputs in TTL gates be connected? Why?

22. How does a low-power TTL circuit differ from a standard TTL circuit?

23. How is the current gain in a Darlington circuit increased?

24. What is an active pull-down circuitry?

25. How does a Schottky diode connected between the base and the emitter of a BJT prevent the transistor from going into saturation?

26. What is the function of the diode connected between the output and ground of any high-speed TTL gate? What is the function of the diodes connected between the inputs and ground?

27. What is a tristate gate?

28. What happens when the outputs of two standard TTL gates, one at logic low and the other at logic high, are conected together?

29. If the output of an open-collector gate realizing $Y = \overline{A + B}$ is connected to the output of another open-collector gate realizing $Y = \overline{C \cdot D}$, what function is realized at the common output?

30. The function $Y = \overline{A \cdot B + C \cdot D}$ can be realized using an AOI gate as well as two 2-input open-collector NAND gates. Which is preferred? Why?

PROBLEMS

* 1. Derive the fan-out condition for the BJT inverter, similar to Equation (4.5), in which the base resistor R_B appears explicitly. Show that

$$(a) \quad N \leq \frac{\beta[V_{CC} - V_{BE\,(sat)}]}{[V_{CC} - V_{CE\,(sat)}]} - \frac{RB}{Rc}$$

$$(b) \quad V_{IH} = V_{BE\,(sat)} + \frac{R_B[V_{CC} - V_{CE\,(sat)}]}{\beta R_C}$$

Use overdrive factor $k = 1$.

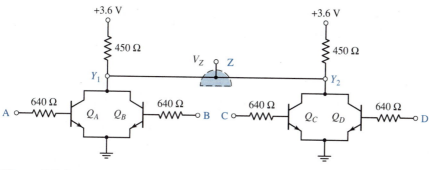

Figure P4.5

2. Determine the logic high fan-out with a noise margin of at least 0.5 V for the BJT inverter of Figure 4.1.

* 3. Determine the switching delays in a BJT inverter (Figure 4.1) using the following circuit and device parameters: $R_B = 10$ kΩ, $R_C = 1$ kΩ, $V_L = 0$ $V_H = 4$ V, $\beta_F = 100$, $\beta_R = 0.5$, $\tau_F = 0.2$ ns, and $\tau_R = 15$ ns. Average junction capacitances are: $C_{E1} = 0.3$ pF and $C_{C1} = 0.12$ pF in the off state, and $C_{E2} = 0.8$ pF and $C_{C2} = 0.12$ pF during conduction. Compare the calculated values with those from PSpice simulation using the approximate zero-bias capacitances given in Figure 4.5.

4. Show that for a two-input RTL NOR gate shown in Figure 4.8 operating as an inverter with $V_{CC} = 3.6$ V, $R_B = 640$ Ω, and $R_C = 450$ Ω, the logic high threshold voltage V_{IH} is
   a. 0.924 V with both of its inputs connected together, and
   b. 1.05 V when one of the inputs is at logic low. Use a β of 20.

5. Determine the truth table for the circuit shown in Figure P4.5 and verify that the logic implemented is Z = $\overline{A + B + C + D}$.
   If only one of the four inputs is at logic high (=3 V), what is the collector current of the saturated transistor? Compare this current with that of a four-input RTL NOR gate (similar to Figure 4.8 with four transistors paralleled together, each with a 640 Ω base resistor) in which only one input is at logic high. What is the minimum β required in each case?

6. Calculate the power dissipated by an RTL inverter with $V_{CC} = 3.6$ V, $R_B = 640$ Ω, and $R_C = 450$ Ω for
   a. the the logic high input of 3.6 V, and
   b. the logic low input of 0.1 V.
   Compare these values with those for the so-called low-power RTL gate which has $R_B = 1.5$ kΩ and $R_C = 3.6$ kΩ.

7. Figure P4.7 on page 214 shows a simplified RTL NOR circuit. Verify its logic operation. Calculate the base and collector currents if
   a. all inputs are high at 5 V, and
   b. only one input is high at 5 V while others are at 0.
   Compare the switching response of this circuit with that of the NOR circuit of Figure 4.8.

8. Another possible RTL circuit is shown in Figure P4.8. Analyze the circuit and determine the logic function and voltage levels at Y for input voltages of 0 and 5 V at A and B. Assume a β of at least 100. What is the effect of loading?

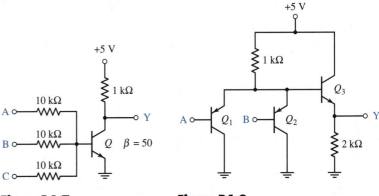

Figure P4.7   **Figure P4.8**

9. Calculate the logic low fan-out of the DTL gate shown in Figure P4.9.

10. Given $R_1 = \gamma R$ and $R_2 = (1 - \gamma)R$, where $R = 3.75 \text{ k}\Omega$, express the fan-out of the IC version of the DTL gate shown in Figure 4.13 in terms of γ and β. Determine the fan-out for $\beta = 50$ and $\gamma = 1, 0.2$, and 0.

11. Referring to Figure 4.14, obtain the voltage table relating the inputs A, B, C, and D, and the output Z, where $A = X, B = \overline{Y}, C = \overline{X}$ and $D = Y$. Show that the circuit implements the logic equivalence function, $Z = 1$ if $X = Y$. If the output Z is connected to both the inputs of Figure 4.13, determine the collector current of Q_4 for $V_A = 4$ V, and $V_B = 0.1$ V.

12. Obtain a MicroSim PSpice simulation of the switching response of the DTL circuit of Figure 4.13 using 2N2222A transistors and 1N914 diode. Use the models available for the active devices in the MicroSim PSpice library.

13. Determine V_{IL}, V_{IH}, and the noise margins for the high-threshold DTL gate shown in Figure 4.15.

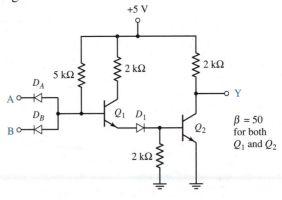

Figure P4.9

14. Show that for a standard TTL gate (Figure 4.19) at logic high output, the maximum fan-out occurs when Q_4 saturates. Calculate this fan-out and the corresponding V_{OH}.

15. Calculate the power dissipated in the circuit of Figure 4.19 for (a) $V_i = 0.1$ V and (b) $V_i = 3$ V. Assume a similar circuit connected as load. Use $\beta_F = 50$ and $\beta_R = 0.1$.

16. Repeat Problem 4.15 for the Schottky (S series) TTL NAND gate (Figure 4.24 with Q_1 to Q_5 replaced by Schottky transistors).

17. Calculate the power supply currents for logic high and logic low inputs in the ALS TTL gate of Figure 4.32. Assume a similar gate connected as load. Compare the calculated values with PSpice (static) simulation results.

18. Calculate the current in the diode D_{1Y} in Figure 4.33 after connecting X and Y together. Assume the output X was at logic low and Y at logic high prior to connecting them together.

*19. Figure P4.19 shows a circuit for pulse amplitude addition. The two inverters are from the open-collector TTL family. Sketch the voltage waveform at Z. Calculate the power supply current I_S.

20. If G_1 and G_2 in Figure P4.20 are open-collector TTL gates, state the logic input combination for which the LED will light. Assuming a diode drop of 1.6 V when on, calculate the diode current.

21. Determine the range of values for R_C to be used with two open-collector gates whose outputs are wired together (Figure 4.40). The wired-AND output drives a single TTL gate with $I_{IH} = 40$ μA (sink) and $I_{IL} = 1.6$ mA (source). Each driver has $I_{OH} = 250$ μA and $I_{OL} = 16$ mA. Recommended output voltages are: $V_{OH} = 2.4$ V and $V_{OL} = 0.2$ V. (These values correspond to 7401 two-input NAND gates.)

22. A simplified TTL inverter is shown in Figure P4.22. Calculate the power supply currents for $V_i = 0.1$ V and $V_i = 4$ V. Assume $\beta_F = 50$ and $\beta_R = 0.2$. State the mode of operation for each transistor in both cases.

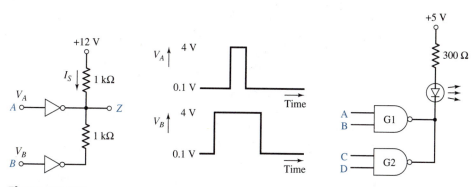

Figure P4.19

Figure P4.20

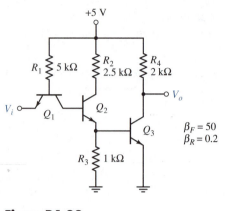

Figure P4.22

23. Figure P4.23 shows a simplified TTL AND circuit. Verify the logic operation and calculate the power supply currents for logic low and high outputs at no load. Assume $\beta_F = 50$ and $\beta_R = 0.2$.

*24. Figure P4.24 shows a TTL circuit in which saturation is controlled by the feedback transistor Q_2. State how saturation of Q_3 is prevented. Obtain a PSpice simulation for the switching response of the circuit with and without Q_2. Use parameters given in Figure 4.5 with $\beta_R = 0.2$

25. What is the function of the diodes in the TTL circuits of Figure 4.47?

*26. Another TTL circuit for LSI implementation is shown in Figure P4.26. Simulate the circuit using PSpice and determine the average propagation delay with (a) $R_1 = 10$ kΩ, and (b) $R_1 = 4$ kΩ and $R_2 = 10$ kΩ. Explain any difference in delays.

*27. A low-voltage, low-power Schottky TTL circuit for LSI implementation is shown in Figure P4.27. Determine the power dissipation when both inputs are at (a) 0.2 V, and (b) 1 V.

28. Verify the logic functions realized by the circuits in Figure 4.49.

29. *Experiment*: Static and dynamic characteristics of a BJT inverter are studied in this experiment. Connect the circuit shown in Figure 4.1 using 2N2222A.

    a. Vary input voltage V_i from 0 to 5 V and obtain the voltage transfer characteristics.

    b. Determine the logic levels and noise margins from the VTC. Calculate these values using measured (or manufacturer's typical) values of β_F and compare experimental results.

    c. Apply a square wave input of -5 V to 5 V amplitude with on and off intervals of 10 μs each. Measure the turn-on and turn-off delays at the output. Vary the dc offset of input so that the low level is changed to -2 V, while keeping the high level at 5 V, and measure the delays. Repeat the delay measurement for zero low level.

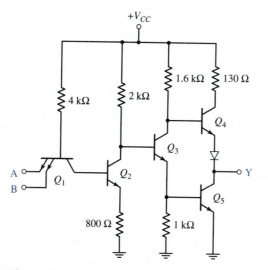

Figure P4.23

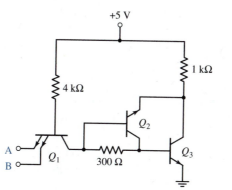

Figure P4.24

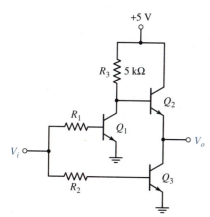

Figure P4.26

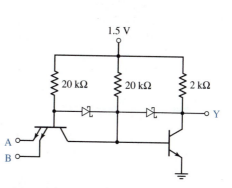

Figure P4.27

   d. Using manufacturer's data or the following MicroSim PSpice model parameters in the charge-control model, calculate the approximate switching delays.[3]

.model Q2N2222A NPN (Is = 14.34f Xti = 3 Eg = 1.11 Vaf = 74.03

Bf = 255.9 Ne = 1.307 Ise = 14.34f Ikf = .2847 Xtb = 1.5

Br = 6.092 Nc = 2 Isc = 0 Ikr = 0 Rc = 1 Cjc = 7.306p Mjc = .3416

Vjc = .75 Fc = .5 Cje = 22.01p Mje = .377 Vje = .75 Tr = 46.91n

Tf = 411.1p Itf = .6 Vtf = 1.7 Xtf = 3 Rb = 10)

Use average voltages in each range of operation to determine the average capacitances given by Equations (3.78) and (3.79). Compare the calculated and experimental delays.

   e. Using Equation (4.38) and (4.39), calculate the value of a speed-up capacitor for -5 V to 5 V input. With this value of capacitor across R_B, determine the switching delays. How do the delays vary as the low level of input is changed as in (*c*) above?

30. *Experiment:* DTL circuit characteristics are studied in this experiment. Connect the discrete DTL circuit shown in Figure 4.10 using the 2N2222A transistor and 1N914 diodes with only one input, A. Use $R_1 = 5$ kΩ, $R_2 = 10$ kΩ, and $R_C = 1$ kΩ, and return R_2 to ground ($V_{BB} = 0$). The following diode model, extracted from PSpice, may be used for switching response analysis.

.model D1N914 D(Is = 10f N = 1 Rs = .1 Ikf = 3 Eg = 1.11

Cjo = 1p M = .3333 Vj = .75 Fc = .5 Isr = 100p Nr = 2 Bv = 100

Ibv = 100u Tt = 5n)

   a. Determine the transfer characteristics for V_i from 0 to 5 V.

   b. Compare the theoretical VTC and logic threshold voltages with the experimental results.

   c. Apply a square wave input of 0 to 5 V amplitude with on and off intervals of 100 μs each. Measure the turn on and turn off delays at the output. Change R_2 to (*i*) 1 kΩ, (*ii*) 50 kΩ, and (*ii*) ∞. Measure the delays in each case.

   d. From Equation (4.39) for the excess base charge stored and using average base discharge current of $[V_{BE(\text{sat})}/2R_2]$, estimate the turn-off delays for each value R_2 used. For $R_2 = \infty$, use the diode reverse current for discharging the base. Compare the estimated delays with experimental values.

---

[3] These parameters are available in the MicroSim PSpice evaluation disk.

31. *Experiment:* TTL inverter characteristics are studied in this experiment. Connect the circuit shown in Figure 4.19 for an inverter using 2N2222A transistors and 1N914 diodes.

   a.   Determine the transfer characteristics for V_i from 0 to 5 V.

   b.   Using the parameters for the transistor, calculate the breakpoints and logic threshold voltages for the inverter and compare with experimental values.

   c.   Keep the output of the inverter at logic low and connect a 1 kΩ potentiometer to the 5 V supply. The variable resistance simulates the load at low output. Vary the potentiometer until the output transistor enters active mode and the output voltage is about 0.2 V. Obtain the load (sinking) current and determine the static logic low fan-out using measured base current and β values. Measure the input current at logic low output.

   d.   Keep the output of the inverter at logic high and connect a 1 kΩ potentiometer to ground. Vary the load resistance until the output voltage drops by 0.4 V. Measure the load (sourcing) current and determine the static logic high fanout using the current measured at logic high input in (c) above.

   e.   Calculate the theoretical fan-out values at low and high outputs and compare with experimental values.

chapter

5

CURRENT-MODE LOGIC FAMILIES

INTRODUCTION

In addition to the saturation logic families, which we considered in Chapter 4, two other bipolar logic families are commercially available: the *emitter-coupled logic* (ECL), and the *integrated injection logic* (IIL or, more commonly, I^2L) families. These families realize logic functions by steering a constant current from one part of a circuit to another. For this reason the two families are called *current-mode logic* (CML) families. Currently, the ECL family has the fastest switching speed among commercially available digital ICs. However, an ECL circuit also requires a relatively large silicon area and dissipates high power; for these reasons, its use was predominantly at the SSI and MSI level until a modified version of ECL, referred to as the *emitter-function logic* (EFL), was developed in 1972. The I^2L family, on the other hand, is a low-power, high-density logic family suitable for LSI and VLSI implementation. At present, both CML families are used in bipolar LSI and VLSI systems.

Although both ECL and I^2L operate by switching currents, the improvement in the delay-power product in each is achieved by a distinct mechanism. The ECL circuit eliminates the turn-off delay of saturated transistors by operating in the active mode. (Recall the Schottky TTL circuits from Chapter 4, in which saturation of a BJT is prevented by the use of a Schottky diode.) Thus the ECL forms a family of nonsaturating logic, unlike the RTL, DTL, and TTL families. To reduce the large power dissipation resulting from the active mode of operation, input logic levels are limited to a narrow voltage range. Currently, EFL cells—modified ECL circuits in LSI and MSI implementations—achieve a delay-power product in the range of 10 to 15 pJ. The I^2L family, on the other hand, operates by saturating one or more input transistors. Power dissipation is significantly reduced by limiting the logic levels to saturation voltages $V_{BE(\text{sat})}$ and $V_{CE(\text{sat})}$. I^2L currently has an average delay-power product of less than 1 pJ per gate in memory and other VLSI circuit implementations.

This chapter presents the analysis of ECL circuits at the gate level, and introduces basic configurations for VLSI realizations. Because of the high switching speed of ECL circuits, there are special considerations for the interconnection of ECL circuits, which are briefly discussed. Next, we consider some simple configurations of the I^2L family and its improved versions, which use

Schottky barriers. The chapter concludes with a discussion of interfacing between saturating and nonsaturating logic families.

5.1 BASIC EMITTER-COUPLED LOGIC (ECL) CIRCUIT

The basic building block of an ECL gate circuit is the emitter-coupled differential amplifier circuit shown in Figure 5.1. An understanding of the linear operation of this circuit in amplifying the difference between the two base voltages,
V_i and V_R, is essential to the analysis of ECL gate circuits. We will assume that
the two transistors Q_1 and Q_2 have identical characteristics, so that for the same
base-emitter voltage $V_{BE(active)}$, they both carry the same terminal currents. This
implies that both transistors have the same reverse saturation currents according
to Equations (3.14)–(3.16). The resistors R_E and R_C are so chosen that for
$V_i = V_R$, both transistors are operating in the active mode and $I_{C1} = I_{C2}$.

Using Equation (3.16) and neglecting the collector-base junction reverse
saturation current I_{CS}, the collector currents in Q_1 and Q_2 are given by

$$I_{C1} \approx \alpha I_{ES} \exp\left(\frac{V_{BE1}}{V_T}\right) \tag{5.1}$$

and

$$I_{C2} \approx \alpha I_{ES} \exp\left(\frac{V_{BE2}}{V_T}\right) \tag{5.2}$$

where α is the forward, common-base current gain, and the forward-bias
voltage V_{BE} for each emitter-base junction is sufficiently large, that is,
$V_{BE} \geq 4V_T$.

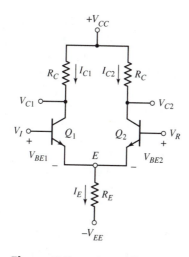

Figure 5.1   Basic differential
amplifier

To express the sum of the collector currents in terms of the total emitter current $I_E = I_{E1} + I_{E2}$, we can write

$$I_{C1} + I_{C2} = \alpha(I_{E1} + I_{E2}) = \alpha I_E \tag{5.3}$$

Next, the difference ΔV between the two base voltages is given by

$$\Delta V = V_{BE1} - V_{BE2} = V_i - V_R \tag{5.4}$$

From Equations (5.1)–(5.4), the collector currents are related to the difference in the input (base) voltages as

$$I_{C1} = \left[\frac{\alpha I_E}{1 + \exp(-\Delta V/V_T)} \right] \tag{5.5}$$

$$I_{C2} = \left[\frac{\alpha I_E}{1 + \exp(\Delta V/V_T)} \right] \tag{5.6}$$

and their ratio is given by

$$\frac{I_{C1}}{I_{C2}} = \exp\left(\frac{V_{BE1} - V_{BE2}}{V_T} \right) = \exp\left(\frac{\Delta V}{V_T} \right) \tag{5.7}$$

Now we consider the three cases of V_i relative to V_R. In the first case, for $V_i = V_R$, $V_{BE1} = V_{BE2}$ and $\Delta V = 0$. Hence, from Equation (5.7), $I_{C1} = I_{C2}$, and the collectors are at the same voltage:

$$V_{C1} = V_{CC} - I_{C1} R_C = V_{C2}$$

This is the quiescent, or operating, state of the circuit.

In the second case, the base voltage of Q_1 is increased slightly above V_R while V_{B2} is held constant at V_R so that $V_i > V_R$. Then, since ΔV is positive, I_{C1} rises and I_{C2} falls according to Equations (5.5) and (5.6). The collector voltage V_{C1}, therefore, drops from its quiescent value while V_{C2} increases.

In the third case, V_i is decreased from V_R by a small amount so that $V_i < V_R$. ΔV is now negative, and hence, we see that the situation is the reverse of the previous case: I_{C2} increases while I_{C1} decreases, resulting in a drop in V_{C2} and a rise in V_{C1}.

Figure 5.2 shows the variations in the collector currents and voltages as a function of ΔV. Observe that for $\Delta V = V_{BE1} - V_{BE2} = V_i - V_R = \pm 120 \ mV$, the collector current ratio is approximately 101 according to Equation (5.7) (using $v_t = 2.6 \ mv$ at room temperature). For this variation in the input voltage V_i, the common emitter voltage V_E changes only by ± 120 mV. Thus, the total emitter current $I_E = I_{E1} + I_{E2} = (V_E - V_{EE})/R_E$ is essentially constant. We observe, therefore, that the hundredfold increase in the collector current of $Q_1(Q_2)$ demanded by the 120 mV rise (fall) in V_i relative to V_R effectively turns $Q_2(Q_1)$ off, and $Q_1 \ (Q_2)$ carries all of I_E. As a result of the change in collector currents, $V_{C1}(V_{C2})$ drops from its quiescent value and $V_{C2} \ (V_{C1})$ rises to V_{CC}. Collector voltage $V_{C2} = V_{CC} - I_{C2} R_C$, and the output voltage $V_{C2} - V_{C1}$ are, therefore, *in phase* with change in V_i while $V_{C1} = V_{CC} - I_{C1} R_C$ and $V_{C1} - V_{C2}$ are *out of*

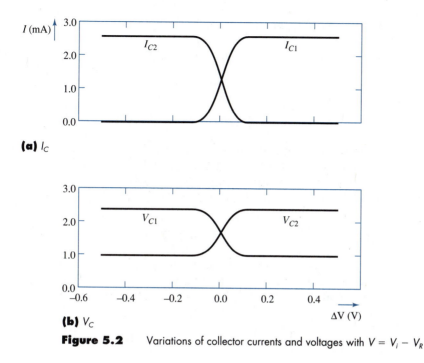

(a) I_C

(b) V_C

Figure 5.2   Variations of collector currents and voltages with $V = V_i - V_R$

phase. Since the output voltage $V_{C1} - V_{C2}$ is proportional to the input voltage difference $V_i - V_R$, the circuit of Figure 5.1 is commonly referred to as the basic *differential amplifier*.

The following example illustrates the changes in output voltage and the common emitter current for changes in the input voltage.

Using the values $R_C = 1$ k Ω, $R_E = 2$ k Ω, $V_R = 1$ V, $V_{CC} = 5$ V $= V_{EE}$, and $\beta = 50$ in Figure 5.1, determine the emitter and collector currents and voltages for $V_i = V_R$, and for $V_i = V_R \pm 120$ mV.

Example 5.1

Solution

In the quiescent state of $V_i = V_R = 1$ V, and using $V_{BE} = 0.7$ V,

$$V_E = V_i - V_{BE1} = 0.3 \text{ V}$$

This gives

$$I_E = \frac{V_E + V_{EE}}{R_E} = 2.65 \text{ mA}$$

which is the quiescent emitter (supply) current.

With matched Q_1 and Q_2,

$$I_{E1} = I_{E2} = \frac{I_E}{2} = 1.33 \text{ mA}$$

and

$$V_{C1} = V_{C2} = V_{CC} - I_{C1}R_C = V_{CC} - I_{E1}\left(\frac{\beta}{\beta + 1}\right)R_C = 3.7 \text{ V}$$

which is the quiescent collector voltage.

When $V_i = V_R + 0.12 = 1.12$ V so that $\Delta V = 120$ mV, the emitter voltage $V_E = 0.42$ V. At this emitter voltage, Q_2 is effectively cut off. Emitter current I_{E1}, therefore, now constitutes the total emitter current I_E given by

$$I_E = I_{E1} = \frac{V_i - V_{BE1} + V_{EE}}{R_E} = 2.71 \text{ mA}$$

Since this current is within about 2 percent of the quiescent value, I_E is essentially constant. We therefore calculate the collector currents using Equations (5.5) and (5.6) (with $V_T = 26$ mV) as

$$I_{C1} = \frac{\alpha I_E}{1 + \exp(-\Delta V/VT)} = \frac{\alpha I_E}{1.01} \approx 2.63 \text{ mA}$$

and

$$I_{C2} = \frac{\alpha I_E}{1 + \exp(\Delta V/VT)} = \frac{\alpha I_E}{102} \approx 26\mu\text{A}$$

Note that I_{C2} is negligibly small, showing that Q_2 is practically off and Q_1 carries the total emitter current I_E. Hence the collector voltages change to

$$V_{C1} = 5 - 2.63 \times 1 = 2.37 \text{ V} \quad \text{and} \quad V_{C2} \approx 5 \text{ V}$$

If $V_i = V_R - 0.12 = 0.88$ V, ΔV becomes -120 mV and Q_1 is turned off. Hence $V_E = V_R - V_{BE2(act)} = 0.3$ V, $I_E = 2.65$ mA (quiescent value) and $I_{C2} \approx \alpha I_E \approx 2.6$ mA.

The collector voltages become $V_{C1} = 5$ V and $V_{C2} \approx 2.4$ V.

The preceding example shows that for a change in input voltage of 120 mV relative to V_R, Q_1 (or Q_2) is turned off while the other transistor carries all of the quiescent total emitter current I_E. Neither transistor, however, saturates while conducting the entire emitter current. Therefore, the switching from cutoff to conduction of Q_1 or Q_2 does not involve any storage delay. This is the principle of fast switching in ECL circuits.

In a logic inverter application of the emitter-coupled differential amplifier of Figure 5.1, V_i is the logic input voltage and V_R is a reference voltage. If the input voltage switches from a low voltage of $V_i = V_R - 120$ mV to a high of $V_i = V_R + 120$ mV, the output (collector) voltage V_{C1} switches from a high of V_{CC} to a low of approximately $V_{CC} - I_E R_C$, causing the inverter action. Therefore, $V_i \leq V_R - 120$ mV may be designated logic low, and $V_i \geq V_R + 120$ mV logic high. By connecting transistors in parallel with Q_1 (as in the case of the RTL gate circuits, Section 4.2), it is possible to realize the logic NOR function using the output V_{C1}. The in-phase output V_{C2} of the same circuit can then be used to obtain logic OR output. For driving other similar ECL gates, however, the output voltage levels must be shifted for compatibility with the low input voltages. Note that the allowable input voltage range of $V_R - 120$ mV $\leq$ $V_I \leq V_R + 120$ mV limits the input noise margins.

From the above discussion of the basic building block of ECL logic gates, we note the following features. Because of the active mode of operation of the differential amplifier, there is no storage delay in switching between on (conducting) and off states of the transistors in the differential pair. Thus, the ECL circuits achieve significant improvement in switching speed over the saturating logic families of Chapter 4. Output logic transition (from the state of Q_1 on, Q_2 off, to Q_1 off, Q_2 on), as the input switches from above V_R to below V_R, causes the total emitter current I_E to switch almost entirely from Q_1 to Q_2. For this reason the differential amplifier operating as a switching circuit is called a *current switch*. As a switching circuit, it is also known as a *voltage comparator* because of its current switching action by the voltage difference $\Delta V = V_i - V_R$. Note that the logic circuit functions by steering a constant current from one side of the differential pair to the other side; for this reason it is known as current-mode logic[1]. Since the current drawn from the negative power supply remains virtually constant—varying by less than 2 percent for a logic state transition—very little surge current occurs in the supply. As a result, the power supply needs to supply essentially constant current, which simplifies its design. Additionally, the absence of current spikes in the supply prevents ringing in the supply lines. This is an important consideration in the interconnection of high-speed switching circuits.

The next two sections consider the IC implementation of currently available ECL gate circuits.

5.2 BASIC INTEGRATED CIRCUIT ECL OR/NOR GATE

The circuit schematic of the earliest ECL OR/NOR gate in the integrated circuit form incorporating the basic differential amplifier of Figure 5.1 is shown in Figure 5.3a. This IC was introduced by Motorola in 1962 as the MECL I series. Transistors Q_3 and Q_4 serve as emitter-follower buffers to shift the collector voltage levels V_{C1} and V_{C2}, so that succeeding (load) ECL gates can be driven between cutoff and active modes. The reference voltage $V_R = V_{BB}$ is selected such that logic low and high levels are symmetric about V_R, and Q_1 and Q_2 operate in the active mode without exceeding their power dissipation ratings.

---

[1] Strictly speaking, current mode logic [Reference 10] does not have the emitter-follower output stage (see the following section). ECL is commonly classified as CML based on the current switching in the differential amplifier section.

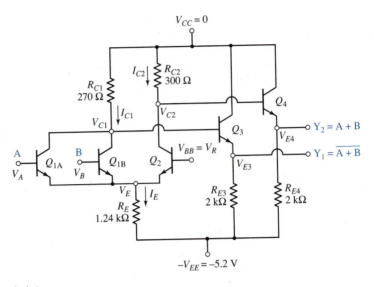

(a) Circuit

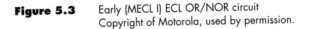

(b) Logic Symbol

Figure 5.3   Early (MECL I) ECL OR/NOR circuit
Copyright of Motorola, used by permission.

The circuit implements two-input OR/NOR logic function. If V_A or V_B is above V_R, Q_{1A} or Q_{1B} conducts and brings the voltage V_{C1} below zero ("low"). Since Q_2 is off, its collector is at $V_{C2} = V_{CC} = 0$ ("high"). When both V_A and V_B are below V_R, Q_{1A} and Q_{1B} are turned off and Q_2 conducts. V_{C1}, therefore, is zero, while V_{C2} becomes negative. Q_3 and Q_4 operate in the active mode and lower V_{C1} and V_{C2} by the base-emitter junction voltage drops. Hence the positive logic outputs at the emitters of Q_3 and Q_4 realize $Y_1 = \overline{A + B}$ and $Y_2 = A + B$. Figure 5.3b shows the circuit symbol for the OR/NOR gate. The following example illustrates the calculation of logic levels.

Example 5.2   For typical supply voltages of $V_{CC} = 0$ and $V_{EE} = -5.2$ V, determine the logic levels and the reference voltage for the circuit shown in Figure 5.3a using $R_{C1} = 270\ \Omega$, $R_{C2} = 300\ \Omega$, $R_E = 1.24$ kΩ, and $R_{E3} = R_{E4} = 2$ kΩ. (In Section 5.2.5 we consider the advantage of grounding the positive supply V_{CC} instead of the negative supply.)

Solution

For $V_A = V_B < V_{BB}$, Q_{1A} and Q_{1B} are off and $V_{C1} = V_{CC} = 0$.
  Neglecting the drop in R_{C1} due to the base current of Q_3, then, the output voltage at the emitter of Q_3 is

$$V_{Y1} = V_{E3} = V_{C1} - V_{BE3(active)} = -0.7 \text{ V}$$

Hence, the logic high output voltage is $V_{OH} = -0.7$ V.
  If V_A (or V_B) is at this logic voltage of -0.7 V, Q_{1A} (or Q_{1B}) conducts and brings V_E to

$$V_E = V_A - V_{BE3(active)} = -1.4 \text{ V}.$$

Hence,

$$I_E = \frac{V_E + V_{EE}}{R_E} = \frac{-1.4 + 5.2}{1.24} = 3.06 \text{ mA}$$

With Q_2 off, and neglecting the base current of Q_{1A} (or Q_{1B}), $I_{C1} \approx I_E = 3.06$ mA, and

$$V_{C1} = -I_{C1} R_{C1} = -3.06 \times 0.27 = -0.83 \text{ V}$$

The logic low output voltage, therefore, is

$$V_{OL} = V_{E3} = V_{C1} - V_{BE3(active)} = -1.53 \text{ V}$$

where, again, we neglect the drop in R_{C1} due to the base current of Q_3.
  Since we obtained the logic low and high output voltages using input voltages that are below and above V_{BB}, we set the reference voltage at the midpoint of the two logic levels as

$$V_R = V_{BB} = \frac{V_{OH} + V_{OL}}{2} = -1.11 \text{ V}$$

If succeeding gates have the same reference voltage of -1.11 V, then the outputs of the present gate can be connected directly to their inputs.

  Besides serving to shift the voltage levels at the collectors of Q_1 and Q_2, the two emitter follower circuits—consisting of Q_3 and R_{E3}, and Q_4 and R_{E4}—provide current driving capability for loads. With large current gain β they have low output resistance ($\approx [r_\pi + R_C]/\beta$) and high input resistance ($\approx \beta R_L$, where, typically, the load resistance $R_L \ll R_{E3}, R_{E4}$). Hence Q_3 and Q_4 can drive high-current loads while isolating the collectors C_1 and C_2 from the loads.
  Using the reference voltage and the output logic voltages we have calculated, we next determine the transfer characteristics and the noise margins of the MECL I series OR/NOR gate.

5.2.1   VOLTAGE TRANSFER CHARACTERISTICS

Voltage transfer characteristics for the ECL gate of Figure 5.3 are obtained between the input voltage $V_i = V_A$ (or V_B), and the two output voltages V_{Y1} and V_{Y2}, assuming all the transistors have high current gain β. A β of 100 to 400, which is typical in IC transistors, allows us to neglect the base current in the anlysis. Further, we assume the base-emitter junction voltage to be constant at 0.7 V throughout the active mode of operation.

For $V_i = V_A = V_B \leq V_{OL} = -1.53$ V and $V_R = -1.11$ V, Q_{1A} and Q_{1B} are off and Q_2 is on. Hence,

$$V_E = V_R - V_{BE2(\text{active})} = -1.81 \text{ V}$$

and

$$I_E = \frac{V_E + V_{EE}}{R_E} = \frac{-1.81 + 5.2}{1.24} = 2.73 \text{ mA}$$

Since $I_{C1} \approx 0$, and $I_{C2} \approx I_E$,

$$V_{C1} = V_{CC} = 0$$

and

$$V_{C2} = V_{CC} - I_{C2} R_{C2} = 0 - 2.73 \times 0.3 = -0.82 \text{ V}$$

Hence,

$$V_{Y1} = V_{C1} - V_{BE3(\text{active})} = -0.7 \text{ V} = V_{OH}$$

$$V_{Y2} = V_{C2} - V_{BE4(\text{active})} = -1.52 \text{ V} = V_{OL}$$

As V_i is increased from V_{OL}, output voltage V_{Y1} and V_{Y2} remain at their logic levels until either V_A or V_B reaches $V_R - 0.12 = -1.23$ V.

If V_A is kept constant at V_{OL}, then, for $V_i = V_B > -1.23$ V, Q_{1B} conducts while Q_{1A} stays in cutoff. I_{C1} increases from zero and I_{C2} decreases, according to Equations (5.5) and (5.6). In the interval -1.23 V $\leq V_i \leq -0.99$ V, therefore, V_{C1} becomes more negative and V_{C2} rises toward zero. Consequently, V_{Y1} drops toward V_{OL} and V_{Y2} increases toward V_{OH} as shown in Figure 5.4a.

For $V_i = V_B = -0.99$ V $> V_R$, Q_2 is essentially off. Hence, $V_{Y2} = -0.7$ V. Since Q_{1B}, is on, $V_E = -1.69$ V, and

$$I_E \approx I_{C1} = \frac{V_E + V_{EE}}{R_E} = \frac{-1.69 + 5.2}{1.24} = 2.83 \text{ mA}$$

Therefore,

$$V_{Y1} = -I_{C1} R_{C1} - 0.7 = -2.83 \times 0.27 - 0.7 = -1.46 \text{ V}$$

For V_i above -0.99 V, current changes very little from its already negligible value in Q_2; Q_{1B}, however, conducts more current since V_E rises with $V_i = V_B$. Therefore, V_{C1} and hence V_{Y1} decrease with a slope of R_{C1}/R_E, while V_{Y2} stays at logic high level.

Note the symmetry in the transfer characteristics of Figure 5.4a except for $V_i > V_R + 120 = -0.99$ V. If V_i is increased much higher than -0.7 V, collec-

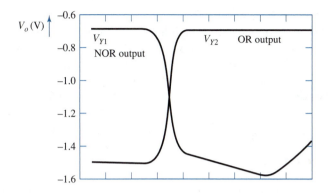

(a) Voltage Transfer Characteristics

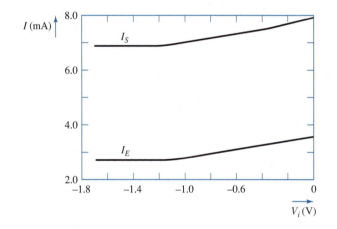

(b) Power Supply and Total Emitter Currents

Figure 5.4   Transfer characteristics of the OR/NOR circuit of Figure 5.3

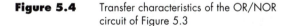

tor current I_{C1} continues to increase and V_{C1} steadily drops towards V_E. Eventually Q_{1B} saturates, which corresponds to the knee of the NOR output in Figure 5.4a. Thus the symmetry in the circuit configuration is not entirely reflected in the voltage transfer characteristics. Problem 5.2 considers the saturation of Q_{1B} and the calculation of the corresponding V_i and V_{Y1}. Since the collector-base junction of Q_{1B} is forward-biased in saturation, an increase in V_i beyond the saturation of Q_{1B} causes V_{C1} to rise as

$$V_{C1} = V_i - V_{BC1} = V_i - 0.7 \text{ V}$$

Hence the NOR output voltage increases with unity slope:

$$V_{Y1} = V_i - 1.4 \text{ V}$$

Since the voltage V_i required to saturate any of the input transistors is much higher than the permissible logic high level, no transistor is driven to saturation during the normal mode of operation. Figure 5.4b shows the emitter current I_E and the power supply current I_S as a function of the input voltage V_i.

Noise Margin   From the piecewise linear transfer characteristics of Figure 5.4a, we may approximate the input logic levels as $V_R \pm 0.12$, or

$$V_{IL} = -1.23 \text{ V} \quad \text{and} \quad V_{IH} = -0.99$$

The unity slope criterion, $dV_o/dV_i = 1$, for the OR output gives $V_{IL} = -1.198$ V, $V_{OL} = -1.493$ V, $V_{IH} = -1.022$ V, and $V_{OH} = -0.727$ V (see Problem 5.4).

Using the approximate values of $V_{OL} = -1.52$ V and $V_{OH} = -0.7$ V, the noise margins are

$$NM_L = V_{IL} - V_{OL} = -1.23 + 1.52 = 0.29 \text{ V}$$

$$NM_H = V_{OH} - V_{IH} = -0.7 + 0.99 = 0.29 \text{ V}$$

The input transition width is 0.24 V, since

$$V_{IL} = V_R - 0.12 \quad \text{and} \quad V_{IH} = V_R + 0.12 \quad \text{and} \quad V_{tw} = V_{IH} - V_{IL}$$

The logic swing V_{ls} is given by

$$V_{ls} = V_{OH} - V_{OL} = 0.82 \text{ V}$$

These values are extremely small compared to the corresponding TTL voltages. However, the noise margins are the same for low and high levels by the choice of the reference voltage as $V_R = (V_{OH} + V_{OL})/2$. The low values for the logic swing and the corresponding noise margins demand strict adherence to the logic levels at the input. Thus, care must be exercised in interconnecting ECL circuits to minimize noise overriding logic signals. Additionally, since the logic levels are strongly dependent on the reference voltage, variation in V_R due to temperature must be reduced. We consider temperature stabilization of the reference voltage in Section 5.3, under modified ECL circuit configurations. Sections 5.2.5 and 5.5 describe noise minimization by the use of negative supply and proper load termination.

The advantage of the low logic swing is that the load (and stray) capacitance needs to charge or discharge to the small voltage of $V_{ls} = 0.82$ V. With the low output resistance of the emitter followers, this results in faster switching between V_{OH} and V_{OL}.

5.2.2   PROPAGATION DELAY

With all the transistors operating in the active mode, storage delay in an ECL circuit is eliminated. Hence, the average propagation delay t_P of an unloaded circuit such as the MECL I OR/NOR gate of Figure 5.3 is on the order of 4 ns. This delay is primarily caused by the everpresent parasitic capacitance at the output terminals, which must be charged and discharged. Detailed analysis that includes the junction capacitances of each of the transistors and the capacitances at the collectors to ground of the differential pair can be performed to determine the propagation delays exactly [Reference 12]. Since load and wiring capacitances are usually predominant over junction capacitances, we can obtain a good approximation by using the lumped load capacitance C_L.

In general, the high-to-low propagation delay in an ECL gate is slightly more than the low-to-high delay. As we saw earlier, charging of C_L from V_{OL} to V_{OH} via the output resistance of $R_o \approx [r_\pi + R_C]/\beta$ is rapidly completed by the active pull-up of the emitter follower transistor. For this reason, the low-to-high propagation delay t_{PLH} is only on the order of a few nanoseconds. When the input changes state, however, the load capacitance C_L cannot discharge from V_{OH} to V_{OL} instantaneously. Hence, the emitter of the output transistor momentarily stays at V_{OH} while the base voltage has decreased from its previous value of zero. This causes the emitter follower to cut off, resulting in the discharge of C_L via the relatively large emitter circuit resistance R_E. Thus, the high-to-low transition occurs using the passive pull-down resistor R_E. Since there is only the small voltage of $V_{ls} = 0.82$ V to discharge, however, high-to-low delay t_{PHL} is also small. Of course, both t_{PHL} and t_{PLH} depend on the amount of load capacitance C_L. The following example illustrates the approximate calculation of the two delays.

Determine the approximate switching delays of the ECL circuit of Figure 5.3 based on charging and discharging time-constants for a 10 pF load capacitance. Assume $\beta = 100$ and $r_\pi = 1000\ \Omega$.

Example 5.3

Solution

High-to-low transition: With Q_3 (or Q_4) momentarily off, load capacitance C_L charges from an initial voltage of -0.7 V toward $V_{EE} = -5.2$ V with a time-constant of $\tau = R_E C_L = (2k\Omega) \times (10pF) = 20$ ns. Hence,

$$V_o(t) = 4.5 \exp\left(\frac{-t}{20\ \text{ns}}\right) - 5.2$$

At $t = t_{PHL}$, V_o reaches the logic low level of -1.52 V.

Therefore, $t_{PHL} = 20 \ln\left(\frac{4.5}{3.68}\right) = 4$ ns.

Low-to-high transition: The emitter follower is now in the active mode with an output resistance of $R_o \approx (r_\pi + R_C)/\beta \approx 13\ \Omega$. Thus, C_L discharges from -1.52 V toward -0.7 V with $\tau = (13\Omega) \times (10pF) = 130$ ps as $V_o(t)$ rises

$$V_o(t) = -0.82 \exp\left(\frac{-t}{130\ \text{ps}}\right) - 0.7$$

Theoretically, V_o reaches $V_{OH} = -0.7$ V at $t = \infty$. In practice, however, we may consider $V_o = -0.75$ V, for example, close to V_{OH}. Therefore, the low-to-high delay time t_{PLH} needed for C_L to reach $V_o = -0.75$ V is

$$t_{PLH} \approx 130 ln\left(\frac{0.82}{0.05}\right) ps \approx 0.36\ \text{ns}$$

The above value of t_{PLH} is a rather coarse approximation which is obtained by assuming linear charging of C_L throughout the interval. Initially, however, the emitter-follower transistor conducts heavily due to the large base-emitter junction voltage, and C_L charges rapidly; the emitter current tapers off as the output voltage rises. Because of this nonlinear behavior, the rise time is usually approximated by $t_r \approx 2.2\tau = 0.286$ ns (Reference 12).

Figure 5.5 shows the MicroSim PSpice simulation results for the transient response of Example 5.3. The response at the NOR output using only the charging and discharging of C_L in the MicroSim PSpice model is shown in Figure 5.5b. Figure 5.5c shows the simulated response using a more accurate and realistic model for the transistors, in which the junction capacitances and ohmic resistances of the transistors are included. The total equivalent capacitance from the base of the emitter follower to ground, which includes substrate and junction capacitances, contributes to increased delay t_{PLH}. With a large time-constant $\tau = R_E C_L$ during high-to-low transition, however, the effect of collector capacitance is negligible on t_{PHL}. We discuss the normal loading of ECL gates in Section 5.2.4 on fan-out.

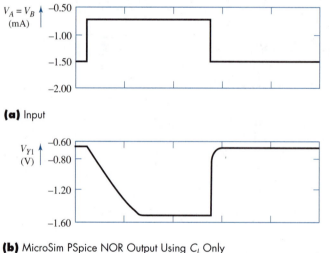

(a) Input

(b) MicroSim PSpice NOR Output Using C_L Only

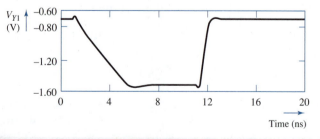

(c) MicroSim PSpice NOR Output Using C_L and Junction Capacitances

Figure 5.5   Transient response of the circuit of Figure 5.3

5.2.3   POWER DISSIPATION

Power dissipation per gate in the ECL circuits is much higher than in the saturating logic families we considered in Chapter 4. This is to be expected, since all the transistors are operating in the active mode, resulting in almost constant high supply currents. The following example demonstrates the calculation of power dissipation.

For the MECL I circuit of Figure 5.3, calculate the static power required at logic high and low levels. | **Example 5.4**

Solution

The power supply current I_S in either logic state is given by

$$I_S = I_E + I_{E3} + I_{E4} \tag{5.8}$$

where I_{E3} and I_{E4} are the currents in the output emitter followers. Since one of the outputs is at logic high while the other is at logic low, the sum of the emitter currents, $I_{E3} + I_{E4}$, is a constant. We have seen earlier that the total emitter current remains essentially constant in going from logic low to high (see Figure 5.4b). Hence, we can determine the power dissipation using the currents calculated at either logic state.

For a logic high input of -0.7 V at either A or B, Q_2 is off and, as calculated in Example 5.2,

$$I_E = 3.06 \text{ mA}$$

With $V_{E3} = V_{Y1} = -1.53$ V,

$$I_{E3} = \frac{V_{E3} + V_{EE}}{R_{E3}} = \frac{-1.53 + 5.2}{2} = 1.84 \text{ mA}$$

Since $V_{E4} = V_{Y2} = -0.7$ V,

$$I_{E4} = \frac{V_{E4} + V_{EE}}{R_4} = \frac{-0.7 + 5.2}{2} = 2.25 \text{ mA}$$

Thus,

$$I_S = 3.06 + 1.84 + 2.25 = 7.15 \text{ mA}$$

and

$$P_D = I_S V_{EE} = 37.2 \text{ mW}$$

Note that the supply current and the power are slightly less if the logic high input goes to the minimum of $V_{IH} = -0.99$ V. At V_A or $V_B = -0.9$ V, for example, $I_S = 7.01$ mA and $P_D = 36.5$ mW. For both inputs at logic low, $I_S = 6.82$ mA and $P_D = 35.5$ mW, which are the lowest possible values. The average total power dissipation, therefore, is $(P_{D(\min)} + P_{D(\max)})/2 = (35.5 + 37.2)/2 = 36.4$ mW per gate. This gives a delay–power product of approximately 145 pJ at an average propagation delay of 4 ns.

A separate circuit for obtaining a constant reference voltage is available for use with the MECL I series gates. Later ECL versions incorporate the reference circuitry within the same chip that contains the logic circuitry (Section 5.3). In either case, one must consider power dissipation of the reference circuitry when determining total gate dissipation. However, with more than one gate per chip and all gates sharing the same reference voltage source, the overall dissipation per gate increases only marginally.

From Figure 5.4b and the power calculations in Example 5.4 we see that an unloaded ECL gate has a relatively constant supply current and gate power dissipation, independent of logic state. We also note that with the active mode of operation the supply current remains constant during logic state transition, since the emitter current I_E is simply diverted from one side to the other of the differential pair. The absence of supply current transients, unlike in a TTL circuit, eliminates supply voltage spikes. Hence the supply for ECL circuits need not be bypassed as frequently as in TTL circuits.

We further observe that the MECL I circuit dissipates considerably more power at 36 mW than the standard TTL circuit. With a significantly faster switching response, however, the delay-power product of the circuit at 145pJ is approximately the same as that of the standard TTL circuit.

5.2.4   FAN-OUT

The static, or dc, fan-out based on a small voltage change at the NOR or OR output of an ECL gate is quite high. This is to be expected, since the base current of the high-β input transistors operating in the active mode is negligibly small. Of greater importance is the output circuitry, which can drive large currents with its active pull-up emitter follower. With the input transistors in cutoff at logic low, we determine the gate fan-out by the allowable voltage drop at logic high. (This is in contrast with TTL fan-out, which we determine by the change in logic low output voltage.) Static fan-out is calculated in the following example.

Example 5.5 | Calculate the static fan-out for the circuit of Figure 5.3. Assume that the NOR output voltage can drop from its nominal (high) value of -0.7 V by $\Delta V = 50$ mV. Use $\beta = 100$.

Solution

Figure 5.6 shows the NOR (Y_1) output of Figure 5.3 connected to load gates. When Y_1 is at logic low, the input transistors of all the N load gates are off. Hence the load current I_L is negligibly small, resulting in virtually unlimited fan-out.

When Y_1 is at logic high of -0.7 V, the input (load) transistors conduct, and the sum of the base currents $I_L = NI_{BL}$ causes the voltage V_{E3} to drop. This drop is given for this example as $\Delta V = 50$ mV. (This is similar to determining the logic high fan-out of a TTL gate; see section 4.4.5.) Note that at logic high output, transistors Q_{1A} and Q_{1B} are off and V_{C1} is zero; hence, V_{E3} cannot drop very much below -0.7 V in active mode. Therefore, ΔV is the drop across R_{C1}

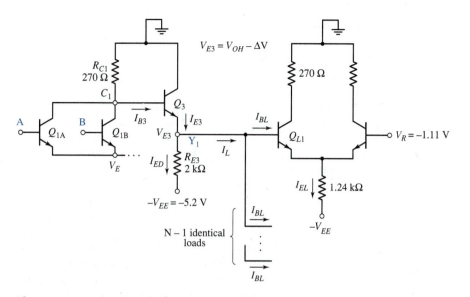

Figure 5.6   NOR output of an MECL I gate driving load gates

due to the base current I_{B3} in Q_3. Clearly, then, the allowable voltage ΔV depends on the current gain β of Q_3.

Referring to Figure 5.6, the base current of each load in the active mode is given by $I_{BL} = \dfrac{I_{EL}}{\beta + 1}$. Thus,

$$I_L = NI_{BL} = \frac{NI_{EL}}{\beta + 1} \tag{5.9}$$

where

$$I_{EL} = \frac{-0.7 - \Delta V - 0.7 + 5.2}{1.24} = \frac{3.8 - \Delta V}{1.24} \tag{5.10}$$

The emitter current of the driver at $V_{E3} = -0.7 - \Delta V$ is

$$I_{E3} = I_L + I_{ED} = \frac{NI_{EL}}{\beta + 1} + \frac{-0.7 - \Delta V + 5.2}{2}$$

or

$$I_{E3} = \frac{N(3.8 - \Delta V)}{1.24(\beta + 1)} + \frac{(4.5 - \Delta V)}{2} \tag{5.11}$$

From the base circuit of Q_3,

$$I_{E3} = (\beta + 1)I_{B3} = \frac{(\beta + 1)\,\Delta V}{R_{C1}} = \frac{(\beta + 1)\,\Delta V}{0.27} \tag{5.12}$$

Combining Equation (5.11) and (5.12) and using $\beta = 100$ and $\Delta V = 50$ mV, we obtain

$$N = 550$$

which is the gate fan-out. For $\beta = 50$, the gate fan-out reduces to $N = 121$.

Note that with $\Delta V = 50$ mV, which is the drop in V_{Y1}, there is still some noise margin available at logic high.

From the foregoing example it is clear that the fan-out for dc or signals at low switching rate is considerably high. In contrast, for signals switching at high speed, parasitic capacitances at the load terminals severely degrade the output waveform as we saw in Section 5.2.2. An MECL I gate typically presents a capacitance of approximately 3 pF at its input. Therefore, to avoid excessive distortion and delay in the output waveform, the useful dynamic fan-out is typically limited to between 10 and 15. Such a small fan-out is a limitation compared with the saturating logic families (Chapter 4) or the MOS logic families (Chapter 7). With *proper termination* (see Section 5.5), however, ECL circuits are capable of driving transmission line loads without loss of switching speed in terms of propagation delay and rise and fall times. Also, with complementary outputs available at the two emitter followers, signal transmission over long distances can be accomplished without additional line driver or receiver circuitry.

5.2.5 ADVANTAGE OF THE NEGATIVE SUPPLY VOLTAGE

It is essential to minimize the amount of noise being transmitted to the output terminals of ECL circuits in order to ensure logic operation within the small logic swing. To see how the choice to ground the positive terminal of the power supply helps in noise reduction, consider the simple ECL circuit of Figure 5.7a where both terminals of the supply are left floating. The noise voltage in the supply V_{SS} (resulting from inadequate filtering and from switching transients caused by other circuitry, for example) is represented by the series source v_n. The two output voltages v_{o1} and v_{o2} are shown with reference to the positive and the negative terminals of the power supply.

For $V_i < V_R$, Q_1 is off and the circuit is simplified as shown in Figure 5.7b. From the mesh that includes the base-emitter junction, the small-signal voltage equation is

$$v_n = i_{b3}[R_{C1} + r_\pi + (\beta + 1)R_{E3}] \tag{5.13}$$

Solving for the base current i_{b3},

$$i_{b3} = \frac{v_n}{R_{C1} + r_\pi + (\beta + 1)R_{E3}} \tag{5.14}$$

Therefore, v_{o1}, which is the change in the output voltage V_{o1} due to V_n, is given by

$$v_{o1} = (R_{C1} + r_\pi)i_{b3} = \frac{v_n(R_{C1} + r_\pi)}{R_{C1} + r_\pi + (\beta + 1)R_{E3}} \tag{5.15a}$$

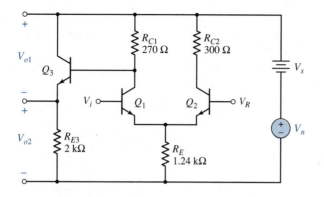

(a) ECL Circuit with Floating Supply

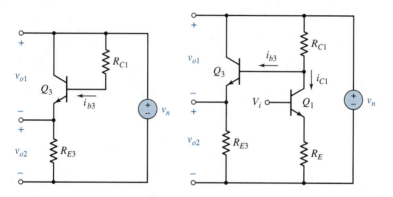

(b) Circuit with Q_1 Off $(V_i < V_R)$ **(c)** Circuit with Q_1 on $(V_i > V_R)$
Figure 5.7   Power supply noise reduction

For $\beta \gg 10$ and $R_{E3} > R_{C1}$, v_{o1} may be approximated as

$$v_{o1} \approx \frac{v_n(R_{C1} + r_\pi)}{(\beta + 1)R_{E3}}$$

(5.15b)

Similarly, v_{o2} is given by

$$v_{o2} = R_{E3}i_{e3} = \frac{v_n(\beta + 1)R_{E3}}{R_{C1} + r_\pi + (\beta + 1)R_{E3}}$$

(5.16a)

which may be approximated as

$$v_{o2} \approx v_n$$

(5.16b)

For the values of $\beta = 100$, $r_\pi = 100\ \Omega$, $R_{E3} = 2\ \text{k}\Omega$, and $R_{C1} = 270\ \Omega$ in the circuit of Figure 5.3, Equations (5.15a) and (5.16b) give

$$v_{o1} \approx 0.0063v_n \quad \text{and} \quad v_{o2} \approx 0.9938v_n$$

which justify our approximations.

Evidently, at logic low input, v_{o1} has negligible amount of power supply noise voltage.

For $V_i > V_R$, Q_2 is off and the output circuit is as shown in Figure 5.7c. Neglecting the small change in V_{CE1}, we write

$$[R_{C1} + r_\pi + (\beta + 1)R_{E3}]i_{b3} + R_{C1}i_{c1} = v_n$$

and

$$R_{C1}i_{b3} + [R_{C1} + R_E]i_{c1} = v_n$$

Solving for i_{b3}, we get

$$i_{b3} = \frac{v_n R_E}{R_{C1}R_E + (R_{C1} + R_E)[(\beta + 1)R_{E3} + r_\pi]} \tag{5.17}$$

as the small-signal current due to v_n.

Therefore,

$$v_{o1} \approx \frac{v_n R_{C1}[R_E + r_\pi + (\beta + 1)R_{E3}]}{[(\beta + 1)R_{E3} + r_\pi](R_{C1} + R_E) + R_{C1}R_E} \tag{5.18}$$

and

$$v_{o2} \approx \frac{v_n(\beta + 1)R_{E3}R_E}{[(\beta + 1)R_{E3} + r_\pi](R_{C1} + R_E) + R_{C1}R_E} \tag{5.19}$$

Again, for $\beta \gg 10$, these equations may be further approximated by

$$v_{o1} \approx \frac{v_n R_{C1}}{(R_{C1} + R_E)}$$

and

$$v_{o2} = \frac{v_n R_E}{(R_{C1} + R_E)}$$

Using the circuit parameters of Figure 5.3, Equations (5.18) and (5.19) result in

$$v_{o1} \approx 0.18v_n \quad \text{and} \quad v_{o2} \approx 0.82v_n$$

Again we see that the supply noise appears more in v_{o2} than in v_{o1}.

It is evident from the above results that using the positive terminal of the power supply as reference attenuates noise voltage significantly at the output.

5.3   MODIFIED ECL CIRCUIT CONFIGURATIONS

A disadvantage of the basic ECL circuit of Figure 5.3 is the requirement for a second power supply for the reference voltage V_R. To avoid the additional supply source, the MECL II family derives the reference voltage internally from the V_{EE} supply as shown in Figure 5.8.

As with all transistors in the ECL family, the reference transistor Q_R is operating in the active mode. If we assume the current gain β of Q_R is high

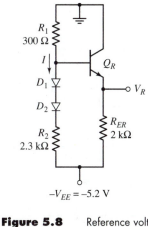

Figure 5.8   Reference voltage derived from V_{EE} supply

enough to justify neglecting the base current in the circuit of Figure 5.8, current I is given by

$$I = \frac{V_{EE} - V_{D1} - V_{D2}}{(R_1 + R_2)} \qquad \textbf{(5.20)}$$

The base voltage $V_{BR} = -IR_1$ is

$$V_{BR} = -\frac{(V_{EE} - V_{D1} - V_{D2})R_1}{(R_1 + R_2)} \qquad \textbf{(5.21)}$$

Using a diode drop of 0.7 V for V_{D1} and V_{D2}, we get, for the given resistances,

$$I = 1.46 \text{ mA} \quad \text{and} \quad V_{BR} = -0.438 \ V$$

With 0.7 V for the active mode base-emitter junction of Q_R, therefore, the reference voltage available at the emitter is

$$V_R = V_{ER} = -1.14 \ V$$

Note that with this voltage at the emitter, $V_{CE} = 1.14$ V, which indeed keeps Q_R in the active mode. With approximately the same current in the emitter as in the diodes, the forward junction voltages are identical. Variations of the junction voltages with temperature are also the same.

Several gates in the MECL II series of logic circuits share the same reference source within a chip. This is possible because the emitter follower in active mode can drive a large current without dropping its voltage. Recall further that the base current of the high-β transistors is negligible. Additionally, it takes a relatively small silicon area and few extra steps to obtain a transistor in integrated circuits compared to a pair of resistors with a precise ratio. The two

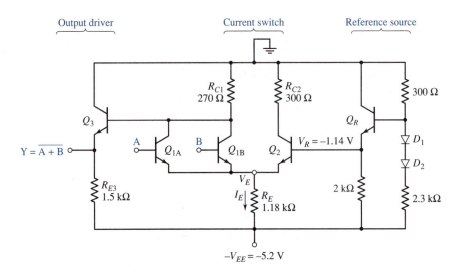

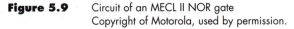

Figure 5.9   Circuit of an MECL II NOR gate
Copyright of Motorola, used by permission.

diodes are used for temperature compensation of the reference supply, as discussed later in this section.

Figure 5.9 shows the complete circuit of a typical MECL II NOR gate. Using $V_R = -1.14$ V as determined earlier, when $V_A = V_B < V_R$, Q_{1A} and Q_{1B} are off and the logic high voltage at the emitter of Q_3 is

$$V_3 = V_Y = -0.7 \text{ V} = V_{OH}$$

For V_A or V_B at the logic high level of -0.7 V,

$$V_E = -1.4 \text{ V} \quad \text{and} \quad I_E = 3.22 \text{ mA}$$

If we assume a minimum β of 50, the above values represent a logic high input current I_{IH} of $3.22/51 \approx 63$ μA. The voltage at the base of Q_3 is $V_{B3} = V_{C1} = -I_E R_{C1} = -0.93$ V. Hence, the logic low output voltage is

$$V_3 = V_Y = V_{B3} - 0.7 = 1.63 \text{ V} = V_{OL}$$

The Motorola MC1010 circuit of the MECL II series is a quad two-input NOR gate, with all four gates sharing the same reference source shown in Figure 5.9. With 14 pins in the chip package, only the NOR output of each gate is available externally. In calculating the supply current for each gate, therefore, one must consider only one-fourth of the reference circuit current.

Note that with no base resistors connected to the input terminals, unused inputs must be tied to the lowest potential in the circuit—usually the V_{EE} supply—to prevent noise pickup and extraneous switching.

Temperature Stability of the Reference Voltage   As with power supply noise, variation of the transfer characteristics with temperature must be kept to a minimum to ensure safe operation over a range of temperature. Since the noise margin and the logic swing depend on the reference voltage, changes in V_R due to temperature can be used effectively to achieve compensated overall circuit behavior. The reference supply circuit of Figure 5.8 ensures close symmetry in logic levels and stable transfer characteristics over a wide range of temperature, as explained below.

Figure 5.10 shows the circuit for determining changes in output logic levels due to temperature variation. In this figure, variation in each of the forward-biased pn junction voltages is denoted by δV V/°C. Note that the voltage source δV represents the junction voltage *dropping* with temperature by approximately 2 mV/°C.

Consider first the OR output voltage V_{OR}. When V_i is at logic high, Q_2 is off; therefore V_{OR} changes by δV, which is the base-emitter junction voltage of Q_4. Denoting the change in voltage by ΔV, we have

$$\Delta V_{OR} = \Delta V_{OH} = \delta V \tag{5.22}$$

In this case, variation in the reference voltage does not affect the OR output voltage.

For $V_i < V_R$, change in the reference voltage, ΔV_R, affects the collector current of Q_2 and, consequently, the output voltage. The reference voltage changes by

$$\Delta V_R = \Delta V_{B5} - \Delta V_{BE5} = \frac{-2\delta V R_1}{R_1 + R_2} + \delta V \tag{5.23}$$

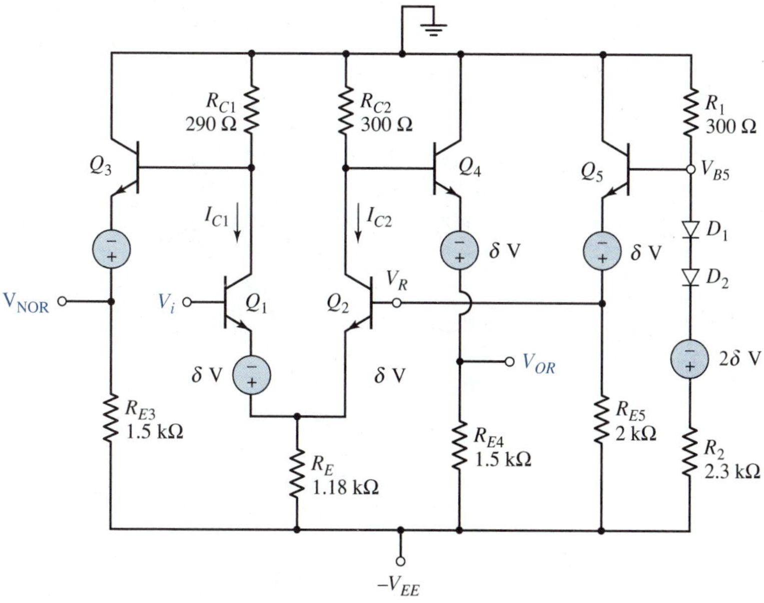

Figure 5.10   Circuit for determining changes in output logic voltages due to temperature variation

using superposition of the δV sources. Note also that the derivation of Equation (5.23) neglects the resistance changes in the diodes and assumes unity gain for the emitter follower.

For the resistor values shown in Figure 5.10, Equation (5.23) gives

$$\Delta V_R = 0.77 \; \delta V$$

Using $\delta V = 2$ mV/°C, this gives an *increase* of 1.54 mV/°C for the reference voltage from its value at room temperature.

The change in the collector current of Q_2 is

$$\Delta I_{C2} = \frac{\Delta V_R + \delta V}{R_E} = \frac{2 \; \delta V R_2}{R_E(R_1 + R_2)} \tag{5.24}$$

Therefore, the OR output at logic low changes by

$$\Delta V_{OR} = \Delta V_{OL} = \Delta I_{C2} R_{C2} + \delta V = \frac{2 \; \delta V R_2 R_{C2}}{[R_E(R_1 + R_2)]} + \delta V \tag{5.25}$$

Using the given resistor values we get

$$\Delta V_{OR} = \Delta V_{OL} = 0.55 \; \delta V \tag{5.26}$$

From Equations (5.22) and (5.26) it is clear that the changes in the OR output voltage with temperature are approximately symmetrical.

For the NOR output, in a similar manner, it is readily shown that the changes in voltage are

$$\Delta V_{OH} = \delta V \tag{5.27}$$

and

$$\Delta V_{OL} = \delta V(1 - R_{C1}/R_E) = 0.73 \; \delta V \tag{5.28}$$

We see from Equations (5.27) and (5.28) that with temperature the NOR output voltage also changes symmetrically from its nominal values.

Comparing the average change in the OR logic voltage of $0.775 \; \delta V$, which is the midpoint of logic swing, with the change in the reference voltage of $\Delta V_R = 0.77 \; \delta V$, it is clear that the variations in the output voltages are approximately symmetrical about the reference voltage. This is also true with the NOR output whose midpoint of logic swing is $0.87 \; \delta V$. The noise immunities for logic low and high levels are therefore the same over a range of temperatures. Figure 5.11

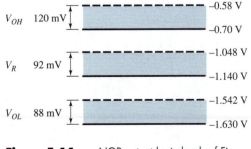

Figure 5.11   NOR output logic levels of Figure 5.9 at 25°C and at 85°C.

shows the logic levels for the NOR output of Figure 5.10 at room temperature and at 85°C.

An additional advantage of using the active circuit for reference voltage generation, as seen from the above equations, is that the voltage variations due to temperature depend on the ratio of resistors instead of their absolute values. In integrated circuits it is relatively simpler to obtain resistors with a given ratio than with given values.

Further developments in achieving higher speed in ECL integrated circuits are discussed next.

5.3.1 THE MECL III AND 10K/10KH FAMILIES OF GATES

Figure 5.12 shows the basic circuit of the MECL III family of logic gates introduced by Motorola in 1968. Two significant differences are evident in this circuit compared with the MECL II family we considered in Section 5.2. The first is the inclusion of the pull-down resistors from each input to V_{EE} supply, and the second is the absence of the passive pull-down resistors in the emitter followers. Open emitter output and the large (50 kΩ) resistor at the base of an input permit transmission line interconnections, which are necessitated by the circuit's high *edge speed* of typically 1 ns.[2] The base pull-down resistor provides a path for the base leakage current to V_{EE} supply if the input is left floating.

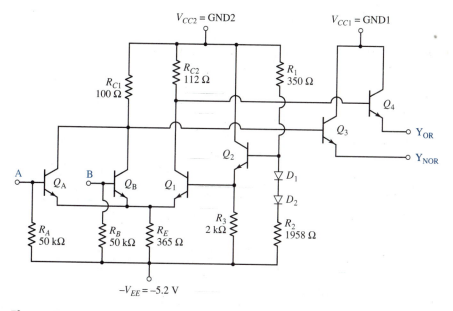

Figure 5.12   Circuit of a basic MECL III OR/NOR gate
Copyright of Motorola, used by permission.

---

[2] Edge speed is the time interval between 20 percent and 80 percent of voltage levels. This is also sometimes referred to as the rise or fall time in manufacturers' data books.

Therefore, an unused input can be safely left open, which may be considered at logic low ($\approx -V_{EE}$).

The MECL III circuit, with a propagation delay and edge speed of 1 ns, is the fastest of the earlier ECL families. Because of its high speed, however, it requires special wiring considerations and its applications are restricted to very high speed communication and computer systems. A slower version of the same configuration, which is easier to use at a slightly increased propagation delay of 2 ns, was introduced as the MECL 10K series by Motorola in 1971. As Figure 5.13 shows, the 10K series basic gate circuit is identical to the MECL III series in circuit configuration, but has larger resistance values. More than the resistor values, however, the differences in transistor geometries, which are not evident from the circuits, contribute to the difference in switching speed. We shall focus our attention on the analysis of the circuit for the 10K series gate.

Logic Levels   Because of the high switching speed requirement, transistors and diodes in the high-speed ECL circuits have smaller dimensions and hence low saturation current I_{ES}. As a result, at high operating currents the forward junction voltages are higher than 0.7 V. At $I_E = 4$ mA with $I_{ES} = 10^{-15}$ A, for example $V_{BE} \approx 0.754$ V. Hence, we will use a forward-biased junction voltage of 0.75 V instead of 0.7 V in the calculations.

From Figure 5.13 we obtain the following voltages using $V_{EE} = -5.2$ V. The reference voltage V_R is given by

$$V_R = V_{B2} - V_{BE2} = \frac{-(5.2 - 1.5)0.907}{4.98 + 0.907} - 0.75 \approx -1.32 \text{ V}$$

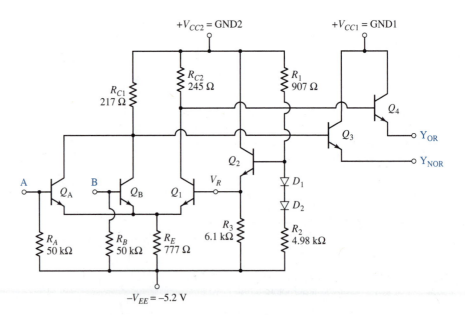

Figure 5.13   Circuit of a basic MECL 10 K OR/NOR gate
Copyright of Motorola, used by permission.

This is the typical value specified by the manufacturer.

For V_A and V_B both at logic high voltage, Q_1 is off and hence

$$V_{C1} = V_{B4} \approx 0.$$

If the emitter of Q_4 is connected to the input of an identical gate, a path is provided for the emitter current via the base resistor of the load gate. Therefore, the output voltage at logic high is

$$V_{OR} = V_{E4} = V_{OH} = V_{C1} - V_{BE4} = -0.75 \text{ V}$$

If V_A or V_B is at this logic high voltage, the input transistor conducts and brings the common emitter voltage to -1.5 V. The emitter and the collector currents, then, are

$$I_C \approx I_E = \frac{-1.5 + 5.2}{0.777} = 4.76 \text{ mA.}$$

Hence,

$$V_{C1} = -I_{C1}R_{C1} = -1.03 \text{ V}$$

and

$$V_{E3} = V_{NOR} = V_{OL} = -1.78 \text{ V}$$

For V_A and V_B below $V_R = -1.32$ V, Q_1 conducts and

$$V_{C1} = \frac{-(-1.32 - 0.75 + 5.2)0.245}{0.777} = -0.99 \text{ V}$$

Therefore $V_{OR} = V_{OL} = -1.74$ V

The above values are approximate based on zero current in all bases and in the collector of the cutoff transistor. When a load is connected to the output, V_{OH}, for example, decreases slightly due to the drop in collector resistor R_{C1} (or R_{C2}), which arises from the finite base current. Taking the loading effect into account, manufacturers provide typical values of $V_{IL} = -1.4$ V, $V_{IH} = -1.2$ V, $V_{OL} = -1.7$ V, and $V_{OH} = -0.9$ V.

If we assume that $V_{IL} = V_R - 100$ mV and $V_{IH} = V_R + 100$ mV, we have

$$V_{IL} = -1.42 \text{ V} \quad \text{and} \quad V_{IH} = 1.22 \text{ V}$$

Typical noise margins are

$$N_{ML} = -1.4 - (-1.7) = 0.3 \text{ V} \quad \text{and} \quad N_{MH} = -0.9 - (-1.2) = 0.3 \text{ V}$$

Power dissipation per gate is about 25 mW. At 2 ns average propagation delay, this gives a delay-power product of 50 pJ, which is comparable to that of the Schottky TTL family (see Table 4.2).

Fan-out, as we saw also in Section 5.2.4, is limited to 10 identical 10K series gates to prevent excessive degradation in switching response.

The MECL 10K (as well as the MECL III) series uses separate ground connections for the output circuit and logic circuit gates to minimize the effect of voltage transients occurring in other parts of a system. Because of high speed of operation the emitter followers must switch high currents for rapid charging and discharging of load capacitances. As a result, collectors Q_3 and Q_4 carry high switching currents, which cause voltage transients in the supply line for V_{CC1} (ground). On the other hand, V_{CC2} supplies essentially constant current to the differential pair and the reference voltage circuit. (Recall that almost the same current is switched between Q_1 and Q_A—or Q_B—during logic transition.) Hence, by separating the two supplies, or grounds, this circuit configuration prevents the coupling of voltage spikes from V_{CC1} to V_{CC2}. As we have seen, this is an essential requirement because of the low noise margins. In a system of ECL logic circuits, separate wires for V_{CC1} and V_{CC2} are brought from all circuits and connected externally to the same system ground.

Modified ECL circuits in the MECL 10K series, called the *MECL 10KH* series, have typical propagation delays of 1 ns with the same noise margins as the 10K series. The circuit configuration and fabrication technology used in the 10KH series parallel those of the 100K series, which we consider next.

5.3.2 THE MECL 100K FAMILY OR/NOR Gate

The MECL 100K series of gates incorporates advanced fabrication technologies to achieve switching speeds in the GHz range. In addition, this series employs compensating circuits to obtain transfer characteristics that are stable with power supply and temperature variations.

The MECL 100K family accomplishes stability of transfer characteristics with power supply and temperature by using (1) a constant current source in place of the common-emitter resistor in the differential pair, and (2) a reference voltage source that is stable with temperature and supply variations. Examples of two commonly used current source circuits in ICs are shown in Figure 5.14. The following numerical example illustrates the stability of these sources.

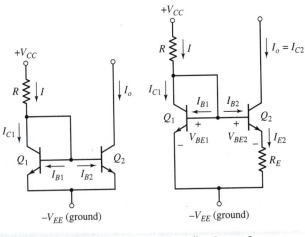

(a) High Current Mirror   **(b)** Widlar Current Source

Figure 5.14   Current sources

Calculate the current I_o in the circuit of Figure 5.14a for $R = 2.5$ kΩ, $R_E = 1$ kΩ, and $V_{CC} = 5$ V. Assume both transistors are identical with $\beta = 100$.

Example 5.6

If β varies by ± 20 percent over the operating temperature range, how much does I_o vary in each circuit?

Solution

In both circuits, the diode-connected transistor Q_1 operates in the active mode with $V_{BC} = 0$. Since the transistors are identical and $V_{BE1} = V_{BE2}$, then $I_{B1} = I_{B2}$ and $I_{C1} = I_{C2}$ in Figure 5.14a. The source current I is given by

$$I = 2I_{B1} + I_{C1} = I_{C1}\left(1 + \frac{2}{\beta}\right) \tag{5.29}$$

With the emitters at ground potential, I is also given by

$$I = \frac{(V_{CC} - V_{BE})}{R} \tag{5.30}$$

Therefore,

$$I_{C1} = I_o = \left(\frac{\beta}{\beta + 2}\right)\left(\frac{V_{CC} - V_{BE}}{R}\right) \tag{5.31}$$

Observe that I is present even if Q_2 is disconnected. This current is reflected $(I_o = I_{C1} \approx I)$ in Q_1. Thus, for a fixed I, I_o remains the same under varying load conditions. Using the given values and $V_{BE} = 0.7$ V, we get $I_o = 1.6863$ mA for a nominal β of 100.

For $\beta = 80$, we obtain $I_o = 1.6781$ mA and for $\beta = 120$, $I_o = 1.6918$ mA. These values show that for a change of 20 percent in β due to temperature I_o changes by less than 0.5 percent.

The circuit in Figure 5.14b is known as the *Widlar current source*. The reference current I is given by Equation 5.30.

$$I = \frac{(V_{CC} - V_{BE1})}{R}$$

Neglecting the base currents I_{B1} and I_{B2}, then,

$$I \approx I_{C1} = \frac{(V_{CC} - V_{BE1})}{R} \tag{5.32}$$

The ratio of the collector currents is given by

$$\frac{I_{C1}}{I_{C2}} = \exp\left(\frac{V_{BE1} - V_{BE2}}{V_T}\right) \tag{5.7}$$

We may approximate the voltage across R_E, given by $I_{E2}R_E = V_{BE1} - V_{BE2}$, by

$$V_{BE1} - V_{BE2} \approx I_{C2}R_E = I_o R_E \tag{5.33}$$

From the last two equations, the source current I_o is given by the transcendental equation

$$I_o = \left(\frac{V_T}{R_E}\right) \ln\left(\frac{I_{C1}}{I_o}\right) \tag{5.34}$$

For the given values, $I \approx I_{C1} = 1.72$ mA, and, after a few trials of Equation (5.34), we arrive at

$$I_o = 80 \ \mu A$$

Equation (5.34) shows that the source current I_o is virtually independent of current gain β. Variation of the base-emitter voltage V_{BE} with temperature, however, causes I_o to vary slightly. The combined variation in I_o due to variations in R, R_E, and V_{BE} with temperature can be made positive, zero, or negative in an IC [Reference 5].

From Equation (5.33) it is clear that I_o can only be a small value for any practical resistance R_E, if the base-emitter junction voltages are closely matched. For obtaining useful values for I_o, the design usually makes the emitter areas different, so that the result is different saturation currents (I_{ES1} and I_{ES2}) and, consequently, a large ratio of I_{C2}/I_{C1}. (Note that in Equations (5.7) and (5.33) we assume that $I_{ES1} = I_{ES2}$. In practice, I_o is specified and the integrated circuit is fabricated with values for R_E and R determined according to Equations (5.32)–(5.34).)

The current sources shown in Figure 5.14 are widely used in systems of ECL circuits. Figure 5.15 depicts the circuit configuration of the MECL 100K series ECL gate incorporating temperature-compensated reference voltage and current source.

Transistors Q_A, Q_B, Q_1, Q_2, and Q_3 form the ECL gate, with Q_4 supplying a constant current. The bias network consisting of Q_5 through Q_{11} and resistors R_3 through R_9 provide stable voltages V_{CS} for the current switch and V_R for the reference transistor Q_1. Stability of the current switch voltage V_{CS} (referred to the V_{EE} supply) is established by Q_7, Q_9, and Q_{10}. If R_4 is zero, voltage across R_5 is given by (see Equation (5.33))

$$V_{R5} = V_{BE7} - V_{BE9} \tag{5.35}$$

It can be shown [Reference 5] that V_{R5} has a positive temperature coefficient if Q_7 is operating at a higher current level relative to Q_9.

Neglecting all base currents, the voltage at the emitter of Q_8 relative to $-V_{EE}$ is

$$V_{E8} = \frac{R_6}{R_5} V_{R5} + V_{BE10}$$

Therefore, the current switch voltage V_{CS} is given by

$$V_{CS} = V_{E8} + V_{BE8} - V_{BE6} = \frac{R_6}{R_5} V_{R5} + V_{BE10} \tag{5.36}$$

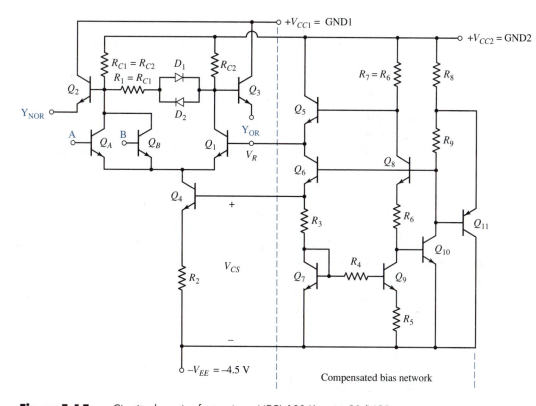

Figure 5.15   Circuit schematic of a two-input MECL 100 K series OR/NOR gate
Copyright of Motorola, used by permission.

where it is assumed that Q_8 and Q_6 are matched. With a positive temperature coefficient for the first term on the right (V_{R5}) and a negative temperature coefficient for the second term (V_{BE10}), Equation (5.36) shows that V_{CS} is invariant with temperature.[3] R_4 is used to compensate for variations in the current gain β of Q_7 and Q_9.

The reference voltage V_R, which is referred to ground (V_{CC2}), is given by

$$V_R = -V_{B5} - V_{BE5}$$

With $R_7 = R_6$,

$$V_{B5} = -\frac{R_6}{R_5} V_{R5} \tag{5.37}$$

[3] This is known as the *band-gap reference method*, since the stable reference voltage is proportional to the silicon band-gap voltage.

Therefore,

$$V_R = -\frac{R_6}{R_5} V_{R5} - V_{BE5} \tag{5.38}$$

From Equation (5.38), which is similar to Equation (5.36), we note that V_R is also invariant with temperature.

From Equation (5.36), observe that to keep the current switch voltage V_{CS} insensitive to variations in the V_{EE} power supply, V_{BE10} must be made independent of supply changes. Since V_{BE} strongly depends on collector current, a constant current in Q_{10} ensures stability of V_{CS} with changes in V_{EE}. Further, a constant V_{BE10} results in constant currents in Q_8 and Q_6; this makes V_{CS} invariant with power supply variations. Current in Q_{10} is regulated by the pnp transistor Q_{11} and the resistors R_8 and R_9. If current through R_9 increases slightly due to an increase in V_{EE}, the emitter-base voltage V_{EB11} rises, which, in turn, increases the collector current. Thus, the additional current is shunted through the collector of Q_{11} while the current in Q_{10} is maintained at a constant level. Constant current in Q_6 and Q_8 ensures a constant current in Q_5. As a result, V_R is independent of variations in V_{EE}.

With V_{CS} compensated for temperature and power supply variations, let us consider the changes in V_{BE4} with temperature. Since V_{BE} decreases with increasing temperature ($\Delta V \approx -2$ mV/°C), the switch current increases in Q_4 by $(-\Delta V/R_2)$. If the input is at logic low so that Q_1 is conducting, there is an increase in voltage drop across R_{C2} at $-\Delta V(R_{C2}/R_1)$. For $R_{C2} \approx R_2$, thus, V_{B3} drops by approximately ΔV V/°C. However, V_{BE} of the OR output transistor Q_3 drops with temperature at ΔV V/°C. Therefore, the OR output at logic high is stable with temperature. With the input transistors off, V_{B2} is at ground voltage and hence D_1 is conducting. Since the diode voltage V_{D1} drops with temperature at the same rate of -2 mV/°C, it compensates for the variation of V_{BE2} with temperature. With $R_1 = R_{C2}$, the temperature-dependent drop across R_{C2} is compensated by the drop in R_1. Hence, at the emitter of Q_2, the NOR output voltage at logic low is unaffected by temperature variations. In a similar manner, temperature compensation at logic high input occurs at both the outputs, Y_{NOR} and Y_{OR}.

Figure 5.16 shows the voltage transfer characteristics of the fully compensated 100K series ECL gates specified by the manufacturers. For comparison, the figure also shows the characteristics of uncompensated ECL gates.

The power supply $-V_{EE}$ in the 100K series gates is changed to -4.5 V to reduce power dissipation in the circuit.

As with the MECL III series and the 10K series, advanced fabrication technology and the compensated circuit configuration of the MECL 100K series result in improved performance. Typical propagation delay of the 100K series is about 750 ps. With a power dissipation of about 40 mW at -4.5 V supply, the delay-power product is 30 pJ. Table 5.1 shows a comparison of typical characteristics of the MECL series logic circuits.

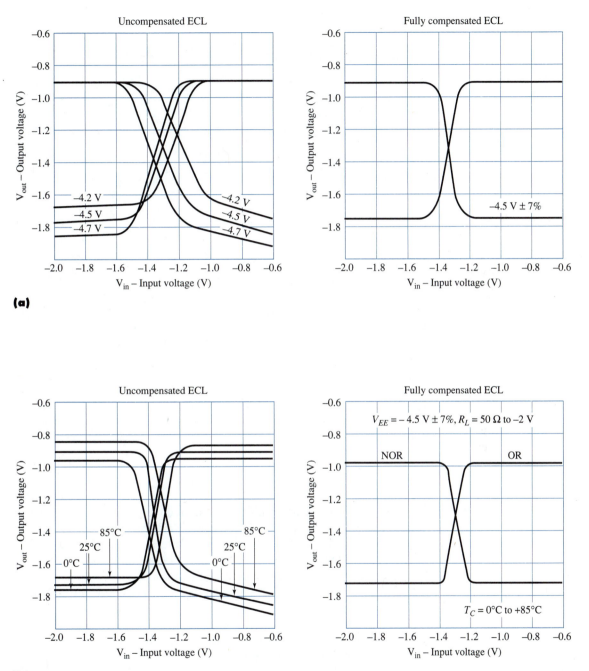

Figure 5.16   Voltage transfer characteristics of uncompensated and compensated ECL gates. [Reference 7] Reprinted with permission of National Semiconductor Corporation.

Table 5.1   Comparison of MECL series circuits

	MECL II	MECL III	MECL 10K	MECL 10KH	MECL 100K
Logic swing	0.9 V	0.8 V	0.8 V	0.8 V	0.8 V
Propagation delay	5 ns	1 ns	2 ns	1 ns	0.75 ns
Power per gate	35 mW	60 mW	25 mW	25 mW	40 mW
Delay-power product	175 pJ	60 pJ	50 pJ	25 pJ	30 pJ

5.4   ECL CIRCUITS FOR LSI AND VLSI IMPLEMENTATION

Because of the high power dissipation of ECL circuits, large scale circuit designs realize complex logic functions using modified forms of ECL configuration. These forms, which are currently implemented in high-speed LSI and VLSI bipolar systems, make use of collector and emitter dotting. Dotting of transistor terminals results in wired logic functions with reduced circuit complexity and power dissipation. Alternatively, ECL gates are stacked together with a single current source in what is known as *series gating. Emitter function logic* is another alternative structure, in which the basic ECL circuit is partitioned for efficient AND and OR realization without using the output emitter follower.

5.4.1   WIRED LOGIC AND GATING OF ECL CIRCUITS

The active pull-up and passive pull-down output circuit of an ECL gate allows many outputs to be wired together to accomplish different combinational functions. Outputs of ECL circuits at the gate level, for example, can be tied together as shown in Figure 5.17. Output Z in Figure 5.17a realizes logic OR of Y_1 and Y_2 in a *wired OR* configuration, which is equivalent to the logic function shown in Figure 5.17b. Note that R_E is common to the emitters of both output transistors. This is known as *emitter-dotting,* or *implied-OR*. Emitter-dotting effectively increases the number of inputs, or fan-in, of the gate and reduces wiring and propagation delay. Additionally, a circuit can realize different logic functions by connecting together OR/NOR outputs from different gates with a single resistor R_E. Therefore, one can implement a network of ECL gates using only ECL OR/NOR gates with minimum fan-in and wired-OR configuration. Of course, individual outputs that are wired together are no longer available. Each gate for use in wired-OR form must have an additional output transistor and emitter resistor for each individual output needed.

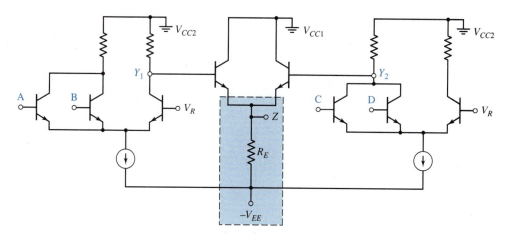

(a) Circuit

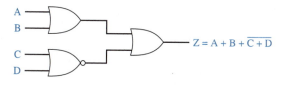

$Z = A + B + \overline{C + D}$

(b) Equivalent Logic

Figure 5.17 Wired-OR of ECL gate outputs

Individual logic functions are not needed at the output when designing complex functions of several variables in a large system. In such cases, the designer can simplify the ECL circuitry to reduce the size and the number of pins in the chip. The simplified circuit, with reduced wiring and fewer interconnections, retains the fast response of the ECL circuits. Clearly, the speed advantage arises from the design based at the transistor level instead of the multilevel ECL gates. Two approaches are used to modify the internal structure of ECL circuitry at the transistor level.

The first modification, known as *wired-AND*, or *collector-dotting* of ECL circuits, is shown in Figure 5.18a. Collectors C_1 and C_2 realize $A + B$ and $C + D$ individually. When wired together, the conducting transistor Q_1 or Q_2 pulls the voltage down and the AND function is realized at the output Z. The equivalent logic shown in Figure 5.18b requires three two-input ECL NOR gates. A second output, $(\overline{A + B}) \cdot (\overline{C + D})$, can be obtained by tying the remaining two collectors to the base of another output transistor. Alternatively, logic functions $(A + B) \cdot (\overline{C + D})$ and $(\overline{A + B}) \cdot (C + D)$ are available by cross-connecting the four collectors. However, only one pair of output functions is available without buffering the collector voltages. Up to eight collectors may be wired together without reducing performance.

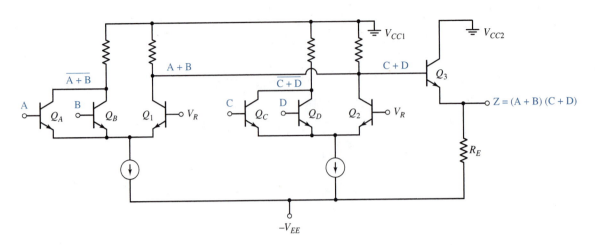

(a) Circuit

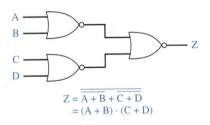

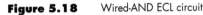

$$Z = \overline{\overline{A+B} + \overline{C+D}}$$
$$= (A + B) \cdot (C + D)$$

(b) Equivalent Logic

Figure 5.18   Wired-AND ECL circuit

Collector- and emitter-dotting may also be combined to realize logic functions. Figure 5.19 shows the circuit for implementing exclusive-OR of two variables. The dotted collectors of Q_3 and Q_4 realize the AND function, $A \cdot \overline{B}$ and $\overline{A} \cdot B$, while the dotted emitters accomplish the OR function at the output. Observe that only one pair of collector resistors is used at the common collector nodes. Q_5 limits current through R to V_{BE5}/R when A is low and B is high. Similar current limiting may also be used at the collectors of Q_1 and Q_3.

The second method of realizing multivariable logic functions is to switch a constant current source through a variety of paths in a relay-like structure. This approach, known as *series-gating*, uses a vertical stack of cascaded transistors to realize AND and NAND functions. Figure 5.20 shows an example of basic series gating realization. Voltage at the base of Q_3 is low if Q_1 or Q_2 conducts. With the sharing of the collector resistor R_{C2}, Q_2 conducts if both C and D are low regardless of A and B. When C (or D) is at logic high, Q_1 conducts I_o via Q_C (or Q_D) if both A and B are low. Hence $(A + B) \cdot (C + D)$ is realized at the base of Q_3, which is level-shifted at the emitter of Q_3. For the NAND function I_o goes through Q_A (or Q_R) and Q_C (or Q_D) only if one input in each pair is at logic

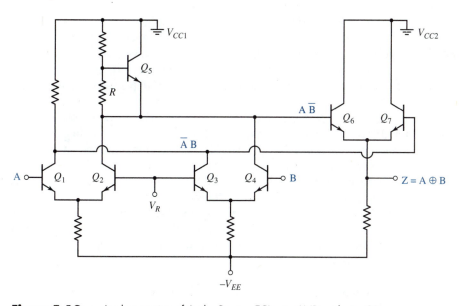

Figure 5.19 Implementation of A plus B using ECL wire-AND and wire-OR

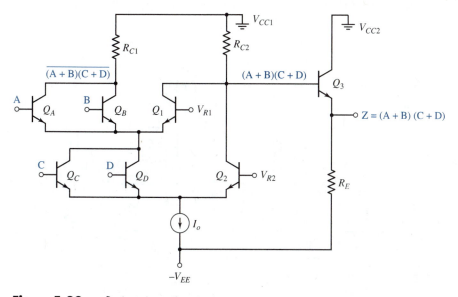

Figure 5.20 Basic series gating structure

high. A complete path, however, is available for I_o regardless of input logic states. Note further that the voltage level at the bases of Q_C, Q_D, and Q_2 must be lower than that at Q_A, Q_B, and Q_1 for I_o to flow. Hence, the inputs are lowered in level before applying at C and D, and the reference voltage V_{R1} is correspondingly reduced from the normal bias voltage V_{R2}, both by a diode drop.

In general, series-gating requires a smaller number of transistors to implement a logic function, compared with the emitter- or collector-dotted configuration. A stack of no more than three transistors, however, are used in series-gating to reduce buildup of junction capacitances and performance degradation. Figure 5.21 shows an example of an ECL full adder, implemented using series-gating.

As we see in wired-logic and series-gated circuits, emitter follower buffers in LSI and VLSI logic systems are needed only at the final outputs that drive other circuitry. A *current-mode logic* (CML) circuit eliminates the intermediate emitter followers in a complex ECL circuit to achieve simpler circuitry and reduced power dissipation. In addition, CML employs series gating to realize multilevel logic functions. Figure 5.22 shows an example of CML realization, which may be compared with the series-gated ECL structure shown in Figure 5.20. As with series-gated ECL structures, the designer must carefully choose reference voltages and current division ratios in CML realizations to obtain compatible logic levels.

Collector-dotting and series-gating are effectively used to build large ECL/CML systems that take advantage of reduced wiring and chip area. *Gate arrays* with transistor-level logic design capability are available from several manufacturers. Transistors and resistors in these arrays can be freely interconnected in collector- and emitter-dotting and series-gating structures.

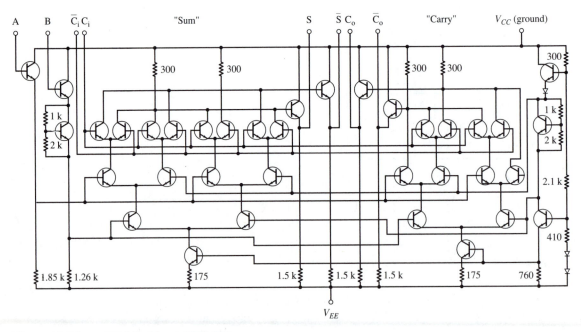

Figure 5.21   ECL full adder © 1970 *IEEE*.
Reprinted by permission from the *IEEE*, see Reference 1.

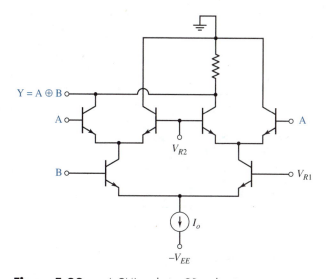

Figure 5.22   A CML exclusive-OR realization

5.4.2   EMITTER-FUNCTION LOGIC (EFL)

Emitter-function logic uses the noninverting (OR) output at the collector of the reference transistor (Q_2 in Figure 5.3, for example) and realizes AND and OR functions in two levels. With the base of the reference transistor at fixed voltage V_R, only the collector voltage changes as input changes. For the input transistors (Q_{1A} and Q_{1B} in Figure 5.3) realizing inverting (NOR) function, however, both input and output change simultaneously in opposite directions. Capacitance between the base and the collector of these transistors must be charged twice as much as that of the reference transistor. (Note that with fixed V_R, Q_2 operates in common-base configuration with its collector voltage changing by the logic swing. In an inverting circuit, the base-collector voltage changes by twice the iogic swing.) Noninverting output at the collector of the reference transistor, therefore, has faster response than the inverting output at the collector of the input transistors. As a result, ECL-type logic systems with low propagation delays can be designed by minimizing the use of inverters. Systems such as shift registers and multiplexers, for example, can be designed with few inverters.

Since the inverted outputs of ECL circuits are unused, the EFL circuits eliminate the input transistors associated with them; instead, inputs are directly applied to the emitter of the reference transistor. Figure 5.23 shows a basic EFL OR gate circuit derived from an ECL gate circuit. Diode D in Figure 5.23b prevents saturation of Q_1. Inputs A and B, which are derived from preceding EFL gate outputs, form wired-OR output to control the operation of Q_1. Note that R_E forms the emitter-follower resistor for the driving gate. By setting $V_R = -V_{BE}/2$, logic high and low levels are the same as in ECL circuits at $-V_{BE}$ and $-2V_{BE}$ at output. At such low voltages the power supply voltage can be as

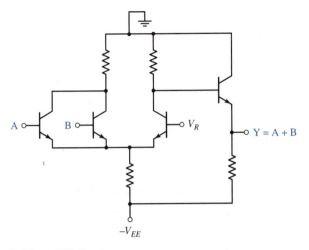

(a) Basic ECL Circuit

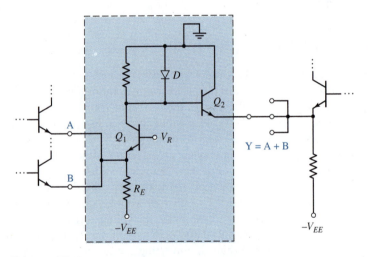

(b) Basic EFL Circuit

Figure 5.23   Comparison of ECL and EFL gates

low as $-2V_{BE}$. This ability of low voltage operation of the EFL circuits results in considerably less power dissipation than that in ECL circuits.

In addition to low power operation, EFL circuits have a simple configuration, with large fan-in and multiple outputs of the same function. Multiple emitters at the input transistor Q_1 of the AND gate in Figure 5.24, for example, increase the fan-in. Output $Y = A \cdot B$ is available at each emitter of Q_2. A design can realize logic functions in the product-of-sum form by connecting more than one input (in wired-OR form) at each of the emitters of Q_1 (see Problem 5.16).

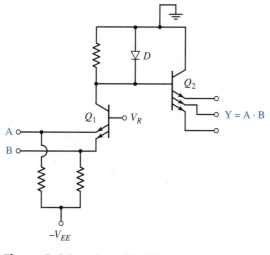

Figure 5.24   Basic EFL AND gate

An EFL circuit realizes complex logic functions using two levels. Figure 5.25 shows an example of two-level EFL realization. The output of this circuit is a limited form of product-of-sum function. Current I_o in R_E is steered through Q_1 (A or B) or Q_2, depending on the voltage level at A or B relative to V_{R1}. This current is further controlled by the voltages at C and D. That is, if the voltage at C is below V_{R2}, Q_3 can supply current to Q_1 (A or B). Now if either A or B is high (voltage above V_{R1}), I_O completes the path via Q_1 (A or B). Hence, the

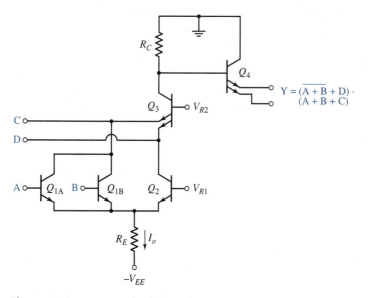

Figure 5.25   A two-level EFL realization

collector of Q_3 goes low and output is at logic low. For C at high (voltage above V_{R2}), the collector of Q_3 rises to zero regardless of the logic level at D. If both A and B are at logic low, Q_2 can conduct and lower the voltage at the collector of Q_3 only if D is low. For logic level compatibility with other EFL circuits, $V_{R2} = V_{R1} + V_{BE(active)}$. The logic threshold levels are then above and below V_{R1}. Note the similarity of the EFL circuit with the ECL series-gating structure shown in Figure 5.20. The use of multiemitter transistors in EFL circuits, however, results in simpler circuit configuration and higher gate density.

Further developments in ECL-related logic families combine low voltage Schottky devices and CMOS devices along with innovative circuit techniques for high speed and low power dissipation. These are presented in the October issues of the *IEEE Journal of Solid-State Circuits*, starting in 1970.

5.5   WIRING CONSIDERATIONS FOR ECL CIRCUITS

As with high-speed TTL circuits, interconnecting wires and printed circuit boards behave as transmission lines at ECL switching speeds, which approach tens of gigahertz. When the length of an interconnecting open wire (line) is 15 cm, for example, propagation delay along the line is about 0.5 ns. Since this delay is comparable to typical MECL III, MECL 10 KH, and MECL 100 K gate delays, multiple reflections are caused from open wires. At edge speeds of about 1 ns, these reflections deteriorate the overall waveform—with excessive ringing at both logic levels—and result in false system operation. Open wiring is therefore limited to about 3 to 5 cm in high-speed ECL circuits. For the MECL I and the MECL 10 K series with an edge speed of about 4 ns, open wires of up to 15 cm may be used. In circuits requiring longer interconnections, waveform reflections are minimized when the load resistance (at the receiving end) or the source resistance (at the sending end) is matched to the impedance of the connecting wires, which behave as transmission lines.

Interconnection using a transmission line of known characteristic impedance Z_o may be terminated with a resistance $R_L = Z_o$ as shown in Figure 5.26a. This termination, known as *parallel termination,* is returned to a supply voltage of $V_{TT} = -2$ V. This is necessary because standard ECL output circuits do not have internal pull-down resistors. The connection of R_L to -2 V ensures that the output transistor remains in the active mode at the logic low output level. Since the output resistance R_o of the emitter follower stage is low, typically less than 10 Ω, ECL circuits can drive low impedance loads with currents as high as 500 mA. At $V_{TT} = -2$ V, quiescent currents in an MECL 10 K gate driving a 50 Ω coaxial cable ($R_L = 50$ Ω), for example, are (Figure 5.26b):

Logic high: $I_{TT} \approx 22$ mA at $V_{OH} = -0.75$ V
Logic low: $I_{TT} \approx 6$ mA at $V_{OL} = -1.7$ V

We must note that the output resistance, and hence the logic levels, vary slightly with the base-emitter voltage and the load current. The variation in the output

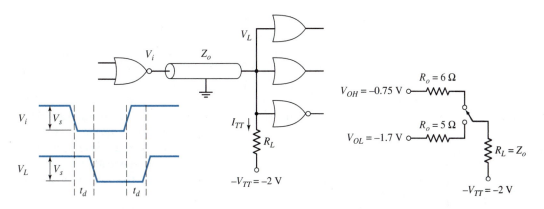

(a) Using a Separate Supply V_{TT}

(b) Approximate Equivalent Circuit

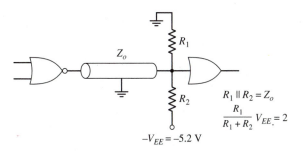

(c) Using V_{EE}

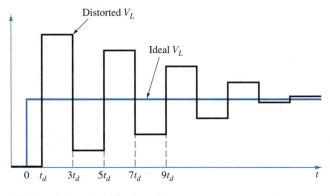

(d) Waveform at Load for $R_L > Z_o$

Figure 5.26   Parallel termination

resistance (approximately from 5 Ω to 10 Ω) is neglected in the above calculation of quiescent currents. At V_{TT} above 2 V, the currents will be higher, and this
will increase the static power dissipation.

An ECL circuit can drive a number of loads using a single parallel termination with loads distributed along the line (Figure 5.26a). Propagation delay of

the signal at the load, however, increases with the length of the line while the signal itself is received undistorted.

We can eliminate the need for a second power supply V_{TT} by using the equivalent circuit shown in Figure 5.26c. The Thevenin equivalent circuit realizes $R_T = Z_o$ and $V_T = -2$ V for $R_1 = 2.6\, Z_o$ and $R_2 = R_2/1.6$.

When the resistance R_L is exactly equal to the characteristic impedance Z_o, all the energy received from the source end is absorbed and no reflection occurs at the load end. Such an interconnection is called a *properly terminated* transmission line. If R_L is not matched with Z_o exactly, as with the Thevenin equivalent circuit in Figure 5.26c, for example, reflections from the load end affect the received signal waveform. The reflected voltage V_r at the load is given by [Reference 17]

$$V_r = r_l V_i \qquad \textbf{(5.39)}$$

where V_i is the incident (source) voltage and

$$r_l = \frac{R_L - Z_o}{R_L + Z_o} \qquad \textbf{(5.40)}$$

is the reflection coefficient at the load. The initial voltage V_L at the load after a delay of t_d along the line, therefore, is

$$V_L = V_i + V_r = (1 + r_l)V_i \qquad \textbf{(5.41)}$$

For $R_L > Z_o$, r_l is positive and Equation (5.41) shows that the received signal voltage at $t = t_d$ is higher than the source voltage. This high voltage is reflected back at the source end with a reflection coefficient of r_s given by

$$r_s = \frac{R_o - Z_o}{R_o + Z_o} \qquad \textbf{(5.42)}$$

For $R_o \approx 6\ \Omega$ and $Z_o = 50\ \Omega$, we have a negative coefficient $r_s \approx -0.8$. Therefore, after another delay of $2t_d$, V_L goes below the initially transmitted voltage. Now this low voltage travels to the source end and the process continues until the voltage at the load V_L settles down to the initially transmitted voltage. Figure 5.26d shows the damped oscillatory waveform V_L for a step change at the output of a driving gate. The ringing waveform for V_L is caused by the finite reflection coefficient r_l, which arises from imperfect matching of R_L to Z_o. The undershoot in V_L may cause faulty logic operation, while the overshoot may drive the input transistor to saturation. Thus it is important that the load voltage oscillations are kept within the noise margins of the driven gates. To limit the overshoot and undershoot to within ± 15 percent of the source voltage, for example, R_L must be in the range

$$\frac{1.15}{0.85} > \frac{R_L}{Z_o} > \frac{0.85}{1.15}$$

or

$$1.35Z_o > R_L > 0.74Z_o \qquad \textbf{(5.43)}$$

Note that when the resistance at the load R_L is chosen in accordance with Equation (5.43,) the output transistor remains in active mode as the logic level switches from low to high.

Parallel termination can also be used in a printed circuit interconnection with a grounded backplane, which typically has a characteristic impedance of 150 Ω. A coaxial cable, however, offers ready impedance-matching, with its well-defined characteristic impedance. Additionally, a coaxial cable has the advantages of minimizing signal attenuation and crosstalk.

Another method of termination, shown in Figure 5.27, is the *series termination*. Series termination matches the source resistance at the sending end to the characteristic impedance of the line. Since the output resistance R_o of an ECL gate is small, matching requires that $R_s + R_o = Z_o$. With this matching, the reflection coefficient at point B is zero. Therefore, a state change from V_{OH} to V_{OL} at point A appears at half the amplitude, $\Delta V = (V_{OH} - V_{OL})/2$ initially. At the load end C, with the large pull-down resistance of the driven gate, $r_l \approx 1$. Hence, the attenuated voltage step reaching the load end after the delay t_d is almost doubled at C to regenerate the change at A.

A number of gates may be connected at the receiving end as a lumped load. The total number of gates is limited to about 10, however, by the drop in voltage in R_s due to the input currents in the logic high state. For driving more than 10 gates, several lines may be connected to the output of the driving gate, each in series with a resistor R_s.

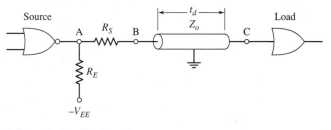

(a) Termination Configuration

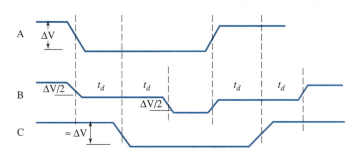

(b) Waveforms

Figure 5.27   Series termination

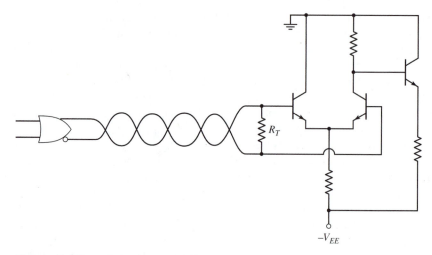

Figure 5.28   Twisted-pair transmission

Minimizing Crosstalk   In the absence of a properly terminated coaxial cable, a twisted pair of wires offers better control of impedance and faster propagation than does straight wire transmission. Additionally, a twisted pair line, when driven differentially, provides maximum noise immunity. *Crosstalk* is any unintended signal coupled from the adjacent circuitry or signal path by a small stray capacitance or mutual inductance. At ECL speeds, crosstalk can be of a large amplitude and can cause faulty switching operation. If a receiver, or load, is driven differentially, crosstalk noise, which is common to both lines, is rejected. With true and complemented (OR and NOR) outputs available from an ECL gate, it is easy to accomplish this differential transmission. At the receiving end, the line is properly terminated before driving a differential amplifier. The twisting of wires, two or more twists per inch, regularly reverses their relative positions. Any induced signal from nearby signal paths has equal and opposite polarity in the wires, while any capacitive pickup is equal in both wires. Thus, their effect is minimized at the differential amplifier output. Figure 5.28 shows a twisted-pair line driving a differential receiver, such as an MECL 10 K series receiver gate. Shielded, twisted-pair wires are available in a variety of sizes and impedances. These are used to distribute commonly used signals, such as clock waveforms, within a system; they are also used for transmission of signals in the presence of high power transients such as between instrument cabinets.

5.6   INTEGRATED INJECTION LOGIC

Integrated injection logic (I²L) combines the low voltage swing of the ECL family and the saturation of the RTL family to achieve the goal of high density at low power. I²L was developed simultaneously by IBM and Philips as an alternative to the TTL family in 1972. The level of power dissipation of the popular TTL family at 1 to 10 mW per gate precludes its usage in LSI systems. A standard TTL gate array consisting of 1000 gates, for example, will dissipate over 10 W of power. Low-power or low-power Schottky TTL gates, on the other hand, require about 1 W, which is still excessive for the allowed maximum chip area of 25 mm^2. ECL gates, at 25 to 60 mW per gate, are even less suitable for large systems. (Note that along with a large surface area, high power dissipation requires special cooling arrangements in large systems.)[4] For high-density logic systems, therefore, power dissipation must be reduced significantly. The I^{2}L family achieves this by modifying the RTL family of saturation gates to switch a built-in current source into different paths of gating transistors, similar to the ECL. To reduce the logic swing and chip area, passive resistors are eliminated. The resulting I^{2}L circuit has low component count, occupies less silicon area, and dissipates low power at low logic levels. I^{2}L, also known as the *merged transistor logic* (MTL) family, competes with the high-density CMOS family in LSI systems such as memories and logic arrays with a packing density of about 5000 gates in an area of 25 mm^2.

In the following sections we discuss the evolution of the I^{2}L and modified I^{2}L family of gates from the basic *direct-coupled transistor logic* (DCTL) gate.

5.6.1   BASIC DCTL GATE

A DCTL gate, shown inside the dashed lines in Figure 5.29, is an RTL NOR gate (Figure 4.8) in which the base resistors are eliminated to reduce component count and power dissipation. The logic levels without the base resistors are the saturation base-emitter and collector-emitter voltages of approximately 0.8 V and 0.1 V. Clearly, these low levels are not compatible with the TTL or other standard logic families. In LSI and VLSI systems where very few nodes need to be interfaced with other families, however, this gate structure has great applications. As with the RTL, the system design must consider the problem of current-hogging by load transistors due to mismatch of junction voltages. Another problem is that load transistors in a DCTL gate are driven heavily into saturation, resulting in long turn-off delays. In addition, collector reverse saturation current of a driver transistor at high temperatures may become large enough to lower the already low output voltage V_{OH}. An I^{2}L circuit, which is a modified version of the basic DCTL circuit, alleviates these problems while retaining the high density and low power dissipation.

5.6.2   STANDARD I^{2}L GATE

If we move the collector resistor R in the DCTL gate of Figure 5.29 to the bases of the load gates, the transitional structure shown in Figure 5.30 results. Since R is now associated with the base of the load transistors, the supply voltage is

---

| [4] ECL LSI circuits used in fast computers, for instance, require cooling fins.

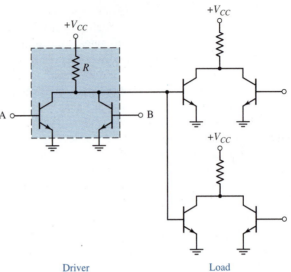

Driver   Load

Figure 5.29   Two-input DCTL NOR gate

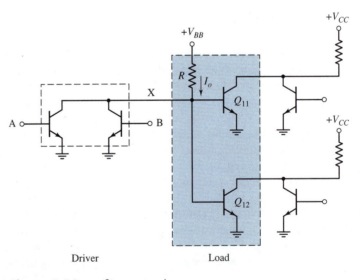

Driver   Load

Figure 5.30   I²L transitional structure

denoted by V_{BB}. Note that the collector resistors for the load gates are now part of succeeding gates or the external interface circuit. Current through R is steered into the collector of the driver at logic low output, or the bases of the load transistors at logic high output. This current I_0, which is different at logic low and high levels, may be obtained from an active current source. Further, the two transistors Q_{11} and Q_{12}, whose bases are at the same potential and whose emitters are at ground voltage, may be replaced by a single multicollector transistor.

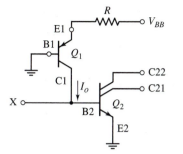

Q_2 combines Q_{11} and Q_{12}
of Fig. 5.26

(a) Multicollector transistor and current source

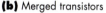

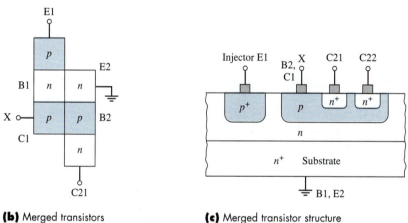

(b) Merged transistors

(c) Merged transistor structure

Figure 5.31   Basic I²L structure

Figure 5.31a shows the multicollector load transistor and the current source implemented by a pnp transistor. The source current I_0 is now virtually constant at

$$I_0 = \frac{\alpha(V_{BB} - V_{EB1})}{R}$$

Observe that the voltage at the input X, which is the collector of Q_1, is at 0.1V at logic low and 0.8V at logic high. Thus, the collector-base junction of the pnp transistor Q is always forward-biased. Consequently, Q_1 operates near or in saturation and $V_{EB1} \approx 0.81$. Still, it can be shown that the source current I_0 given above is valid at both input logic levels.

With common B_1 and E_2, and common C_1 and B_2, the pnp current source transistor and the npn load (switching) transistor can be merged together as Figure 5.31b shows. Implementation of the merged transistor is shown in Figure 5.31c. Note, from the structure, that the grounded-base pnp transistor Q_1 is a *lateral* device. Because of the lateral structure, current gain α of Q_2 is low, with a typical value of 0.8.

The structure of the multicollector npn transistor Q_2 in Figure 5.31c looks like a conventional multiemitter transistor; however, the lightly doped n diffusion region of the emitter, which is normally the collector, is grounded through the substrate. Further, in contrast with the pnp transistor, the geometry of the npn transistor shows it is a *vertical* device. The multicollector npn transistor, therefore, operates by injecting carriers from its lightly doped n region upward into its heavily doped n region. Because of the small area of the collectors, each of which is surrounded by the large emitter region, the current gain β of Q_2 in its normal mode of opertation is low. This current gain, which is referred to by β_R, the reverse active mode current gain in a conventional transistor, is called β_U for Q_2 operating in the *active upward mode*. The heavily doped n^+ substrate gives a typical β_U of 5. This value of β is adequate to keep Q_2 in saturation with up to five collectors. (Note that the saturation base current of Q_2 is the same as each of its collector currents.)

Merging of the complementary transistors as shown in Figure 5.31c, which gives the name *merged transistor logic,* reduces the number of metal interconnections. Note from the structure that there are a total of four separate regions (excluding multiple collectors) needed to form the two transistors. This integration reduces the chip area and the number of processing steps. Observe further that most of the current leaving from the emitter of the pnp transistor is injected directly into the base of the npn transistor. For this reason, the emitter of the pnp transistor is called the *injector* and the integrated gate structure is referred to as the *integrated injection logic.*

An inverter cell using merged transistors and its representation are shown in Figure 5.32a and b. To reduce chip area and to provide a range of switching delay-power products, the biasing resistor for the current source is connected external to the cell. The logic low input is 0.1 V, which is the saturation voltage of the preceding multicollector transistor (Figure 5.32c). At this input voltage, the base current of Q is zero, and the driving transistor sinks the injector current I_0. Since Q is turned off, its collectors are at logic high level. With a succeeding cell connected to each collector, the logic high level at output (for each of the collectors) is the saturation base-emitter voltage of 0.8 V. If the input is at 0.8 V, Q is driven to saturation, and the collectors are at 0.1 V. We must note that these voltages are applicable only in a cascaded system of I^2L circuits. The logic NOR function is realized by connecting collectors of inverter cells together.

From the above discussion we observe the following features of the I^2L circuits: (1) simpler circuit configuration (cf. TTL and ECL inverter configurations), (2) the merging of npn switching transistor and pnp current source transistor, and (3) the combining of several switching transistors in a single multicollector transistor. All of these features contribute to reduced silicon area per gate in LSI systems. In addition, an I^2L circuit achieves higher packing density by sharing the injector current with a large array of gates.

As an example of injector current sharing and implementing logic functions, consider the half-adder circuit shown in Figure 5.33a. To realize the circuit using I^2L inverter cells, the sum S and the carry C are derived using NOR gates. Figure 5.33b shows the I^2L realization, where the injector current for each switching transistor is obtained from the same supply voltage. Each npn transis-

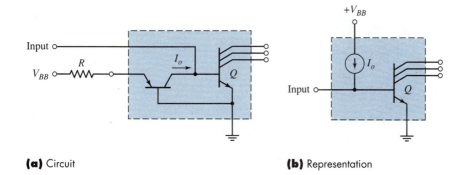

(a) Circuit   **(b)** Representation

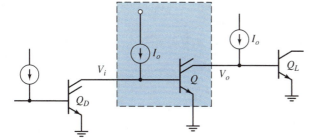

(c) Cell in a Cascaded System

Figure 5.32   I²L inverter cell

tor Q_1–Q_6, with its pnp current source, is an inverter, similar to the cell shown in Figure 5.32. By tying the collectors of Q_1 and Q_2 together, for example, $\overline{A + B}$ is obtained at the common node. Since the emitters of all pnp transistors are at the same potential and the bases are grounded, the circuit can use a single external resistor for all the current sources. This leads to common (p⁺) emitters and common (grounded) bases (n) for all the pnp transistors, which can be realized using a single rail of p⁺ region run along the length of the chip. Such a layout combines merged transistors and results in a density of 200 gates/mm², compared with 20 gates/mm² for the standard TTL family. Another significant advantage of the merging of transistors is that no isolation is needed between npn and pnp transistors. This saves chip area and masking steps for interconnection. Figure 5.33c shows a possible I²L layout for the half-adder circuit. More npn switching transistors can be added on either side of the p⁺ rail to realize other logic functions, each carrying the same fraction of the total supply current. Note the compact two-dimensional rectangular form of this layout, which is relatively simple to design. This form is also more amenable to computer-automated design than are three-dimensional layouts requiring complex interconnections.

Fabrication of all the pnp transistors on the same chip results in excellent tracking of their base-emitter junction voltages. In this way, current-hogging

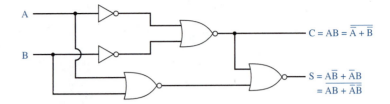

(a) NOR Gate Realization

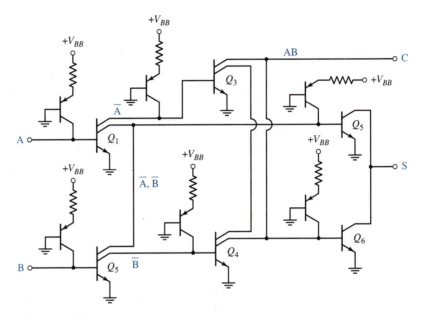

(b) I²L Realization

Figure 5.33   I²L implementation of half-adder circuit

by any of the pnp transistors is minimized. Example 5.7 illustrates the power dissipation per gate in a shared injector current I²L system.

Example 5.7 | Calculate the injector current and the average power dissipation per gate in an array of 1000 I²L gates operating at a power supply of 5 V and using a 20 Ω external resistor.

Solution

With a forward-bias voltage of 0.8 V for the emitter-base junction of the saturated pnp transistor, the total current from the supply is

$$I_T = \frac{V_{BB} - V_{EB}}{R} = \frac{5 - 0.8}{20} = 0.21 \text{ A}$$

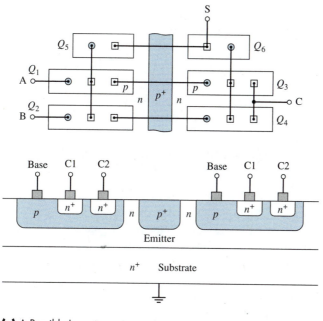

(c) A Possible Layout

Figure 5.33   (continued)

Therefore, current per gate using $\alpha = 0.8$ is

$$I_0 = \frac{0.21 \times 0.8}{1000} A = 168\mu A$$

Total power dissipated in the gate array is

$$P_T = I_T\, V_{EB(\text{sat})} = 0.168 \text{ W}$$

Hence the average power dissipated per gate is

$$P_G = 168 \ \mu W$$

This value of power per gate in a large I^{2}L array is about two orders of magnitude less than that for a standard TTL or ECL gate.

5.6.3   CHARACTERISTICS OF STANDARD I^{2}L GATES

As we discussed earlier, output logic levels in a standard I^{2}L gate driving other I^{2}L gates are the saturation emitter-base and emitter-collector voltages; hence, $V_{OH} = 0.8$ V and $V_{OL} = 0.1$ V. At the input, using a piecewise linear model, the npn switching (inverter) transistor is off if $V_i = V_{BE} < 0.6$ V. Hence, $V_{IL} = 0.6$ V. Since the base-emitter junction voltage at the edge of saturation is

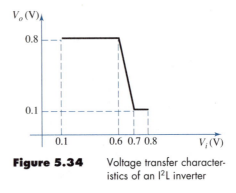

Figure 5.34   Voltage transfer characteristics of an I²L inverter

0.7 V, $V_{IH} = 0.7$ V. Figure 5.34 shows the piecewise linear voltage transfer characteristics of an I²L inverter in a cascade of gates.

From the above voltages, we find the noise margins and the logic swing as

$$NM_L = 0.6 - 0.1 = 0.5 \text{ V} \quad \text{and} \quad NM_H = 0.8 - 0.7 = 0.1 \text{ V}$$

$$V_{ls} = 0.8 - 0.1 = 0.7 \text{ V}$$

Clearly the logic swing and the noise margin are low compared to corresponding TTL values; however, these low values do not cause any problem within a single chip. Since an entire I²L subsystem such as a shift register, digital-to-analog converter, or memory is normally built as a single MSI/LSI chip, there is little noise voltage generated within the chip, and hence, the low noise margin does not affect internal performance. To increase the noise margins when interfacing with external circuitry, TTL circuits, for example, are used as buffers at the input and output of an I²L system.

Fan-Out   Fan-out of an I²L gate depends on the forward (upward) current gain β of the switching (vertical) transistor. With N collectors in the switching transistor, the multicollector transistor must remain in saturation while sinking a total current of NI_0 at logic low output. Since the base current is I_0, then $\beta_U \geq N$, or $N \leq \beta_U$. For a typical β_U of 5, two to five collectors are generally used in most I²L circuits. Fan-out, therefore, is two to five. As with noise margins, this low fan-out is not a limitation in a complete I²L system fabricated in a single chip. Additionally, wire-ANDing of collector outputs (collector-dotting) can be used when driving a single input from different I²L gates.

Switching Speed and Power-Delay Product   The switching speed of an I²L gate is typically that of a saturated BJT inverter. However, at low levels of collector currents (generally less than 100 μA for the vertical npn transistors), the removal of excess minority carriers from the base contributes less to the propagation delay than does the charging and discharging of the load transistor junction and the parasitic capacitances. For a fixed capacitance, as in a given I²L subsystem, the delay time can be reduced by injecting a higher value of current. As we see in Figure 5.32, the external resistance R may be reduced for a higher injector current I_0. Alternatively, a higher I_0 can be obtained at a higher supply voltage V_{BB} when using the on-chip resistor. While it decreases the charging time of load capacitance, however, higher injector current also causes increased power dissipation. This feature of varying the propagation delay with the injector current enables the user to trade dissipation for switching speed, depending on the application. At currents above 100 μA, both the delay time *and* the dissipation increase, due to the series resistance and the stored charge in the base region. Currents lower than 1 μA, on the other hand, cause too large delays (in the μs range) to be of practical use.

The inverse relationship between power dissipation and delay for injector currents of approximately 1 to 100 μA results in a constant delay-power product of 0.1 to 2 pJ. Typical delay time for this current range is 1 ns to 10 μs. The following example compares the performance of I²L gates with TTL gates.

Using a typical delay-power product of 0.2 pJ, calculate the propagation delay of the I²L gate array of Example 5.7. If the array is operated at a current of $I_0 = 10 \ \mu$A per gate, what is the propagation delay?
Compare the power dissipation for the two delays with that of the TTL family at corresponding delays.

| **Example 5.8**

Solution

Delay time at 0.2 pJ with $I_0 = 168 \ \mu$A current is

$$\frac{0.2 \ \text{pJ}}{168 \ \mu\text{W}} = 1.19 \ \text{ns}$$

This is at a power level of 168 μW per gate.

At $I_0 = 10 \ \mu$A, power per gate is $P_G = (I_o/\alpha)V_{EB} = (10 \ \mu\text{A}/0.8) \times 0.8\text{V} = 10\mu\text{W}$, and the delay is $0.2\text{pJ}/10\mu\text{W} = 20$ ns.

At the expense of increased propagation delay, power dissipation for this example is reduced from 168 μW to 10 μW per gate. With less dissipation, the total supply current drops from 210 mA to $I_T = 1000 \times 10\mu\text{A}/0.8 = 12.5$ mA. At a 5 V supply, then, a 400 Ω resistor must be used for the slow application of the gate array. (If the bias resistance is fixed at 20 Ω, as in Example 5.7, the resulting supply voltage of 250 mV will be too low to be practical.)

To obtain about a 1 ns delay in the TTL technology, the 74AS Series TTL, which dissipates about 8 mW per gate, must be used (Table 4.2). For 20 ns, a standard or low-power Schottky TTL performs better, with a 10 ns delay. Power dissipation, however, is several orders greater: at 10 mW for the standard TTL and 2 mW for the 74 LS Series. Note further that the size of the TTL gate array chip is about 10 times larger than the I²L chip.

From the above discussion it is clear that an I²L circuit can operate over a range of supply voltage and power dissipation levels, and the user can readily choose a value based on the switching-speed requirement. This choice is generally not available in other bipolar logic families. Improved technology and circuit structures have extended the useful current to a range of 1 nA to 1 mA, while keeping the delay-power product at as low as 30 fJ.

With its high density and low delay-power product, I^2L technology is used in such LSI and VLSI systems as 16-bit microprocessors, memories, and data converters. Other advantages offered by the I^2L family are: operation using a wide range of power supply voltage and current, ready interfacing with the popular TTL family (Section 5.7), and ability to combine other bipolar logic families as well as analog bipolar circuits in the same chip as the I^2L circuit. A major limitation of the I^2L technology is its speed at the high end compared to Schottky TTL and ECL families. Another limitation arises from the simple collector-dotting, which realizes the NOR function using a small silicon area; other mutivariable functions may require more interconnections, and hence more chip area may be needed than when using the TTL or ECL families. In the next section we consider combining Schottky metal barriers with standard I^2L technology to improve speed.

5.6.4   SCHOTTKY I^2L AND STL GATES

One approach to improving the switching speed of an I^2L gate is to reduce the logic swing so that the output transistor junction capacitance can be charged faster. A Schottky diode connected to each collector of the switching transistor as shown in Figure 5.35a increases the logic low output voltage from $V_{OL} = V_{CE(sat)} \approx 0.1$ V to $V_{OL} = V_{CE(sat)} + V_{sb} \approx 0.1 + 0.4 = 0.5$ V[5]. As a result, the load capacitance needs to charge by only 0.3 V (from $V_{OL} = 0.5$ V to $V_{OH} = 0.8$ V) when the output switches from low to high. The reduced logic swing decreases switching delay and hence improves power-delay product by a factor of two to five. This circuit is also known as a *Schottky coupled transistor logic* (SCTL).

Instead of several n^+ diffusion regions forming the collectors of the switching transistor as in the standard I^2L gate, a single, lightly doped collector region is formed by ion implantation for the Schottky I^2L gate. Multiple collector contacts are made using platinum or palladium metal barriers as shown in Figure 5.35b. In addition to reducing the logic swing, the Schottky diodes provide isolation between collectors of the switching transistor. Hence, the collector region is more compact, resulting in increased packing density for the Schottky I^2L gate.

We can achieve further reduction in propagation delay and delay-power product by preventing saturation of the switching transistor. A Schottky diode connected across the collector-base junction of the npn transistor, similar to that in a Schottky TTL gate, clamps the low voltage at the collector to 0.4 V. Without

---

[5] Note that unlike in a Schottky TTL gate, where the diode is connected across the emitter-base junction, the series Schottky diode in an I^2L gate does not prevent saturation.

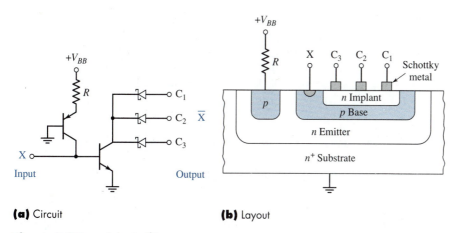

(a) Circuit   **(b)** Layout

Figure 5.35   Schottky I²L gate

saturation, the storage delay is now reduced. The Schottky clamp can be combined with the Schottky collectors to further reduce logic swing. The resulting
circuit is the *Schottky transistor logic* (STL) gate, shown in Figure 5.36. With
the pnp current source and the npn Schottky transistor, the circuit is also known
as *complementary constant-current logic* (C³L).

If V_{SC} and V_{SI} are the forward voltages of the clamp (base) and isolation
(collector) Schottky diodes, the logic low voltage is given by

$$V_{OL} = V_{BE(sat)} - V_{SC} + V_{SI}$$

With $V_{OH} = V_{BE(sat)}$, the logic swing is

$$V_{ls} = V_{SC} - V_{SI}$$

This shows that for useful logic swing, the forward voltage of the clamp diode
must be more than that of the isolation diode; thus, the Schottky barriers must
be of different metals. Using aluminum with 0.5 V forward drop for the clamp
and titanium with 0.3 V for the isolation diode, a logic swing of 0.2 V can be
obtained. If platinum-silicon having 0.4 V drop is used for the clamp, the swing

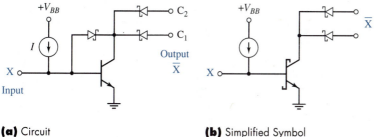

(a) Circuit   **(b)** Simplified Symbol

Figure 5.36   Schottky transistor logic gate

can be as low as 0.1 V. Fabrication of STL gates with two different metals, however, is complex, and the yield is low. Additionally, dielectric isolation is required between the switching transistor and the injector current source, and this increases the junction capacitance. The advantage with low voltage swing is that a delay-power product of 0.2 pJ is achieved in STL gates at a minimum delay of 2 ns.

5.6.5   INTEGRATED SCHOTTKY LOGIC GATE

Instead of using two different Schottky barriers to prevent saturation of the switching transistor in the SCTL circuit, *integrated Schottky logic* (ISL) uses a merged pnp transistor as shown in Figure 5.37. At the onset of saturation of the npn transistor Q_1 (at logic high input), its collector-base junction voltage forward-biases the emitter-base junction of the pnp transistor Q_2. Since $V_{CE1} = V_{BC2} > 0$, Q_2 operates in forward active mode, drawing part of the source current I away from the base of Q_1. Q_1, therefore, does not go into heavy saturation.

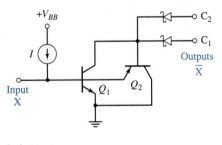

(a) Circuit

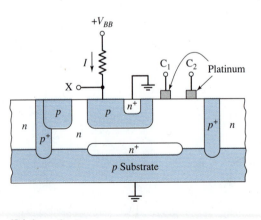

(b) Cross Section

Figure 5.37   Integrated Schottky logic gate

Table 5.2   Summary of I²L technologies

	Minimum Delay, ns	Delay-Power, pj	Maximum Density, gates/mm²	Current, per gate, µA	Logic Swing, V	Complexity
Standard	3.5	1–2	1000	0.001–1000	0.7	Low
STL	2	0.2	250	100–200	0.1–0.2	High
ISL	0.7	0.1–1.5	250	200	0.2	Medium

Unlike in the standard I²L circuit, Q_1 in the ISL circuit is a vertical npn device with its emitter on the top layer (Figure 5.37b); hence, the device operates in the normal (downward) mode, with a higher current gain β. Storage delay is therefore reduced. The pnp transistor Q_2 is a composite of a lateral device and a vertical (parasitic) device with a common collector. With the grounded emitter of Q_1 on top, another pnp transistor for the current source cannot be merged with Q_1. Instead, a 5 kΩ resistor, which takes up more area than a transistor, is formed. Also, this design needs two layers of metallization to ground the top emitter and to bring out the collectors (see the ground connections Figure 5.37b). The packing density of the ISL is therefore lower than that of the standard I²L. An advanced ISL that uses only the vertical pnp transistor for saturation control improves packing density and switching delay.

Use of the pnp transistor in place of the Schottky barrier to clamp the collector voltage reduces speed relative to the STL gates. However, ISL, which can be implemented using standard processing technology, is faster than standard I²L by a factor of at least 5. ISL has a delay-power product of 0.1 to 1.5 pJ, with a delay of 2 to 5 ns.

Table 5.2 shows a comparison of performance characteristics of different I²L technologies.

5.7   INTERFACING SATURATION AND CURRENT-MODE LOGIC FAMILIES

When a design calls for connecting the output of a logic circuit (driver) from a given family to the input of another logic family (load), the designer must take care in matching the voltage and current characteristics of the two circuits. An interface or buffer circuit whose input characteristics match the output characteristics of the driver, while its output characteristics match the input characteristics of the load, ensures compatibility of the load and the driver. In general, we need an interface circuitry between two different logic families because of the lack of certain subsystems in a given family. For example, the serial output of a

TTL shift register may need to be interfaced with an ECL gate for high-speed processing. If the speed of the combined system in such cases is to be high, the switching delays of the driver and the load families must be approximately the same. Additionally, the delay contributed by the interface circuit must be as low as possible. On the other hand, interfacing may be needed between a low-speed family and a high-speed family. For example, an ECL counter may need to be interfaced with a standard TTL circuit that drives a display. In such cases, a buffer circuit is necessary to translate the logic voltage levels. In the following sections we discuss interfacing between logic families of compatible speeds.

5.7.1   TTL-TO-ECL INTERFACING

If the number of ECL circuits is small in the system, translation from the TTL to the ECL level can be accomplished using a single positive supply as shown in Figure 5.38. Although this requires all the ECL circuits to operate with positive supply voltage, good noise rejection can be achieved with isolation of noise generated by the TTL supply. The resistor divider consisting of R_1, R_2, and R_3 translates TTL output levels to ECL input voltages of approximately ± 0.7 V relative to the reference voltage. The propagation delay of the circuit, which depends on the wiring delay and stray capacitance, is typically 1 ns.

   In large ECL systems using a separate negative supply we can use the basic circuit shown in Figure 5.39. In this circuit, TTL input voltages at A and B are shifted to compatible ECL voltages at V_b by the diode-resistive divider circuit. It is readily shown that the diodes D_A and D_B perform the AND function on the TTL inputs and the ECL output is $Y = A \cdot B$. When either of the inputs is at logic low (i.e., $0 \leq V_i \leq V_{IL} = 0.1$ V), V_P is clamped at 0.8 V and V_b is -1.97 V.

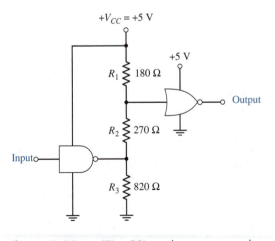

Figure 5.38   TTL to ECL translator using a single supply

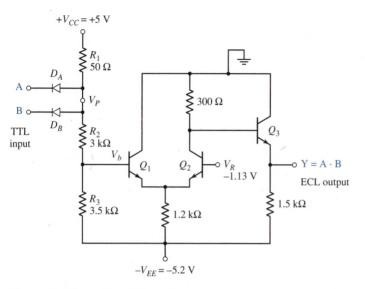

Figure 5.39   TTL to ECL translator

With $V_R = -1.13$ V, then, Q_1 is off and Q_2 conducts. Output voltage at Y can be shown as -1.54 V, which corresponds to ECL logic low level. If both the inputs are at TTL high, the diodes are off and the base of Q_1 is at 0.25 V. At this voltage, Q_1 is on and Q_2 is off. Hence Y is at logic high at $=0.7$ V. Note that with no collector resistor, Q_1 is prevented from going into saturation.

Translator circuits for high-speed Schottky TTL outputs use Schottky diodes at the TTL front end and temperature-compensated reference voltage and stabilized current source for the ECL output. Additionally, with their open-emitter differential outputs similar to the 10K series, these circuits can be used as differential line drivers for TTL outputs. The use of a common ground and separate supplies isolates noise generated in the TTL part of the circuit from the ECL supply line. MECL 10124 is a quad TTL-to-ECL translator circuit with a typical propagation delay of 3.5 ns. High-speed versions of the translator in the 10KH and 100K series have typical delays of 2 ns and 1 ns, respectively, at no load.

5.7.2   ECL-TO-TTL INTERFACING

Level shifting from an ECL output to drive TTL inputs can be performed using the circuit shown in Figure 5.40. With more than one input, A and B, this circuit also realizes the NOR function of ECL-compatible signals while shifting the output to TTL levels. The NOR output voltage at the collector of Q_2 is inverted by the saturated inverter Q_3. If input A or B is at logic high (voltage above $V_R = -1.3$ V), Q_2 is turned off. This saturates Q_3, and the output is at 0.1 V. When

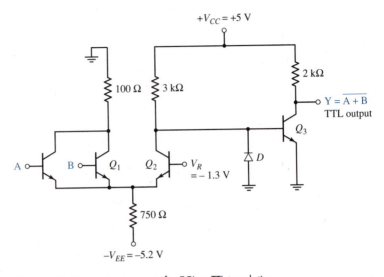

Figure 5.40   Basic circuit for ECL to TTL translation

both inputs are at logic low (voltage below V_R), Q_2 conducts, and its collector voltage becomes negative. Q_3 is therefore turned off and the output becomes 5 V. The diode clamps the negative base voltage of Q_3 to -0.7 V and damps out any ringing present in the circuit. The MECL 10125 is a quad ECL-to-TTL translator with differential inputs. This chip incorporates circuitry for a stabilized reference voltage and current source and provides totem-pole output. A typical propagation delay of 4.5 ns and a fan-out of 10 are specified by the manufacturer. High-speed versions in the MECL 10KH and 100K series have delays of 2.5 ns and 2 ns, respectively.

5.7.3   TTL-I²L INTERFACING

Since both TTL and I²L are saturation logic circuits, in principle no interfacing circuit is needed. However, the low operating current level of the I²L requires a simple modification at the output of the TTL circuit or the input of the I²L circuit. With its low power operation, it is most appropriate to interface I²L with a low-power Schottky TTL (74 LS) input and output using the circuitry shown in Figure 5.41.

The resistance R_I must be a large value such that, at logic low output of the I²L gate, Q_{I1} remains in saturation. Since the 74 LS TTL input (Figure 4.29a) sources a current of 225 μA into Q_{I1}, R_I is given by

$$R_I \geq \frac{V_{BB} - 0.1}{I - 225\ \mu A}$$

where a β_U of 1 is assumed for the vertical transistor Q_{I1}. Note that the value of injector current I for the I²L gate must be above 225 μA. Such a large current

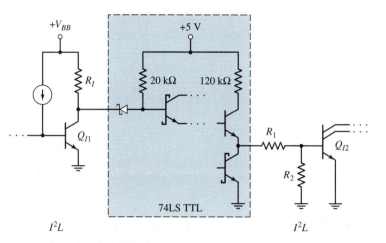

I^2L

I^2L

Figure 5.41   TTL-I^2L interfacing circuit

is required to achieve a propagation delay comparable to the 74 LS gate. At the logic high output of the I²L gate, R_I supplies I_{IH} for the 74 LS gate. Parasitic capacitance of the large resistor R_I may affect the switching response of the combination. The resistor divider at the output of the 74 LS gate limits the base drive of Q_{I2} at logic high to the injector current of I.

Although ECL-I²L interfacing circuitry can be fabricated on the same chip as an I²L circuit, wide mismatch in speed does not warrant its usage.

SUMMARY

An emitter-coupled logic circuit is a symmetrical BJT circuit which realizes logic NOR at the collectors of the coupled transistors. A constant current is switched from one path to another during logic transition. The ECL circuit achieves high switching speed by operating the coupled transistors in the active mode. ECL circuits have typical propagation delays on the order of a few nanoseconds. Low logic swing reduces power dissipation due to the active mode of operation to about 30 mW. Because of low swing and noise margin, compensating circuitry for temperature and power supply variations is required. With emitter follower drivers and complementary outputs, ECL circuits can drive high currents into terminated transmission lines without adversely affecting signal integrity.

Modified ECL circuit configurations can be used to realize complex logic functions in VLSI systems. Emitter-and collector-dotting and series-gating of ECL circuits reduce transistor count and improve speed in VLSI systems. Emitter-function logic circuits realize AND and OR functions using multiemitter input and output transistors. EFL circuits have fast response time, reduced device count, and low power dissipation compared to ECL circuits.

Integrated-injection logic is another bipolar saturation logic available for implementing MSI and LSI systems. Similar to ECL, the I^2L switches a constant injector current into a driver or input circuit. Power dissipation is reduced to microwatts by operating switching transistors in saturation or cutoff.

The I^2L family achieves a large packing density in several ways: simpler circuit configuration, the merging of npn switching and pnp current source transistors, multiple collectors for the switching transistors, and a common injector rail for all current sources. I^2L circuits can operate over a range of power supply voltages. An I^2L circuit can obtain propagation delays ranging from a few nanoseconds to microseconds by operating at different currents. Packing density of I^2L circuits can be as high as 1000 gates/mm^2.

The switching speed and delay-power product of a standard I^2L circuit are improved by using Schottky barrier contacts to reduce logic swing and/or to control saturation. I^2L circuits using Schottky barriers have propagation delays of under a nanosecond and dissipate a few milliwatts of power.

Circuits that interconnect different logic families must translate voltage levels for logic compatibility. Some interfacing circuits for TTL-ECL and TTL-I^2L perform basic logic functions along with level translation.

REFERENCES

1. Garret, L. S. "Integrated Circuit Digital Logic Families." Parts I, II, and III, *IEEE Spectrum*, October, November, December, 1970.

2. Grinich, V. H., and H. G. Jackson. *Introduction to Integrated Circuits.* New York: McGraw-Hill, 1975.

3. Hamilton, D. J., and W. G. Howard. *Basic Integrated Circuit Engineering.* New York: McGraw-Hill, 1975.

4. Taub, H., and D. Schilling. *Digital Integrated Electronics.* New York: McGraw-Hill, 1977.

5. Glaser, A. B., and G. E. Subak-Sharpe. *Integrated Circuit Engineering: Design, Fabrication, and Applications.* Reading, Mass: Addison-Wesley, 1977.

6. Muroga, S. *VLSI System Design, When and How to Design Very-Large-Scale Integrated Circuits.* New York: Wiley, 1982.

7. *F100K ECL User's Handbook.* Mountain View, Calif: Fairchild Camera and Instrument Corporation, 1982.

8. Blood, W. R., Jr. *MECL System Handbook.* Motorola Semiconductor Products Inc., 1983.

9. Hodges, D. A., and H. G. Jackson. *Analysis and Design of Digital Integrated Circuits.* New York: McGraw-Hill, 1988.

10. Treadway, R. L. "DC Analysis of Current Mode Logic." *IEEE Circuits and Devices Magazine,* March 1989, pp. 21–35.

11. Wilson, G. R. "Advances in Bipolar VLSI." *Proc. IEEE* 78, no. 11 (November 1990), pp. 1707–1719.

12. Elmasry, M. I. *Digital Bipolar Integrated Circuits.* New York: Wiley Interscience, 1983.

13. Skokan, Z. E., "Emitter Function Logic—Logic Family for LSI." *IEEE Journal of Solid-State Circuits* SC-8, no. 5 (October 1973), pp. 356–61.

14. Lynn, D. K., C. S. Meyer; and D. H. Hamilton. "*Analysis and Design of Intergrated Circuits.*" New York, McGraw-Hill, 1967.

15. Sedra, A. S., and K. C. Smith. *Microelectronic Circuits.* Philadelphia: Saunders College Publishing, 1991.

16. Altman, L. (ed.) *Large Scale Integration.* Electronics Book Series, New York: McGraw-Hill, 1976.

17. Millman, J., and Tamb, H. *Pulse, Digital, and Switching Waveforms.* New York: McGraw-Hill, 1965.

REVIEW QUESTIONS

1. What is the advantage of constant power supply current in ECL circuits?

2. What is the disadvantage of the active mode of operation in ECL circuits?

3. How is the input transition width in an ECL circuit determined?

4. If the reference voltage in an ECL circuit is at ground voltage, approximately what are the input logic levels?

5. On what circuit/device parameters does the static fan-out depend?

6. Why is the dynamic fan-out of an ECL gate limited to about 10?

7. What is the advantage of using $-V_{EE}$ over $+V_{CC}$?

8. How is the reference voltage derived from the supply voltage $-V_{EE}$?

9. How does a constant current source help in the transfer characteristics?

10. What is emitter-dotting in ECL circuits? What logic function does it perform? Does it increase propagation delay?

11. What is proper temination for transmission of ECL outputs? Why is it needed?

12. How does an IC inverter configuration of the I^2L family differ from that of an RTL family?

13. In what ways does an I^2L inverter differ from ECL and TTL inverters?

14. Why is the β of the switching transistor in an I^2L gate lower than that of a normal npn transistor?

15. What determines the logic levels in an I^2L gate?

16. What is the typical logic low noise margin of an I^2L gate? Why is this value not a problem in I^2L systems?

17. What determines the fan-out of an I²L inverter? If the switching transistor has a β of 4, what is the fan-out? How many collectors can the switching transistor have?

18. How does the delay-power product of an I²L system differ from that of other families?

19. What logic function is realized by collector-dotting of inverters in an I²L system?

20. How does a Schottky diode at the collector of a switching transistor improve speed in an I²L gate?

21. How is saturation of switching transistors controlled in Schottky-coupled and integrated Schottky logic gates?

PROBLEMS

1. Determine the currents I_{C1}, I_{C2}, and I_E, and the voltages V_{C1} and V_{C2} in the circuit shown in Figure P5.1 for (a) $V_i = 0$ (b) $V_i = -0.1$ V, and (c) $V_i = 0.1$ V. Assume $\beta = 50$ for both transistors.

2. If V_A is kept constant at logic low of -1.51 V in the circuit of Figure 5.3, calculate the input voltage V_B at which Q_{1B} saturates. Assume $V_{CE(sat)} = 0.1$ V and $V_{BE(sat)} = 0.8$ V and use $V_R = -1.11V$.

* 3. In the circuit of Figure 5.3, calculate transition width and noise margins based on input voltages such that (a) $I_{C1} = 0.05\ I_E$ and $I_{C2} = 0.95\ I_E$ and (b) $I_{C1} = 0.95\ I_E$ and $I_{C2} = 0.05\ I_E$. Assume I_E is constant at its quiescent value corresponding to $V_i = V_R$.

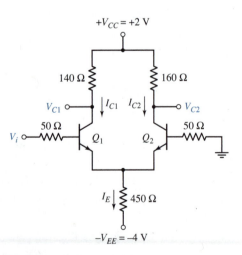

Figure P5.1

* 4. Derive the logic levels for the OR output in Example 5.2 using the slope criterion of $\dfrac{dV_0}{dV_i} = 1$.

5. Calculate the static logic high fan-out for the circuit of Figure 5.3, assuming that V_{OH} of the driving gate can drop from -0.7 V to -0.8 V. Use $\beta = 50$.

6. For compatibility with other ECL gates, logic low voltages at the NOR and OR outputs must be the same. By equating the low voltages at Y_1 and Y_2 in Figure 5.3, show that the collector resistors have a ratio given by

$$\frac{R_{C2}}{R_{C1}} = \frac{V_{IH} - V_{BE} + V_{EE}}{V_R - V_{BE} + V_{EE}}$$

Determine the ratio for $V_R = -1.3$ V, $V_{IH} = -1.2$ V, and $V_{EE} = 5.2$ V. If $R_{C1} = 300\ \Omega$ and the input logic levels are symmetrical about V_R, determine the output low level and the noise margins.

7. Calculate the average power supply current and power dissipation for a single MECL II NOR gate shown in Figure 5.9.

8. Verify the logic voltage changes with temperature for the NOR output given in Equations (5.27) and (5.28).

9. Determine the logic low and high voltages at the OR output of Figure 5.9 at 25°C and at 80°C.

10. Aside from the power supply noise consideration, an ECL circuit can operate satisfactorily at any supply voltage if $V_{CC} + V_{EE}$ is the same. Determine the reference voltage and the logic levels in the circuit of Figure 5.9 for $V_{CC} = 3$ V (instead of zero) and $-V_{EE} = -2.2$ V.

11. Determine the value of R in the current source circuit of Figure 5.14a for obtaining a current of $I_o = 3$ mA. Use $V_{CC} = 0$ and $V_{EE} = -5.2$ V, $V_{BE} = 0.75$ V, and $\beta = 100$. What is the percentage change in I_0 if β varies from 70 to 130?

*12. Run a MicroSim PSpice simulation for the ECL circuit of Figure 5.9 in which the emitter resistor R_E is replaced with a constant current source of 3 mA. Determine the transfer characteristics and logic levels.

13. Verify that the series-gated ECL circuit in Figure P5.13 realizes $Y = A \oplus B \oplus C$. Show which transistors are on for Y at logic high.

14. Figure P5.14 shows part of an ECL octal decoding tree that realizes all eight three-variable functions at the collectors at the top level. Complete the circuit and indicate the function realized at each collector at top level.

*15. For $V_R = -0.3$ V and $-V_{EE} = -5$ V, obtain a MicroSim PSpice simulation of the switching response for the EFL AND gate (single output) shown in Figure 5.24. Use emitter resistors of 2 kΩ (for both input and output) and a collector resistor of 600 Ω. Terminate the output with 2 kΩ resistor to $-V_{EE}$. Note that the inputs, which are obtained from preceding EFL gates, cannot be below -1 V. Use the 2N2222A transistor model.

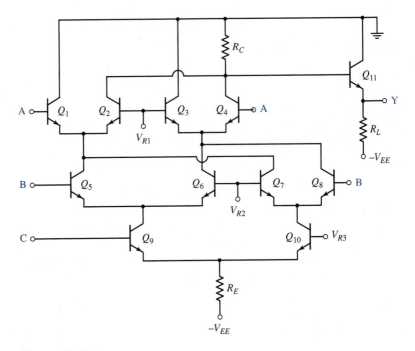

Figure P5.13

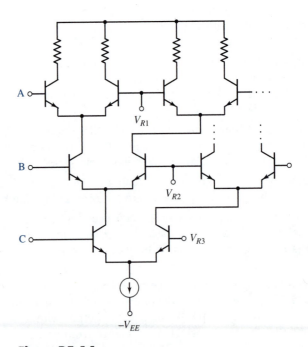

Figure P5.14

16. Determine the logic functions at the outputs of EFL circuits shown in Figure P5.16.

17. Figure P5.17 shows a three-variable I²L decoder, similar to the ECL octal decoding tree of Problem 5.14. (Injector currents are not shown.) Complete

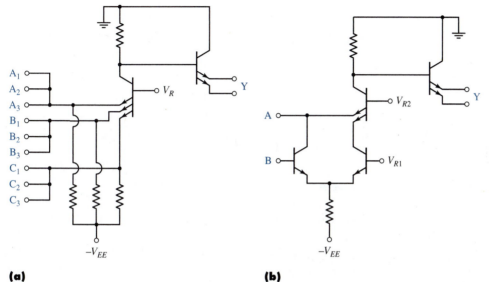

(a)

(b)

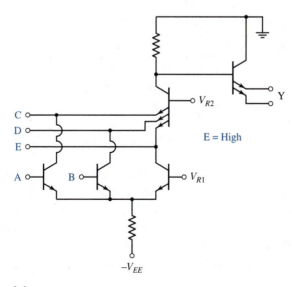

(c)

Figure P5.16

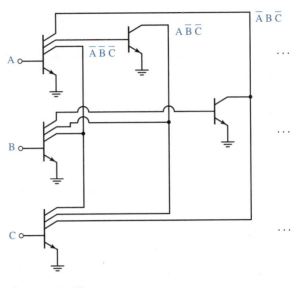

Figure P5.17

the circuit and show the logic functions available at each output. What is the maximum number of collectors required?

18. If the injector current I_0 in Problem 5.17 above is 20 μA, calculate the resistance R needed from a 5 V power supply. What is the total power dissipated in the circuit? At a power-delay product of 1 pJ, what is the propagation delay at each of the outputs?

19. Determine the input and output voltages of the single supply TTL-ECL translator shown in Figure 5.38. Neglect base currents in the ECL circuit and logic high output current of the TTL circuit. Assume the standard TTL and MECL 10K Series ECL (Figure 5.13).

*20. Run a MicroSim PSpice simulation for the TTL-ECL translator shown in Figure 5.39 and obtain its transfer characteristics. Use 2N2222A transistors and 1N914 diodes.

21. Determine the average power dissipated by the ECL-TTL translator of Figure 5.40.

*22. Perform a MicroSim PSpice simulation of the switching response of the ECL-TTL translator of Figure 5.40 for a single input. Use 2N2222A transistors and 1N914 diode.

*23. The circuit of the MECL 10K Series TTL-ECL translator is shown in Figure P5.23. Determine V_R, V_b, and V_{y1} for $V_i = 0.4$ V and $V_i = 4.3$ V.

24. *Experiment:* Static and dynamic characteristics of the basic discrete ECL circuit are studied in this experiment.
    a. Connect the circuit of Figure 5.3 with a single input using 2N2222A transistors or, preferably, using a transistor array such as CA3086.

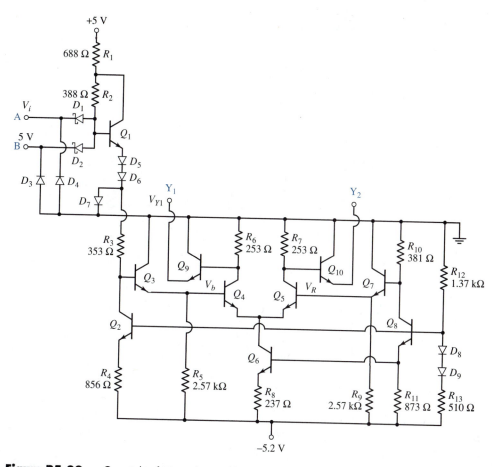

Figure P5.23   Copyright of Motorola, used by permission.

b. Using a reference supply voltage of -1.2 V, obtain the transfer characteristics at the NOR and OR outputs.

c. Calculate the logic levels and noise margins from the transfer characteristics.

d. Applying a square wave input with amplitudes in accordance with the
   input logic levels calculated above, determine the propagation delays at
   both the outputs. Use a period of 1 μs for the square wave. To avoid
   ringing in the waveforms, short wires must be used.

e. Verify the transfer and the switching characteristics by running MicroSim PSpice simulations. Use transistor parameters from manufacture's data books or MicroSim PSpice models.

chapter 6

INTRODUCTION TO METAL-OXIDE-SEMICONDUCTOR FIELD-EFFECT TRANSISTORS

INTRODUCTION

The *metal-oxide-semiconductor field-effect transistor* (MOSFET) is the single most widely used device in the fabrication of large-scale integrated circuits. The MOSFET is a three-terminal device, with only one type of carrier involved in current conduction; thus, it is a unipolar device.[1] Depending on the type of device, the carrier may be either the majority electron or hole. Current in a MOSFET depends on the control of the majority carriers, which are available in a channel, by an applied electric field. The device, therefore, behaves as a voltage-controlled current source.

MOSFET technology is the basis for VLSI circuits because of its many advantages over bipolar technology. The principal advantage of MOSFETs over BJTs is their small size: a MOSFET occupies typically less than one-third the area of a BJT. Second, the fabrication of MOSFET circuits requires fewer and less complex steps than does the fabrication of BJT circuits. Hence, circuits using MOSFETs are less expensive to manufacture and have higher density per chip area. The third advantage of MOSFETs is their low power requirement compared with BJTs. MOSFET devices dissipate much less power during dynamic operation and almost negligible amounts under static conditions. Because of these advantages, MOSFET LSI and VLSI circuits are used increasingly in battery-powered and portable systems. Also, the versatility of the MOSFET circuits extends to applications in imaging and signal processing.

A major limitation of MOSFET logic devices is their relatively low speed of operation, due to large lumped capacitance at the output node. Nonlinear capacitances that depend on fabrication processes and device geometries arise be-

---

[1] Although the substrate, or body, of the device is also available as the fourth external terminal, its effect is neglected in approximate calculations (see Section 6.5).

tween each of the four terminals in a MOSFET. With improvements in the fabrication technology and reduction in device dimensions, however, the speed difference between MOSFETs and BJTs is narrowing. The second limitation of MOSFET circuits arises from their low current-driving capability. High-current loads such as data busses and transmission lines cannot be driven without degradation in performance. Special MOSFETs at slightly lower speeds are available for such high-power loads.

Another unipolar device that operates similarly to a MOSFET is the *gallium arsenide* (GaAs) *metal-semiconductor field-effect transistor* (MESFET). Replacing silicon with the compound semiconductor GaAs increases the speed of operation. This is because the electrons in GaAs have a significantly higher mobility and travel faster than the electrons in silicon. Recent developments in the complex fabrication of MESFETs show that GaAs devices may perform at speeds beyond those of the silicon ECL circuits.

This chapter begins with a brief discussion of the structure and basic characteristics of the different types of MOSFETs. Modeling and simple circuit configurations for use in logic circuits are considered next. The chapter concludes with an introduction to the MESFET structure and model.

6.1   STRUCTURE AND OPERATION OF **MOSFETS**

There are two broad types of MOSFETs available: the enhancement-type MOSFET, and the depletion-type MOSFET: Depending on the type of majority carrier present in the channel, a MOSFET is further identified as an n- or p-channel device. We consider the structure and the operation of each of these types in the following sections.

6.2   ENHANCEMENT-TYPE NMOSFET

The *enhancement-type MOSFET* (E-MOSFET) is the most widely used device in the fabrication of LSI circuits. The simplified structure of an n-channel enhancement-type MOSFET, abbreviated E-NMOSFET, is shown in Figure 6.1. (The reason for the term *enhancement* will become clear shortly.) The device has two heavily doped n-type regions (concentrations of at least $10^{18}/cm^3$) diffused in a p-type silicon substrate, or body, which provides physical strength. External terminals, called the *source* (S) and the *drain* (D), are brought out with aluminum contacts made with these two n-type regions. A thin (40 nm to 100 nm) layer of silicon dioxide, an insulator, is grown on the surface between the source and the drain. A layer of aluminum is deposited on the oxide insulator, and the third terminal, called the *gate* (G), is brought out. Formation of the transistor with the metal (Al) gate, the insulating oxide layer (SiO_2), and the semiconductor below the oxide layer gives rise to the name *metal-oxide-*

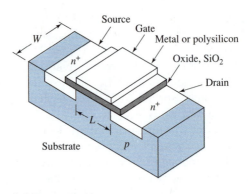

(a) Perspective View

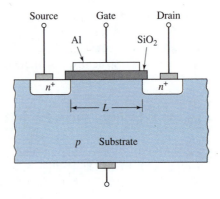

(b) Cross Section

Figure 6.1   Simplified structure of an enhancement-type NMOSFET

semiconductor transistor. Modern MOSFET devices use heavily doped polycrystalline silicon (commonly referred to as *polysilicon* or *poly*) for the gate electrodes. While undoped polysilicon has a high resistivity, the electrical behavior of doped (n- or p-type) polysilicon is similar to that of a metal. Additionally, polysilicon has the same melting temperature as single-crystal silicon. Hence, the entire silicon wafer can be subjected to the necessary heat treatment after forming the gate. The *substrate* or *body* terminal, B, is also brought out, although in most applications its role may be negligible if the substrate-to-source voltage V_{SB} is small (see Section 6.5).

6.2.1   PHYSICAL OPERATION

Under zero-bias conditions, or with the gate left open, the body and the source, and the body and the drain form two pn junctions connected back to back. Hence the device is normally off, with negligible (diode reverse saturation) current.

Note that the gate is completely insulated from the source and the drain terminals (with a resistance of 10^{10} Ω to 10^{15} Ω) because of the oxide layer. For this reason, the MOSFET is sometimes referred to as the *insulated gate field-effect transistor* (IGFET), or *metal-insulator-semiconductor* (MIS) transistor.

Operation of the NMOSFET requires the creation of a conducting path, or *channel*, between the n^+ regions of the source and the drain. Connecting the source and the body together and applying a positive voltage to the gate establishes an electric field between the gate and the body. This field repels the majority holes in the p substrate from under the gate region and causes a depletion region. As the gate voltage is increased, the increasing field attracts electrons from the two n^+ regions of the source and the drain into the region directly below the oxide layer. For a sufficient number of electrons to accumulate beneath the gate terminal and form a conducting channel (*inversion layer*) between the source and the drain,[2] the positive gate voltage ($V_{GS} > 0$) must be above a certain minimum. This minimum positive voltage required for the formation of an *induced n channel* in the NMOSFET is called the *threshold voltage, V_T*. The induced channel has a length L (below 1 μm to 6 μm), which is the distance between the source and the drain, and a width W (up to 500 μm); see Figure 6.1. Since the holes are repelled by the positive gate voltage, the region below the channel and adjacent to the n^+ wells is depleted of charges. The threshold voltage V_T is a parameter of the MOSFET and depends on the doping concentrations of the drain, the source, and the substrate, and the capacitance between the gate and the body. Typically, V_T is between 0.5 V and 3 V for integrated circuit NMOSFETs.

If, after forming the channel, the drain is kept at a slightly positive voltage relative to the source ($V_{DS} > 0$), electrons from the source end reach the drain causing positive current flow from drain to source. Note that once the channel is created by applying a positive voltage above V_T between the gate and the body, current can flow from source to drain, or drain to source, depending on which terminal is more positive relative to the other. In circuit applications, therefore, the source and the drain can be interchanged; they can be identified only after establishing a voltage difference between them. This reversibility of the two terminals is possible because of the symmetrical doping of the source and the drain regions, unlike in a BJT. The term *field-effect* arises from the fact that the channel characteristics are determined by the *electric field* established between the gate and the body of the device; in addition, current flow between the drain and the source is due to the field created along the channel.

If the gate voltage is increased above the threshold voltage, the number of induced electrons in the channel increases and the channel becomes deeper (see Figure 6.2a). Consequently, the drain-to-source resistance decreases (or, equivalently, the channel conductivity increases), and more current flows for the same (small) positive voltage applied to the drain. The channel, and hence the drain current, are thus *enhanced* for positive gate voltage. The device is, there-

---

[2] Note that the surface of the substrate just below the oxide layer has an accumulation of electrons; hence, this layer has conductivity opposite to that of the p substrate.

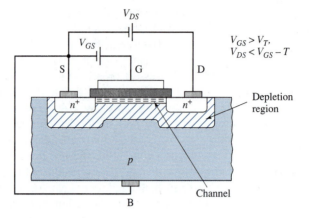

$$V_{GS} > V_T,$$
$$V_{DS} < V_{GS} - T$$

(a) Induced (Continuous) Channel Formation: $V_{GS} > V_T$, V_{DS} Small

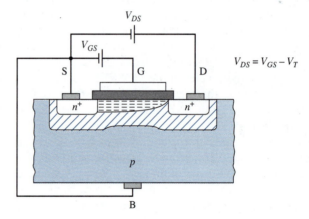

$$V_{DS} = V_{GS} - V_T$$

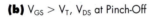

(b) $V_{GS} > V_T$, V_{DS} at Pinch-Off

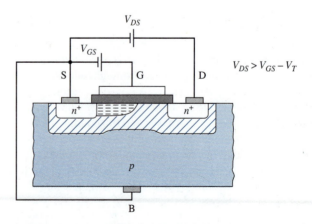

$$V_{DS} > V_{GS} - V_T$$

(c) $V_{GS} > V_T$, V_{DS} Above Pinch-Off

Figure 6.2   Operation of an enhancement-type NMOSFET

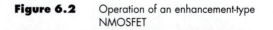

fore, called an *enhancement-type* MOSFET. We shall use the abbreviation E-NMOSFET to denote an n-channel enhancement-type MOSFET.

The above mode of operation of the MOSFET, where an increase in the gate voltage causes an increase in the drain current for small drain voltages, is called variously the *ohmic region,* the *voltage-controlled resistance region,* the *triode region* (for its similarity to the triode characteristics of a bygone era), or the *linear region;* all of these names refer to the fact that the channel resistance decreases linearly with increasing gate voltage. We note further that, for small V_{DS}, the drain current increases linearly for a given V_{GS}. This linear increase in the drain current with the drain-to-source voltage arises because of the increased electric field between the drain and the source.

Increasing the drain voltage while keeping the gate voltage constant at or above V_T decreases the voltage between the gate and the body near the drain; that is, $V_{GD} = V_{GS} - V_{DS}$ decreases as V_{DS} is raised. This widens the depletion region and makes the conducting channel shallower around the drain and near the channel-drain junction, as shown in Figure 6.2b. Further increase in the drain voltage causes a constriction in the channel near the drain. Since the source-to-gate voltage is kept constant, the channel depth does not change at the source end. The channel resistance and the drain current, therefore, increase nonlinearly with the drain voltage. Eventually the channel is *pinched off* near the drain and the drain-to-source resistance becomes very high (≈ 300 kΩ to 500 kΩ). At this point the MOSFET is said to be operating in the *pinch-off* mode. Note that the channel pinch-off occurs near the drain when the gate-to-drain voltage, $V_{GD} = V_{GS} - V_{DS}$, is below the threshold voltage V_T. This is because the voltage V_{GD} at the drain end of the channel is now below the voltage V_T needed to form the channel; that is, $V_{GS} - V_{DS} < V_T$, or equivalently, $V_{DS} > V_{GS} - V_T$ at pinch-off. For drain voltages above pinch-off, the depletion region near the drain end extends toward the source, thereby reducing the channel length (see Figure 6.2c); but the number of electrons arriving from the source, and hence the drain current, remain the same as at $V_{GD} = V_T$. For sufficiently large drain-to-source voltage (i.e., $V_{DS} > V_{GS} - V_T$), there is a large electric field with very low electron density in the depletion region near the drain. Electrons leaving the source, therefore, are accelerated toward the drain while their number remains the same. The drain current thus *saturates* at the value reached at the onset of pinch-off when $V_{DS} = V_{GS} - V_T$. The saturation drain-to-source voltage depends on the gate-to-source voltage.[3] An increase in the drain current after saturation can only be accomplished by an increase in the depth of the channel near the source, which requires an increase in the gate voltage.

Since the drain current is constant for a given gate voltage, this region of operation of the MOSFET is known as the *saturation* or *pinch-off region,* where $V_{DS} > V_{GS} - V_T$. Drain current does increase slightly with increase in V_{DS} after

---

[3] The following water flow analogy for the saturation of the drain current may be given [Reference 2]. If S is a reservoir which drains into D via a connecting hose, increase in flow rate occurs when D is lowered in level relative to S. The rate "saturates" at a value determined by the hose diameter. Further increase in rate arises only with a larger hose.

pinch-off. As V_{DS} increases beyond $V_{GS} - V_T$, the depletion region widens near the drain, and the pinch-off point extends toward the source, thereby reducing the effective channel length. The reduction in channel length, known as *channel-length modulation,* contributes to an increase in drain current because of increased field [Equation (6.13)]. This effect is neglected in discrete MOSFETs where the change in channel length relative to L is negligible.

At high drain voltage (about 20 V), the depletion region near the drain extends completely to the source. With the channel narrowed, the gate loses control and the drain current increases rapidly. This breakdown, which is reversible, is the *punch-through* of the MOSFET. Punchthrough is a limitation in short-channel ($L < 3$ μm) devices. At even higher drain voltages (50 V to 100 V) relative to the substrate, the large reverse bias causes *avalanche breakdown* of the pn junction. In this process, holes from the substrate form a drift component of current, leading to a rapid increase in the drain current. More severe, irreversible damage occurs when the gate voltage exceeds the insulating oxide breakdown voltage. At a breakdown field strength of 7 MV/cm, an oxide thickness of 50 nm can only withstand 35 V. For a margin of safety, gate voltage is limited to about 10 V. Because of the high input (gate) impedance, however, this voltage can be easily reached by the accumulation of a small amount of static charge. For this reason, MOSFET devices are protected against static charge buildup by the use of input clamping diodes.

6.2.2   STATIC CHARACTERISTICS

Since the induced channel in an E-NMOSFET is not formed until the gate voltage reaches the threshold voltage V_T, (with the source and the body at ground potential), the current-voltage (I_D-V_{DS}) characteristic in Figure 6.3 shows zero current for $V_{GS} \leq V_T$. For a given $V_{GS} > V_T$, the MOSFET begins to conduct. As the drain voltage V_{DS} increases, the drain current I_D increases linearly. This is the *ohmic region* defined by

$$V_{DS} \leq V_{GS} - V_T \quad \text{or} \quad V_{GD} \geq V_T \qquad \textbf{(6.1)}$$

The drain current I_D in this region, where V_{DS} is small, is given by the average induced charge (due to electrons) flowing from drain to source per unit time as

$$I_D = \frac{Q_C}{\tau} \qquad \textbf{(6.2)}$$

where Q_C is the induced charge and τ is the average transit time of induced electrons. Since I_D is the drift current due to the field from source to drain, the transit time is related to the drift velocity as

$$\tau = \frac{\text{Channel length}}{\text{Drift velocity}} = \frac{L}{v_d} \qquad \textbf{(6.3)}$$

where the drift velocity, from Equation (2.2), is given by

$$v_d = \mu_n E = \frac{\mu_n V_{DS}}{L} \qquad \textbf{(6.4)}$$

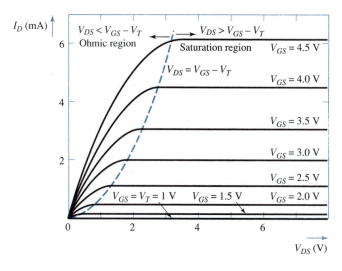

Figure 6.3 I_D-V_{DS} characteristic of an enhancement-type NMOSFET

For simplicity, the electric field magnitude E, given by $E = V_{DS}/L$ (Equation (2.22)), is assumed uniform along the length of the channel L in the above equation. μ_n is the mobility of electrons in the inversion layer, or surface mobility. This mobility, which describes the motion of electrons in the channel with surface scattering, is smaller than the mobility of electrons inside the bulk semiconductor.

From Equations (6.2)–(6.4), the drain current is given by

$$I_D = \frac{Q_c \mu_n V_{DS}}{L^2} \tag{6.5}$$

The gate and the body of the MOSFET form the plates of a parallel plate capacitor with the oxide insulator (SiO$_2$) as dielectric. Hence, the charge Q_c stored in the channel is given by

$$Q_c = V_G C_g = V_G\left(\frac{A\epsilon}{t_{ox}}\right) = V_G WL\left(\frac{\epsilon}{t_{ox}}\right) \tag{6.6a}$$

where

C_g = the gate-to-channel capacitance
A = the area of the plate, $A = WL$
t_{ox} = the thickness of the oxide
ϵ = the permittivity of the gate oxide

The voltage V_G in the above equation is the gate-to-channel, or plate, voltage. Since the channel of induced charges is formed when V_{GS} is above V_T, the gate capacitance has a plate voltage of $V_G = V_{GS} - V_T$ everywhere along the gate from the source to the drain for $V_{DS} = 0$. For $V_{DS} > 0$, the source-to-channel voltage increases along the length, from 0 at the source to V_{DS} at the drain. As a result, V_G varies from $V_{GS} - V_T$ to $V_{GS} - V_T - V_{DS}$ along the length of the plate. If we assume that the channel voltage increases linearly, the average, or effective, plate voltgage becomes

$$V_G = V_{GS} - V_T - \frac{V_{DS}}{2}$$

The charge Q_c in Equation (6.6a), then, becomes

$$Q_c = \left(V_{GS} - V_T - \frac{V_{DS}}{2} \right) WL\left(\frac{\epsilon}{t_{ox}} \right) \qquad \textbf{(6.6b)}$$

Combining Equations (6.5) and (6.6), we find the drain current in the ohmic region as

$$I_D = \left(\frac{1}{2} \right)\left(\frac{W}{L} \right)\left(\frac{\mu_n \epsilon}{t_{ox}} \right) [2(V_{GS} - V_T) V_{DS} - V_{DS}^2] \qquad \textbf{(6.7a)}$$

or

$$I_D = (1/2)(W/L) \, \mu_n \, C_{ox}[2(V_{GS} - V_T) V_{DS} - V_{DS}^2] \qquad \textbf{(6.7b)}$$

where $C_{ox} = \epsilon/t_{ox}$ is the gate capacitance per unit area.

Equation (6.7) is usually rewritten as

$$I_D = k[2(V_{GS} - V_T)V_{DS} - V_{DS}^2], \, V_{DS} \leq V_{GS} - V_T \quad \text{and} \quad V_{GS} > V_T \qquad \textbf{(6.8)}$$

The constant k in the above equation is called the *conduction parameter* or *device transconductance parameter.*[4] The device-dependent parameter k is given by

$$k = (1/2) \, \mu_n \, C_{ox}(W/L), \, A/V^2 \qquad \textbf{(6.9)}$$

where

μ_n = Mobility of electrons in the induced channel

$C_{ox} = \epsilon/t_{ox}$, the gate oxide capacitance per unit area

$\epsilon = \epsilon_r \epsilon_o$ = Permittivity of the gate oxide

ϵ_o = Permittivity of free space

ϵ_r = Dielectric constant of $SiO_2 = 4$

t_{ox} = Thickness of the oxide

W = Induced channel width

L = Induced channel length

For a given fabrication process, the oxide thickness is a constant; thus, the conduction parameter k depends on the width-to-length, or the *aspect*, ratio

---

[4] Most textbooks define k, the device transconductance parameter, without the constant $(1/2)$. With the constant $(1/2)$ included, we prefer to use the term conduction parameter for k in Equation (6.9) to distinguish from the device transconductance parameter used in other texts.

W/L. The device conduction parameter is usually expressed in terms of the *process transconductance parameter, k'* as

$$k = \left(\frac{1}{2}\right) k' \left(\frac{W}{L}\right)$$

where *k'* is given by

$$k' = \mu_n C_{ox} = \frac{\mu_n \epsilon}{t_{ox}} \qquad \qquad \textbf{(6.10)}$$

Note that *k'* is a constant for a given process. Typical values for the conduction parameter *k* range from 10 μA/V^2 to 1 mA/V^2.

As the drain voltage V_{DS} increases, the device enters the *saturation* or *pinch-off* region. This region is defined by

$$V_{DS} \geq V_{GS} - V_T \qquad \qquad \textbf{(6.11)}$$

We note that at $V_{DS} = V_{GS} - V_T$ in Equation (6.8), the drain current reaches the maximum. At this V_{DS}, the channel voltage at the drain end becomes zero. Hence, the charge stored at the drain end, according to our simplified model, is zero.

Physically, the drain current increases with V_{DS} and the peak current occurs when the channel is pinched off at the drain end; further increase in drain voltage beyond the pinch-off voltage, $V_{DS} = V_{GS} - V_T$, reduces the induced channel length. The drain voltage in excess of $V_{GS} - V_T$ appears across the depletion region near the drain and creates a short region of high electric field; as a result, the free carriers in the narrowed inversion region are accelerated across the depletion region by this field. Note that as V_{DS} increases, the voltage across the inversion region saturates at the pinch-off voltage, $V_{GS} - V_T$, while the field in the depletion region increases. The drift velocity V_d of the channel electrons, however, does not increase with increasing field as given by Equation (6.4); instead, velocity saturation occurs due to the inverse dependence of mobility μ_n at high electric fields ($E > 10^5$ V/cm). The net effect is a fairly constant drain current evaluated at $V_{DS} = $ Peak channel voltage $= V_{GS} - V_T$ in Equation (6.8). This saturation current in the pinch-off region is given by

$$I_D = k(V_{GS} - V_T)^2, \qquad V_{DS} \geq V_{GS} - V_T \quad \text{and} \quad V_{GS} > V_T \qquad \textbf{(6.12)}$$

Because of the square-law dependence of the drain current with the gate voltage, the above model is referred to as the *square-law model*. Equation (6.12) is also known as the first-order *gradual channel approximation*. It yields fairly accurate results in hand calculations for channels in discrete MOSFETs. For short channels ($L < 1$ μm), as encountered in modern VLSI devices, more sophisticated models that take velocity saturation into account are available [Reference 2]. MicroSim PSpice, for example, has four levels of MOSFET models, two of which consider short-channel effects.

The effect of channel-length modulation, that is, the reduced effective channel length in pinch-off mode, is to increase the field across the pinched-off region of channel. With the surface mobility of channel electrons increasing with field, this results in a slight increase in drain current in the pinch-off mode (see Figure 6.15). Although the effective channel length can be calculated [Reference 3], the small increase in drain current is usually approximated by the empirical relation

$$I_D = k(V_{GS} - V_T)^2(1 + \lambda V_{DS}) \tag{6.13}$$

where λ is called the *channel-length modulation factor*. Higher-level models calculate the effective channel length in pinch-off for small-geometry devices. λ, which has the units of $1/V$, is in the range of 0.005 to 0.03. Note the similarity of channel-length modulation to base-width modulation in the BJT due to the Early effect. As with the Early voltage V_{AF} (Section 3.8), λ can be measured by extending the pinch-off region plots in the I_D-V_{DS} characteristics back to the $-V_{DS}$ axis. The voltage at which the lines intersect corresponds to $-1/\lambda$.

The current equations, Equations (6.8) and (6.12), show that for a given fabrication process and a typical value of $V_T = 1$ V, drain currents above 10 mA can be obtained only with a large *aspect ratio* W/L, and hence, a large silicon area. Moreover, MOSFETs with the same threshold voltage but with different current capabilities can be fabricated on the same chip, with different aspect ratios. The aspect ratio, therefore, is an important design parameter for discrete MOSFETs.

In the case of VLSI design, the aspect ratio is crucial because of its role in determining the conduction parameter and the threshold voltage. It is shown [Reference 3] that if all the dimensions for a MOSFET including those vertical to the silicon surface are *reduced* and the doping densities are *increased* by a factor of $\alpha > 1$, then the conduction parameter increases by α while the aspect ratio remains the same. Furthermore, the propagation delay, supply current, and supply and threshold voltages are reduced by α; the dc power consumption is therefore reduced by α^2 and, consequently, the delay-power product for switching circuits decreases by α^3. Since the channel length-to-width *ratio* is unchanged, a faster MOSFET circuit dissipating lower power and occupying smaller size than the original device is possible for $\alpha > 1$. This is the principle of *scaling of MOSFETs* for VLSI design. In the last two decades, scaling has been applied successfully to achieve dense circuit layouts with minimum feature size (internal dimensions) down from about 10 μm to less than 0.5 μm. As a result, gate propagation delays have been reduced from 10 ns to less than 400 ps. Physical limitation to feature size arises at the fundamental limit of about 0.3 μm. Attendant with the smaller dimensions, however, are the functional limitations caused by the smaller supply and logic voltages, and higher current densities; also, difficulties in the fabrication and interconnection of small-sized devices prevent scaling from being carried out to its full extent.

Figure 6.3 also shows the boundary between ohmic and pinch-off regions. With the separation between the two operating regions given by $V_{DS} = V_{GS} - V_T$, we plot the dividing line using

$$I_D = kV_{DS}^2 \tag{6.14}$$

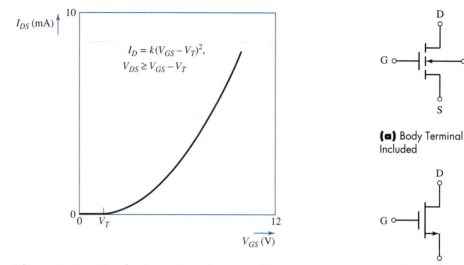

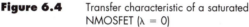

Figure 6.4 Transfer characteristic of a saturated NMOSFET ($\lambda = 0$)

(a) Body Terminal Included

(b) Body Connected to Most Negative Potential

Figure 6.5
Circuit symbols for the E-NMOSFET

This parabolic equation determines the locus of pinch-off drain voltage for different gate voltages.

Figure 6.4 shows the transfer characteristic $I_D - V_{GS}$, for an NMOSFET operating in the saturation region. (In this figure channel-length modulation λ is neglected.) Note that the device is operating as a nonlinear voltage-controlled current source, with zero current for V_{GS} below V_T.

6.2.3 CIRCUIT SYMBOL

Figure 6.5 shows two of the most commonly used circuit symbols for the E-NMOSFET. Figure 6.5a shows explicitly the body terminal. The absence of a continuous physical channel between the source and the drain terminals is indicated by the broken line from D to S. The second symbol (Figure 6.5b) is used when the body is connected to either the source or the most negative potential in the circuit. The arrowhead on the source shows the direction of positive current leaving the source terminal.

Determine the drain current I_D and the drain voltage V_D for the circuit shown in Figure 6.6 for an input voltage V_i of (a) 0.5 V, (b) 2 V, and (c) 5 V. What is the voltage V_i at which $V_{DS} = V_i$?

| **Example 6.1**

The E-NMOSFET has a threshold voltage of $V_T = 1$ V and a conduction parameter of $k = 100$ μA/V^2.

Analytical Solution
(a) $V_i = 0.5$ V: Since $V_i < V_T$, the MOSFET is off. Therefore, $V_D = V_{DS} = V_{DD} = 5$ V and $I_D = 0$.

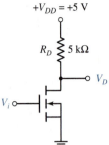

+V_{DD} = +5 V

R_D 5 kΩ

V_D

V_i

Figure 6.6
Circuit for Example 6.1

(b) $V_i = 2$ V: Since V_i is only slightly greater than V_T, we shall assume the MOSFET is operating in the saturation region. (Note that for V_{GS} slightly above V_T, I_D saturates at low V_{DS}—see Figure 6.3.)

The drain current, from Equation (6.12), is

$$I_D = k(V_{GS} - V_T)^2 = 0.1(2 - 1)^2 = 0.1 \text{ mA}$$

The drain voltage is

$$V_{DS} = V_{DD} - I_D R_D = 5 - 0.1 \times 5 = 4.5 \text{ V}$$

At this drain voltage the MOSFET is indeed operating in the saturation region, as we assumed.

(c) $V_i = 5.0$ V: At this large value of V_{GS} ($= V_{DD}$ = maximum possible V_{DS}), the device must be in the ohmic mode of operation. The drain current is given by Equation (6.8):

$$I_D = k[2(V_{GS} - V_T)V_{DS} - V_{DS}^2] = 0.1[2(5 - 1)V_{DS} - V_{DS}^2]$$

$$= 0.8V_{DS} - 0.1V_{DS}^2$$

Hence

$$V_{DS} = V_{DD} - I_D R_D = 5 - 5[0.8V_{DS} - 0.1V_{DS}^2]$$

Rearranging,

$$V_{DS}^2 - 10V_{DS} + 10 = 0$$

which gives $V_{DS} = 1.13$ V and 8.87 V. Clearly, V_{DS} cannot be 8.87 V.

Hence the drain voltage $V_{DS} = 1.13$ V, and since $V_{DS} < V_{GS} - V_T = 4$ V, our initial assumption is valid. The drain current I_D is

$$I_D = \frac{5 - V_{DS}}{5} = 0.77 \text{ mA}$$

For $V_i = V_{GS} = V_{DS}$, we have $V_{DS} > V_{GS} - V_T$. Hence the MOSFET is operating in the saturation region.

With current I_D given by

$$I_D = k(V_{GS} - V_T)^2 = k(V_{DS} - V_T)^2$$

the drain voltage is

$$V_{DS} = 5 - 0.5(V_{DS} - 1)^2$$

Solving, $V_{DS} = V_{GS} = 3$ V. (This gives a current of $I_D = 0.4$ mA.)

$V_{DS} = V_{GS}$ is used as one of the critical voltages in MOSFET inverters (see Chapter 7).

Graphical Solution

The above solutions can be obtained graphically from the I_D-V_{DS} characteristics for $V_{GS} = 0$, 2 V, and 5 V, and the load line for $R_D = 5$ kΩ as shown in Figure 6.7. The load line equation is $V_{DD} = V_{DS} + I_D R_D$, which describes the I_D-V_{DS}

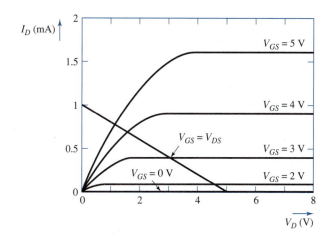

Figure 6.7   Graphical solution to Example 6.1

variation for a given supply voltage V_{DD} and external resistance R_D. Intersection of the line with the I_D-V_{DS} characteristic at a given V_{GS} gives the operating point. Note in Figure 6.7 the critical voltage $V_{DS} = V_{GS}$ occurring at 3 V.

In the above example, the dc resistance offered by the MOSFET between its source and drain terminals varies from $R_{DS} = V_{DS}/I_D = 45$ kΩ for $V_i = 2$ V to $R_{DS} = 1.45$ kΩ for $V_i = 5$ V. For $V_i < V_T$, the theoretical drain-to-source resistance is infinity. This large variation in resistance R_{DS} as voltage $V_{GS} = V_i$ switches from below V_T to above V_T ($V_i = V_{DD}$) makes the circuit of Figure 6.6 useful in switching applications.

6.3   DEPLETION-TYPE NMOSFET (D-NMOSFET)

A depletion-type NMOSFET (D-NMOSFET) has a structure identical to that of an enhancement-type NMOSFET, with the difference that it has a lightly doped n-channel implanted between the source and the drain. Figure 6.8 shows the basic structure and commonly used circuit symbols of a depletion-type NMOS-FET. The symbol in Figure 6.8c is the same as that for the E-NMOSFET (Figure 6.5b). In circuits employing both enhancement and depletion devices, the symbol shown in Figure 6.8b is used to distinguish D-NMOSFET devices from E-NMOSFET devices. The solid vertical line denotes the presence of a physical channel. Here again the substrate (B) is omitted when connected to the most negative potential (or to the source); instead, an arrowhead leaving the source is again commonly used to represent an NMOS device of either depletion or enhancement type.

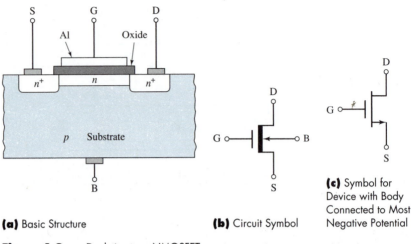

(a) Basic Structure   **(b)** Circuit Symbol

(c) Symbol for Device with Body Connected to Most Negative Potential

Figure 6.8   Depletion-type NMOSFET

Because of the presence of a physical channel, current can flow from drain to source even for zero gate-to-source voltage; hence, the depletion MOSFET is a normally on device. For positive gate voltages the channel depth is enhanced, as in the E-NMOSFET, and the drain current increases. If V_{GS} is negative, some of the electrons in the channel are repelled and a depletion region forms under the gate surface. Since this decreases the depth of the channel, the drain current is reduced. For a small drain-to-source voltage, if the gate voltage is sufficiently negative, the depletion region extends to the entire channel and the current drops to zero. The voltage V_{GS} at which the channel is completely *pinched-off* is the *threshold voltage* V_T of the D-NMOSFET, similar to the threshold voltage of the E-NMOSFET. The threshold voltage for the D-NMOSFET, however, is negative. The device can operate in the *enhancement mode* with $V_{GS} > 0$, or in the *depletion mode* with $V_{GS} < 0$. Currents in the ohmic and the pinch-off regions are given by the same equations as those for the E-NMOSFET, with the difference that $V_T < 0$. Figure 6.9 shows the drain current characteristics of the D-NMOSFET.

Example 6.2

For the circuit shown in Figure 6.10, determine (*a*) the drain voltage $V_D = V_o = V_{DS}$ for $V_i = 0$, and (*b*) the input voltage V_i when the drain is at 1 V. The MOSFET parameters are: $k = 0.5$ mA/V^2 and $V_T = -2$ V.

Solution

(*a*) For $V_i = V_{GS} = 0$, assume saturation mode of operation for the MOSFET. Then the drain current is

$$I_D = 0.5[0 - (-2)]^2 = 2 \text{ mA}$$

which gives a drain voltage of

$$V_{DS} = 5 - 2 \times 1 = 3 \text{ V}$$

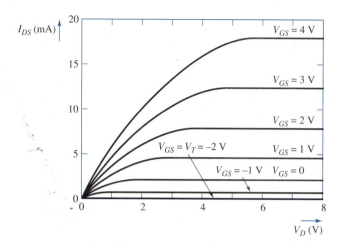

Figure 6.9 Drain current characteristics of a depletion-type NMOSFET

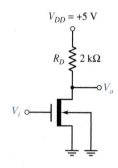

Figure 6.10
Depletion-type
NMOSFET example

Since $V_{DS} > V_{GS} - V_T = 2$ V, the MOSFET is operating in the saturation region as assumed.

(b) At the small voltage of $V_{DS} = 1$ V, we may assume operation of the device in the ohmic region. Then

$$I_D = 0.5[2(V_i + 2)1 - 1] = (V_i + 1.5)\text{mA}$$

From the circuit

$$I_D = \frac{5 - 1}{2} = 2 \text{ mA}$$

Hence $V_i = 0.5$ V, which justifies the assumed region of operation.

These results for voltages and currents can be verified graphically using the depletion NMOSFET characteristics shown in Figure 6.9.

6.4 P-CHANNEL MOSFETS

The p-channel MOSFETs (PMOSFETs) were the first to be used in LSI circuits because of their relative simplicity in fabrication. However, their slow switching speeds due to the lower mobility of the majority holes is a disadvantage compared with NMOSFETs. Figure 6.11 shows a PMOSFET of the enhancement type (E-PMOSFET) and its circuit symbol. To form an induced channel of holes in the region between the drain and the source, the gate must be at a lower potential than the source; hence the threshold voltage V_T is negative. The drain must be at a lower voltage than the source so that it can draw holes from the channel to constitute drain current. The channel is pinched off at the drain end

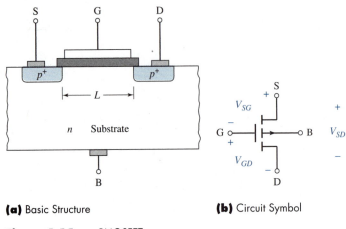

(a) Basic Structure   **(b)** Circuit Symbol

Figure 6.11   PMOSFET

when the gate-to-drain voltage is above V_T. Therefore, the condition for saturation of the drain current in the PMOSFET is

$$V_{GD} \geq V_T, \quad \text{or} \quad V_{SD} \geq V_{SG} + V_T \qquad (6.15)$$

Since $V_T < 0$, Equation (6.15) may be rewritten, in accordance with Equation (6.11), as

$$V_{SD} \geq V_{SG} - |V_T| \qquad (6.16)$$

for pinch-off.

The current in the ohmic region is given by

$$I_D = k[2(V_{SG} - |V_T|)V_{SD} - V_{SD}^2] \qquad (6.17)$$

for

$$V_{SD} \leq V_{SG} - |V_T| \quad \text{and} \quad V_{SG} > |V_T|$$

and, in the pinch-off region, by

$$I_D = k(V_{SG} - |V_T|)^2 \qquad (6.18)$$

for

$$V_{SD} \geq V_{SG} - |V_T| \quad \text{and} \quad V_{SG} > |V_T|$$

These current equations are similar to those of Equations (6.8) and (6.12), with the reversal of terminals. The *V-I* characteristics of the PMOSFET, shown in Figure 6.12, are, therefore, identical to those of an NMOSFET.

Since the mobility of holes is only about half as much as that of electrons, the conduction parameter of a PMOSFET is approximately half of that of an NMOSFET for the same aspect ratio, *W/L*.

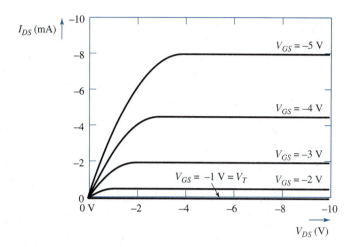

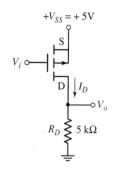

Figure 6.13
Circuit for Example 6.3

Figure 6.12   PMOSFET characteristics

Analyze the PMOSFET circuit shown in Figure 6.13 and determine (a) V_o and   **Example 6.3**
I_D for $V_i = 0$, (b) V_o and I_D for $V_i = 4$ V, and (c) V_i for $V_o = 1$ V. The transistor
has $V_T = -1$ V and $k = 50$ μA/V².

Solution

Note that the source of the PMOSFET is at 5 V while the drain is connected to
ground via the 5 $k\Omega$ resistor. Hence $V_i = V_G$ and $V_o = V_D$.
(a) For $V_i = 0$, $V_{SG} = V_S - V_G = V_S - V_i = 5$ V.
At this high voltage (much above the threshold voltage magnitude) the device
conducts heavily and $V_{SD} = V_S - V_D = V_S - V_o$ is likely to be small. Hence,
assuming ohmic mode of operation, we have

$$I_D = V_o/R_D = (5 - V_{SD})/5 = 0.05[2(5 - 1)V_{SD} - V_{SD}²]$$

Solving, we get $V_{SD} = 2$ V (discarding the other solution of $V_{SD} = 10$ V), $I_D =$
0.6 mA and $V_o = 0.6 \times 5 = 3$ V. Note that the calculated source-drain voltage
satisfies the ohmic mode of operation.
(b) At $V_i = 4$ V, $V_{SG} = 5 - 4 = 1$ V. Since the channel is barely formed, $I_D =$
0, $V_{SD} \approx 5$ V and $V_o \approx 0$.
(c) At $V_0 = 1$ V, $V_{SD} = 5 - 1 = 4$ V and $I_D = 1/5 = 0.2$ mA. Assuming
pinch-off mode of operation, we have

$$I_D = 0.05(5 - V_i - 1)² = 0.2 \text{ mA}$$

Hence, $V_i = 2$ V so that, for saturation,
$V_{SD} = 4 > V_{SG} - |V_T| = 5 - 2 - 1$. (The other solution, $V_i = 6$ V, is
discarded. Why?)

6.5   EFFECT OF SOURCE-TO-SUBSTRATE POTENTIAL: THE BODY EFFECT

The purpose of connecting the body terminal B to the source terminal S is to apply a constant zero-bias across the source–body (n^+–p) junction so that the npn BJT action between S, B, and D in an NMOSFET is prevented. If, as in integrated circuits, the body is connected to the most negative potential in the circuit, there results a reverse-bias voltage, $V_{SB} > 0$. This reverse bias draws electrons from the channel, and the channel depth is reduced. That is, the substrate voltage causes the body to act as another gate. This is known as the *body effect.* To create a channel of the same depth, therefore, the gate voltage must be increased. The effect of the source-to-body potential is thus to increase the threshold voltage needed to produce a strong channel. The dependence of the threshold voltage on V_{SB} is approximately given by

$$V_T(V_{SB}) \approx V_{T0} + \gamma \sqrt{V_{SB}} \tag{6.19}$$

where V_{T0} is the threshold voltage for $V_{SB} = 0$, and γ is a constant called the *body effect coefficient,* which depends on the doping level of the substrate and the thickness of the gate oxide, Typically, γ has a value of 0.3 to 0.5 $\sqrt{V}$.

A more accurate modification to the threshold voltage is given by [see Reference 2, for example]

$$V_T(V_{SB}) = V_{T0} + \gamma \left(\sqrt{2|\phi_F| + V_{SB}} - \sqrt{2|\phi_F|} \right) \tag{6.20}$$

where ϕ_F is the bulk Fermi potential. ϕ_F, which is typically -0.3 V, depends on the temperature and the substrate doping concentration.

We can use the circuit shown in Figure 6.14a to observe the effect of body-source reverse bias on the threshold voltage of a MOSFET. With gate and body connected together, that is, $V_{DS} = V_{GS}$, the pinch-off condition $V_{DS} \geq V_{GS} - V_T$ is satisfied when the device is conducting. Hence, the plot of $\sqrt{I_D} = V_{GS}$ in saturation (Equation (6.12)) shows zero current for $V_{GS} \leq V_T$. As the body bias voltage V_{SB} is increased, the threshold voltage increases; hence, for the same drain current, higher gate voltage is needed as we see in Figure 6.14b.

Figure 6.15 shows the drain characteristics of the E-NMOSFET shown in Figure 6.3 with body bias and channel-length modulation. At $V_{SG} = 4$ V, substrate bias of -3 V relative to the source causes pinch-off of the channel at a smaller drain voltage. In addition, the drain current with $V_{SB} = 3$ V is decreased due to a weaker channel.

The body effect causes a shift in the voltage transfer characteristics of logic circuits. A similar effect is preset in p-channel devices when the n-type body is forward-biased relative to the p^+ source.

Example 6.4   Determine the source voltage V_S in the circuit shown in Figure 6.16 for $V_i = 2$ V. Assume $V_T = 1$ V for $V_{SB} = 0$, and $k = 100$ μA/V^2. Neglect the body effect.

If the body-effect coefficient is 0.5, what is V_S?

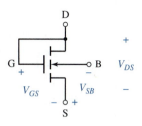

(a) MOSFET in Saturation

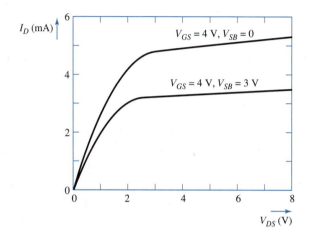

(b) $\sqrt{I_D}$ Variation with V_{GS}

Figure 6.14   Effect of body bias on threshold voltage

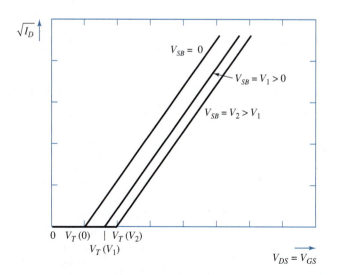

Figure 6.15   Effect of body bias and channel-length modulation on drain characteristics

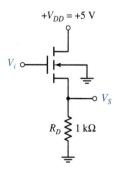

Figure 6.16
Circuit for Example 6.4

Solution

Since the body and the source are not at the same potential, the threshold voltage will be different from the value of V_T given for $V_{SB} = 0$. If we neglect the body effect, then the drain current I_D, assuming pinch-off mode, is given by Equation (6.12):

$$I_D = k(V_{GS} - V_T)^2$$

Since

$$V_{GS} = V_G - V_S = V_i - (V_{DD} - V_{DS}) = V_{DS} - 3,$$

and

$$V_{DS} = V_{DD} - I_D R_D,$$

we have, for V_{DS},

$$V_{DS} = 5 - 0.2(V_{DS} - 4)^2 \times 1$$

Solving,

$$V_{DS} = 4.854 \text{ V}, -1.854 \text{ V}$$

Since V_{DS} cannot be negative, $V_{DS} = 4.854$ V and $V_S = 0.146$ V. Note that this drain voltage satisfies the condition for pinch-off mode.

With the body effect: Since $V_{SB} = V_S$, the approximate threshold voltage with body effect is given by

$$V_T(V_S) \approx V_{T0} + \gamma \sqrt{V_S} \qquad \qquad \textbf{(6.19)}$$

where the threshold voltage for $V_{SB} = V_S = 0$ is 1 V. Hence, the drain current in pinch-off, using Equations (6.18) and (6.19) is

$$I_D = k[V_i - V_S - V_{T0} - \gamma\sqrt{V_S}]^2$$

Also,

$$I_D = \frac{V_S}{R_D}$$

Combining the above equations with the values given results in

$$V_S = 0.2[1 - V_S - 0.5\sqrt{V_S}]^2$$

Substituting $x = \sqrt{V_S}$, we have the following fourth-order equation.

$$x^4 + x^3 - 6.75x^2 - x + 1 = 0$$

The four solutions for x are -3.063, -0.456, 0.327, and 2.192. If we use only the positive solutions, the corresponding values for V_S are 0.11 and 4.81. Of these, $V_S = 4.81$ V is invalid for the given input. Therefore, $V_S \approx 0.11$ V.

From Equation (6.19), we find that the threshold voltage for $V_{SB} = V_S = 0.11$ V is $V_T(V_{SB}) = 1.17$ V. Since V_{SB} is small, the body effect causes only a slight change in the threshold voltage.

The MicroSim PSpice simulation (see Section 6.6) uses the more accurate model given by Equation (6.20), and gives a source voltage of $V_S = 0.136$ V and a threshold voltage of 1.04 V. Evidently, a computer-aided analysis and a higher-order model of physical behavior are required for more accurate results than we can obtain by hand calculations.

Since the order of magnitude is the same using the approximation given by Equation (6.18) or, more simply, with the body effect neglected, we can conclude that approximate but rapid hand calculations are facilitated by assuming $\gamma = 0$.

Although we have assumed no inversion layer being present in a MOSFET when the gate voltage is below the threshold voltage, in practice a weak channel is formed. This channel of low density of carriers gives rise to a small amount of predominantly diffusion current, called *subthreshold current*. Operation of MOSFETs in the subthreshold region, where the current is exponentially related to the gate voltage above the thermal voltage, is typically utilized in low-voltage logic and memory circuits. For currents above microampere range, however, subthreshold operation is usually negligible.

6.6   MOSFET Equivalent Circuits

One can obtain a complex model of a MOSFET using the drain and the gate characteristics under all modes of operation. For rapid hand calculations, however, simplified equivalent circuits are commonly used based on signal level and frequency of operation. In linear circuit applications at low frequencies, for example, the square-law model yields a simple equivalent circuit with high enough accuracy. Large-signal variations, as one encounters in switching applications, are analyzed with reasonable accuracy using lumped terminal capacitances. In this section we discuss the equivalent circuits for hand calculations and for more accurate MicroSim PSpice simulation.

6.6.1.   Circuits for Hand Calculations

The small-signal equivalent circuit of a MOSFET consists of a dependent current source that accounts for the drain current as a function of gate-source voltage and the incremental drain-source (channel) resistance, as shown in Figure 6.17. This model is valid for low-frequency operation, where the device capacitances (see below) have negligible effect. In addition, the device operates with a small change (ΔV_{GS}) in gate-source voltage from its quiescent point (V_{GS}). In linear amplifier circuits, for example, the MOSFET is biased to operate in the saturation region. In such cases, $|\Delta V_{GS}| \ll V_{GS} - V_T$, to ensure operation entirely in the saturation region. The transconductance g_m describes the change in drain

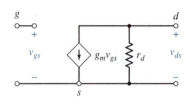

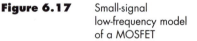

Figure 6.17   Small-signal
low-frequency model
of a MOSFET

current for change in gate-source voltage at a constant drain-source voltage; g_m
is, therefore, the slope (at the operating point) of the transfer characteristic I_D - V_{GS}
(see Figure 6.4, for example).

From Equation (6.12),

$$g_m = \frac{\Delta I_D}{\Delta V_{GS}}\bigg|\Delta V_{DS} = 0$$

becomes

$$g_m = 2k(V_{GS} - V_T) \qquad \text{(6.21a)}$$

or

$$g_m = 2\sqrt{kI_D} \qquad \text{(6.21b)}$$

in the saturation region, where I_D is the quiescent drain current. Note that
channel-length modulation is neglected. If the body effect is present
($|V_{SB}| \neq 0$), g_m given in the above equation must be modified using $V_T(V_{SB})$
[Equation (6.19) or (6.20)] for more accurate results.

The dynamic, ac, or incremental drain resistance r_d is the slope of the I_D-V_{DS}
characteristic at the operating point. Clearly, $r_d = \infty$ in the saturation region, if
we assume I_D is constant. Practical values for r_d, because of channel-length
modulation [see Equation (6.13)], range from tens of kilo ohms to hundreds of
kilo ohms.

From Equation (6.13),

$$r_d = \frac{\Delta V_{DS}}{\Delta I_D} = \frac{1}{k(V_{GS} - V_T)^2\lambda}, \qquad V_{DS} > V_{GS} - V_T \qquad \text{(6.22a)}$$

or

$$r_d \approx \frac{1}{\lambda I_D} \qquad \text{(6.22b)}$$

The following example illustrates the linear application of a MOSFET.

Example 6.5   In the MOSFET amplifier circuit shown in Figure 6.18, the input voltage v_i
varies by ± 0.5 V. Determine the drain voltage change and the gain of the
amplifier if the coupling capacitor (C_c) effect is negligible at the frequency of

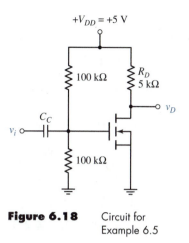

Figure 6.18 Circuit for
Example 6.5

operation. The MOSFET parameters are: $V_T = 1$ V and $k = 100$ μA/V^2. Neglect channel-length modulation.

Solution

The quiescent, or operating, point corresponding to $v_i = 0$ has, by voltage division,

$$V_{GS} = \frac{100 \times 5}{200} = 2.5 \text{ V}$$

At this gate voltage, the device is in saturation, with

$$I_D = 0.1(2.5 - 1)^2 = 0.225 \text{ mA}.$$

Therefore,

$$V_{DS} = 5 - 5 \times 0.225 = 3.875 \text{ V},$$

which verifies the assumption of pinch-off mode.

At the operating point of $(V_{DS}, I_D) = (3.875 \text{ V}, 0.225 \text{ mA})$,

$$g_m = 2[\sqrt{0.1 \text{ mA/V}^2 \times 0.225 \text{ mA}}] = 0.3 \text{ mA/V}$$

and $r_d = \infty$, if channel-length modulation is neglected. This results in the small-signal equivalent circuit shown in Figure 6.19.

From Figure 6.19, the time-varying component of drain voltage, v_d, is given by

$$v_d = -g_m v_{gs} R_D = -1.5 v_{gs} = -1.5 v_i$$

where v_{gs} denotes the time-varying component of the applied gate voltage.

Hence, the voltage gain of the amplifier is $A_v = v_d/v_i = -1.5$, and the drain voltage changes by $1.5 \times 1 = 1.5$ V for a total input change of 1 V.

Note that as v_i changes from 0 to -0.5 V, the total instantaneous gate voltage v_{GS} is reduced from 2.5 V to 2 V. At this gate voltage, the MOSFET is still in saturation with a total drain current of

$$i_D = 0.1(2 - 1)^2 = 0.1 \text{ mA}$$

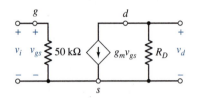

Figure 6.19   Small-signal equivalent circuit for Figure 6.18

and the drain (dc + instantaneous) voltage is

$$v_{DS} = 5 - 5 \times 0.1 = 4.5 \text{ V}$$

Similarly, as v_i goes from 0 to 0.5 V, v_{GS} is increased from 2.5 V to 3 V, and the drain current becomes

$$i_D = 0.1(3 - 1)^2 = 0.4 \text{ mA}$$

giving

$$v_{DS} = 5 - 5 \times 0.4 = 3 \text{ V}$$

We can obtain the same results graphically from Figure 6.7, which shows the load line for $R_D = 5 \text{ k}\Omega$ superimposed on the MOSFET characteristics.

In switching applications, where the signal swings are large, the mode of operation for a MOSFET undergoes changes; hence, static, or dc, resistance in each mode is important in determining loading, power dissipation, and switching delays.

The next example illustrates the magnitude of dc resistances offered by two series-connected MOSFETs, one used as driver and the other as active load.

Example 6.6 | Calculate the voltage V_0, the current I_D, and the resistances between the source and the drain terminals of each MOSFET in the discrete circuit shown in Figure 6.20 for (a) $V_i = 0$, (b) $V_i = 2$ V, and (c) $V_i = 4$ V. The threshold voltage of the E-NMOSFET is $V_{TE} = 1$ V, and that for the D-NMOSFET is $V_{TD} = -2$ V. The conduction parameter for each device is 10 $\mu A/V^2$.

Solution

(a) $V_i = 0$: Since the gate-to-source voltage for the E-NMOSFET M_E is below V_T, M_E is off. The D-NMOSFET M_D, on the other hand, is on, with $V_{GS} = 0$. With M_E off, however, there is no current from the supply through M_E and M_D; that is, M_D is operating in the ohmic region with zero current. Hence, the voltage $V_o = V_{DSD} = V_{DD} = 5$ V and the current $I_D = 0$.

With zero current in M_E, resistance $R_{DE} = \infty$ between the source and the drain. For M_D, resistance R_{DD} in the ohmic region is zero. Under static conditions, the (dc) resistances are, therefore,

$$R_{DE} = \infty \quad \text{and} \quad R_{DD} = 0$$

(b) $V_i = 2$ V: With $V_{SB} = 0$ for both the devices, no body effect is present. Assuming the pinch-off region of operation for M_E, and the ohmic region for M_D, the drain current in M_E is

$$I_{DE} = k_E(V_i - V_{TE})^2 = 10 \ \mu A$$

Since M_D carries the same current in ohmic mode,

$$I_{DD} = k_D[2(-V_{TD})(V_{DD} - V_o) - (V_{DD} - V_o)^2] = 10 \ \mu A$$

where $V_{DSD} = V_{DD} - V_o$ and $V_{GSD} = 0$ are used in Equation (6.8).

Solving the quadratic equation resulting from the above equation gives $V_o = 4.73$ V and 1.27 V. Of these two values, only $V_O = 4.73$ V satisfies the assumed ohmic mode for the depletion device. Therefore, for $V_i = 2$ V,

$$V_o = 4.73 \text{ V} \quad \text{and} \quad I_D = 10 \ \mu A$$

Hence, the dc resistances are given by

$$R_{DE} = \frac{4.73 \text{ V}}{10 \ \mu A} = 473 \text{ k}\Omega \quad \text{and} \quad R_{DD} = \frac{0.27 \text{ V}}{10 \ \mu A} = 27 \text{ k}\Omega$$

(c) $V_i = 4$ V: For this large input, M_E is likely to be in the ohmic mode with a small drain voltage, and, consequently, M_D will be in the saturation mode with large V_{DSD}. The current in the saturated device is

$$I_{DD} = k_D(-V_{TD})^2 = 40 \ \mu A$$

Equating the above to the current in M_E, we get

$$I_{DE} = k_E[2(V_i - V_{TE})V_o - V_o^2] = 40 \ \mu A$$

Solving the quadratic equation, we have $V_O = 5.23$ V or 0.76 V. Hence, for $V_i = 4$ V,

$$V_o = 0.76 \text{ V} \quad \text{and} \quad I_D = 40 \ \mu A$$

These values verify the initial assumption about the modes of operation for both the devices.

The dc resistances are given by

$$R_{DE} = 0.76 \text{ V}/40 \ \mu A = 19 \text{ k}\Omega \quad \text{and} \quad R_{DD} = 4.24 \text{ V}/40 \ \mu A = 106 \text{ k}\Omega$$

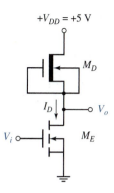

Figure 6.20
Circuit for Example 6.6

As seen in the above example, the dc resistance of the enhancement-type device (the driver) switches from ∞ for $V_i = 0$V to 19 kΩ for $V_i = 4$ V. The resistance of the depletion-type device (the load) varies in a complementary manner, switching from zero to 106 kΩ. Because of the large series resistance offered by one of the MOSFETs, the current in the circuit is extremely small at both input voltages. This unique feature of the MOSFET circuit is used to advantage in designing low-power switching circuits with no external resistances.

For large-signal variations, the capacitances between the electrodes in a MOSFET, shown in Figure 6.21, become significant. Capacitances between the gate and the other three terminals depend on the gate oxide capacitance, which is intrinsic to the device operation, and the area, WL. The gate-to-substrate capacitance C_{gb} when the channel is uniform, for example, is WLC_{ox}. This capacitance is split approximately equally between the drain and the source in a uniform channel; that is, in the ohmic region. When the channel is pinched off at the drain end, the gate-to-drain capacitance becomes zero. Only a small parasitic capacitance due to the overlap of the gate oxide above the drain region contributes to C_{gd}, which is on the order of 1 to 10 fF. Therefore, in the saturation region C_{gd} is negligibly small. At the source end, C_{gs} is given by $C_{gs} \approx (2/3)WLC_{ox}$, in saturation. This value of C_{gs} also neglects the capacitance due to the overlap of the gate oxide above the source region. (The overlap capacitance may be included in C_{gs} and C_{gd} by modifying the length of the channel to include the overlap.) In the ohmic region both C_{gs} and C_{gd} depend on voltages V_{DS} and V_{GS} in a nonlinear manner. For small V_{DS}, however, both are approximated by $(1/2)WLC_{ox}$. Combining the three capacitances, we use the following first-order approximation to the lumped gate capacitance in all modes of operation.

$$C_g = C_{gs} + C_{gd} + C_{gb} \approx WLC_{ox} \qquad \textbf{(6.23)}$$

Capacitances between the substrate and the channel, C_{bs} and C_{bd}, are the reverse-biased pn junction capacitances. As in junction diodes, these capacitances depend on the body-channel voltage, area, and concentration levels (see Section 2.4).

Other higher-order effects include the bulk resistances of the drain and the source, and the resistances and capacitances of the reverse-biased pn junctions

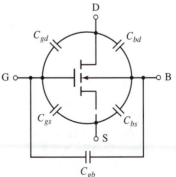

Figure 6.21   MOSFET capacitances

(between the drain and the substrate and between the source and the substrate). Additionally, one must consider the effect of temperature on the parameters at high operating temperatures. Since the mobility of the carriers varies approximately inversely with temperature, both the conduction parameter and the threshold voltage change with temperature. Similar to junction voltage change in a diode, threshold voltage changes with temperature in the range of approximately -2 mV/°C to -5 mV/°C. The conduction parameter k varies linearly with the mobility of carriers, which changes approximately as $T^{-(3/2)}$. Because of the exponential decrease of mobility with temperature, k decreases much more than V_T. Combining the slow decrease of V_T and the fast decrease of k with temperature, the overall effect is a general decrease in drain current.

6.6.2 MicroSim PSpice Model

The MicroSim PSpice model includes five levels for simulating MOSFET current-voltage characteristics. One selects a model level depending on the minimum feature size (from below 0.3 μm to above 4 μm), and availability of processing parameters, geometry, and measured characteristics. Model parameters in level 1 (the square-law model) include the dimensions of the channel, transconductance parameter, zero-bias threshold voltage, body-effect coefficient, channel-length modulation factor, and all of the capacitances we discussed in the previous section. Other levels use processing parameters or experimental values to compute the dc parameters such as threshold voltage. Level 1 uses simpler model equations requiring fewer parameters; drain current, for example, is calculated using 8 parameters in Level 1 while 21 or more are needed for other levels, which include higher-order effects. The Level 1 model, which is the default level, yields high enough accuracy with the analytic equations used in this text. Hence, MicroSim PSpice simulations in the text use only the Level 1 model. Figure 6.22 shows the dc model of an NMOSFET employed

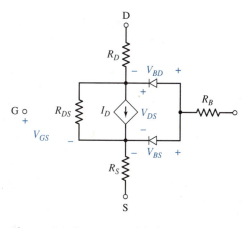

Figure 6.22   Dc model of an NMOSFET
used in MicroSim PSpice

in MicroSim PSpice. By specifying the process parameters and doping densities, one can evaluate the model parameters in MicroSim PSpice. Additionally, some parameters, such as the bulk (contact) resistances, may be included in the model statement to override calculated values.

Table 6.1 lists some of the MicroSim PSpice parameters used in Level 1. For more detailed lists of parameters in all five levels, one can refer to MicroSim PSpice user's guide [Reference 8].

Table 6.1   Some MicroSim PSpice level 1 (default level) parameters for MOSFET

Symbol	Name	Parameter	Units	Default Value	Typical Value
V_{TO}	VTO	Threshold Voltage	V	0	1.0
k'	KP	Process Transconductance parameter	A/V^2	2E-5	2E-5
λ	LAMBDA	Channel-length modulation coefficient	V^{-1}	0	0.02
γ	GAMMA	Bulk threshold parameter	V$^{1/2}$	0	0.5
t_{ox}	TOX	Oxide thickness	m	—	1E-7

In addition to the above parameters, others such as channel length and width (whose default values may be set), resistances of the terminals (default = 0), and junction and terminal capacitances per unit area (default = 0) may be specified in the device statement of MicroSim PSpice files. When length and width are specified, parasitic capacitances and terminal resistances are scaled by the area. For identical devices paralleled together (to obtain high current driving capability, for example), a multiplier M can be included to simulate scaled parameters for the device.

The process transconductance parameter KP in the MicroSim PSpice model is defined as KP $= \mu C_{ox}(W/L)$. Since the conduction parameter k used in this text is defined by Equation (6.9) as $k = (\frac{1}{2})\mu C_{ox}(W/L) = (\frac{1}{2})k'(W/L)$, the value of k' used in the text must be doubled for use as KP in the MicroSim PSpice simulation. Similarly, when default length L and width W are used, simulation must be carried out with KP $= 2k$ where k is the conduction parameter used in the text. Dc parameters such as KP may also be computed in Level 1; to do so, one must specify the process parameters such as substrate doping and oxide thickness, and must explicitly state Level $= 1$ in the model statement. Note further that the channel-length modulation factor, if specified, is used as a multiplier, $(1 + \lambda V_{DS})$, in both ohmic and saturation regions of operation in MicroSim PSpice. This ensures mathematical continuity of the current at the boundary of ohmic and saturation regions with negligible error.

The following example shows a PSpice simulation for Example 6.5.

```
* NMOSFET Simulation
; Example 6.7
ME      2   1    0    0  NE;  Enhancement
VI      1   0              ;  Input
MD      3   2    2    2  ND;  Depletion
VS      3   0    DC   5V
.DC     VI  0    5V   0.1V
.PROBE
.MODEL NE NMOS(VTO= 1 KP=20E − 6); 2*k, with default L = W = 100 micron
.MODEL ND NMOS(VTO= −2 KP=20E − 6); 2*K, with default L = W = 100 micron
.END
```

 Input File

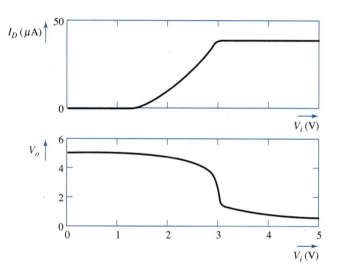

(b) Transfer Characteristics

Figure 6.23   MicroSim PSpice simulation of the circuit in Figure 6.20

Simulate the circuit of Figure 6.20 using MicroSim PSpice and obtain the V_i-V_o and V_i-I_D characteristics.

Example 6.7

Solution

Figure 6.23a shows the simulation input file. Note that the default length and width are used for both MOSFETs. Figure 6.23b shows the transfer voltage and current characteristics.

We consider different configurations of MOSFET logic circuits in Chapter 7.

6.7   Structure and Operation of GaAs MESFETs

Gallium arsenide is a III-V compound semiconductor formed of gallium, a rare element in the third column of the periodic table, and arsenic, a chemically active element in the fifth column of the periodic table.[5] The compound GaAs has electronic properties superior to those of silicon. The principal advantages of GaAs are the high mobility of electrons (5 to 10 times that in silicon) at low electric fields, high drift velocity (twice that in silicon), and high resistivity (up to 10,000 times that of silicon). Such high mobility and drift velocity contribute to the high transconductance and improved high-frequency performance of devices made of GaAs. High resistivity—in the range of 10^6 to 10^8 $\Omega \cdot$ cm— makes a GaAs substrate useful in providing isolation between devices with minimal processing steps in high-density integrated circuits. Furthermore, the semi-insulating substrate produces very low parasitic capacitance, thus improving the propagation of high-speed signals. Because of these advantages, field-effect transistors using GaAs in place of silicon have been in use in high-frequency microwave applications (in the gigahertz range). In their most commonly used form, GaAs transistors have a structure similar to that of MOSFETs but without the oxide insulation between the gate and the channel (see Figure 6.24). These transistors are known as *metal-semiconductor field effect transistors* (MESFETS). Digital circuit applications of GaAs MESFETs began in 1974, with SSI frequency-divider circuits.

A major limitation of the GaAs IC technology is the difficulty in achieving defect-free devices on GaAs because of problems encountered in the bulk growth of the compound semiconductor. For complex circuits, the yield is much lower than that using silicon, and this increases the cost. In addition, with the low mobility of holes in GaAs (400 cm²/V · s as opposed to 8000 cm²/V · s for electrons), the p-channel GaAs devices are not as advantageous; hence, complementary logic circuits, which are widely used in silicon MOSFET logic systems to reduce static power, are not effective with GaAs. On the other hand, the speed availability of the silicon devices, which have been the mainstay of semiconductor electronics over four decades, is nearing the fundamental limits of technology at around 10 Gbits/s of data rate. GaAs devices with improved technology are promising to offer speeds beyond this rate.

Two major GaAs technologies, *the planar ion-implanted depletion-type MESFET* and *the enhancement-type MESFET,* are available for digital circuit applications with better fabrication yield and reliability than others. A Schottky barrier diode fabricated on GaAs is another high-performance device that is extensively used in digital circuits. (Silicon Schottky barrier diodes are considered in Chapter 2.) Both depletion-type and enhancement-type MESFETs have been used in MSI and LSI systems. This section introduces the structure and operation of the depletion- and enhancement-type GaAS MESFET devices.

| [5] Some other two-element or binary III-V semiconductors are AlP, GaP and InP.

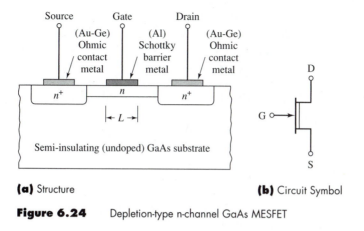

(a) Structure

(b) Circuit Symbol

Figure 6.24   Depletion-type n-channel GaAs MESFET

6.7.1   N-CHANNEL DEPLETION-TYPE MESFET (D-MESFET)

Figure 6.24 shows the structure of a GaAs n-channel depletion-type MESFET, which was the first GaAs device to be used at the LSI level. An active layer (about 0.3 μm) of n dopping (10^{17}/cm^3) is grown on an undoped (semi-insulating) substrate of GaAs. Two ohmic contacts (with heavy n doping of about 10^{18}/cm^3), separated by about 5 μm, are made for the source and the drain terminals. A Schottky barrier gate electrode is formed as a rectifying contact by depositing a film of aluminum (or a trilayer of titanium, platinum, and gold) on the n-doped active layer (see Figure 6.24) between the source and the drain contacts. The length of the gate electrode in direct contact with the active layer defines the channel length L, presently less than 0.5 μm. The short channel reduces carrier transit (drift) time and contributes to high-frequency operation. In LSI circuits, all devices are made of the same length L with different widths W for design flexibility. Because of the semi-insulating nature of the substrate, isolation between devices is provided by undoped regions of the substrate.

With the active n-channel present, the operation of a D-MESFET is very similar to that of a silicon D-NMOSFET. At zero gate-source and drain-source voltages, the built-in voltage of the Schottky barrier metal and the GaAs channel depletes part of the channel symmetrically below the gate. For a negative gate voltage V_{GS}, the thickness of the depletion region stretches further into the channel. At gate voltage $V_{GS} < V_T < 0$, the Schottky barrier depletion region extends through the entire depth of the channel and the channel is pinched off. This pinch-off occurs at a slightly higher threshold voltage V_T, due to the Schottky barrier voltage, than in an insulated-gate D-NMOSFET. V_T for a depletion-type GaAs MESFET is in the range of -0.3 V to -3 V. For a small positive drain-source voltage V_{DS} ($\ll$1V), the conductivity of the channel is modulated by the gate voltage $V_{GS} > V_T$. With $V_{GS} > V_T$, for small V_{DS} the channel behaves as a linear resistor causing an increase in drain current with V_{GS}, much like a MOSFET in the ohmic region. As V_{DS} is increased, the reverse bias across the gate-to-channel junction increases at the drain end. Therefore, there is no charge flow from the channel to the gate. With increasing voltage drop along the channel, the depletion region near the drain widens and constricts the channel. At $V_{DS} = V_{DS(\text{sat})}$ the drain current saturates and remains relatively constant, as in a MOSFET. This behavior agrees well with experimental drain characteristics for MESFETs with gate length longer than channel depth. With short gate lengths ($L \leq 1$ μm), however, the velocity of channel electrons saturates at a much lower drain-source voltage due to mobility saturation. As a result, the drain current I_D saturates at a much smaller V_{DS} (below $V_{GS} - V_T$) than in a corresponding MOSFET. This effect is modeled in the $I_D - V_{DS}$ relationship using an empirical function of V_{DS}. At V_{GS} around 0.6 V, the Schottky barrier gate junction is sufficiently forward-biased to cause significant forward current in the gate. This is in contrast to the silicon D-MOSFET operation where the gate is insulated from the channel by the oxide layer.

6.7.2   N-CHANNEL ENHANCEMENT-TYPE MESFET (E-MESFET)

GaAs enhancement-type MESFETs (E-MESFETs), although not as mature in technology and yield as the depletion-type devices, are available as normally off devices. These operate in a manner similar to the silicon E-MOSFETs, but the MESFETs have a lightly doped n-channel that is totally depleted by the built-in potential of the Schottky junction. Application of a positive gate voltage above threshold voltage forward-biases the gate junction and reduces the thickness of the depletion region. As with D-MESFETs, a high forward gate voltage of above about 0.6 V causes a gate current. The threshold voltage of the E-MESFETs is typically in the range of 0.1 V to 0.3 V.

Another type of GaAs MESFET is fabricated by separating a heavily doped channel (AlGaAs) from the undoped semi-insulating substrate (GaAs) by a thin layer of undoped spacer (AlGaAs). The spacer separates the channel electrons from the donor ions so that the interaction between them is reduced. With reduced scattering effects from ionized donors, electron mobility is increased to as high as 2×10^6 cm^2/V · s at low temperatures. Such a MESFET is called a *high-electron-mobility transistor* (HEMT). Because of its multilayer structure with doped and undoped layers, the HEMT is also known as the *modulation-doped FET* (MODFET). HEMTs offer an ultrafast switching speed of around 10 ps, with power dissipation of below 1 mW.

Other GaAs devices being developed for digital logic applications include heterojunction bipolar transistors. These devices offer far greater switching speeds compared to depletion- or enhancement-type GaAs MESFETs. Their fabrication technology, however, is still maturing.

6.8 MESFET Models

There are several analytical models available for describing the operation of MESFETs. Because of the low operating voltages, simplified dc models only approximately represent the MESFET behavior in the region below saturation. In this section, we consider two commonly used models that are suitable for hand calculations. These models are also available for simulation in MicroSim PSpice.

6.8.1 Curtice and Raytheon Models

The drain current-voltage relationship in the *Curtice model is* given by

$$I_D = 0 \quad \text{for} \quad V_{GS} < V_T \tag{6.24a}$$

$$I_D = \beta(V_{GS} - V_T)^2(1 + \lambda V_{DS}) \tanh(\alpha V_{DS}) \quad \text{for} \quad V_{GS} > V_T \tag{6.24b}$$

where β represents the transconductance parameter, A/V^2 per unit area. λ is the empirical channel-length modulation factor to account for the increase in drain current in saturation, which is similar to the behavior in silicon MOSFETs. The coefficient α is determined by fitting to experimental data. (Note that α can be determined from the slope of the I_D-V_{DS} graph at $V_{DS} = 0$.) The value of α ranges from 0.3 (for a long channel, $L \approx 20 \ \mu m$) to about 4 (for $L \leq 1 \ \mu m$).

Since the hyperbolic tangent function is monotonic, it may be approximated by a polynomial depending on the argument. This is the approximation that is used in the *Raytheon model.* It gives the drain current for $V_{DS} < 3/\alpha$ as

$$I_D = \frac{\beta(V_{GS} - V_T)^2(1 + \lambda V_{DS})\left[1 - \left(\dfrac{\alpha V_{DS}}{3}\right)^3\right]}{1 + b(V_{GS} - V_T)} \tag{6.25a}$$

and, for $V_{DS} \geq 3/\alpha$,

$$I_D = \frac{\beta(V_{GS} - V_T)^2(1 + \lambda V_{DS})}{1 + b(V_{GS} - V_T)} \tag{6.25b}$$

The constant b, called the *doping tail extending parameter,* is used to model the effects of velocity saturation. Equation (6.25) improves computation time for simulation at the expense of accuracy compared to Equation (6.24). Figure 6.25 shows the simulated characteristics of an n-channel GaAs depletion-type MESFET using the two models in PSpice.

In addition to the above models, an early approximate square-law model, similar to that representing a MOSFET in the ohmic and the saturation regions, is found satisfactory for noncritical applications. In this model the drain current is given by [Reference 5]

$$I_D = \beta[2(V_{GS} - V_T)V_{DS} - V_{DS}^2](1 + \lambda V_{DS}) \tag{6.26}$$

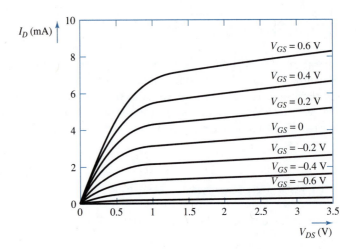

(a) Using Curtice Model

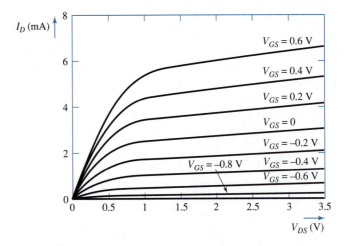

(b) Using Raytheon Model

Figure 6.25 Simulated I_D-V_{DS} characteristics of a GaAs MESFET

in the ohmic region, where $V_{DS} < V_{GS} - V_T$ and $V_{GS} \geq V_T$.

In the saturation region,

$$I_D = \beta(V_{GS} - V_T)^2 (1 + \lambda V_{DS}) \tag{6.27}$$

where $V_{DS} \geq V_{GS} - V_T$ and $V_{GS} \geq V_T$.

Note that compared with the Curtice model [Equation (6.24b)], Equation (6.27) assumes $\tanh(\alpha V_{DS}) = 1$.

Although Equations (6.26) and (6.27) are much simpler to use, calculated currents in the ohmic region differ from those measured experimentally. The error occurs because of deficiencies in the model, namely, (1) the model does not include velocity-saturation effects, and (2) it assumes current saturation and pinch-off occurring at $V_{DS} = V_{GS} - V_T$. In addition, the square-law model works well only for an impulse doping profile in the channel [see Reference 5]. Also, convergence of solution may not result in the vicinity of saturation as predicted by the model. In spite of these limitations, the simplified model may be used for rapid solutions where exact results are not crucial to understanding the behavior of a MESFET circuit. (Although the voltages in the model equations refer to the internal junction voltages, we shall use the terminal voltages for hand calculations.)

6.8.2   MicroSim PSpice Model

MicroSim PSpice uses a model for the GaAs MESFET with the schematic shown in Figure 6.26. The diodes in this model represent the Schottky barrier junctions between gate and source, and gate and drain. Since the internal junction voltages for these diodes are different from the terminal voltages, the model includes ohmic resistances R_D and R_S to account for the difference. These resistances are inversely proportional to the channel width. The dependent current source models the channel conductivity modulation by V_{GS} and V_{DS}.

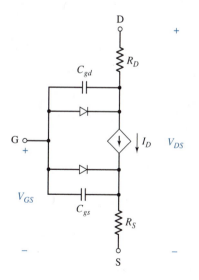

$$I_D = B(V_{GS} - V_T)^2 (1 + \lambda V_{DS}) \tan h \, (\alpha V_{DS})$$

Figure 6.26   MicroSim PSpice model for GaAs MESFET

Both Curtice (Level = 1) and Raytheon (Level = 2) models are available for calculating the drain current in MicroSim PSpice. [A third model, called TriQuint model, is also available as Level = 3; see Reference 8.]

Table 6.2 lists the MicroSim PSpice parameters commonly used with unit area ($L = W = 1$ μm). Parameters shown with an asterisk under the area column indicate that their values are scaled by the area parameter specified in the device statement. For a device of area $1 \times W$ μm^2, for example, the transconductance parameter used for simulation will be βW, where β is specified in the model statement. Resistances and capacitances are similarly scaled by the area parameter (see Example 6.8). Values shown under "Typical Value" are those used to obtain the characteristics shown in Figure 6.25 with an area of 1×20 μm^2. Note that, unlike with other device models, the drain and source resistances must be nonzero to obtain meaningful simulation results.

Table 6.2   MicroSim PSpice GaAs MESFET model parameters

Symbol	Name	Parameter	Units	Default	Area	Typical Value
Level	Level	Model level		1		
V_T	VTO	Threshold voltage	V	−2.5		−1
α	ALPHA	Saturation voltage parameter	V^{-1}	2		3
b	B	Doping tail extending parameter (Level = 2 only)	V^{-1}	0.3		0.3
β	BETA	Transconductance parameter	A/V^2	0.1	*	7E-5
λ	LAMBDA	Channel-length modulation	V^{-1}	0		0.1
R_D	RD	Drain resistance	Ω	0	*	1600
R_S	RS	Source resistance	Ω	0	*	1600
C_{gs}	CGS	Zero-bias gate-source junction capacitance	F	0	*	5E-14
C_{gd}	CGD	Zero-bias gate-drain junction capacitance	F	0	*	5E-14
V_{bi}	VBI	Gate junction built-in potential	V	1		0.8
I_S	IS	Gate junction saturation current	A	1E-14	*	3E-16

GaAs Schottky diodes are modeled similarly to silicon Schottky diodes. As with the MESFETs, the series resistance due to contact and bulk n-type GaAs material must be included in the ideal diode model for useful simulations.

The following example analyzes a resistive MESFET inverter similar to the MOSFET configuration we considered in Example 6.2. In this example, the MOSFET square law model is used because of its simplicity in hand calculations.

Example 6.8

Determine the output voltage V_o in the circuit shown in Figure 6.27 for (a) $V_i = 0$ V and (b) $V_i = 0.4$ V. M_D has an area of 50 μm^2. Use the typical values shown in Table 6.2 for the parameters of the MESFET with unit area.

Solution

(a) For $V_i = 0$ V, the depletion MESFET is on. To simplify the calculations we shall assume the MESFET is in saturation, that is, $V_{DS} \geq V_{GS} - V_T$, and we shall use the approximate current equation [Equation (6.27)]:

$$I_D = \beta(V_{GS} - V_T)^2(1 + \lambda V_{DS})$$

From the circuit,

$$I_D = \frac{2 - V_{DS}}{250}$$

Solving the above equations, we have

$$V_o = V_{DS} = 1.034 \text{ V} \quad \text{and} \quad I_D = 3.86 \text{ mA}$$

The value for V_{DS} verifies our assumption of $V_{DS} \geq V_{GS} - V_T$. The Raytheon model, which includes a correction factor of $[1 + b(V_{GS} - V_T)]$, gives $V_{DS} = 1.243$ V and $I_D = 3.03$ mA.

(b) For $V_i = 0.4$ V, the MESFET conducts heavily, dropping significant voltage across R_D; hence, the device is likely to be operating in the ohmic region. Again, using the form of MOSFET current in the ohmic region,

$$I_D = \beta[2(V_{GS} - V_T)V_{DS} - V_{DS}^2](1 + \lambda V_{DS}) \qquad \textbf{(6.26)}$$

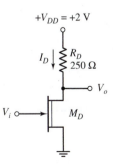

Figure 6.27
Resistive inverter using a MESFET

and

$$I_D = \frac{2 - V_{DS}}{250}$$

we have

$$V_o = V_{DS} = 0.706 \text{ V} \quad \text{and} \quad I_D = 5.18 \text{ mA}$$

Note that the $V_{DS} < V_{GS} - V_T$.

Figure 6.28 shows the MicroSim PSpice simulation for the behavior of the MESFET using the Curtice model and the graphical solution for Example 6.8. The results are: for $V_i = 0$ V, $V_o = 1.22$ V and $I_D = 3.11$ mA; for $V_i = 0.4$ V,

```
* GaAs D-MESFET CHARACTERISTICS-Curtice model
; Example 6.8
B        2    1    0           g1d 50
VD       1    0
VD       2    0
.DC      VD   0    3.5   0.1   VG − 1 0.6 0.6 0.2
.PROBE
.MODEL g1d gasfet (VTO= −1 alpha=3 beta=73-5
+ lambda=0.1 rd=1600 rs=1600 is=3e−16 vbi=0.8 level=1)
.END
```
(a) Simulation

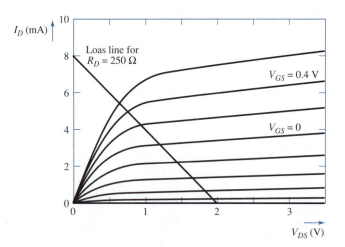

(b) Graphical Solution

Figure 6.28   MicroSim PSpice simulation of circuit in Example 6.8

$V_o = 0.76$ V and $I_D = 4.91$ mA. These values show a reasonable agreement with the results of calculations using the simplified MOSFET model.

Note that the transconductance parameter β is specified for unit area in the device statement. This value is scaled by the area factor of 50 given for the device parameter for the device in Example 6.8. Other values in the .model statement may also be appropriately scaled in the MicroSim PSpice before simulation.

The next example illustrates a basic direct-coupled FET logic (DCFL) inverter. In this circuit, an E-MESFET (source) drives a D-MESFET (load), similar to the complementary MOSFET circuit we considered in Example 6.6.

Determine the output voltage V_o and the drain current I_D in the circuit shown in Figure 6.29 for (a) $V_i = 0$, (b) $V_i = 0.2$ V, and (b) $V_i = 0.5$ V. Use the following parameters for the depletion-type MESFET: $V_{TD} = -0.5$ V, $\alpha = 2.5$, $\beta = 80\mu$A/V^2, $\lambda = 0.05$, $R_D = R_S = 1\ \Omega$. For the enhancement-type MESFET, use the same parameters with $V_{TE} = 0.1$ V, and $\beta = 500\ \mu$A/V^2.

Example 6.9

Solution

(a) For $V_i = 0$, the E-MESFET M_E is off while the D-MESFET M_D is on, with $V_{GS} = 0$. Hence,

$$I_D = 0 \quad \text{and} \quad V_o = V_{DD} = 1.2 \text{ V}$$

(b) For $V_i = 0.2$ V, which is only slightly above $V_{TE} = 0.1$ V, M_E may be in saturation with M_D supplying a small current in its ohmic region.

Using the simplified MOSFET model [Equation (6.27)] for M_E,

$$I_D = \beta(V_i - V_T)^2 = 5(1 + 0.05V_o) \ \mu\text{A}$$

For M_D, in the ohmic region.

$$I_D = \beta[2(V_{GS} - V_T)V_{DS} - V_{DS}^2](1 + \lambda V_{DS})$$
$$= 80[(2 - V_o) - (2 - V_o)^2][1 + 0.05(2 - V_o)]\mu\text{A}$$

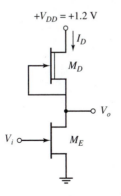

$+V_{DD} = +1.2$ V

I_D

M_D

V_o

V_i

M_E

Figure 6.29 Circuit for Example 6.9

Solving the two equations, we get

$$V_o = 1.07 \text{ V} \quad \text{and} \quad I_D = 5.26 \ \mu\text{A}$$

The above voltage satisfies the assumed models of operation for the two MES-FETs.

The MicroSim PSpice simulation results given in Figure 6.30 show $I_D = 5.22 \ \mu\text{A}$ and $V_o = 1.1$ V.

(c) For $V_i = 0.5 \text{ V} > V_{TE}$, it is reasonable to assume that M_E is in the ohmic region and M_D in saturation. Using the simplified current equations for the two devices, we have

$$20[1 + 0.05(2 - V_o)] = 500(0.8V_o - V_o^2)(1 + 0.05V_o)$$

The only meaningful solution to the above equation is $V_o = 59.2$ mV. This gives a current of $I_D = 21.94 \ \mu\text{A}$.

The MicroSim PSpice solution for this case (Figure 6.30) is $V_o = 106$ mV and $I_D = 21 \ \mu\text{A}$. As stated earlier, neglect of the voltage drops in the bulk resistances of the source and the drain terminals, and the model inaccuracy at low voltages, contribute to the large error in V_{DS} using the simplified model.

From the above examples we see that the MESFETs can be used in place of the MOSFETs in logic circuits with the same configuration but at low voltages.

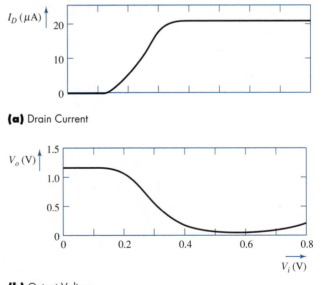

(a) Drain Current

(b) Output Voltage

Figure 6.30   MicroSim PSpice simulation of circuit in Example 6.9

To avoid gate currents, however, we must restrict operation to significantly low voltages. Low-voltage operation, on the other hand, contributes to faster charging and discharging of load capacitances in switching circuits. In addition, we can accomplish on-chip interconnection of a large number of circuits in LSI and VLSI systems without requiring large voltage swings. High switching speed, low-voltage operation, and the relative ease in isolating different devices in a large system are definite advantages of the MESFET logic circuits.

SUMMARY

This chapter introduced the metal-oxide-semiconductor field-effect transistors and the metal-semiconductor field-effect transistors, both operating with only one type of charge carriers. An n-channel enhancement-type MOSFET (normally off device) has a p substrate to which the drain and the source terminals are attached. The channel, which is formed in the space between the drain and the source terminals, is separated from the gate terminal by a thin insulating layer of oxide. For normal circuit operation the substrate (body) is connected to the lowest (most negative) potential. Positive voltage above a minimum of V_T, the threshold voltage, applied between the gate and the channel (body) draws electrons toward the gate and forms a continuous conducting channel between the source and the drain. Terminal current from drain to source is caused by a positive potential at the drain relative to the source. As the small drain-to-source voltage V_{DS} is increased, the drain current I_D increases linearly. For a fixed gate-to-source voltage V_{GS}, an increase in V_{DS} decreases the gate-to-channel voltage near the drain. This extends the depletion region. When $V_{GD} = V_{GS} - V_{DS} < V_T$, the depletion region near the drain end becomes deep enough to pinch off the channel. The drain current then levels off.

Depletion-type MOSFETs with n-channels are normally on devices with a lightly doped n-channel connecting the drain and the source. With $V_T < 0$, these devices operate in a manner similar to the enhancement-type MOSFETs.

Both depletion- and enhancement-type MOSFETs with p-channels are also available. Since the mobility of holes is smaller than that of electrons, PMOSFETs operate at lower speeds compared with NMOSFETs.

The higher-order effects of MOSFETs include the dependance of the threshold voltage on the reverse-bias voltage between the body and the source. In addition, the narrowing of the effective channel in pinch-off increases the electric field and causes a slight increase in drain current.

Because of the insulated gate, gate current in a MOSFET is extremely small, and the device is voltage operated. Static operation of MOSFET logic circuits offers large resistances, and hence, draws low power supply currents.

Gallium arsenide MESFETs are fabricated from a substrate of the semi-insulating compound semiconductor GaAs; they are similar to the silicon MOSFET, but with a doped channel. The gate terminal is formed by a Schottky metal barrier with the channel. Electrons in GaAs have high mobility and drift velocity, which enable higher speed of operation for MESFETs compared to silicon devices. Of the many evolving technologies of GaAs devices, enhancement- and depletion-type MESFETs are relatively more mature than others. These two types of devices operate in a manner similar to MOSFETs but at lower voltages.

Summary of MOSFET and MESFET Equations   Cut-off region:

$$V_{GS} < V_T \qquad I_D = 0$$

Ohmic/linear/voltage-variable resistance region:

$$V_{DS} \leq V_{GS} - V_T \quad \text{or} \quad V_{GD} \geq V_T$$

$$I_D = k[2(V_{GS} - V_T)V_{DS} - V_{DS^2}]$$

Saturation/pinch-off region:

$$V_{DS} \geq V_{GS} - V_T$$

$$I_D = k(V_{GS} - V_T)^2 \quad \text{Without channel-length modulation}$$

$$I_D = k(V_{GS} - V_T)^2(1 + \lambda V_{DS}) \quad \text{With channel-length modulation}$$

The above equations are applicable for GaAs MESFETs as a first-order model for hand calculations. The conduction parameter k is replaced by the transconductance parameter β, and the channel-length modulation must be included for MESFETs.

REFERENCES

1. Penney, W. M., and L. Lau (eds.). *MOS Integrated Circuits: Theory, Fabrication, Design, and Systems Applications of MOS LSI.* New York: Van Nostrand Reinhold, 1972.

2. Muller, R. S., and T. I. Kamins. *Device Electronics for Integrated Circuits.* New York: Wiley, 1986.

3. Uyemura, J. P. *Fundamentals of MOS Digital Integrated Circuits.* Reading, MA: Addison-Wesley, 1988.

4. Hodges, D. A., and H. G. Jackson. *Analysis and Design of Digital Integrated Circuits.* New York: McGraw-Hill, 1988.

5. Long, S. I., and S. E. Butner. *Gallium Arsenide Digital Integrated Circuit Design.* New York: McGraw-Hill, 1990.

6. Sedra, A. S., and K. C. Smith. *Microelectronic Circuits.* Philadelphia: Saunders College Publishing, 1991.

7. Tuinenga, P. W. *SPICE: A Guide to Circuit Simulation and Analysis Using PSpice.* Englewood Cliffs, NJ.: Prentice Hall, 1995.

8. *PSpice Circuit Analysis User's Guide,* Irvine, CA: MicroSim Corporation, 1991.

REVIEW QUESTIONS

1. Why is there no gate current in the operation of a MOSFET?
2. What are the regions of operation of a MOSFET? How are they defined?
3. What is the difference in physical structure between an E-NMOSFET and D-NMOSFET? How is the channel in an E-MOSFET formed?
4. How does the threshold voltage differ between an E-NMOSFET and a D-NMOSFET?
5. Why are NMOSFETs preferred over PMOSFETs?
6. Define the aspect ratio of a MOSFET. What is its significance?
7. How does saturation of drain current occur in an NMOSFET?
8. What is channel-length modulation in a MOSFET?
9. What are the significant elements in the dc model of a MOSFET in saturation?
10. What is the body effect in a MOSFET?
11. What causes speed improvement in GaAs devices compared to Si devices?
12. How does the gate formation in a MESFET differ from that in a MOSFET?
13. What causes gate current in a MESFET?
14. Why is the input voltage range smaller in a MESFET compared to that in a MOSFET?
15. Why are PMESFETs not preferred over NMESFETs?

PROBLEMS

1. For the circuit shown in Figure P6.1, determine (*a*) V_i and V_o at which the MOSFET is at the boundary between the pinch-off and ohmic mode, and (*b*) V_o for $V_i = 5$ V. Assume $V_T = 1$ V and $k = 100$ $\mu A/V^2$.

2. For the circuit shown in Figure P6.2, determine (*a*) V_i for $V_o = 3$ V, and (*b*) V_o for $V_i = 4$ V. Given: $k_L = 50$ $\mu A/V^2$, $k_D = 100$ $\mu A/V^2$, and $V_{TL} = V_{TD} = 1$ V. What are the dc resistances of the MOSFETs in (*a*) and (*b*)? If the body of M_L is at ground potential, recalculate V_i and V_o for (*a*) and (*b*). Use the approximate body-effect equation with a γ of 0.5.

3. For the circuit shown in Figure P6.3, calculate (*a*) V_i for $V_o = 3$ V, and (*b*) V_o for $V_i = 2$ V. Use $k_L = 50$ $\mu A/V^2$, $k_D = 100$ $\mu A/V^2$, and $V_{TL} = V_{TD} = 1$ V. What are the dc resistances of the MOSFETs in (*a*) and (*b*)?

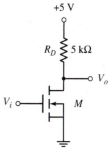

Figure P6.1

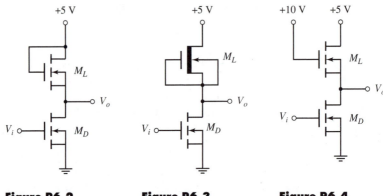

Figure P6.2   **Figure P6.3**   **Figure P6.4**

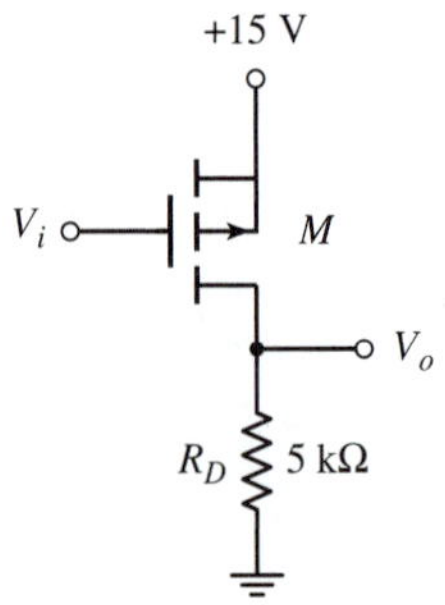

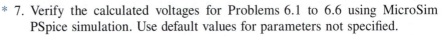

4. For the circuit shown in Figure P6.4, calculate (*a*) V_i for $V_o = 3$ V, and (*b*) V_o for $V_i = 4$ V. Find the power supply current for both these cases. Given: $k_L = 100 \ \mu\text{A/V}^2$, $k_D = 1$ mA/V², $V_{TL} = -2$ V, and $V_{TD} = 1$ V.

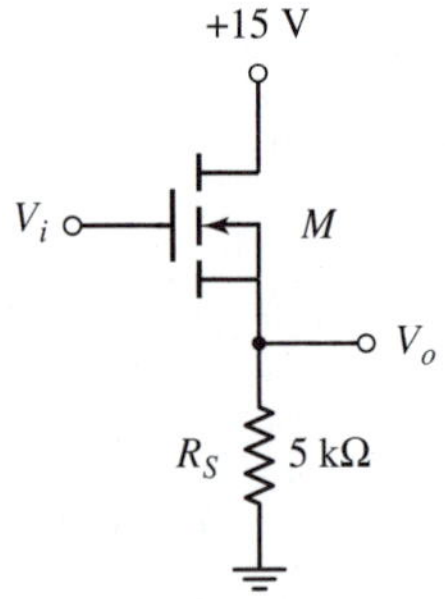

5. For the PMOS circuit shown in Figure P6.5 with the MOSFET parameters of $V_T = -1.5$ V and $k = 30 \ \mu\text{A/V}^2$, calculate (*a*) V_o for $V_i = 3$ V, (*b*) V_o for $V_i = 12$ V, and (*c*) V_i for $V_o = 7$ V. Express the relationship for saturation of the MOSFET in terms of V_i and V_o. At what voltage is $V_o = V_i$?

6. Repeat Problem 6.5 for the NMOS circuit shown in Figure P6.6. The MOSFET parameters are: $V_T = 1.5$ V and $k = 30 \ \mu\text{A/V}^2$.

* 7. Verify the calculated voltages for Problems 6.1 to 6.6 using MicroSim PSpice simulation. Use default values for parameters not specified.

8. For the MOSFET amplifier circuit shown in Figure P6.8, determine the voltage gains, v_d/v_i and v_s/v_i. The MOSFET parameters are: $V_T = 1.5$ V

Figure P6.5

Figure P6.6

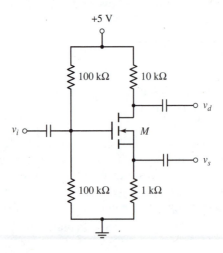

Figure P6.8

and $k = 30 \ \mu A/V^2$. The coupling capacitors have negligible reactance at the frequency of operation.

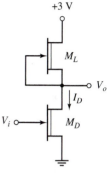

Figure P6.10

9. Determine the drain voltage and current in Figure 6.26 for $R_D = 500 \ \Omega$ and (*a*) $V_i = 0$, and (*b*) $V_i = 0.4$ V. Use typical values given in Table 6.2 for the MESFET. Verify your results graphically using the characteristics shown in Figure 6.28b.

*10. The circuit shown in Figure P6.10 is an unbuffered FET Logic inverter. Using simplified current equations, determine the output voltage V_o and the current I_D for (*a*) $V_i = -0.5$ V, (*b*) $V_i = 0$, and (*c*) $V_i = 0.5$ V. Assume M_D and M_L are identical depletion MESFETs with parameters shown under "Typical Values" in Table 6.2.

Verify your results using Curtice and Raytheon models in a MicroSim PSpice simulation.

11. *Experiment:* In this experiment, some of the MOSFET parameters are extracted for NMOS and PMOS devices in the array, CD 4007UB. CD 4007UB consists of three pairs of E-NMOS and E-PMOS devices, with the gates of each complementary pair connected together as shown in Figure P6.11a.

   a. Connect the circuit shown in Figure P6.11b using one of the NMOS-FET devices. Wire the body terminal to the source so that $V_{SB} = 0$ and thus there is no body effect. To determine the threshold voltage $V_T = V_{T0}$, increase the gate voltage V_{GS} gradually from zero until the drain voltage begins to drop from $V_D = V_{DD} = 10$ V. (The drain current at this point may be about 1 μA.)

   b. For each gate voltage V_{GS} from 2 V to 8 V in steps of 1 V, measure I_D as a function of V_{DS} by varying the supply voltage V_{DD} from 0 to 10 V.

   c. Measure the threshold voltage with body effect using the circuit for (*a*) above and a negative supply connected to the body. Use $V_{SB} = -V_B = 2$ V, 4 V, and 6 V.

   d. From the data for (*b*) above, obtain a plot of the drain characteristics I_D-V_{DS} as a function of V_{GS}. Using the drain current in the saturation region, where I_D is fairly constant with V_{DS}, plot $\sqrt{I_D}$ against V_{GS}. From the approximate straight-line plot, determine the device transconductance parameter k [Equation (6.12)]. From the slope of the $I_D - V_{DS}$ plot in the saturation region for a given V_{GS}, determine the channel-length modulation factor [Equation (6.13)]. If the slope is too small to measure, extend the drain current plots to the $-V_{DS}$ axis and determine $-1/\lambda$ = the intersecting voltage.

   e. From the data for (*c*) above, plot the measured threshold voltage as a function of $\sqrt{V_{SB}}$. Using Equation (6.19), determine the approximate body effect coefficient γ.

   f. Wire the PMOS circuit shown in Figure P6.11c. For measuring V_{T0}, connect the source and the body terminals together. Keep $-V_{SS} = -10$ V. Starting at gate voltage of $V_G = 0$, gradually decrease

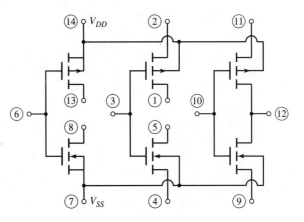

(a) Functional Diagram of CD 4007U
Courtesy Harris Semiconductor.

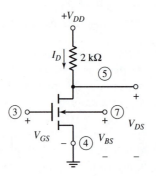

(b) NMOS Circuit for Measurement   **(c)** PMOS Circuit for Measurement

Figure P6.11

V_G. Note down $V_{T0} = V_{GS}$ when the drain voltage V_D begins to rise above $-V_{SS}$. Repeat (*b*) for $V_{GS} = -2$ V to -8 V in steps of -1 V. Repeat (*c*) for $V_{BS} = V_B = 2$ V, 4 V, and 6 V.

g. As with the NMOS device, plot the data and calculate the transconductance parameter, the channel-length modulation factor, and the approximate body-effect coefficient for the PMOS device.

7

MOSFET LOGIC CIRCUITS

INTRODUCTION

In this chapter we study logic circuits using silicon MOSFET and gallium arsenide MESFET devices. LSI and VLSI circuits using NMOSFETs have been available since the 1970s. While PMOSFETs are equally useful, they were overshadowed by the high-speed and high-density NMOSFET technology since the beginning of the MOSFET evolution in digital ICs. Complementary MOS-FET (CMOS) circuits use both NMOS and PMOS transistors, with reduced power dissipation and improved static and dynamic performance. Although CMOS circuits first appeared in the 1960s, difficulty in fabricating both n-channel and p-channel devices on the same chip and the development of bipolar digital circuits with high speed (TTL and ECL) and low power (I^2L) left them in the background. But with advances in processing technology in the 1980s, CMOS has reemerged as the leading technology for SSI to VLSI applications, with high density, low power, and superior performance characteristics. The switching speed of CMOS circuits, however, remained behind that of bipolar circuits, especially for large fan-out or for capacitive loads. More recently, bipolar output stages, with their high current-driving capabilities, have been merged with CMOS circuits to combine the advantages of both bipolar and CMOS technologies. The resulting BiCMOS technology, which appeared in the mid 1980s, is rapidly becoming the leading technology for high-speed and high-density LSI and VLSI applications.

Parallel with the silicon CMOS and bipolar technologies, integrated circuits fabricated in gallium arsenide began to appear in the early 1980s. With higher electron mobility at lower electric field in GaAs than that in silicon, GaAs devices are capable of operating at frequencies in the gigahertz range. Digital ICs based on the GaAs MESFET technology has been developed over the past decade for applications in high-speed computers and optical communications systems. Because of the maturing of technology, many different digital circuit configurations have evolved that have delays in the tens of picoseconds range. Due to processing complexity and low yield, however, GaAs technology is presently limited to selected MSI and LSI functions only.

In this chapter we analyze logic circuit configurations using NMOS, CMOS, BiCMOS, and GaAs MESFETs. The chapter concludes with a brief description of interfacing circuits for compatible bipolar and MOS logic families, and a comparison of the different logic families studied so far.

7.1 NMOSFET INVERTERS

Logic inverters and gates can be realized using NMOSFETs exclusively, somewhat analogous to using only npn transistors. Unlike the bipolar transistor, however, NMOSFETs are available as depletion type (D-NMOSFET) or enhancement type (E-NMOSFET), both of which can be operated in saturation (pinch-off) or linear (ohmic) mode. As a result, several NMOS inverter configurations are possible. We study the analysis of commonly used NMOS inverter circuits in this section.

7.1.1 NMOSFET INVERTER WITH A RESISTIVE LOAD

The simplest MOSFET inverter uses an enhancement-type NMOSFET (E-NMOSFET) and a resistor as shown in Figure 7.1a. The body and the source of the MOSFET are connected together so that the body effect is prevented. The input logic signal is applied to the gate, ($V_i = V_{GS}$) and the output is taken at the drain ($V_o = V_{DS}$).

For input V_i below the (positive) threshold voltage V_T of the E-NMOSFET, the device is off and $V_o = V_{DD}$. At the logic high level of input, that is, for $V_i \approx V_{DD}$ ($>V_T$), the transistor is on and the output is nearly at ground (source) voltage. We can obtain the complete voltage transfer characteristic (VTC) of the circuit for $0 < V_i < V_{DD}$, shown in Figure 7.1c, from the drain characteristics and the load line as shown in Figure 7.1b. Analytically the VTC is derived based on the mode of operation of the MOSFET.

As input V_i is increased from zero, conduction of the MOSFET begins, and the output (drain) voltage drops from V_{DD} at $V_i = V_T$. Thus, the first critical point in the characteristic occurs at $V_i = V_T$ and $V_o = V_{DD}$. Since the induced channel from drain to source has barely formed, there is little drain current. Hence, $V_{DS} = V_o > V_{GS} - V_T$, and the device is operating in the pinch-off, or saturation, mode. Further rise in V_i increases the drain current, and the output voltage drops as

$$V_o = V_{DD} - I_D R_D \tag{7.1}$$

In the saturation region, the drain current I_D is given by

$$I_D = k(V_{GS} - V_T)^2 \quad \text{or} \quad I_D = k(V_i - V_T)^2 \tag{7.2}$$

Solving Equations (7.1) and (7.2), we have the input-output relationship after the first critical point as

$$V_o = V_{DD} - kR_D(V_i - V_T)^2 \tag{7.3}$$

Equation (7.3) shows that the output voltage V_o drops in a nonlinear manner due to the nonlinear increase of the drain current with the input voltage V_i. When

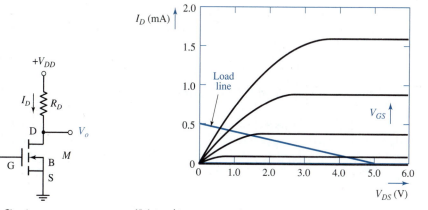

(a) Circuit   **(b)** Load Line

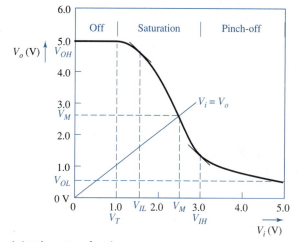

(c) Voltage Transfer Characteristic

Figure 7.1   NMOSFET inverter with a passive resistor load

V_o is sufficiently low so that the condition for pinch-off, that is, $V_{DS} \geq V_{GS} - V_T$ or $V_o \geq V_i - V_T$, is no longer satisfied, the MOSFET enters the ohmic mode of operation.

The second critical point in the VTC occurs when the MOSFET is at the boundary between the saturation and ohmic modes. Output voltage V_o at this critical point is obtained using $V_o = V_i - V_T$ in Equation (7.3).

Increasing V_i above the second critical point keeps the transistor in the ohmic mode. The drain current in the ohmic mode is given by

$$I_D = k[2(V_i - V_T)V_o - V_o^2] \tag{7.4}$$

Since this is also nonlinear, there is a second nonlinear region in the VTC beginning at $V_o = V_i - V_T$. Using Equations (7.1) and (7.4), the output voltage in this region is given by

$$V_o = V_{DD} - kR_D[2(V_i - V_T)V_o - V_o^2] \qquad \textbf{(7.5)}$$

All three regions are identified in the transfer characteristics shown in Figure 7.1c.

Since the VTC shows a continuous variation of V_o with V_i, there are no clear breakpoints, as in the case of the VTC of a TTL circuit. Logic levels V_{IL}, V_{IH}, V_{OL}, and V_{OH} are, therefore, determined based on the slope criterion $dV_o/dV_i = -1$ for the inverter. The following example illustrates the calculation of logic levels.

Example 7.1

Determine the critical voltages and the logic levels for the NMOSFET inverter of Figure 7.1a. The MOSFET has $V_T = 1$ V and $k = 100 \ \mu A/V^2$. The load resistor is $R_D = 10 \ k\Omega$ and the supply voltage is $V_{DD} = 5$ V.

Solution

The first critical point is at $V_i = V_T = 1$ V and $V_o = V_{DD} = 5$ V. The MOSFET is off until $V_i = 1$ V.

For V_i above 1 V, the MOSFET is in pinch-off mode. From Equation (7.3), the output voltage in the pinch-off region is given by

$$V_o = 5 - (V_i - 1)^2 \qquad \textbf{(7.6)}$$

At the transition between the pinch-off and ohmic modes, $V_o = V_i - 1$. Hence the drain current at this voltage becomes

$$I_D = k(V_i - V_T)^2 = 0.1(V_i - 1)^2 \text{ mA}$$

or $$I_D = 0.1V_o^2 \text{ mA} \qquad \textbf{(7.7)}$$

and the output voltage is given by

$$V_o = 5 - V_o^2$$

Solving the quadratic equation, we have $V_o = 1.79$ V and -2.79 V. Discarding the negative value, the transition between pinch-off and ohmic modes occurs at

$$V_i = 2.79 \text{ V} \quad \text{and} \quad V_o = 1.79 \text{ V}$$

The logic low input voltage V_{IL} is obtained using Equation (7.6) and the slope requirement $dV_o/dV_i = -1$.

From Equation (7.6) we have

$$dV_o/dV_i = -2(V_i - 1)$$

Equating the slope to -1 at $V_i = V_{IL}$ we get $V_{IL} = 1.5$ V.

It is readily seen that at this input voltage, output V_o from Equation (7.6) is

$$V_o = 5 - (1.5 - 1)^2 = 4.75 \text{ V}$$

The above input and output voltages justify our assumption of pinch-off mode for the MOSFET.

Although V_{OH} is the output voltage corresponding to $V_i = V_{IL}$, the nominal output voltage of $V_o = V_{DD}$ is considered the logic high level. Note that the MOSFET is off at this output voltage.

Hence,
$$V_{OH} = 5 \text{ V}$$

At logic high input, the MOSFET is in the ohmic mode. Therefore, we calculate V_{IH} using Equation (7.5) and the slope criterion. From Equation (7.5),

$$V_o = 5 - [2(V_i - 1)V_o - V_o^2] \qquad \textbf{(7.5a)}$$

Thus,
$$dV_o/dV_i = [2V_o + 1 - 2V_i]/2V_o$$

Equating the slope to -1 at $V_i = V_{IH}$ results in

$$V_o = \frac{V_{IH}}{2} - 0.25 \qquad \textbf{(7.8)}$$

Substituting Equation (7.8) into Equation (7.5a) and solving, we get the logic high input and the corresponding output voltages:

$$V_{IH} = 3.08 \text{ V} \quad \text{and} \quad V_o = 1.29 \text{ V}$$

These voltages verify that the MOSFET is in the ohmic mode.

Since V_{OH} is defined as $V_{DD} = 5$ V, V_{OL} is the output voltage of an identical inverter (load) with $V_i = V_{OH} = 5$ V. From the quadratic equation resulting from Equation (7.5a), the physically meaningful solution is

$$V_{OL} \approx 0.6 \text{ V}$$

At this output voltage, a succeeding (load) inverter will be turned off and the logic levels are regenerated.

From the results for Example 7.1, we can determine the noise margins, the transition width, and the logic swing:

$$NM_L = V_{IL} - V_{OL} = 0.9 \text{ V} \quad \text{and} \quad NM_H = V_{OH} - V_{IH} = 1.92 \text{ V}$$

$$V_{tw} = V_{IH} - V_{IL} = 1.58 \text{ V}$$

$$V_{ls} = V_{OH} - V_{OL} = 4.4 \text{ V}$$

Other points in the VTC that are often useful are the transition points at which the MOSFET is changing from one mode of operation to the other, and the point where the input and the output voltages are equal.

The first transition occurs when the MOSFET begins to conduct. For the circuit in Example 7.1, this transition is at $V_i = V_T = 1$ V and $V_o = V_{DD} = 5$ V.

The second transition occurs when the MOSFET goes from pinch-off to ohmic mode. At this point $V_o = 1.79$ V, and $V_i = 2.79$ V. At $V_i = V_o = V_M$, the MOSFET is in the pinch-off mode. From Equation (7.3),

$$V_M = V_{DD} - kR_D(V_M - V_T)^2 \qquad \textbf{(7.9)}$$

which, for the data given in Example 7.1, gives $V_M = 2.56$ V. V_M is known as the *gate threshold voltage,* since it separates low and high levels at input. Figure 7.1c shows the $V_i = V_o$ line and the point V_M.

Although the logic swing of the MOSFET inverter with a passive load is quite high, the VTC has a large transition width. It can be shown, in general, using Equations (7.3) and (7.5), that the transition width is given by

$$V_{tw} = 2\sqrt{\frac{V_{DD}}{3kR_D} - \frac{1}{kR_D}} \qquad \textbf{(7.10)}$$

Therefore, for a given conduction parameter k, a smaller transition width, and hence a sharper VTC, can be achieved using a larger load resistance R_D. Figure 7.2a shows the voltage transfer characteristics of Figure 7.1 for $R_D = 5$ kΩ, 10 kΩ, and 100 kΩ. Observe from Equations (7.3) and (7.5) that a large R_D causes low output voltages in both pinch-off and ohmic regions.

A relatively large load resistance is also required to limit the power dissipation in the circuit. Note, from Figure 7.2b, that the drain current, which is also the power supply current, decreases with increasing load resistance. However, a large resistor in the MOS technology occupies much more area than a MOSFET device and requires a relatively complex process to realize. Reducing

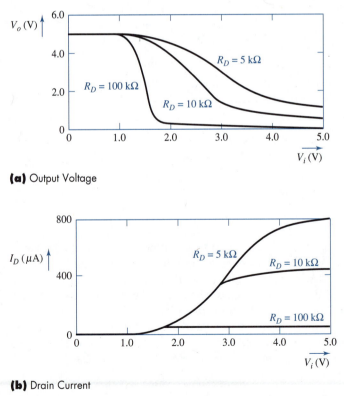

(a) Output Voltage

(b) Drain Current

Figure 7.2   Transfer characteristics of the E-NMOSFET inverter for different load resistances

the value of the conduction parameter k to reduce the current, on the other hand, makes the device long and narrow for a given technology. (Recall that k is directly proportional to the aspect ratio, W/L.) For these reasons the E-NMOSFET inverter with a passive load resistor is not commonly employed in IC logic gates.

7.1.2 NMOSFET INVERTER WITH SATURATED E-NMOSFET LOAD

It is relatively more advantageous to fabricate a second MOSFET than a large resistor for the inverter of Figure 7.1. Figure 7.3 shows an E-NMOSFET inverter employing another E-NMOSFET device as a load. With the gate tied to

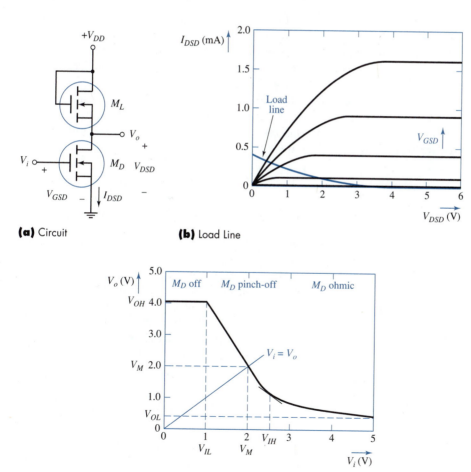

(a) Circuit

(b) Load Line

(c) VTC

Figure 7.3   NMOSFET inverter with saturated enhancement-type load

the drain, the load device M_L is always operating in the pinch-off mode. (Why?) The substrate in an IC logic gate is common to all devices and therefore it is connected to ground (or the most negative) potential. Consequently the body effect will be present, as a result of the nonzero source-to-body voltage (which varies with logic levels) for the load device. But, to simplify the analysis, we shall assume that two discrete devices are available and that their substrates are connected to their respective source terminals, so that $V_{SB} = 0$ for both M_L and M_D.

Since the load MOSFET M_L is in the pinch-off mode, its I-V characteristic is given by

$$I_{DL} = k_L(V_{DSL} - V_{TL})^2 \tag{7.11a}$$

where the subscript L denotes the load device. Since $V_{DSL} = V_{DD} - V_o$, the load line for the inverter is

$$I_{DL} = k_L(V_{DD} - V_o - V_{TL})^2 \tag{7.11b}$$

Figure 7.3b shows the load line characteristic of the device M_L given in Equation (7.11b) superimposed on the I_D-V_{DS} characteristics of the driver device M_D. The intersection of the driver characteristics with the load line gives the output voltage $V_o = V_{DSD}$ as the input voltage $V_i = V_{GSD}$ is varied. The voltage transfer characteristic is shown in Figure 7.3c.

As with the resistive load, the critical voltages are determined analytically from the drain currents of M_D and M_L. For the $V_i < V_{TD}$ the driver transistor M_D is off, and the drain current is zero. Since the load current is the same as the driver current, from Equation (7.11b),

$$I_{DL} = I_{DD} = k_L(V_{DD} - V_o - V_{TL})^2 = 0$$

Hence,

$$V_o = V_{OH} = V_{DD} - V_{TL}$$

At this output voltage, the load device M_L is conducting negligible current while operating in the pinch-off mode. Also, since a voltage of $V_{GS} = V_{DS} = V_T$ is needed to keep M_L pinched off, the output voltage V_o cannot rise above $V_{DD} - V_{TL}$. If the body of M_L is connected to ground, as in IC inverters, the threshold voltage will be higher than V_{TL}, and V_{OH} will be even smaller.

For $V_i > V_{TD}$, the driver M_D conducts and goes into pinch-off mode. We calculate the steady-state voltage V_o for a given V_i, in the pinch-off mode of M_D, by setting the drain currents of the two devices equal.

$$I_{DD} = k_D(V_{GSD} - V_{TD})^2 = I_{DL} = k_L(V_{DSL} - V_{TL})^2$$

or
$$k_D(V_i - V_{TD})^2 = k_L(V_{DD} - V_o - V_{TL})^2 \tag{7.12}$$

Solving for V_o, we get

$$V_o = V_{DD} - V_{TL} - \sqrt{\frac{k_D}{k_L}}(V_i - V_{TD}) \tag{7.13}$$

which shows a linear variation for V_o with V_i.

The slope of the transfer characteristic, from Equation (7.13), is a constant given by

$$\frac{dV_o}{dV_i} = -\sqrt{\frac{k_D}{k_L}} \qquad \text{(7.14)}$$

in the region where M_D and M_L are in saturation. Although this is useful in linear amplifier applications, it gives a rather wide transition region for the inverter. The slope with which V_o falls with V_i can be increased, and, consequently, the transition width narrowed with a large value for k_D/k_L. Since the conduction parameter is proportional to the aspect ratio W/L, the slope is given by

$$\frac{dV_o}{dV_i} = -\sqrt{\frac{k_D}{k_L}} = -\sqrt{\frac{\mu C_{ox}(W/L)_D}{\mu C_{ox}(W/L)_L}} = -\sqrt{K_R} \qquad \text{(7.15)}$$

where

$$K_R = \frac{k_D}{k_L} = \frac{(W/L)_D}{(W/L)_L} \qquad \text{(7.16)}$$

In Equation (7.15) the process transconductance parameter ($k' = \mu_n \epsilon / t_{ox}$) is the same for both devices. If the body effect of M_L is included, the slope dV_o/dV_i is slightly altered by the bulk threshold parameter [Reference 1]. K_R, which is the ratio of the aspect ratios of M_D to M_L (when the process transconductance parameters are the same for both devices), is known as the *geometry ratio*.

Equation (7.15) shows that the VTC can be made steeper in the transition region using a MOSFET with a large geometry (short and wide) for the driver and a small one (long and narrow) for the load device. Clearly the geometry ratio K_R must be above unity for practical inverters with small transition widths. Typically, a value of 10 is used for K_R in LSI inverters. Figure 7.4 shows the transfer characteristics for different values of K_R.

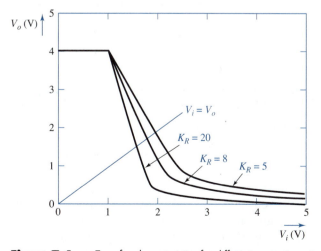

Figure 7.4   Transfer characteristics for different geometry ratios

Since the slope dV_o/dV_i is a constant, the breakpoint V_{IL} is reached when M_D begins to conduct. Thus, $V_{IL} = V_{TD}$. Note that with a constant slope given by Equation (7.15), $dV_o/dV_i = -1$ may not be possible for a given value of K_R.

Further increase in the voltage V_i in the pinch-off mode increases the drain current, and $V_{DS} = V_o$ drops. For $V_o < V_i - V_{TD}$, M_D enters the ohmic mode. We calculate the output voltage in the ohmic region of the driver by setting the drain currents equal for M_D and M_L, and using $dV_o/dV_i = -1$. Equating the drain currents, we have

$$I_{DD} = k_D[2(V_i - V_{TD})V_o - V_o^2] = I_{DL} = k_L(V_{DD} - V_o - V_{TL})^2 \quad \textbf{(7.17)}$$

Using $dV_o/dV_i = -1$, Equation (7.17) gives

$$V_i = V_{TD} + 2V_o + \frac{k_L}{k_D}(V_o + V_{TL} - V_{DD}) \quad \textbf{(7.18)}$$

Solving Equations (7.17) and (7.18), the logic high input voltage V_{IH} and the corresponding V_o are given by

$$V_{IH} = \frac{(V_{DD} - V_{TL})(2 + 1/K_R)}{\sqrt{1 + 3K_R}} + V_{TD} - \frac{V_{DD} - V_{TL}}{K_R} \quad \textbf{(7.19)}$$

$$V_o = \frac{V_{DD} - V_{TL}}{\sqrt{1 + 3K_R}} \quad \textbf{(7.20)}$$

When $V_i = V_{OH}$, the driver is in the ohmic mode; hence, we can determine the logic low output voltage V_{OL} from Equation (7.17). Again, the dependence of V_{TL} on V_o—the body effect—makes Equation (7.17) a fourth-order equation for accurate results. Neglecting the body effect gives a reasonable approximation to V_{OL}.

The critical, or gate threshold, voltage $V_M = V_i = V_o$ occurs with M_D in saturation. From Equation (7.12),

$$k_D(V_M - V_{TD})^2 = k_L(V_{DD} - V_M - V_{TL})^2$$

or

$$V_M = \frac{V_{DD} - V_{TL} + V_{TD}\sqrt{K_R}}{1 + \sqrt{K_R}} \quad \textbf{(7.21)}$$

The second critical point is the input voltage at which M_D goes from pinch-off to ohmic mode. Using Equation (7.12) and $V_{DS} = V_{GS} - V_T$ for the driver, it can be shown that this point occurs at

$$V_i = \frac{V_{DD} + V_{TD}(1 + K_R) - V_{TL}}{1 + K_R} \quad \textbf{(7.22)}$$

These equations simplify the calculation of critical and threshold voltages by neglecting the body effect in the load transistor M_L. In IC inverters, however, the body effect causes an increase in the threshold voltage of M_L. With a higher V_{TL}, V_{OH} drops significantly and the noise margin suffers. In addition, a large geometry ratio (large M_D and small M_L) is required to achieve a steep VTC. Although the silicon area occupied by M_L is small compared to that required by a large passive resistor load (Figure 7.1), other circuit configurations overcome the problem of low V_{OH}.

7.1.3 NMOSFET INVERTER WITH LINEAR E-NMOSFET LOAD

As we saw in Section 6.2.1, when the gate-to-source voltage of an NMOSFET is large, so that

$$V_{DS} < V_{GS} - V_T$$

is satisfied, the MOSFET operates in its ohmic, or linear region, much like a passive resistor. Hence, an E-NMOSFET biased in the linear region can be used as a load in an inverter. Figure 7.5a shows an NMOSFET inverter with a linear

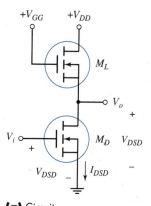

(a) Circuit

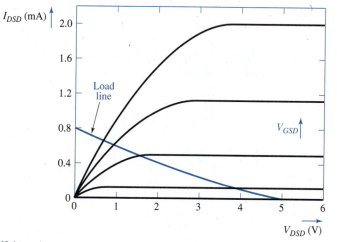

(b) Load Line

Figure 7.5   NMOSFET inverter with linear E-NMOSFET load

enhancement-type NMOSFET M_L as load. The linear mode of operation for M_L is ensured using a second power supply $V_{GG} > V_{DD} + V_T$. Note again that the body and the source of M_L are connected together to prevent the body effect, which is possible only with discrete devices.

Static analysis of this circuit can be performed graphically by constructing the load line on the characteristics of the driver transistor M_D. The current-voltage relationship of the load transistor M_L is given by

$$I_{DL} = k_L[2(V_{GSL} - V_{TL})V_{DSL} - V_{DSL}^2]$$

Since $V_{GSL} = V_{GG} - V_o$, and $V_{DSL} = V_{DD} - V_o$,

$$I_{DL} = k_L(V_{DD} - V_o)(2V_{GG} - V_o - 2V_{TL} - V_{DD}) \qquad \textbf{(7.23)}$$

This load line characteristic is superimposed on the I_D-V_{DS} characteristics of the driver transistor M_D as shown in Figure 7.5b. From Equation (7.23) and Figure 7.5b it is clear that the load line is strictly not linear except for small values of V_o relative to V_{GG}. Still, the curvature of the line is less than that for the saturated NMOSFET. Also, with a large V_{GG} the load transistor M_L can operate in the linear region, with $V_{DS} = 0$ and $I_{DL} = 0$. Consequently, there is no drop across M_L and the logic high output voltage V_{OH} is the full supply voltage V_{DD} when M_D is off.

Figure 7.6 shows the voltage transfer characteristic of the linear load inverter. We determine the critical voltages, as with the saturated E-NMOSFET load, based on the equality of the drain currents I_{DL} and I_{DD} at any input voltage V_i. The following example illustrates the analysis of the inverter.

Example 7.2	**D**etermine the logic levels and critical voltages for the inverter using the linear mode E-NMOSFET load shown in Figure 7.5a. Use $V_{TD} = V_{TL} = 1$ V, $k_D = 125$ μA/V^2, $k_L = 12.5$ μA/V^2, $V_{GG} = 10$ V, and $V_{DD} = 5$ V. Neglect body effect in M_L.

What are the noise margins for this inverter?

Solution

For $V_i < V_{TD} = 1$ V, the driver M_D is off and, with the load device M_L operating in the ohmic region with zero current, we have

$$V_{OH} = V_{DD} = 5 \text{ V}$$

The first critical voltage where the driver transistor goes from off to saturation is $V_i = V_{TD} = 1$ V and $V_o = 5$ V.

For $V_i > V_{TD}$, M_D goes into the saturation mode. Equating the saturation drain current of M_D to the linear drain current of M_L, we get

$$k_D(V_i - V_{TD})^2 = k_L[2(V_{GG} - V_o - V_{TL})(V_{DD} - V_o) - (V_{DD} - V_o)^2] \quad \textbf{(7.24a)}$$

which, for the given circuit parameters and $K_R = k_D/k_L = 10$, becomes

$$10(V_i - 1)^2 = 2(9 - V_o)(5 - V_o) - (5 - V_o)^2 \qquad \textbf{(7.24b)}$$

Equation (7.24) describes the region in Figure 7.6a where M_D is in saturation.

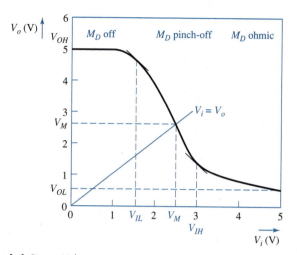

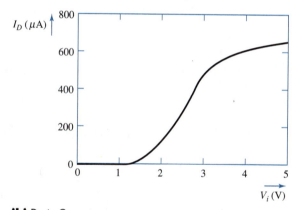

(a) Output Voltage

(b) Drain Current

Figure 7.6   Transfer characteristics of the linear load inverter

We determine V_{IL} from Equation (7.24) and the slope condition, $dV_o/dV_i = -1$. Equating the slope to -1 using Equation (7.24) gives

$$V_i = V_{IL} = 1.9 - 0.1\, V_o \qquad \textbf{(7.25)}$$

Solving Equations (7.24b) and (7.25), we get $V_{IL} = 1.42$ V, and the corresponding output voltage $V_o = 4.78$ V.

As $V_i = V_{GSD}$ is increased above

$$V_i = V_o + V_{TD}, \quad \text{or} \quad V_{DSD} = V_{GSD} - V_{TD} \qquad \textbf{(7.26)}$$

M_D goes from the pinch-off to the linear mode of operation. The second critical point at which M_D makes a transition in its mode of operation is found from

Equations (7.24a) and (7.26). Solving Equations (7.24a) and (7.26), this transition occurs at

$$V_i = 2.87 \text{ V} \quad \text{and} \quad V_o = 1.87 \text{ V}$$

For $V_i > 2.87$ V, M_D is operating in the ohmic region. Equating the drain currents in the ohmic region for both M_D and M_L,

$$k_D[2(V_i - V_{TD})V_o - V_o^2]$$
$$= k_L[2(V_{GG} - V_o - V_{TL})(V_{DD} - V_o) - (V_{DD} - V_o)^2] \quad \textbf{(7.27a)}$$

or

$$10[2(V_i - 1)V_o - V_o^2] = 2(9 - V_o)(5 - V_o) - (5 - V_o)^2 \quad \textbf{(7.27b)}$$

Equation (7.27) describes the region for V_i above the transition voltage in the VTC.

Taking the derivative of Equation (7.27b) and equating it to -1, we get

$$V_i = V_{IH} = 2.1V_o + 0.1 \quad \textbf{(7.28)}$$

Solving Equations (7.27b) and (7.28) gives

$$V_{IH} = 3.14 \text{ V} \quad \text{and} \quad V_o = 1.45 \text{ V}$$

Note that $V_o = V_{DSD} < V_{GSD} - V_{TD}$, so that M_D is in its ohmic mode.

For $V_i = V_{OH} = 5$ V, which is above the transition voltage for M_D, the logic low output voltage V_{OL} is obtained from Equation (7.27b) as

$$V_{OL} = 0.72 \text{ V}$$

The gate threshold voltage $V_M = V_i = V_o$, using Equation (7.24b), is given by

$$V_M = 2.59 \text{ V}$$

The results of the linear load inverter data are: $V_{IL} = 1.42$ V, $V_{IH} = 3.14$ V, $V_{OH} = 5$ V, $V_{OL} = 0.72$ V, $V_M = 2.59$ V, and $V_i = 2.87$ V for M_D going from saturation to ohmic mode.

The noise margins are:

$$NM_L = 0.70 \text{ V} \quad \text{and} \quad NM_H = 1.86 \text{ V}$$

The above values are approximate since we have neglected the body effect for M_L. Figure 7.6a shows the logic and critical voltages.

From the above example it is seen that the logic high output voltage V_{OH} is higher than that for the saturated enhancement-type MOSFET load. However, note that the linear MOSFET load needs a second power supply. It is preferable to implement systems that operate with a single power supply to reduce cost, size, power dissipation, and availability. Also, a second power supply requires additional interconnections and more chip area when implemented in IC form. Also, it can be shown that for the same slope in the transition region of the VTC as for the saturated NMOS load, the linear load inverter requires a much larger geometry ratio. For these reasons linear load NMOS circuits find limited applications in MOS LSI circuits.

7.1.4 NMOSFET INVERTER WITH DEPLETION-TYPE LOAD

The fourth type of NMOSFET inverter uses a normally on depletion-type MOSFET as an active load, as shown in Figure 7.7a. Although the D-NMOSFET requires additional processing steps to fabricate on the same chip (because of the physical channel) as the E-NMOSFET, the greatly improved performance of

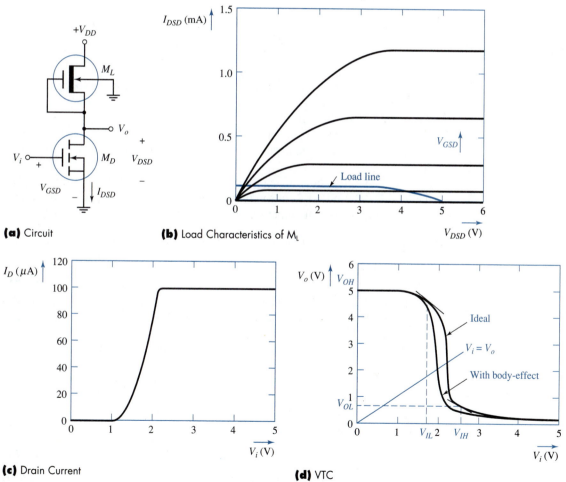

(a) Circuit   **(b)** Load Characteristics of M_L

(c) Drain Current   **(d)** VTC

Figure 7.7   NMOSFET inverter with depletion-type load

this configuration has led to a wide variety of modern NMOSFET circuit applications.

Figure 7.7b shows the combined load characteristics of the D-NMOSFET M_L, with $V_{GS} = 0$, and the I_D-V_{DS} characteristics of the E-NMOSFET M_D. Since the load device M_L is on for $V_{GS} = 0$, there is no voltage drop in the conducting channel from drain (V_{DD}) to source (output) for zero current. Hence the output voltage is $V_o = V_{DD}$ when the driver M_D is off. For $V_i > V_{TD}$, M_D conducts while M_L is in saturation over a large range of V_i, as explained below. The saturated D-NMOSFET M_L approximates a constant-current source, as shown by the load line in Figure 7.7b and the drain current in Figure 7.7c. This behavior of the load device results in the sharp voltage transfer characteristics as shown in Figure 7.7d.

With $V_{TL} < 0$ and $V_{GS} = 0$, the saturation condition for the depletion device M_L is given by

$$V_{DSL} \geq V_{GSL} - V_{TL},$$

or

$$V_{DD} - V_o \geq |V_{TL}| \tag{7.29}$$

As the input voltage V_i is increased above the driver threshold voltage V_{TD}, M_D conducts and V_o falls. The drain-source voltage of the load, $V_{DSL} = V_{DD} - V_o$, therefore, rises. As a result, the saturation condition in Equation (7.29) is satisfied for V_i increasing much above V_{TD}. (Note that the body effect of M_L raises V_{TL}, and V_{TL} becomes less negative; hence, saturation is not affected by the body effect.)

For V_i slightly above V_{TD} so that $V_o \approx V_{DD}$, the driver is in saturation while the load D-NMOSFET is in the ohmic (linear) mode. Equating the drain currents of M_D and M_L, the input-output voltage relationship is given by

$$k_D(V_i - V_{TD})^2 = k_L[2(V_{GSL} - V_{TL})V_{DSL} - V_{DSL}^2]$$

or

$$k_D(V_i - V_{TD})^2 = k_L[2|V_{TL}|(V_{DD} - V_o) - (V_{DD} - V_o)^2] \tag{7.30}$$

The range of input in Equation (7.30) corresponds to current I_{DL} increasing with decreasing V_{DSD} in the load line characteristics of Figure 7.7b. Output V_o in this range drops, due to increasing drop across M_L.

As V_o decreases from V_{DD}, either M_D or M_L changes mode of operation. If

$$V_o = V_{DSD} \leq V_i - V_{TD} \tag{7.31}$$

is reached first, the enhancement driver M_D goes from saturation to ohmic mode while the depletion load M_L continues in ohmic mode.

If, instead,

$$V_{DD} - V_o \geq |V_{TL}|, \quad \text{or} \quad V_o \leq V_{DD} - |V_{TL}| \tag{7.32}$$

occurs first, the load device M_L enters saturation mode while the driver M_D continues in saturation. We determine the transition point for M_L going from ohmic to saturation using the boundary condition in Equation (7.32) and the drain current given by Equation (7.30). Then we can obtain the second transition point for M_D going from saturation to ohmic mode using Equation (7.31).

Assuming M_L enters saturation mode first, the output at the boundary in Equation (7.32) is given by

$$V_o = V_{DD} - |V_{TL}| \tag{7.33}$$

If M_D is still in saturation at this V_o, from Equation (7.30)

$$V_i = V_{TD} + \frac{|V_{TL}| k_L}{k_D}$$

or, using the geometry ratio $K_R = k_D/k_L$,

$$V_i = V_{TD} + \frac{|V_{TL}|}{K_R} \tag{7.34}$$

(For more accurate results V_{TL} must include the body effect.)

With V_i and V_o given by Equations (7.33) and (7.34), we can now check the conditions in Equations (7.31) and (7.32). Clearly, the threshold voltages and the geometry ratio determine the transition point. The threshold voltages are set during the fabrication process and, in practice, $V_{TD} < |V_{TL}|$. The geometry ratio K_R, therefore, is a design parameter.

Consider the case when Equation (7.32) is satisfied first, that is, when both M_D and M_L are in saturation. The drain current is given by

$$k_D(V_i - V_{TD})^2 = k_L(V_{TL})^2 \tag{7.35}$$

Since the right side of Equation (7.35) is a constant, it shows that both M_D and M_L are in saturation at only one point corresponding to V_o and V_i given by Equations (7.33) and (7.34). Thus, a sharp drop in VTC occurs at this point. With body effect, V_{TL} varies with V_o as [see Chapter 6, Equation (6.20)]

$$V_{TL}(V_o) = V_{TL0} + \gamma[\sqrt{V_o + 2|\Phi_F|} - \sqrt{2|\Phi_F|}] \tag{7.36}$$

where V_{TL0} is the threshold voltage for zero source-body potential. Since $|V_{TL}|$ decreases with falling V_o, the load current decreases slightly. Hence, even with body effect, V_o drops quite rapidly with V_i.

A drop in V_o below $V_{GSD} - V_{TD} = V_i - V_{TD}$ takes the driver to ohmic mode while the load stays in saturation. With low drain current in the driver, the output voltage drops slowly with the input voltage. The equation describing this region is given by

$$k_D[2(V_i - V_{TD})V_o - V_o^2] = k_L(V_{TL})^2 \tag{7.37}$$

If, for a given inverter, M_D goes from saturation to the ohmic mode first while M_L stays in the ohmic mode (i.e., Equation (7.31) is satisfied), the input-output relationship can be similarly derived from the drain currents of the two devices.

Figure 7.7d shows the VTC for the depletion load inverter with and without body effect. The following example illustrates the calculation of the logic and critical voltages for the inverter.

Example 7.3

Determine the logic levels and critical voltages for the D-NMOSFET load inverter of Figure 7.7a with $V_{DD} = 5$ V. Use the following device parameters: $k_D = 75$ μA/V^2, $V_{TD} = 1$ V, $k_L = 25$ μA/V^2, and $V_{TL} = -2$ V.

Solution

For $V_i = V_{GSD} < V_{TD}$, M_D is off while the channel is present in M_L for $V_{GSL} = 0 > V_{TL}$. Therefore, the current in the circuit is zero, and $V_o = V_{OH} = V_{DD} = 5$ V.

As V_i is increased above V_{TD}, M_D goes into saturation while M_L continues in ohmic mode. We find the V_i-V_o characteristic in this region by equating the drain currents [Equation (7.30)]:

$$75(V_i - 1)^2 = 25(5 - V_o)(V_o - 1) \qquad \text{(7.38)}$$

From Equation (7.38), the slope dV_o/dV_i is -1 for

$$V_o = 3 V_i \qquad \text{(7.39)}$$

Solving Equations (7.38) and (7.39), we find $V_{IL} = 1.58$ V, and the corresponding $V_o = 4.73$ V. These values verify the modes of operation assumed for M_D and M_L.

For $V_i > V_{IL}$, assume that Equation (7.32) is satisfied first; that is, M_L goes from ohmic to saturation mode while M_D continues in saturation mode. At the boundary between ohmic and saturation modes for M_L, we have

$$5 - V_o = -V_{TL}, \quad \text{or} \quad V_o = 3 \text{ V}$$

Equating the drain currents at this voltage, the corresponding input is

$$75(V_i - 1)^2 = 25 \times 4, \quad \text{or} \quad V_i = 2.15 \text{ V}$$

Clearly, at these voltages both M_L and M_D are in saturation as we assumed. With the body effect of M_L ignored, at $V_i = 2.15$ V, V_o drops abruptly from 3 V to 2.15 V. It can be readily shown that $V_i = V_o = 2.15$ V is the only point at which both the devices are operating in the saturation mode.

As V_i is increased above V_M, M_D enters ohmic mode while M_L continues in saturation. From Equation (7.37), the VTC is described by

$$3[2(V_i - 1)V_o - V_o^2] = 4 \qquad \text{(7.40)}$$

Again, equating the slope dV_o/dV_i to -1 gives

$$V_o = \frac{V_i - 1}{2} \qquad \text{(7.41)}$$

Solving Equations (7.40) and (7.41) for $V_i = V_{IH}$, we get $V_{IH} = 2.33$ V, and the corresponding output $V_o = 0.67$ V. Again we can verify that M_D and M_L are indeed in their assumed modes of operation corresponding to these voltages.

For $V_i > V_{IH} = 2.33$ V, M_D continues in the ohmic region while M_L stays in saturation. For $V_i = V_{OH} = 5$ V, equating the drain currents, we have

$$V_o = V_{OL} = 0.17 \text{ V}$$

Summarizing the results for the ideal depletion load inverter (without the body effect), we have

$$V_{IL} = 1.58 \text{ V}, \quad V_{IH} = 2.33 \text{ V}, \quad V_{OH} = 5 \text{ V}, \quad V_{OL} = 0.17 \text{ V}, \quad \text{and}$$
$$V_M = 2.15 \text{ V}$$

It can be shown that if the body effect for the depletion load device is included, the results are close to the values obtained above. For example, at $V_i = V_o = V_M$, assuming saturation for both devices and equating the drain currents, we have

$$75(V_M - 1)^2 = 25|V_{TL}(V_M)|^2, \quad \text{or} \quad V_M = 1 + \frac{|V_{TL}(V_M)|}{\sqrt{3}}$$

where $V_{TL}(V_M)$ is given by Equation (7.36). Using typical values of $\gamma = 0.4 \sqrt{V}$ and $2|\Phi_F| = 0.6$ V, the above equation can be iteratively solved to obtain $V_M = 1.96$ V.

The results of Example 7.3 show a far better performance for the depletion-load inverter compared with the saturation or the linear load inverters. As with the other active load inverters, the geometry ratio of the depletion load inverter, $K_R = k_D/k_L$, is an important design parameter. To see this, consider $V_o = V_{OL}$ for $V_i = V_{OH}$. Equating the drain current of the driver, which is in the ohmic mode, to the drain current of the load in saturation, we have, from Equation (7.37),

$$k_D[2(V_{OH} - V_{TD})V_{OL} - V_{OL}^2] = k_L(V_{TL})^2$$

or
$$V_{OL} = (V_{OH} - V_{TD}) - \sqrt{(V_{OH} - V_{TD})^2 - \frac{V_{TL}^2}{K_R}} \quad \textbf{(7.42)}$$

where only the physically meaningful solution for the quadratic equation is used. Equation (7.42) shows that a low value for V_{OL} is obtained using a large geometry ratio. MOSFET logic circuits in which V_{OL} depends on the geometry ratio K_R are referred to as *ratioed* circuits. (Note that K_R also determines the sharpness of the VTC and hence the transition width.) Theoretically, $V_{OL} = 0$ can be achieved with $K_R = \infty$. In practice, however, large K_R requires large silicon area for the enhancement driver device, assuming the same process transconductance parameter (μC_{ox}) for both the driver and the (depletion) load devices. On the other hand, to obtain V_{OH} as close as possible to V_{DD}, a large conduction parameter k_L and a large threshold voltage V_{TL} are needed. We can see this as follows. If the (leakage) current in the enhancement driver device M_D is not negligible for $V_i = 0$, M_L may be approximated by its resistance. Since M_L is operating in the linear mode, its conductance G_L at $V_{GSL} = 0$ is given by the slope,

$$G_L = \frac{dI_{DL}}{dV_{DSL}} = 2k_L[-V_{TL} - V_{DSL}] = 2k_L[|V_{TL}| - (V_{DD} - V_o)]$$

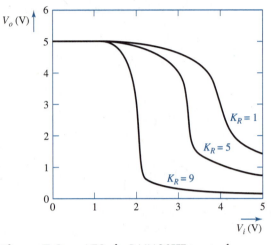

Figure 7.8   VTC of a D-NMOSFET inverter for different geometry ratios

Approximating the drain-source voltage of M_L as $V_{DSL} \approx I_{DL}/G_L$, where I_{DL} is the leakage drain current of the driver M_D, and noting that $V_o = V_{OH}$ for $V_i = 0$, we have

$$V_{OH} = V_{DD} - V_{DSL} \approx V_{DD} - \frac{I_{DL}}{2k_L} \frac{1}{|V_{TL}| - (V_{DD} - V_{OH})} \qquad (7.43)$$

From this iterative equation, it is clear that for $V_{OH} \approx V_{DD}$, k_L must be large when the leakage drain current is significant. Note also that with large k_L, the body effect of M_L has negligible effect on V_{OH}. In addition, a load device with large k_L charges output (load) capacitance with high current and hence results in low propagation delay t_{PLH}, as we will see later.

Instead of increasing the inverter chip size by requiring large sizes for the load and the driver, a designer would choose a relatively high threshold voltage V_{TL} for the load, then determine the driver size from specified V_{OL} and K_R using Equation (7.42). This approach gives a low V_{OL} and a sharp VTC. Figure 7.8 shows the transfer characteristics for different K_R using the same V_{TL}.

The following example illustrates the design of a D-NMOSFET inverter based on V_{OL}.

Example 7.4   Design a D-NMOSFET inverter as shown in Figure 7.7a to obtain $V_{OL} \leq 0.2$ V at a supply voltage of 5 V. The process threshold voltages are: $V_{TD} = 1$ V, and $V_{TL} = -3$ V. Neglect the body effect.

What is the geometry ratio K_R if the body-effect coefficient for the load device is $\gamma = 0.4 \sqrt{V}$ and the Fermi potential is $2|\Phi_F| = 0.6$ V?

Assume in both cases that $V_{OH} \approx V_{DD} = 5$ V; that is, neglect the leakage drain current of the driver.

Solution

Rewriting Equation (7.42), we get the geometry ratio without body effect as

$$K_R = \frac{(V_{TL})^2}{2V_{OL}(V_{OH} - V_{TD}) - V_{OL}^2}$$

For the given values this gives $K_R = 5.77$. Noting that a higher value gives a value for V_{OL} better than specified, we can round up K_R to 6. For this value of K_R, Equation (7.42) gives $V_{OL} = 0.192$ V.

With body effect, the threshold voltage of the load at $V_{SB} = V_{OL} = 0.2$ V is given by, from Equation (7.36),

$$V_{TL}(0.2) = -3 + 0.4(\sqrt{0.8} - \sqrt{0.6}) = -2.952 \text{ V}$$

This results in $K_R = 5.586$. Rounding this again to $K_R = 6$ gives

$$V_{OL} = 4 - \sqrt{16 - \frac{V_{TL}(V_{OL})^2}{6}}$$

which can be iteratively solved to obtain $V_{OL} = 0.186$ V, which is also better than specified.

Figure 7.9 shows a MicroSim PSpice simulation of the VTC for the inverter designed using a minimum size of 50 μm and typical process transconductance parameters.

For purposes of comparison, Figure 7.10 shows the load and the transfer characteristics for the four NMOSFET inverter configurations studied. The devices in all cases have the same threshold voltages and the same geometry ratios. Conduction parameters have been adjusted to obtain the same drain current at logic low output. Clearly the static performance of the depletion-load inverter is superior to that of the other three load configurations. Also, we see later that the depletion-load inverter has a higher speed of operation than the others. For these reasons the D-NMOSFET load configuration is widely used in NMOSFET logic circuits.

7.1.5   SWITCHING RESPONSE

As with other logic families, MOSFET inverter switching delays arise primarily in charging and discharging the capacitance at the output (load) terminal. Sources of the load capacitance are both internal, from the terminals of the driver and the load MOSFETs, and external, from wiring and the load being driven. Figure 7.11a shows the various sources of capacitances for the NMOSFET inverter with a general NMOSFET (active) load driving an identical gate. Depending on the load MOSFET configuration, some of these capacitances become zero. The lumped line capacitance is denoted by C_{line} and the load gate capacitance by C_{in} in Figure 7.11a.

```
* Example 7.4 DEPLETION  LOAD NMOS INVERTER VTC - Design verification
MD      2    1    0    0   NENH   L=50U W = 150U
ML      3    2    2    0   NDEP   L=100U W=50U ; KR = 6
VI      1    0
VDD     3    0 DC 5
.PROBE
.MODEL NDEP NMOS (VTO = -3 GAMMA = 0.4 PHI=0.6 KP = 50E-6)
.MODEL NENH NMOS (VTO = 1 KP = 50E-6)
.DC     VI  0 5V  0.1V
.PRINT DC V(2),ID(MD)
.OPTIONS NOPAGE
.END
```

(a) Simulation File

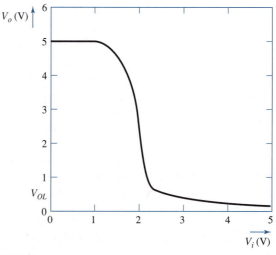

(b) VTC

Figure 7.9   MicroSim PSpice simulation for the inverter in Example 7.4

We can express the capacitances between the gate and the drain in terms of the oxide capacitance per unit area, $C_{ox} = \epsilon/t_{ox}$, and we can obtain the substrate to source and drain capacitances from pn junction depletion capacitances. In general, however, these capacitances are nonlinear functions of the drain-source voltage. Although an estimate of the values for all these capacitances can be made based on the mode of operation and the device and process parameters [Reference 1], it can be shown that the contribution to the output terminal (lumped) capacitance C_L in Figure 7.11b arises, in order of decreasing magnitude, from (1) the input (gate) capacitance C_{in} of the fan-out gates, (2) the

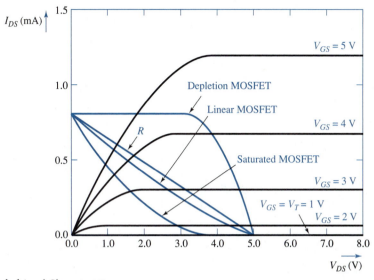

(a) Load Characteristics

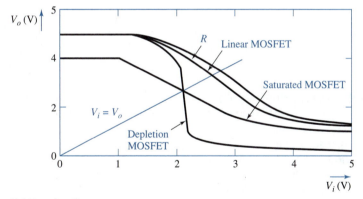

(b) Transfer Characteristics

Figure 7.10 Comparison of NMOSFET inverter characteristics

junction capacitances C_{dbD} and C_{sbL} between the drain and the source to the substrate, (3) the drain-to-gate capacitances C_{gdD} and C_{gdL} of the driver and the load MOSFETs, and (4) the parasitic capacitance of the interconnect line. These contributions to C_L in Figure 7.11b are as given in Equation (7.44).

$$C_L = C_{in} + K_c(C_{dbD} + C_{sbL}) + C_{gdD} + C_{gdL} + C_{line} \qquad \textbf{(7.44)}$$

K_c is the voltage-dependency factor for the drain and the source junction capacitances with the body.

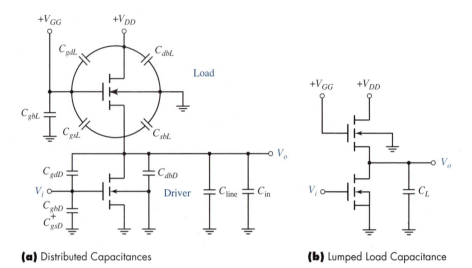

(a) Distributed Capacitances   **(b)** Lumped Load Capacitance

Figure 7.11   Pertaining to switching delays in MOSFET inverters

While MicroSim PSpice can calculate C_L for a cascaded inverter using specified parameters, one may use lumped capacitance on the order of 10 to 50 fF (10^{-15} F) for hand calculation of propagation delays. For each fan-out gate above one, we may add an additional gate capacitance C_{in} of 10 to 15 fF.

Output High-to-Low Delay Time   In the general NMOSFET inverter using the active NMOSFET load shown in Figure 7.11b, let the input switch from low ($V_i = V_{OL}$) to high ($V_i = V_{OH}$) at $t = 0$. Output voltage V_o changes from high to low by discharging the load capacitance C_L via the on device M_D. To simplify calculations we shall assume that the input rise time is negligible and estimate the delay in discharging C_L from $V_o = V_{OH}$ to $V_o = (V_{OH} + V_{OL})/2$, that is, the 50 percent point in the output transition. With $V_{DSD} = V_o = V_{OH}$ and $V_{GSD} = V_i = V_{OH}$, the driver M_D is in saturation. The load M_L may be in saturation or ohmic mode, depending on the configuration used. Referring to Figure 7.12a, the capacitor discharge current is given by

$$i_c = i_{DL} - i_{DD}$$

With a large geometry ratio $K_R = k_D/k_L$ in practical inverters, $i_{DD} \gg i_{DL}$. Hence, to simplify calculations, we shall assume that $i_c \approx -i_{DD}$.

Then, at $t = 0$, we have

$$C_L \frac{dV_o}{dt} = -k_D(V_{OH} - V_{TD})^2 \qquad \textbf{(7.45)}$$

As the capacitor discharges from V_{OH} it reaches a voltage of $V_o = V_{OH} - V_{TD}$ at $t = t_1$ when M_D goes from saturation to the ohmic mode. The time t_1 is deter-

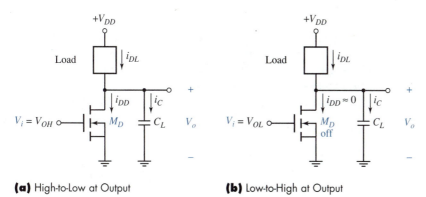

(a) High-to-Low at Output   **(b)** Low-to-High at Output

Figure 7.12   Pertaining to switching delays in MOSFET inverters

mined by integrating Equation (7.45) from $V_o = V_{OH}$ to $V_o = V_{OH} - V_{TD}$. For $t > t_1$, the driver operates in the ohmic mode. If the threshold voltage is such that $V_{OH} - V_{TD} > (V_{OH} + V_{OL})/2$, as is the case in practical inverters, C_L continues to discharge, with a current of

$$i_c = C_L \frac{dV_o}{dt} = -k_D[2(V_{OH} - V_{TD})V_o - V_o^2] \qquad \textbf{(7.46)}$$

At $t = t_2$, V_o reaches the halfway mark in transition. We obtain t_2 by integrating Equation (7.46) from $V_o = V_{OH} - V_{TD}$ to $V_o = (V_{OH} + V_{OL})/2$. The sum of the two delays, $t_1 + t_2$, gives the high-to-low propagation delay time.

Output Low-to-High Delay Time   If the input switches from $V_i = V_{OH}$ to $V_i = V_{OL}$ at $t = 0$, the output at the logic low voltage of $V_o = V_{OL}$ cannot change instantaneously, due to charging of C_L. With $V_{GSD} = V_{OL} < V_{TD}$ at $t = 0$, the driver is off, Thus, the charging current for C_L is supplied by the load device as given by (see Figure 7.12b)

$$i_C = C_L \, dV_o/dt = I_L \qquad \textbf{(7.47)}$$

This equation shows that, depending on the device and configuration for M_L, the load capacitance may be charged faster or slower to reach V_{OH}. Since the load and the driver have different conduction parameters, the charging and discharging currents for C_L are different. Consequently, the rising and falling edges of the output waveform have different delay times. The following examples illustrate the approximate delay calculations for the inverters with resistor and D-NMOS-FET load configurations.

Estimate the high-to low and low-to-high propagation delays for the resistive load inverter of Figure 7.1 with a lumped load capacitance of $C_L = 50$ fF. | **Example 7.5**

Solution

High-to-low delay: From Example 7.1, $V_{OL} = 0.6$ V and $V_{OH} = 5$ V. Thus, the delay t_{PHL} is the time taken by the load capacitor to discharge from 5 V to $5 - (5 - 0.6)/2 = 2.8$ V. Figure 7.13a shows the movement of the operating point as the output makes the transition from V_{OH} to V_{OL}.

At $t = 0$, input $V_i = V_{OH} = 5$ V and output $V_o = V_{OH} = 5$ V. Hence, the MOSFET driver is in saturation. The discharge current, from Equation (7.45), is

$$i_C = -i_D, \quad \text{or} \quad C_L \frac{dV_o}{dt} = -0.1(5 - 1)^2 \text{ mA} = -1.6 \text{ mA}$$

Integrating the above equation from $t = 0$ and $V_o = 5$ V to $t = t_1$ at which $V_o = V_i - V_T = 4$ V, we get

$$t_1 = 31.2 \text{ ps}$$

Now the MOSFET is in the ohmic mode. Therefore, from Equation (7.46), we write

$$C_L \, dV_o/dt = -0.1(8V_o - V_o^2) \text{ mA}$$

Integrating the above equation[1] from $t = 0$ and $V_o = 4$ V, we have

$$(1/8) \ln\left[\frac{V_o}{(8 - V_o)}\right]\Bigg|_4^{V_o} = -2 \times 10^9 t$$

Time t_2 at which $V_o = 2.8$ V is given by $t_2 = 38.7$ ps.

Therefore, $t_{PHL} = t_1 + t_2 = 69.9$ ps

Note, in Figure 7.13a, as the operating point moves from $t = 0$ to $t = t_1$, the driver current is constant at a large value; for $t > t_1$, the MOSFET in the ohmic mode contributes less current to discharge C_L. It is easily verified that the current through R_D is negligible compared with the capacitor discharge current throught the interval $t_1 < t < t_2$.

Low-to-high delay: When the input switches from high to low, the MOSFET driver is off. Thus, the load capacitor is charged by the passive resistor load from $V_{OL} = 0.6$ V to $V_{OH} = 5$ V. With a time-constant of $\tau = RC_L = 0.5$ ns, the output rises as

$$V_o = 5 - 4.4 \exp(-t/\tau)$$

V_o reaches 2.8 V, which is the midpoint of logic swing, at $t = t_{PLH}$, given by

$$t_{PLH} = \tau \ln(2) = 347 \text{ ps}$$

---

[1] $\dfrac{\int dx}{ax^2 + bx} = (1/b) \ln\left[\dfrac{x}{ax + b}\right]$

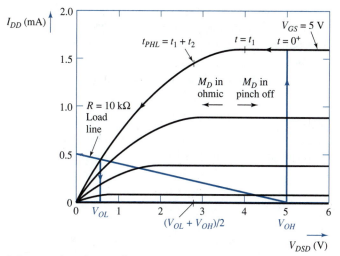

(a) V_i Switching from High to Low

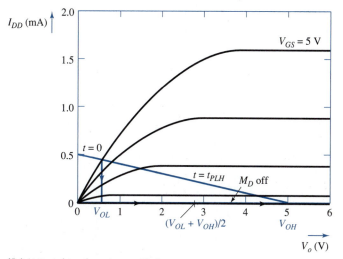

(b) V_i Switching from Low to High

Figure 7.13 Movement of operating point during switching in a resistive load inverter

The average propagation delay t_P is, therefore, given by

$$t_P = \frac{(t_{PHL} + t_{PLH})}{2} = 209 \text{ ps}$$

Figure 7.13b shows the movement of the operating point as the output changes from low to high.

Figure 7.14 shows the MicroSim PSpice simulation of the inverter switching response with only the lumped load capacitor. The simulation results given $t_{PHL} \approx 80$ ps and $t_{PLH} \approx 360$ ps, which are in reasonable agreement with our estimated values.

The preceding example points to a conflicting requirement for the resistor in a passive load inverter. To decrease the low-to-high propagation delay time, clearly, we must reduce the load resistance. But, for reduced power dissipation

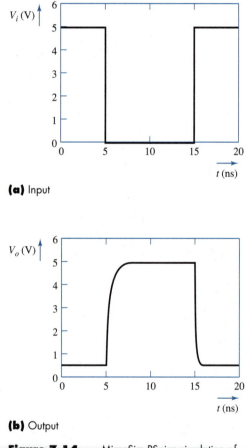

(a) Input

(b) Output

Figure 7.14   MicroSim PSpice simulation of resistive load inverter switching response

and sharp VTC, we must make R_L as large as possible. The next example considers the switching delays in an inverter using a D-NMOSFET load.

| Determine the approximate propagation time delays and the rise and fall times for the D-NMOSFET inverter of Example 7.3, with $C_L = 50$ fF. | **Example 7.6** |

Solution

Propagation delays: From Example 7.3, $V_{OH} = 5$ V and $V_{OL} = 0.17$ V. Hence, t_{PHL} is the time taken for the capacitor C_L to discharge from 5 V to 2.6 V as the input switches from V_{OL} to V_{OH}. Calculation of t_{PHL} is identical to that with a resistive load, since we neglect the current from the load device in both cases.

At $t = 0$, $V_o = V_i = 5$ V. Therefore, the driver M_D is in saturation. Neglecting the current from the D-NMOSFET load (which is in the ohmic mode), the capacitor is discharging with a current of

$$i_C = C_L \frac{dV_o}{dt} = -i_{DD} = -75(5-1)^2 \ \mu A = -1.2 \text{ mA}$$

Integrating,

$$V_o - 5 = -2.4 \times 10^{10} t \qquad \textbf{(7.48)}$$

Thus, $V_o = 4$ V, at which M_D goes from saturation to ohmic mode, is reached at $t_1 = 41.7$ ps.

For $t > t_1$, the discharge of C_L is governed by the ohmic mode drain current of M_D as

$$C_L \, dV_o/dt = -0.075(8V_o - V_o^2) \text{ mA}$$

Integrating the above equation from $V_o = 4$ V, we have

$$(1/8) \ln\left[\frac{8 - V_o}{V_o}\right] = 1.5 \times 10^9 t \qquad \textbf{(7.49)}$$

Hence, $V_o = 2.6$ V is reached at $t_2 = 61$ ps.

Adding,

$$t_{PHL} = t_1 + t_2 = 103 \text{ ps}$$

For input switching from high to low, M_D is off and the charging current for C_L is supplied by M_L. At $t = 0$, $V_o = V_{OL} = 0.17$ V and the load D-NMOSFET is in saturation. Thus, the charging current is given by

$$i_c = C_L \frac{dV_o}{dt} = 25|V_{TL}|^2 \ \mu A \qquad \textbf{(7.50)}$$

where the body effect at $V_{SB} = V_{OL} = 0.17$ V is neglected. As V_o increases to 2.6 V, M_L is still in saturation if the body effect is neglected. Integrating equation (7.50), we have

$$V_o - 0.17 = 2 \times 10^9 t \qquad \textbf{(7.51)}$$

Hence, the time $t_1 = t_{PLH}$ at which $V_o = 2.6$ V is given by

$$t_{PLH} = t_1 = 1.215 \text{ ns}$$

The average propagation delay is therefore given by

$$t_p = \frac{t_{PHL} + t_{PLH}}{2} = 659 \text{ ps}$$

Fall and rise times: Fall time t_f is the time required for the output voltage V_o to drop from $V_{OH} - (V_{OH} - V_{OL})/10 = 4.52$ V to $V_{OL} + (V_{OH} - V_{OL})/10 = 0.65$ V when the input is at V_{OH}.

From Equation (7.48), $V_o = 4.52$ V is reached at $t_{f1} = 20$ ps. After $t = t_{f1}$, V_o continues to drop with M_D still in saturation until $V_o = 4$ V, which is reached at $t = t_i = 42$ ps as we calculated earlier.

For $t > t_i$, Equation (7.49) describes the discharge of C_L. $V_o = 0.65$ is reached at t_3 given by

$$t_3 = \left(\frac{1}{8} \times 1.5\right)\ln\left[\frac{8 - 0.65}{0.65}\right]\text{ns} = 0.202 \text{ ns}$$

Hence we arrive at the fall time as (see Figure 7.15)

$$t_f = t_1 + t_3 - t_{f1} = 224 \text{ ps}$$

Note that a large contribution to the fall time comes from the driver M_D operating in the ohmic mode.

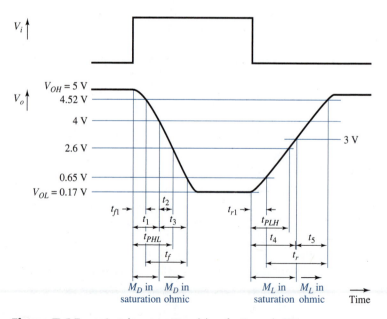

Figure 7.15   Switching transition delays for Example 7.6

Rise time t_r is calculated for V_o to rise from 0.65 V to 4.52 V while V_i is at V_{OL}.

From Equation (7.51), V_o rises from V_{OL} to 0.65 V in t_{r1} given by $t_{r1} = 240$ ps. As V_o increases to $V_o = V_{DD} - |V_{TL}|$, M_L enters the ohmic mode. If we ignore the body effect, the time t_4 at which $V_o = 5 - 2 = 3$ V is, from Equation (7.51), given by

$$t_4 = 1.415 \text{ ns}$$

Now the load device is in the ohmic mode; hence, the charging current is given by

$$C_L \frac{dV_o}{dt} = k_L[2|V_{TL}|(V_{DD} - V_o) - (V_{DD} - V_o)^2] \qquad \textbf{(7.52)}$$

or

$$\frac{dV_o}{4(5 - V_o) - (5 - V_o)^2} = 5 \times 10^8 \, dt$$

Rewriting this equation as

$$\frac{d(5 - V_o)}{4(5 - V_o) - (5 - V_o)^2} = -5 \times 10^8 \, dt$$

and integrating from $V_o = 3$ V, we have

$$(1/4) \ln\left[\frac{V_o - 1}{5 - V_o}\right] = 5 \times 10^8 t$$

Thus we find the time t_5 at which $V_o = 4.52$ V is given by

$$t_5 = 0.996 \text{ ns}$$

If we include body effect for an average source-to-body voltage of $(0.65 + 4.52)/2 = 2.6$ V,

$$V_{TL}(2.6 \text{ V}) = -2 + 0.4\left[\sqrt{2.6 + 0.6} - \sqrt{0.6}\right] = -1.6 \text{ V}$$

Now Equation (7.51) becomes

$$V_o - 0.17 = 12.8 \times 10^9 t$$

The saturation time of M_L, that is, the time at which $V_o = 5 - 1.6 = 3.4$ V, using the new V_{TL}, is

$$t_4' = 2.52 \text{ ns}$$

Charging of C_L with M_L in ohmic mode is [Equation (7.52)] now given by

$$d(5 - V_o)/[3.2(5 - V_o) - (5 - V_o)^2] = -5 \times 10^8 \, dt$$

Integrating,

$$(1/3.2)\ln[(V_o - 1.8)/(5 - V_o)] = 5 \times 10^8 \, dt$$

Hence, $V_o = 4.52$ V is reached at

$$t_5' = 1.08 \text{ ns}$$

We now find the rise time t_r (see Figure 7.15) as

$$t_r = t_4 + t_5 - t_{r1} = 2.17 \text{ ns, without body effect,}$$

and

$$t_r = t'_4 + t'_5 - t_{r1} = 3.36 \text{ ns, with body effect included.}$$

Summarizing the results, we have $t_{PHL} = 103$ ps, $t_{PLH} = 1.215$ ns, $t_f = 224$ ps, and $t_r = 2.17$ ns without body effect, and $t_r = 3.36$ ns with some body effect included. The MicroSim PSpice simulation for the example with body effect, shown in Figure 7.15, gives $t_{PHL} = 222$ ps, $t_{PLH} = 1.58$ ns, $t_f = 273$ ps, and $t_r = 3.2$ ns. Clearly, the inclusion of some body effect of the depletion load MOSFET gives a better estimate for the rise time.

The preceding example makes it clear that the depletion load MOSFET charges the load capacitance with a nearly constant current; hence, the output voltage rises linearly over a wide range. It can be shown that the low-to-high delay time t_{PLH} is inversely proportional to the conduction parameter of the load device (see Problem 7.7). In a resistive load inverter, however, the output voltage increases exponentially with a time-constant of $R_L C_L$. Thus, for the same propagation delay and power dissipation (see the following section), a D-NMOSFET inverter is more advantageous to use than a large resistor.

It is shown empirically that including the rise time of the input waveform, t_r, increases the overall delay time in the mean squared sense as

$$t_{PLH} = \sqrt{(t_{PLH}z)^2 + \left(\frac{t_r}{2}\right)^2}$$

where $t_{PLH}z$ is the delay with zero rise time of input, that is, ideal step input V_i.

In comparing with other logic families, note that the small delay times are obtained in the preceding examples using capacitances that are extremely small. Although the values we have used for C_L are reasonable as on-chip gate input capacitances, large capacitive off-chip loads (in the pF range) will result in excessive delay and degraded dynamic performance. This is a limiting factor in using MOS circuits to drive a transmission line or highly capacitive loads, where bipolar logic families (Chapters 4 and 5), particularly ECL, have an advantage.

7.1.6.   POWER DISSIPATION AND DELAY-POWER PRODUCT

Current in each of the four NMOSFET inverters is negligibly small at logic high output. Although the active load device is on (in ohmic or pinch-off mode) it carries only the leakage current of the off driver. Since the leakage current is many orders smaller than normal operating current, it is of little consequence

```
* Example 7.6 DEPLETION LOAD INVERTER Switching
MD      2   1   0   0   NENH
ML      3   2   2   0   NDEP   ; KR = 3
CL      2   0   0.05pF
VI      1   0   PULSE(5V OV 5n 0.005n 0.005n 10n 20n)
VDD     3   0 DC 5
.PROBE
.MODEL NDEP   NMOS (VTO = -2 KP = 50E-6 GAMMA = 0.4)
.MODEL NENH NMOS (VTO = 1 KP = 150E-6); 2*K for both
.TRAN 1n 25n
.END
```

(a) Simulation File

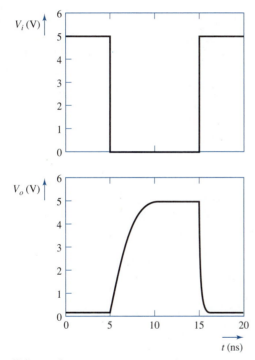

(b) Waveforms

Figure 7.16 Switching transition delays for Example 7.6

in power calculations. But at logic low output, with both the load and the driver devices conducting, the supply (drain) current cannot be ignored. Therefore, static power dissipation occurs only at the output low state. The following example compares the average static power dissipation in the four inverter configurations.

Example 7.7 | Determine the average static power dissipation in the NOMOSFET inverter circuit for each of the following cases: (*a*) resistive load (Figure 7.1a) with $R = 10 \text{ k}\Omega$, (*b*) saturated NMOSFET load (Figure 7.3a) with $k_L = 10 \ \mu\text{A/V}^2$ and $V_{TL} = 1 \text{ V}$, (*c*) linear NMOSFET load (Figure 7.5a) with $V_{GG} = 10 \text{ V}$, $k_L = 10 \ \mu\text{A/V}^2$, and $V_{TL} = 1 \text{ V}$, and (*d*) depletion NMOSFET load (Figure 7.7a) with $k_L = 10 \ \mu\text{A/V}^2$ and $V_{TL} = -2 \text{ V}$. In all cases $V_{DD} = 5 \text{ V}$, $k_D = 100 \ \mu\text{A/V}^2$, and $V_{TD} = 1 \text{ V}$. Neglect the body effect.

Solution

In all four cases, the gate current is zero and the supply current at logic high output is negligible. Hence, the average power dissipation is half the power supplied by the source at logic low output. (Note that at $V_o = V_{OH} \approx V_{DD} = 5 \text{ V}$, the load MOSFET is in the ohmic mode in all cases.) Based on the mode of operation of the load device, supply current at $V_o = V_{OL}$, which corresponds to $V_i = V_{OH}$, is obtained in each case.

Resistive load: At $V_i = V_{OH} = 5 \text{ V}$, the ohmic mode drain current is

$$I_D = 0.1[2(5 - 1)V_{OL} - V_{OL}^2] = \frac{5 - V_{OL}}{10}$$

Solving, $V_{OL} = 0.6 \text{ V}$ and $I_D = 0.44 \text{ mA}$.

Thus, the static power dissipation of the circuit at logic low output is given by

$$P_D = V_{DD} I_D = 2.2 \text{ mW}$$

Saturated MOS load: We determine the current at $V_o = V_{OL}$ by setting the ohmic mode current of the driver and the saturation mode current of the load. Note, however, that $V_i = V_{OH} = V_{DD} - V_{TL} = 4 \text{ V}$. Hence,

$$I_D = 100[2(4 - 1)V_{OL} - V_{OL}^2] = 10(5 - V_{OL} - 1)^2$$

Solving, $V_{OL} = 0.245 \text{ V}$ and $I_D = 0.14 \text{ mA}$.

Thus, $P_D = 0.7 \text{ mW}$.

Linear MOS load: With both the load and the driver MOSFETs operating in the ohmic mode, the drain current is given by Equation (7.27a)

$$I_D = 100[2(5 - 1)V_{OL} - V_{OL}^2]$$
$$= 10[2(10 - 1 - V_{OL})(5 - V_{OL}) - (5 - V_{OL})^2]$$

Solving, $V_{OL} = 0.722 \text{ V}$ and $I_D = 0.525 \text{ mA}$.

Thus, $P_D = 2.63 \text{ mW}$.

Depletion MOS load: With the load MOSFET in saturation and the driver in ohmic mode, the drain current is given by Equation (7.37)

$$I_D = 100[2(5 - 1)V_{OL} - V_{OL}^2] = 10(2)^2$$

Solving, $V_{OL} = 0.05 \text{ V}$ and $I_D = 0.04 \text{ mA}$.

And thus, $P_D = 0.2 \text{ mW}$.

Table 7.1 shows typical values of average power dissipation.

Table 7.1
Typical average static power dissipation for NMOSFET inverters

Load	P_{avg} μW
Resistor	1000
Saturated NMOS	300
Linear NMOS	1000
Depletion NMOS	100

Delay-Power Product   Example 7.7 and Table 7.1 show that NMOSFET inverters dissipate far less static power and have lower delays than BJT logic circuits. We must keep in mind, however, that the delays are low due to much smaller load capacitance than we encounter in BJT circuits. Clearly, the average propagation delay is proportional to (1) the logic swing $V_{ls} = (V_{OH} - V_{OL})$, which may be approximated by the supply voltage V_{DD}, and (2) the load capacitance C_L, which must charge and discharge by V_{ls}. In addition, the charging current in an active NMOSFET load increases with increasing conduction parameter; hence, t_{PLH} and, consequently, the average propagation delay t_p, decrease with k_L. Note further that the low-to-high delay t_{PLH} is, in general, greater than the high-to-low delay t_{PHL}. This is because of the large conduction parameter k_D ($>k_L$) for the driver to achieve a low value for V_{OL}.

For $V_{OL} \approx 0$, we can write the average static power dissipation as approximately proportional to V_{DD}^2. Combining the static power dissipation with the delay, we may express the static delay-power product DP_s of an NMOS inverter as

$$DP_s = k_{dp} C_L V_{DD}^2 \qquad \textbf{(7.53)}$$

where the constant k_{dp} depends on the inverter circuit configuration.

A more important point to consider is the power dissipation during output transitions. This dissipation arises in charging and discharging the load capacitance C_L in each cycle of input clock. Dynamic dissipation is quite significant at high clock rates, that is, when the input clock period T is comparable to $(t_{PHL} + t_{PLH})$. The energy transferred to C_L during the charging interval (at logic high output via the load device) is given by $(1/2) C_L V_{ls}^2$, which is also the energy discharged by C_L (to the driver device).

Therefore, average dynamic power dissipation P_d in each period of T s (with equal on and off intervals) is

$$P_d = \frac{C_L V_{ls}^2}{T} \qquad \textbf{(7.54a)}$$

Note that P_d is the average dynamic power dissipated in the circuit.

Using the frequency of operation $f = 1/T$, this becomes

$$P_d = f C_L V_{ls}^2 \qquad \textbf{(7.54b)}$$

If the average propagation delay is $t_P = (t_{PHL} + t_{PLH})/2$, the dynamic delay-power product is then given by

$$DP_d = (t_P/T) C_L V_{is}^2 = (f t_P) C_L V_{ls}^2 \qquad \textbf{(7.55a)}$$

Writing $f_{max} = 1/(t_{PHL} + t_{PLH}) = 1/2t_P$ for the maximum frequency of operation, we rewrite the above equation as

$$DP_d = \left(\frac{1}{2}\right)\left(\frac{f}{f_{max}}\right) C_L V_{ls}^2 \qquad \textbf{(7.55b)}$$

Equations (7.54) and (7.55) show that the dynamic power dissipation and thus, the dynamic delay-power product increase linearly with the frequency of switching. Therefore, for a more accurate measure of the circuit switching and dissipation performance, one must consider both static and dynamic delay-power products.

If we again use the approximation $V_{ls} \approx V_{DD}$, Equation (7.55) becomes

$$DP_d = \frac{1}{2}(f/f_{max}) C_L V_{DD}^2 \qquad \textbf{(7.56)}$$

Combining Equations (7.53) and (7.56), the total delay-power product DP for an NMOS inverter is

$$DP = C_L V_{DD}^2 \left[k_{dp} + \frac{1}{2}(f/f_{max}) \right] \qquad \textbf{(7.57)}$$

where k_{dp} depends on the inverter configuration. For example, for a resistive load inverter, $k_{dp} \approx \ln(2)$, where it is assumed that $t_{PHL} \ll t_{PLH}$.

Equation (7.57) shows that we can reduce the total delay-power product by reducing C_L or V_{DD} for a given inverter configuration. Since V_{DD} is dictated by system power supply and noise margins, C_L must be reduced by careful wiring and interconnection. Clearly, circuits operating at low supply voltages can have low values of DP. Alternatively, the static product DP_s may be decreased to a negligible value, as is the case with CMOS inverters (Section 7.2).

7.1.7 FAN-OUT

With virtually no input (gate) current drawn by load inverters, an NMOSFET inverter can drive any number of identical circuits without degradation in logic levels under static conditions. Each load inverter, however, presents a significant gate capacitance at the output terminal of the driving gate. As a result, the switching response of the driving gate deteriorates with a large number of load gates. Buffer circuits, which use large-sized devices, increase the current capability of the load (sourcing) and the driver (sinking) MOSFETs to charge and discharge load capacitances.

7.2 COMPLEMENTARY NOSFET (CMOS) INVERTER

Digital integrated circuits that range from a few gates per chip for conventional (random logic) system design to several thousand per chip for custom design are currently available in the Complementary MOSFET (CMOS) technology. The wide popularity of the CMOS technology is due to its virtually zero static power dissipation and large noise margins in either logic state. In addition, CMOS circuits can operate over a range of power supply voltage and at a relatively high speed. For these reasons, CMOS circuits are widely used—in portable battery-powered circuits and systems as well as in large complex systems.

One of the disadvantages of the CMOS technology is the difficulty in fabricating both NMOS and PMOS devices on the same wafer. Since NMOSFETs are formed on a p-type substrate, additional processing steps are needed to form an n-type substrate first for the PMOSFET devices. Also, two threshold voltages, positive for the NMOS and negative for the PMOS, must be adjusted during fabrication. Although with the maturing of technology neither of these is a problem, NMOS technology is simpler for large circuits.

Another disadvantage is that complex logic circuits in the CMOS technology need more transistors than in the NMOS technology. Consequently, CMOS technology has lower chip density. Further, CMOS circuits are susceptible to a condition known as *latch-up,* in which a large supply current is diverted from a CMOS circuit to ground due to parasitic bipolar junction transistors. With advances in CMOS processing technology, all these problems are minimized to a great extent, and the high density and sharp voltage transfer characteristic stand out as clear advantages over many other technologies. In this section we begin our study of CMOS circuits with the basic CMOS inverter.

7.2.1 TRANSFER CHARACTERISTICS

Figure 7.17a shows the circuit of a CMOS inverter, which consists of an NMOS transistor M_N and a PMOS transistor M_P, both of the enhancement type. The substrate of each MOSFET is connected to its source so that no body effect is present. Input V_i is applied to both the gates, and output is taken from both the drains. Figure 7.17b shows a simplified cross section of the inverter with p-type substrate and n-type isolation, or *n-tub* for a PMOSFET device. Polysilicon (in silicon gate technology) or metal interconnections are made for input, output, V_{SS}, and V_{DD} connections. A thick field oxide (not shown in the figure) grown

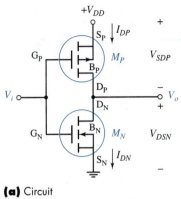

(a) Circuit

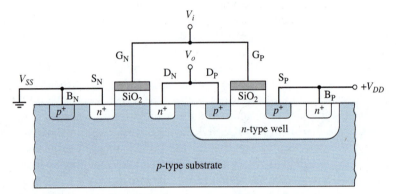

(b) Simplified Cross Section

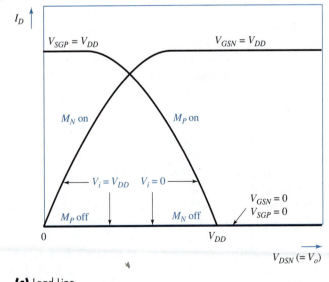

(c) Load Line

Figure 7.17   CMOS inverter

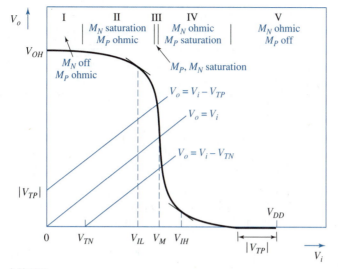

(d) VTC

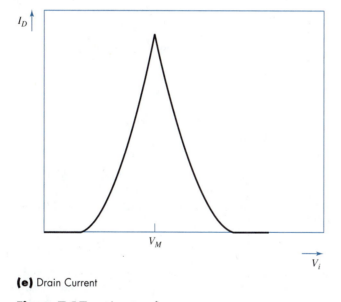

(e) Drain Current

Figure 7.17   (continued)

over the p-type wafer isolates the terminals. Alternatively, CMOS circuits are also built using n-type substrate and p-type isolation regions.

The input voltage V_i controls the gate bias for both M_N and M_P. Hence neither one drives the other; instead, each device operates in a mode complementary to the other.

For perfectly matched M_N and M_P, that is, both having the same magnitude of threshold voltage, and the same conduction parameter, we may qualitatively view the inverter operation as follows. For $V_i = V_{GSN} = 0$, M_N is turned off (no channel formed), while M_P, with $V_{SGP} = V_{DD}$, is turned on (conducting channel present); very little current flows through the off device M_N, and V_o is at logic high level ($\approx V_{DD}$). At $V_i = V_{GSN} = V_{DD}$, M_N is turned on, and M_P is turned off; again, negligible drain current exists and $V_o \approx 0$. Thus, one of the MOSFETs is off in both of the extreme logic levels while the other device provides a path from output node to ground or V_{DD}. The static power dissipation in either logic state is, therefore, negligibly small.

Figure 7.17c shows the load characteristics of the two MOS devices for the two extreme input voltages. For $V_i = 0$, M_N is off with zero drain current as shown by the horizontal line. M_P has the V_{SD}-I_D characteristic corresponding to $V_{SGP} = V_{DD}$. Since $I_{DP} = I_{DN}$, both graphs intersect only at $V_o = V_{DSN} = V_{DD}$. Similarly for $V_i = V_{DD}$, M_N has the V_{DS}-I_D characteristic for $V_{GSN} = V_{DD}$, while M_P has the zero-current horizontal line, both of which intersect only at $V_o = V_{DSN} = 0$.

As with the other MOSFET inverters, the voltage transfer characteristics for the CMOS inverter are obtained by equating the drain currents. For $V_i < V_{TN}$, M_N is off and M_P is on, with zero source-drain current. Hence,

$$V_o = V_{OH} = V_{DD}$$

which is in Region I shown in Figure 7.17d. As V_i is increased above V_{TN}, M_N goes to saturation. V_o is now dropping slightly due to conduction of M_N. The saturation condition for M_P, given by

$$V_{SDP} \geq V_{SGP} - |V_{TP}|$$

or
$$V_{DD} - V_o \geq V_{DD} - V_i - |V_{TP}| \tag{7.58a}$$

which may be rewritten as

$$V_o \leq V_i - V_{TP} \tag{7.58b}$$

is not satisfied. Hence, M_P is operating in the ohmic mode. Figure 7.17d shows the boundary between saturation and ohmic modes of M_P. The VTC in the ohmic mode of M_P with M_N in saturation, which is identified by Region II in Figure 7.17d, is described by $I_{DN} = I_{DP}$, or

$$k_N(V_i - V_{TN})^2 = k_P[2(V_{DD} - V_i - |V_{TP}|)(V_{DD} - V_o) - (V_{DD} - V_o)^2] \tag{7.59}$$

The logic low threshold voltage V_{IL} occurs in this region of VTC.

A further increase in V_i causes V_o to drop further and the saturation condition given in Equation (7.58) is now satisfied. At this point, both M_N and M_p are in saturation and operate as constant current sources. This is Region III in Figure 7.17d; it is described by

$$k_N(V_i - V_{TN})^2 = k_p'(V_{DD} - V_i - |V_{TP}|)^2 \tag{7.60}$$

Clearly, Equation (7.60) is true only at a single voltage. In practice, the drain current of each device increases slightly with increasing drain-source voltage. Still, the two nearly constant-current sources in series cause a large drop in drain voltage with a small increase in input voltage V_i. Thus, the region where both the devices are in saturation is extremely narrow, and the VTC falls sharply at V_i given Equation (7.60).

For input voltage V_i such that

$$V_o \le V_i - V_{TN} \quad \text{or} \quad V_i \ge V_o + V_{TN}$$

M_N operates in the ohmic mode while M_p continues in the saturation mode in Region IV. Figure 7.17d shows the boundary between the saturation and ohmic modes of M_N. V_o falls in the ohmic region of M_N according to

$$k_N[2(V_i - V_{TN})V_o - V_o^2] = k_P(V_{DD} - V_i - |V_{TP}|)^2 \qquad \textbf{(7.61)}$$

$V_i = V_{IH}$ occurs in Region IV as shown in Figure 7.17d.

As V_i is increased above $V_{DD} - |V_{TP}|$, M_p is turned off while M_N continues in the ohmic in Region V. Hence, $V_o = V_{OL} \approx 0$. The critical voltage $V_M = V_i = V_o$, which is the gate threshold voltage, occurs when both M_N and M_p are in saturation. From Equation (7.60) V_M is given by

$$V_M = \frac{V_{TN} + \sqrt{K_R}(V_{DD} - |V_{TP}|)}{1 + \sqrt{K_R}} \qquad \textbf{(7.62)}$$

where $K_R = k_P/k_N$.

As the geometry ratio K_R is increased, V_M shifts to the right in the VTC while the transition remains steep. Thus, the slope of the midregion where both MOS-FETs are in saturation does not change with K_R, unlike in the case of NMOSFET inverters. Note that for a symmetric inverter, that is, one with $V_{TN} = |V_{TP}|$ and $k_P = k_N$, $V_M = V_{DD}/2$.

Observe from the transfer characteristics in Figure 7.17d that the supply voltage must satisfy

$$V_{DD} \ge V_{TN} + |V_{TP}|$$

so that the logic levels will be unambiguously defined. Hence, the inverter operates satisfactorily at any supply voltage above the sum of the two threshold voltage magnitudes. In practice, the low end of V_{DD} is higher than $V_{TN} + |V_{TP}|$ to have some transition width $V_{IH} - V_{IL}$. At the high end, V_{DD} is limited by the drain-source breakdown voltage.

Since one of the devices is off for $V_i < V_{TN}$ and for $V_i > V_{DD} - |V_{TP}|$, the drain current in the circuit is zero in these regions. For V_i satisfying Equation (7.60), both devices are in saturation and the drain (supply) current is a maximum. Figure 7.17e shows the supply current for the CMOS inverter.

Determine the logic threshold voltages and V_M for a CMOS inverter having $V_{DD} = 5$ V, $V_{TN} = -V_{TP} = 1$ V, $k_N = 100$ μA/V^2, and $k_P = 50$ μA/V^2. | **Example 7.8**

Solution

For $V_i < V_{TN} = 1$ V, M_N is off and $V_o = V_{OH} = V_{DD} = 5$ V. For $V_i > V_{TN}$, M_N goes into saturation mode and M_p goes into ohmic mode. From Equation (7.59), the VTC is decribed by

$$2(V_i - 1)^2 = 2(4 - V_i)(5 - V_o) - (5 - V_o)^2 \qquad \textbf{(7.63)}$$

The slope of the VTC is

$$\frac{dV_o}{dV_i} = \frac{V_o - 2V_i - 3}{V_o - V_i - 1} \qquad \textbf{(7.64)}$$

Equating the slope to -1 and using Equation (7.63), V_{IL} and the corresponding output are given by

$$V_{IL} = 1.8 \text{ V} \quad \text{and} \quad V_o = 4.7 \text{ V}$$

To determine V_{IH} in the region where M_N is in ohmic mode and M_P in saturation, from Equation (7.61), we have

$$4(V_i - 1)V_o - 2V_o^2 = (4 - V_i)^2$$

Again, equating the slope dV_o/dV_i to -1, V_{IH} and the corresponding output are given by

$$V_{IH} = 2.54 \text{ V} \quad \text{and} \quad V_o = 0.4 \text{ V}$$

For $V_i = V_{OH} = 5$ V, M_P is in cutoff and M_N is in the ohmic region. Hence, $V_{OL} \approx 0$.

With $K_R = k_P/k_N = 0.5$, from Equation (7.62), $V_i = V_o = V_M = 2.24$ V.

Note that since both MOSFETs are in saturation, the drain current at $V_i = V_o = V_M$ is the maximum at 154.4 μA.

The critical voltages for the inverter are:

$$V_{IL} = 1.8 \text{ V}, \; V_{IH} = 2.54 \text{ V}, \; V_{OH} = 5 \text{ V}, \; V_{OL} = 0 \text{ V}, \text{ and } V_M = 2.24 \text{ V}.$$

Note that from the above example the logic low noise margin $NM_L = V_{IL}$, since $V_{OL} = 0$. Clearly this noise margin is larger than that for the NMOSFET inverters with nonzero V_{OL}.

If PMOS and NMOS devices are symmetric i.e., $(V_{TN} = |V_{TP}|$ and $k_P = k_N)$, it can be shown (see Problem 7.10) that $V_{IL} + V_{IH} = V_{DD}$. To achieve $k_P = k_N$, however, the geometry ratio must be adjusted. Since the conduction parameter is proportional to the mobility of carriers in the induced channel [Equation (6.9)] as

$$k_N = \left(\frac{1}{2}\right)\mu_N C_{\text{ox}}\left(\frac{W}{L}\right)_N \quad \text{and} \quad k_P = \left(\frac{1}{2}\right)\mu_P C_{\text{ox}}\left(\frac{W}{L}\right)_P$$

the geometry ratio K_R is given by

$$K_R = \frac{k_P}{k_N} = \frac{\mu_P(W/L)_P}{\mu_N(W/L)_N} \qquad \textbf{(7.65)}$$

Therefore, for $k_P = k_N$, or $K_R = 1$,

$$\frac{(W/L)_P}{(W/L)_N} = \frac{\mu_N}{\mu_P}$$

With the electron-hole mobility ratio of $\mu_N/\mu_p \approx 2.5$ in silicon, the above equation shows that a minimum area CMOS inverter with symmetrical characteristics must have the PMOSFET with an aspect ratio of 2.5 times that of the NMOSFET. The resulting inverter has $V_{IL} = V_{IH} = V_M$ and maximum noise margins of $V_{DD}/2$ for low and high levels.

Although CMOS inverters are usually designed for symmetrical voltage transfer characteristics, in general the design may be accomplished for a specified critical voltage V_M given by Equation (7.62). Rewriting Equation (7.62) in terms of the geometry ratio, we have

$$\sqrt{K_R} = \frac{V_M - V_{TN}}{V_{DD} - V_M - |V_{TP}|}$$

or

$$\sqrt{\frac{k_P}{k_N}} = \sqrt{\frac{\mu_P(W/L)_P}{\mu_N(W/L)_N}} = \frac{V_M - V_{TN}}{V_{DD} - V_M - |V_{TP}|}$$

Using $\mu_N/\mu_p \approx 2.5$ in silicon,

$$\frac{(W/L)_P}{(W/L)_N} \approx 2.5 \left[\frac{V_M - V_{TN}}{V_{DD} - V_M - |V_{TP}|} \right]^2 \qquad \textbf{(7.66)}$$

Equation (7.66) gives the geometry ratio for the inverter with specified gate and device threshold voltages, if the oxide capacitance C_{ox} is the same for both the the PMOS and NMOS devices. While this straightforward design meets the static transfer characteristic specification of V_M, large aspect ratios resulting from Equation (7.66) increase the silicon area.

Design a CMOS inverter to operate at a supply voltage of 5 V with $V_M = 2$ V, $V_{TN} = 1$ V, and $V_{TP} = -1.5$ V. What are the input logic threshold voltages? | **Example 7.9**

Solution
From Equation (7.66) the geometry ratio is

$$\frac{(W/L)_P}{(W/L)_N} = 2.5 \left(\frac{1}{1.5} \right)^2 = \frac{10}{9}$$

using integer ratios. This gives $k_P/k_N = 4/9$.

Using Equations (7.59) and (7.61) and the unity slope condition, the logic threshold voltages are:

$$V_{IL} = 1.62 \text{ V} \quad \text{and} \quad V_{IH} = 2.23 \text{ V}$$

with corresponding ouput voltages of 4.76 V and 0.33 V, respectively.

Figure 7.18 shows the MicroSim PSpice simulation for the VTC of the circuit as designed above. In this simulation, 5 μm is used for channel length with different widths for the NMOSFET and PMOSFET devices. A process transconductance parameter of $\mu_n\epsilon/t_{ox} = 40$ μA/V^2 is typical for NMOS processes.

```
* Example 7.9 CMOS INVERTER VTC
MN       2    1    0    0   NCH L=5U W=45U
MP       2    1    3    3   PCH L=5U W=50U
VI       1    0
VDD      3    0 DC 5
.PROBE
.MODEL PCH   PMOS (VTO = -1.5 KP = 40E-6)
.MODEL NCH   NMOS (VTO = 1 KP = 100E-6)
.DC      VI 0 5V 0.1V
.PRINT DC V(2),ID(MN)
.OPTIONS NOPAGE
.END
```

(a) Simulation

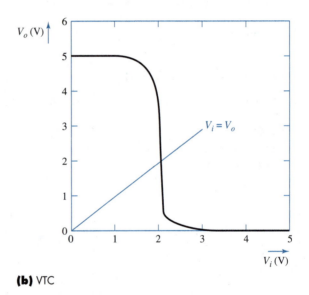

(b) VTC

Figure 7.18   MicroSim PSpice simulation for Example 7.9

In practice, for devices in LSI systems the device threshold voltage magnitudes may be set the same for all so that $V_M = V_{DD}/2$ results in $k_P = k_N$ and $(W/L)_P/(W/L)_N = 2.5$. This gives the minimum-area inverter with symmetrical devices, and VTC.

7.2.2   FAN-OUT AND DELAY-POWER PRODUCT

As with NMOSFET inverters, a CMOS inverter has practically no limit in driving other CMOS inputs under static conditions. The dynamic fan-out is limited by the allowable propagation delay in an application. Typically, input of a standard CMOS inverter in the 4000B series and the 74HC series presents 2 to 5 pF to the driving gate. This results in a propagation delay of a few nanoseconds. Hence, fan-out is usually limited to 10, to prevent excessive delays at the output. For increased fan-out with reduced delay, CMOS buffers may be built by paralleling several inverters at the input and output. Connecting the output nodes together reduces the drain-source resistance, which is typically 1000Ω for an on device and up to several hundred $M\Omega$ for an off device. Thus, the drive current in both logic levels is increased. Clearly, this is an alternative to fabricating large-sized output devices.

Power Dissipation   As seen in the VTC, a CMOS inverter draws negligible current from the supply voltage in either logic state. The leakage current due to finite off resistance of each MOSFET device is typically a few nanoamperes at 5 V supply. Hence static power dissipation is almost zero.

As with NMOSFET inverters, a CMOS inverter draws significant current (of up to several hundred μA depending on load capacitance) during logic transition. Since $V_{OH} = V_{DD}$ and $V_{OL} = 0$, average dynamic power dissipation in the inverter circuit in one period $T \,(= 1/f)$ of a clock is

$$P_{avg} = \left(\frac{1}{T}\right)C_L V_{DD^2} \qquad \textbf{(7.67)}$$

Since power supply current in a CMOS inverter is maximum during transition between logic states (See Figure 7.16e), finite rise and fall times in the input waveform also contribute to power dissipation. However, this contribution becomes significant only with input waveforms having large rise and fall times.

Delay-Power Product   If the average propagation delay is t_P, which is approximately the interval in which C_L is charging, the dynamic delay-power product is given by

$$DP = P_{avg}\, t_P = \left(\frac{t_P}{T}\right)C_L V_{DD^2} \qquad \textbf{(7.68a)}$$

Using $f_{max} = 1/2t_P$ and $f = 1/T$, the above equation becomes

$$DP = \frac{1}{2}\left(\frac{f}{f_{max}}\right)C_L V_{DD^2} \qquad \textbf{(7.68b)}$$

Equation (7.68b) is the same as Equation (7.56) for the dynamic DP of an NMOS inverter. However, for CMOS inverters Equation (7.68) represents the total delay-power product.

Propagation Delay   Charging of C_L via the PMOSFET device and discharging via the NMOSFET device contribute to delays t_{PLH} and t_{PHL}. Figure 7.12, which shows charging and discharging of C_L for a general NMOSFET inverter, is applicable for the CMOS inverter with an E-PMOSMET "load" device. However, unlike with an NMOSFET inverter, the range of voltage for C_L is from 0 to V_{DD}. The second difference in the case of a CMOS inverter is that C_L is usually higher—on the order of 5 to 15 pF. Increased capacitance arises from the paralleling of the PMOSFET gate, and also from reverse-biased input diodes used for protection against electrostatic discharge (see Section 7.2.3). (Note that unlike in NMOSFET inverters, gates of both PMOS and NMOS devices are driven by the input signal; therefore, the driving circuit must charge the capacitance of both the devices.) Also, in symmetrical and in high-speed CMOS inverters, the device widths are large by design, which contribute to large capacitances. As a result, delays in CMOS circuits are usually higher than in comparable NMOSFET circuits. In addition, bus-connected systems add the largest component of capacitance: the interconnect capacitance, on the order of 50–70 pF per meter. Clearly the stray interconnect capacitance causes the maximum delay and hence determines the maximum frequency of operation.

The calculation of CMOS delay times proceeds in a manner similar to that for NMOS inverters. When the input voltage V_i changes from $V_{OH} = V_{DD}$ to $V_{OL} = 0$, the NMOS device M_N is turned off. C_L charges from 0 toward V_{DD} by the PMOS device M_P, which is operating as a constant-current source in the saturation mode. With an initial charging current of

$$i_C = k_P(V_{DD} - |V_{TP}|)^2 \qquad \textbf{(7.69)}$$

V_o rises until it reaches $|V_{TP}|$. At this point, M_P changes its mode of operation to ohmic. We obtain the time t_1 when $V_o = |V_{TP}|$ is reached using constant-current charging as

$$t_1 = \frac{|V_{TP}|C_L}{k_P(V_{DD} - |V_{TP}|)^2} \qquad \textbf{(7.70)}$$

Now C_L charges with a current of

$$i_C = k_P[2(V_{DD} - |V_{TP}|)(V_{DD} - V_o) - (V_{DD} - V_o)^2] \qquad \textbf{(7.71)}$$

The time taken to charge from $V_o = |V_{TP}|$ to the 50 percent point of $V_{DD}/2$ (using the integral in Example 7.5) is given by

$$t_2 = \frac{C_L}{2k_P(V_{DD} - |V_{TP}|)} \ln \frac{3V_{DD} - 4|V_{TP}|}{V_{DD}} \qquad \textbf{(7.72)}$$

Hence the low-to-high propagation delay time $t_{PLH} = t_1 + t_2$ is

$$t_{PLH} = \frac{C_L}{2k_p(V_{DD} - |V_{TP}|)}\left[\frac{2|V_{TP}|}{V_{DD} - |V_{TP}|} + \ln \frac{3V_{DD} - 4|V_{TP}|}{V_{DD}}\right] \qquad \textbf{(7.73)}$$

Similarly, we can obtain the high-to-low propagation delay time during which the NMOS device carries the discharging current of C_L as

$$t_{PHL} = \frac{C_L}{2k_N(V_{DD} - V_{TN})}\left[\frac{2V_{TN}}{V_{DD} - V_{TN}} + \ln \frac{3V_{DD} - 4V_{TN}}{V_{DD}}\right] \qquad \textbf{(7.74)}$$

Note the similarity in the two equations. Since only one of the two MOSFETs is conducting during either interval, the above equations are more accurate than those for the NMOSFET inverters. In the case of a symmetrical CMOS inverter with $k_P = k_N$ and $|V_{TP}| = V_{TN}$, we can readily see that $t_{PHL} = t_{PLH}$, making the symmetry of the inverter complete. For nonsymmetrical inverters, increasing the conduction parameter reduces propagation delays. Since the process transconductance parameter is fixed for a given process, it appears that the aspect ratios of the MOSFETs may be increased for faster switching response. However, aspect ratios increase the size of the inverter and also the capacitance at input and output. Additionally, the noise margin specification may not be met. Thus a trade-off is usually called for between size and performance specifications.

Example 7.10

Determine the propagation delays for the CMOS inverter designed in Example 7.9 if the process transconductance parameter is 50 μA/V^2 for the NMOSFET and 20 μA/V^2 for the PMOSFET, and the lumped capacitance at the output terminal is 10 pF. Use $V_{DD} = 5$ V.

Solution

The conduction parameters are

$$k_N = 50\left(\frac{45}{5}\right) = 450 \ \mu\text{A/V}^2 \quad \text{and} \quad k_p = 20\left(\frac{50}{5}\right) = 200 \ \mu\text{A/V}^2$$

Using $|V_{TP}| = 1.5$ V and $V_{TN} = 1$ V, from Equations (7.70) and (7.72),

$$t_1 = 6.12 \text{ ns} \quad \text{and} \quad t_2 = 4.2 \text{ ns}$$

Hence,

$$t_{PLH} = t_1 + t_2 = 10.32 \text{ ns}$$

With NMOSFET replacing PMOSFET in Equations (7.70) and (7.72), we have

$$t_1 = 1.39 \text{ ns} \quad \text{and} \quad t_2 = 2.19 \text{ ns}$$

Hence,

$$t_{PHL} = t_1 + t_2 = 3.58 \text{ ns}$$

The MicroSim PSpice simulation for the transient response of this example, shown in Figure 7.19, gives $t_{PLH} = 10.45$ ns and $t_{PHL} = 3.72$ ns.

The last two examples point to an important aspect in using simulation for the design of CMOS circuits. As stated earlier, although the design in Example 7.9 accomplished the dc specification of $V_M = 2$ V, the resulting inverter may not satisfy the switching response requirement for the given load capacitance. MicroSim PSpice or any other simulation of the designed circuit provides the expected transient response, based on which one can readily make modifications to the design.

For gates and other SSI circuits manufacturers provide delay equations as functions of load capacitance and supply voltage.

```
* Example 7.10 CMOS INVERTER TRANSIENT RESPONSE
MN      2   1   0   0   NCH L=5U W=45U
MP      2   1   3   3   PCH L=5U W=50U
CL      2   0   10pF
VI      1   0   PULSE (0 5V 10N 0.001N 0.001N 100N 200N)
VDD     3   0 DC 5
.PROBE
.MODEL PCH   PMOS (VTO = -1.5 KP = 40E-6)
.MODEL NCH   NMOS (VTO = 1               KP = 100E-6)
.TRAN 5N 200N
.PRINT DC V(2),ID(MN)
.END
```

(a) Simulation

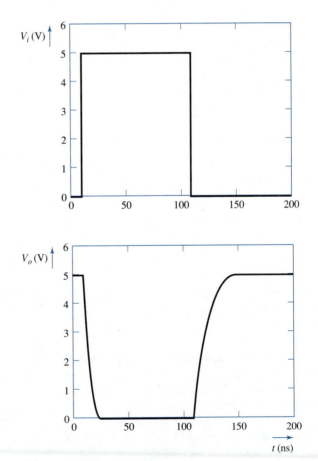

(b) Input and Output Waveforms

Figure 7.19   MicroSim PSpice simulation of transient response for Example 7.10

7.2.3 INPUT PROTECTION AND LATCH-UP

While the extremely high input impedance at the gate of a CMOS inverter is beneficial to the driving circuitry, the gate must be protected against static charge build-up. The insulating gate oxide is extremely thin ($1000 \, \text{Å} = 0.1 \, \mu\text{m}$ for a metal gate and less than $500 \, \text{Å}$ for a polysilicon gate) with a breakdown voltage of 50 to 100 V. If a gate voltage above the breakdown voltage is applied, it ruptures the oxide insulation and causes excessive gate current, resulting in permanent damage. With a gate resistance in the hundreds of mega ohms, stray electrostatic discharge from a person handling the circuits can easily release enough energy to cause breakdown. (Note that with a body capacitance of 100 to 300 pF, a person walking across a waxed laboratory floor or brushing against a garment may generate static voltage in excess of 10 kV. The "charged" body releases an average power of many kilowatts over hundreds of nanoseconds.) To prevent gate current due to high voltages at the input terminal of a CMOS circuit, the circuit includes clamping diodes as shown in Figure 7.20. Diodes D_1 and D_2 at the input conduct and clamp the gate voltage V_i at $-V_{\text{diode}}$ and $V_{DD} + V_{\text{diode}}$. The series resistance R_s, which is typically 1500 Ω for a metal gate and 250 Ω

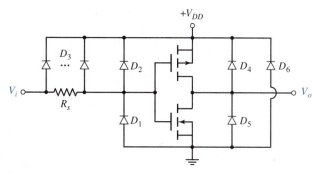

(a) Metal Gate

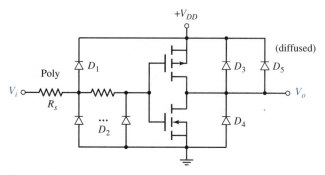

(b) Polysilicon Gate

Figure 7.20   Input and output protection in CMOS circuits

for a polysilicon gate, limits transient gate current. The distributed diode (D_3 in the metal gate and D_2 in the polysilicon gate) is parasitic arising from the fabrication of R_s, which has p diffusion in the n-substrate of the PMOSFET in metal gates (see Figure 7.21).

At the output side, the parasitic diode D_4 in Figure 7.20a (D_3 in Figure 7.20b) conducts when the output voltage exceeds V_{DD}. Similarly, the drain to p-tub diode (D_5 in Figure 7.20a or D_4 in Figure 7.20b) conducts if the voltage goes negative. D_6 arises from the pn junction between the p-substrate (or tub) to the n-tub (or substrate). Note that all three diodes are parasitic.

A polysilicon resistor R_s is used at the input of silicon gate CMOS circuits as shown in Figure 7.20b. Additionally, a diffused diode D_5 may be used for increased protection.

The protection diodes in both metal and silicon gate CMOS circuits are reverse- or zero-biased under normal operating conditions. Hence, their contribution of depletion capacitance increases the total capacitance at the input and output nodes.

Another type of protection in a CMOS circuit is needed to prevent conduction from power supply to ground via a complementary pair of bipolar transistors. Figure 7.21 shows a common layout of a CMOS inverter in which the PMOSFET is fabricated on an n-substrate; on the same n wafer, the NMOSFET is formed in a p-type well. This gives rise to parasitic bipolar junction transistors.[2] These BJTs, the vertical npn and the lateral pnp, form a *thyristor,* or *silicon controlled rectifier,* which is a pnpn composite transistor with three terminals. R_w and R_s are the resistances of the p-well and the n-substrate. The composite transistor conducts current from the external p terminal (connected to V_{DD}) to the external n terminal (connected to ground) when a positive current is injected at the base-collector junction labeled G in Figure 7.21b. The injected current causes an increase in the collector current of the npn transistor Q_N which, in turn, increases the base current and, consequently, the collector current of the pnp transistor Q_P. This results in the cumulative increase of current from the power supply, which is almost shorted to ground. The positive feedback caused by external injection at G continues even after removing the source of injection. The current drain from V_{DD} can be stopped only by disconnecting the supply. In CMOS circuits, the cumulative process is initiated by a short-duration positive voltage at G. In addition, since the output node V_o is common to one of the emitters from each transistor, voltage overshoot (above $V_{DD} + 0.6$ V) or undershoot (below -0.6 V) at the common emitter (V_o) can trigger parasitic conduction. Also, the input protection diodes contribute to the parasitic pn junctions at the p-well and the n-substrate. Thus, a sudden application of V_{DD} or incorrect connection of output to a positive voltage source forward-biases Q_N, and the positive feedback process may begin. The condition in which current from V_{DD} is diverted from logic circuit to ground is known as *latch-up.* In addition to the failure of logic function, a large current from V_{DD} to ground may overheat and cause damage to the chip. Latch-up is most likely to occur in CMOS driver circuits where large voltages arise.

| [2] A similar pair of npn-pnp parasitic transistors also arise in the n-well process where a p substrate is used.

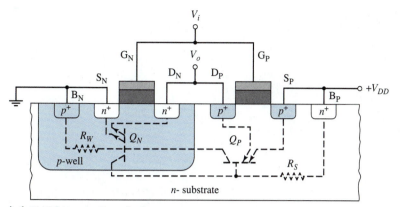

(a) CMOS Inverter Cross Section

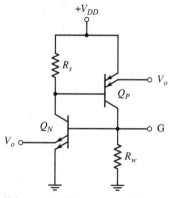

(b) Parasitic BJTs Forming SCR

Figure 7.21   Latch-up in a CMOS inverter

One can minimize the effect of latch-up by reducing the current gains of the parasitic BJTs. Positive feedback does not sustain if the *loop gain,* which is the product of the BJT current gains, β_P and β_N, is below unity. The condition $\beta_P \beta_N < 1$ is achieved by providing heavily doped *guard rings* around PMOS and NMOS devices. Guard rings, or bands, of diffusions—p^+ between the drain of the n-channel and p-well, and n^+ outside the p-well near the drain of the p-channel—isolate the CMOS terminals from parasitic pn junctions. These diffusions reduce both current gains for the BJTs while simultaneously eliminating the parasitic resistances R_w and R_s. Note that with R_w and R_s zero, the BJTs are off. Only at currents much higher than normal operating values, or at higher temperatures, which increase current gains, can latch-up in such circuits occur. Other recent approaches to eliminate latch-up include using an insulating substrate (such as sapphire, instead of a silicon semiconductor), and separating NMOS and PMOS devices using deep trench oxide barriers. [Reference 1]

7.3   MOSFET LOGIC GATES

NMOS and CMOS inverter implementations can be extended to realize other logic functions. This section considers the realization of basic gates using NMOS and CMOS configurations.

7.3.1   NMOS GATES

Basic NMOSFET gates are implemented using enhancement-type drivers and depletion-type loads as depicted in Figure 7.22. In both the circuits, the channel is always present in the normally on load D-NMOSFET (M_L). If either input (A or B) is at logic high in the NOR gate of Figure 7.22a, the E-NMOSFET driver M_{DA} or M_{DB} conducts with low resistance and the output is at logic low. If both inputs are at logic low, the path from V_{DD} to ground is open at Y and the output is at logic high. The number of inputs, or fan-in, can be increased by connecting additional driver E-NMOSFETs in parallel. With all driver MOSFETs identical in geometry, the logic low output voltage V_{OL} is the same regardless of which of the drivers is on. When more than one driver is on, however, the on devices connected in parallel reduce V_{OL} slightly. Note that the paralleling of driver MOSFETs is equivalent to increasing the effective conduction parameter of a single driver in the case of an inverter to the sum of the two driver parameters. Since the worst case (highest) V_{OL} arises when only one of the drivers is conducting, the design of an NMOSFET NOR gate is based on a specified maximum allowable value for V_{OL}. For symmetry, the aspect ratio of each driver MOSFET is chosen to be the same. When all the drivers are on in such a design, the logic low voltage would be better (lower) than the specified value. Clearly this design

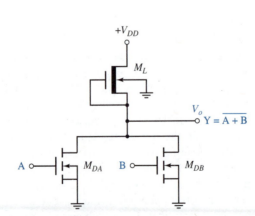

(a) Two-Input NOR

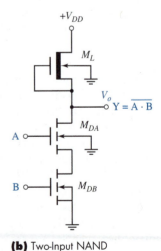

(b) Two-Input NAND

Figure 7.22   Basic NMOSFET gates

assumes equal current-sharing in all the drivers; that is, the logic high voltage at input is the same for all drivers. Otherwise V_{OL} will vary, depending on the magnitude of input voltage.

With the addition of more driver E-NMOSFETs in a NOR gate, capacitance at the output node increases. This increase is caused by the drain-to-body and drain-to-gate capacitances of each added device. In addition, the interconnect capacitance from each driver MOSFET also contributes to the lumped load capacitance. As a result, the propagation delay of an NMOSFET NOR gate is, in general, more than that of an inverter. One must consider this difference in delays when eliminating timing problems ("glitches") in the design of logic circuits using a combination of different gates. The body effect of the drivers, however, is absent in NMOSFET NOR gates.

In the NAND gate of Figure 7.22b, if one of the inputs is at logic low, the path to ground via the cascaded E-NMOSFET devices is broken and the output is at logic high. When all inputs are at logic high, all the driver E-NMOSFETs are on, with low resistance, and the output goes to logic low. For each additional input to increase the fan-in, a driver E-NMOSFET is added in series between the load D-NMOSFET and ground. Unlike in the NOR gate, however, the series combination of N driver MOSFETs increases the on resistance from output node Y to ground, when all inputs are high, to N times that of a single on MOSFET. This results in a higher output voltage V_{OL}. To obtain approximately the same V_{OL} as in an inverter, the conduction parameter of the combination must be adjusted to that of the single driver in an inverter. This is accomplished by increasing the aspect ratio of each of the N drivers by the number of inputs, N. Clearly the size of the NMOSFET NAND gate becomes larger than a NOR gate with the same fan-in and V_{OL}. In addition, the body effect arising from all but the bottom driver complicates the exact calculation of V_{OL}.

As with the NOR gate, capacitance at the output node increases with fan-in in the NAND gate. However, the total load capacitance must be estimated for different cases of input voltages, since each driver capacitance varies with voltage. Figure 7.23 shows a MicroSim PSpice simulation of switching response for a three-input NAND gate operating as an inverter. For comparison, the switching response of the depletion load inverter of Example 7.6 is shown alongside. For the same conduction parameters as in Example 7.6 and for the same output node capacitance, the high-to-low propagation delay is increased from 224 ps to 415 ps, and the fall time is increased from 273 ps to 993 ps. Note also the higher voltage of V_{OL}—from 0.17 V to 0.53 V. (Delays on the rising edge of output are not affected significantly in this case, since all the driver MOSFETs are off.) In practice, if the increased capacitance due to all of the driver E-MOSFETs are included, the delays at both edges would be much higher. Because of this and the increased silicon area, NMOSFET NAND gates are seldom used. With the availability of NOR gates (and inverters) this is not a limitation in NMOS logic implementation.

We can realize more complex combinational logic functions readily by stacking enhancement-type drivers and depletion-type loads. Figure 7.24 shows two examples of NMOSFET gates.

```
* Fig. 7.23 DEPLETION LOAD NAND Switching
ML      1  2  2  0  NDEP  ; KR = 3
MD1     2  3  4  0  NENH
MD2     4  5  6  0  NENH
MD3     6  5  0  0  NENH
CL      2  0  0.05pF
VH      5  0  DC 4.5
VI      3  0  PULSE(5V 0V 5n 0.005n 0.005n 10n 20n)
VDD     1  0  DC 5
.PROBE
.MODEL NDEP NMOS (VTO = -2 KP = 50E-6 GAMMA = 0.4)
.MODEL NENH NMOS (VTO = 1 KP = 150E-6 GAMMA = 0.4)
.TRAN 1n 25n
.PRINT DC V(2), ID(ML)
.OPTIONS NOPAGE
.END
```

(a) Simulation File

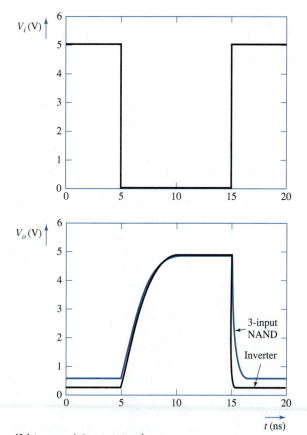

(b) Input and Output Waveforms

Figure 7.23   Comparison of NMOSFET NAND and inverter switching response

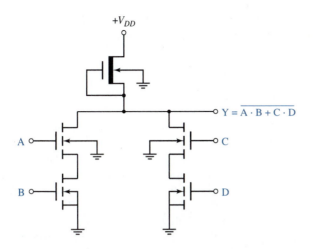

$$Y = \overline{A \cdot B + C \cdot D}$$

(a) AND-OR-Invert Gate

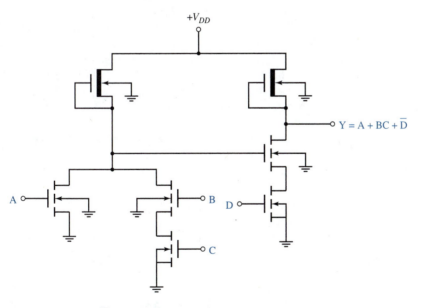

$$Y = A + BC + \overline{D}$$

(b) Y = A + BC + $\overline{D}$

Figure 7.24   Examples of NMOSFET gates

As we discussed earlier, NMOS circuits require smaller chip area and fewer fabrication steps compared with PMOS and CMOS circuits. As a result, high-density NMOS circuits are available as microprocessors, memory units, and other VLSI systems. These circuits are used with external buffering or interface circuits to drive high-current or capacitive (bus-connected) loads. For small-

and medium-scale circuits, however, the lack of drive capability is a serious limitation; hence, NMOSFET gates and other SSI/MSI functions are not available.

7.3.2 CMOS Gates

CMOS gate circuits are implemented in a manner similar to the NMOS circuits by paralleling and/or cascading pairs of PMOS and NMOS devices. Figure 7.25 shows two-input NAND and NOR gates. (Note the simplified symbols used for the NMOS and PMOS devices, implying that the source terminal of each device is connected to the appropriate body or substrate terminal, that is, to V_{DD} for PMOS and ground for NMOS devices.) If input A (B) is low in Figure 7.25a, the NMOS device $M_{N1}(M_{N2})$ is off while the PMOS device $M_{P1}(M_{P2})$ is on; hence, output goes high. When both inputs are high, the cascaded NMOS devices are turned on and form a low-resistance path from output (Y) to ground. With the paralleled PMOS devices off, output becomes logic low.

In the NOR gate of Figure 7.25b, $M_{N1}(M_{N2})$ turns on while $M_{P1}(M_{P2})$ is off when input A (B) is high; thus, output Y is at logic low. If both inputs are low, the paralleled NMOS devices are disconnected from the output node. The cascaded PMOS devices conduct and the output goes to logic high.

For each additional input in both the circuits, one pair of PMOS and NMOS transistors is required, with the PMOS device connected in parallel (series) and the NMOS device connected in series (parallel) for the NAND (NOR) function. Because of the body effect in the cascaded devices, the characteristics of CMOS NAND and NOR gates differ slightly from those of the CMOS inverter.

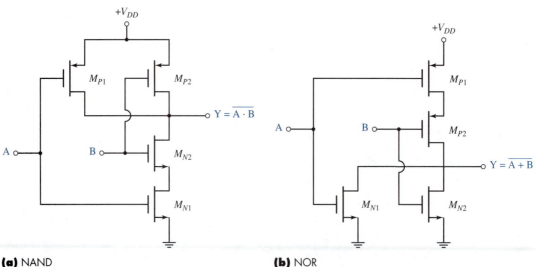

(a) NAND   **(b)** NOR

Figure 7.25   CMOS gates

NAND Gate Characteristics   In the transfer characteristics of a two-input CMOS NAND gate, the input voltage at which the output low-to-high transition occurs depends on which of the two inputs is changing. Since the lower NMOS M_{N1} in Figure 7.25a has no body effect, it requires $V_A = V_{TN0}$ (the zero-bias device threshold voltage) to begin conduction when V_B is at logic high. If, instead, V_A is at logic high, V_B must be above V_{TN}, which is larger than V_{TN0} due to body effect. For gates with three or more inputs, in general, the transfer characteristics shift to the right because of the increased input voltage required (due to body effect) in the NMOSFETs higher in the rung.

The case when both inputs switch simultaneously represents the worst case in terms of the gate threshold voltage $V_M = V_i = V_o = V_Y$. This is because of the higher voltage V_M required to keep both NMOS devices on. V_M is determined based on the common drain current of the NMOS devices, which is equal to the sum of the drain currents in the PMOSFETs. Figure 7.26 shows a two-input NAND gate with both inputs at $V_i = V_M$ and output at $V_o = V_M$. M_{N2} is in saturation with the gate and drain at the same voltage, and $V_{GS2} = V_{DS2} = V_M - V_{DS1}$. Since M_{N1} has $V_{DS1} = V_M - V_{DS2} < V_{GS1} = V_M$, it is in the ohmic mode. (Note that with $V_M = V_{DS1} + V_{DS2}$, V_{DS2} must be below the threshold voltage V_{TN} for M_{N1} to be in saturation.) If both the NMOS devices have the same conduction parameter k_N and threshold voltage V_{TN} and the body effect of M_{N2} is ignored, we have, for the drain current,

$$I_{DN} = k_N(V_M - V_{DS1} - V_{TN})^2 = k_N[2(V_M - V_{TN})V_{DS1} - V_{DS1}^2] \qquad \textbf{(7.75)}$$

For both the PMOSFETs

$$V_{SG} = V_{DD} - V_M = V_{SD}$$

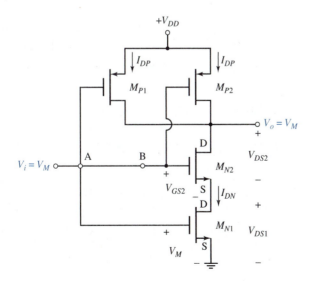

Figure 7.26   Two-input NAND gate at transition with $V_A = V_B = V_M$

Thus, the PMOSFETs are in saturation. Again, if the devices have the same conduction parameter k_P and threshold voltage V_{TP}, they carry the same drain current I_{DP}, which is given by

$$I_{DP} = k_P(V_{DD} - V_M - |V_{TP}|)^2 \tag{7.76}$$

Solving for V_{DS1} in the second half of the equation in Equation (7.75) as

$$V_{DS1} = (V_M - V_{TN})\left(1 \pm \frac{1}{\sqrt{2}}\right)$$

and using the first half of the equation, we get

$$V_M = V_{TN} + \sqrt{\frac{2I_{DN}}{k_N}}$$

Noting that $I_{DN} = 2I_{DP}$ and using Equation (7.76), we have

$$V_M = \frac{V_{TN} + 2\sqrt{\dfrac{k_P}{k_N}}\,(V_{DD} - |V_{TP}|)}{1 + 2\sqrt{\dfrac{k_P}{k_N}}} \tag{7.77a}$$

or, using $K_R = k_P/k_N$ as with the CMOS inverter,

$$V_M = \frac{V_{TN} + 2\sqrt{K_R}\,(V_{DD} - |V_{TP}|)}{(1 + 2\sqrt{K_R})} \tag{7.77b}$$

Equation (7.77b) shows the difference in the gate threshold voltage between the inverter and the two-input NAND gate operating as inverter. The inverter threshold voltage given by Equation (7.62) is identical to Equation (7.77b) except for the factor 2 occurring in the denominator and the numerator. If the devices are symmetrical, that is, $k_P = k_N$ and $V_{TN} = |V_{TP}|$, then

$$V_M = \frac{1}{3}(2V_{DD} - V_{TN}) \tag{7.78}$$

which shows an unsymmetrical VTC, unlike for the inverter. Therefore, the static operation of the NAND gate as an inverter with both inputs tied together is different from that of an inverter even if the MOSFETs have the same characteristics in both cases.

Design   Equation (7.77) provides a basis for the design of CMOS NAND gates. Rewriting Equation (7.77) for $K_R = k_P/k_N$, we get

$$k_P/k_N = \frac{\mu_P(W/L)_P}{\mu_N(W/L)_N} = \left[\frac{V_M - V_{TN}}{2(V_{DD} - V_M - |V_{TP}|)}\right]^2$$

or

$$\frac{(W/L)_P}{(W/L)_N} = 2.5\left[\frac{V_M - V_{TN}}{2(V_{DD} - V_M - |V_{TP}|)}\right]^2 \tag{7.79}$$

which is similar to the CMOS inverter static design equation given in Equation (7.66). The following example illustrates the worst-case dc design.

Design a two-input CMOS NAND gate to meet the worst case (maximum) gate | **Example 7.11**
threshold voltage of 2 V operating at a supply of 5 V. The PMOS and NMOS devices have identical conduction parameters, and $V_{TN} = 1$ V and $V_{TP} = -1.5$ V.

What is the gate threshold voltage if the input A (the lower NMOSFET) is at 5 V? Use process transconductance parameters of 20 $\mu A/V^2$ for the PMOS-FET and 50 $\mu A/V^2$ for the NMOSFET. Neglect body effect.

Verify the values of V_M using MicroSim PSpice simulation.

Solution

From Equation (7.79), we have

$$\frac{(W/L)_P}{(W/L)_N} = 2.5 \left[\frac{(2 - 1)}{2(5 - 2 - 1.5)} \right]^2 = \frac{2.5}{9}$$

Hence,

$$(W/L)_P = 2.5, \quad \text{or} \quad W = 25 \ \mu m \quad \text{and} \quad L = 10 \ \mu m$$

and

$$(W/L)_N = 9, \quad \text{or} \quad W = 45 \ \mu m \quad \text{and} \quad L = 5 \ \mu m$$

may be used. (Usually, channel length of all devices is kept the same. Thus, we may use $W = 90 \ \mu m$ and $L = 10 \ \mu m$ for the NMOS devices.)

For $V_A = 5$ V, let $V_B = V_o = V_{MB}$ be the threshold voltage in Figure 7.25a. At V_{MB}, M_{N1} is in the ohmic mode and M_{N2} is in saturation. Equating the drain current in the NMOSFETs, then, we have

$$k_N[2(V_A - V_{TN})V_{DS1} - V_{DS1}^2] = k_N(V_{MB} - V_{DS1} - V_{TN})^2$$

or

$$8V_{DS1} - V_{DS1}^2 = (V_{MB} - V_{DS1} - 1)^2 \tag{7.80}$$

With the gate at 5 V, M_{P1} is off and there is no current contribution to the NMOSFETs; M_{P2} is in saturation with the drain and gate at V_{MB}. Thus, the drain current in M_{P2} is

$$I_{DP} = (20)(2.5)(3.5 - V_M)^2$$

Equating I_{DP} to the drain current in M_{N2}

$$(20)(2.5)(3.5 - V_{MB})^2 = (50)(9)(V_{MB} - V_{DS1} - 1)^2 \tag{7.81}$$

Eliminating V_{DS1} using Equations (7.80) and (7.81) and solving,

$$V_{MB} = 1.66 \text{ V} \quad \text{and} \quad V_{DS1} \approx 0.045 \text{ V}$$

Clearly, at this low voltage of $V_{SB2} = V_{DS1} \approx 0.05$ V, body effect in M_{N2} is negligible. (With iteration it can be shown that V_{SB2} is approximately 0.05 V, which is still negligible.)

```
* Example 7.11 CMOS NAND VTC
MP1       2   3   1   1   PCH L=10U W=25U
MP2       2   4   1   1   PCH L=10U W=25U
MN1       5   3   0   0   NCH L=5U  W=45U
MN2       2   4   5   0   NCH L=5U  W=45U
RSHORT    3   4   0.0001;  Both inputs shorted
VI        3   0
VDD       1   0   DC 5
.PROBE
.MODEL PCH   PMOS (VTO =
- 1.5 KP = 40E-6 GAMMA = 0.4)
.MODEL NCH   NMOS (VTO = 1 KP = 100E-6 GAMMA = 0.4)

.DC    VI   0 5V 0.1V
.PRINT DC  V(2),ID(MN1)
.OPTIONS NOPAGE
.END
```

(a) Simulation File

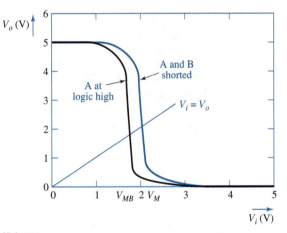

(b) VTC

Figure 7.27　　Simulation of NAND VTC

Figure 7.27 shows the MicroSim PSpice simulation of the NAND VTC with both inputs connected together. With body effect of M_{N2} (at $V_{DS1} \approx 0.26$ V) the worst-case (maximum) gate threshold voltage V_M is 2.06 V. The VTC with input A at logic high (M_{N1} conducting) shows approximately the same threshold voltage as that which we calculated.

Although the above design satisfies the worst-case gate threshold voltage, the circuit has longer switching delays than an inverter. This is because of increased capacitance at the output node. As with NMOSFET gates, higher output capacitance arises primarily from the drain-body capacitance of the increased number of PMOS and NMOS devices. In addition, the low-to-high delay time at the output is dependent on the number of inputs that change from high to low. Clearly, if all the inputs in an N-input NAND gate switch simultaneously from high to low, all N PMOSFETs are turned on, thereby providing high charging current for the lumped output capacitance and reducing the charging delay. If only one of the inputs goes low, current from a single PMOSFET cannot charge as fast, and thus, the delay time increases. On the other hand, for the high-to-low switching at the output, all inputs must be at logic high. The discharge path for the capacitance at the output node is provided by the series combination of all N NMOSFETs. Because of the increased effective resistance to ground (N times the resistance of a single NMOSFET) and the capacitance, the high-to-low delay is more than that for an inverter. However, this delay is constant for a given fan-in. The increase in delays must be considered in the design of CMOS NAND gates in addition to the worst-case gate threshold voltage.

Although we considered only the worst case of both inputs simultaneously switching, other cases of one of the inputs at logic high while the other changes state can be similarly analyzed [see Reference 1].

In general circuit designers adjust the dimensions of the NMOSFETs in the design phase so that, for simultaneous switching of inputs, the charging and the discharging currents are approximately the same. Although this does not ensure equal delays, the drive current for the load will be the same at logic low and high levels. For a two-input NAND gate, the output high (charging or sourcing) current is the sum of the two PMOSFET drain currents, while the discharge (sinking) current is that of one of the two series-connected NMOSFETs. Hence, for identical NMOSFETs and identical PMOSFETs, the conduction parameter k_N must be twice k_P. Or,

$$\left(\frac{W}{L}\right)_N \frac{\mu_N C_{\mathrm{ox}}}{2} = 2\left(\frac{W}{L}\right)_P \frac{\mu_P C_{\mathrm{ox}}}{2}$$

Using $\mu_N/\mu_P \approx 2.5$, the above equation gives

$$\left(\frac{W}{L}\right)_N \approx \frac{\left(\dfrac{W}{L}\right)_P}{1.25}$$

For a fan-in of N, we have

$$\left(\frac{W}{L}\right)_N \approx \frac{N}{2.5}\left(\frac{W}{L}\right)_P \qquad\qquad \textbf{(7.82)}$$

NOR Gate Characteristics   As with the NAND gate, the characteristics of a CMOS NOR gate vary in both static and switching behavior, depending on the number of inputs that change from one logic level to the other. Since the

PMOSFETs are connected in series and the NMOSFETs in parallel, the behavior is exactly opposite that of the NAND gate. The following example illustrates the analysis of a two-input NOR gate, which parallels the NAND gate analysis.

Example 7.12

Determine the logic threshold voltage for the two-input NOR gate of Figure 7.25b if (a) both inputs are tied together, and (b) input A is at ground potential. $V_{DD} = 5$ V, and the devices have the following parameters: $V_{TN} = 1$ V $= -V_{TP}$, $k_N = 100$ μA/V^2, and $k_P = 50$ μA/V^2. Neglect body effect.

Solution

(a) With both inputs tied together, at the threshold voltage $V_A = V_B = V_Y = V_M$, the NMOSFETs are in saturation. With the drain-source junction of the PMOS-FETs at $V_X > V_M$, M_{P1} is in the ohmic mode while M_{P2} is in saturation.

Equating the drain currents of the PMOSFETs, we have

$$k_P[2(V_{SG1} - |V_{TP}|)V_{SD1} - V_{SD1}^2] = k_P(V_{SG2} - |V_{TP}|)^2$$

or

$$50[2(4 - V_M)(5 - V_X) - (5 - V_X)] = 50\,(V_X - V_M - 1)^2 \quad \textbf{(7.83)}$$

Since this current is equal to the sum of the saturation drain currents of M_{N1} and M_{N2}, the second equation is given by

$$2k_N(V_M - V_{TN})^2 = k_P(V_{SG2} - |V_{TP}|)^2$$

or

$$200(V_M - 1)^2 = 50(V_X - V_M - 1)^2 \quad \textbf{(7.84)}$$

Solving Equations (7.83) and (7.84), we get

$$V_M = 1.78 \text{ V} \quad \text{and} \quad V_X = 4.35 \text{ V}$$

Note that with $V_{SB2} = 4.35 - 5 = -0.65$ V, the body effect for M_{P2} is negligible.

(b) With V_A at ground potential and $V_B = V_Y = V_M$, M_{N1} is off and M_{P1} is in the ohmic mode while both M_{N2} and M_{P2} are in saturation. Since the drain current in all three conducting devices is the same, it is given by

$$k_P[2(5 - 1)(5 - V_X) - (5 - V_X)^2]$$
$$= k_P(V_X - V_M - 1)^2 = k_N(V_M - 1)^2 \quad \textbf{(7.85)}$$

where V_X is the new voltage at the junction of the two PMOS devices. Solving the pair of equations resulting from Equation (7.85), we get

$$V_M = 2.11 \text{ V} \quad \text{and} \quad V_X = 4.68 \text{ V}$$

It can be shown (see Problem 7.15) that the worst-case (largest) gate threshold voltage occurs when the input is applied to A (upper PMOS device M_{P1}) while the lower device is conducting with constant $V_{SG} = V_{DD}$, that is, input

B at logic low. As with the NAND gate, this is due to the increased drop from V_{DD} at the source of M_{P2} and body effect. Note that the body effect may become significant in the lower device with more PMOSFETs in series. Figure 7.28 shows the VTC for a two-input NOR gate using MicroSim PSpice simulation for all three cases.

As with a NAND gate, a two-input CMOS NOR gate may be designed for a specified gate threshold voltage for both inputs switching. However, with increased capacitance at the output node due to the drain-source capacitances of the two paralleled NMOS devices, propagation delays are dependent on which

```
* Fig. 7.28 CMOS NOR VTC - Input A at low
MP1      3   2   1   1 PCH
MP2      5   4   3   1 PCH
MN1      5   2   0   0 NCH
MN2      5   4   0   0 NCH
VL       2   0   0
VI       4   0
VDD      1   0   DC 5
.PROBE
.MODEL PCH  PMOS (VTO = -1 KP = 25E-6 GAMMA = 0.4)
.MODEL NCH  NMOS (VTO = 1 KP = 50E-6 GAMMA = 0.4)
.DC     VI  0 5V 0.1V
.PRINT DC V(5),ID(MP1)
.OPTIONS NOPAGE
.END
```

(a) Simulation File

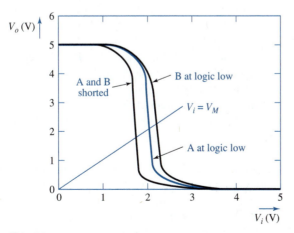

(b) VTC

Figure 7.28   VTC for a two-input NOR gate

input is switching. Because of the increased delays due to higher capacitance, CMOS NOR gates are slower than CMOS NAND gates with the same fan-in. Note that the PMOSFETs connected in series in a NOR gate have higher resistance (lower conduction parameter) than the series-connected NMOSFETs of the same size in a NAND gate. Thus, the charging time for the output capacitance is more in a NOR gate than the discharging time in a NAND gate. If, on the other hand, the sizes of the p-channel and n-channel devices are adjusted for the same charge and discharge current when all inputs are switching simultaneously, the NOR gate occupies a larger silicon area than a NAND gate. For example, for a two-input NAND gate, the conduction parameters must be such that $k_P = 2k_N$, or

$$\left(\frac{W}{L}\right)_P \frac{\mu_P C_{ox}}{2} = 2\left(\frac{W}{L}\right)_N \frac{\mu_N C_{ox}}{2}$$

which gives

$$\left(\frac{W}{L}\right)_P \approx 5\left(\frac{W}{L}\right)_P$$

For a fan-in of N, the aspect ratio of the PMOSFETs is

$$\left(\frac{W}{L}\right)_P \approx 2.5N\left(\frac{W}{L}\right)_P$$

Because of the smaller size and delay times, CMOS NAND gates are preferred over NOR gates for high-speed and high-density applications.

Complex logic functions can be implemented by modifying basic CMOS gates. As we have seen, for each additional fan-in above two, a complementary pair is added with PMOSFET in parallel (series) and NMOSFET in series (parallel) for NAND (NOR) function. AND-OR-inverter and OR-AND-inverter functions are readily realized as shown in Figure 7.29. Observe that for each variable or combination of variables in the AND function, a complementary pair is used, with NMOSFET in series and PMOSFET in parallel. The equivalent switch configurations in Figures 7.29b and 7.29d (the switch representing PMOSFET closing for logic low while that for NMOSFET closing for logic high) show clearly the complementary implementations of the two functions. With NAND, NOR, and inverter realizations available, all complex combinational functions can be directly implemented. However, certain logic functions, such as exclusive OR and equivalence, are realized more efficiently using CMOS transmission gates.

CMOS Transmission Gate   The CMOS transmission gate (TG) (Figure 7.30) is a bidirectional pass gate built using a complementary pair of MOSFETs. As Figure 7.30 shows, the MOSFETs connected in parallel are controlled by the logic signals C and $\overline{C}$. With drain and source interchangeable, the signal flow is bidirectional. Figure 7.30b shows a commonly used symbol for the transmission gate. If we arbitrarily designate input and output nodes as shown

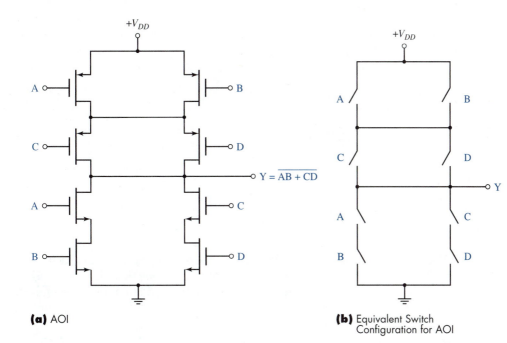

(a) AOI

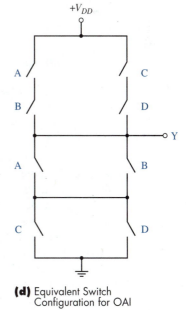

(b) Equivalent Switch Configuration for AOI

(c) OAI

(d) Equivalent Switch Configuration for OAI

Figure 7.29   CMOS AOI and OAI Realizations

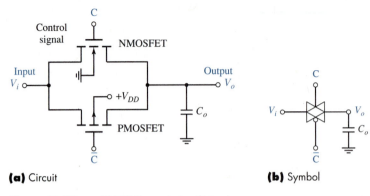

(a) Circuit

(b) Symbol

Figure 7.30   CMOS Transmission Gate

in Figure 7.30, we can obtain the voltage V_o at the output, which has a capacitive load C_o, for different combinations of V_i and V_o. We assume that the logic levels at V_i and at the gates are 0 for low and V_{DD} for high. In addition, for symmetry, we assume $V_{TN} = -V_{TP} = V_T$.

Case I: C = low and V_i = 0   Since the gate voltage is zero the NMOS device is off. For the PMOS, the gate is at V_{DD} ($\overline{C} = 1$) while the drain and the source are at zero voltage if $V_o = 0$; if, on the other hand, $V_o = V_{DD}$, neither the drain nor the source is above V_{DD}. Hence, no channel is formed and the PMOS device is also off. V_o, therefore, stays at its previous value of zero or V_{DD}.

Case II: C = low and V_i = V_{DD}   Again both devices are off and V_o is held at its previous value. Therefore, for C at logic low, input and output are isolated from each other and the output capacitor voltage does not change.

Case III: C = high and V_i = 0   If $V_o = 0$ initially, the NMOS device is turned on. However, no current exists since the drain and the source are at zero voltage. For the PMOS device, all three terminals are at zero, and thus it is off. Therefore, V_o stays at zero.

If V_o was initially at some positive voltage (V_{DD}, as shown in the next case we consider), the input node of the NMOSFET acts as source and there is current flow from output to input via the saturated NMOSFET device. Simultaneously, the output node of the PMOSFET acts as source, providing a second path for current from output to input via the saturated PMOSFET device. Hence, the output capacitor discharges to $V_i = 0$ through the small parallel resistance of the two conducting MOSFETs. As V_o drops to $|V_{TP}| = V_T$, the PMOSFET no longer conducts. The NMOSFET, however, is still on, with its gate at constant voltage V_{DD}. Therefore, C_o discharges completely to zero.

Case IV: C = high and V_i = V_{DD}   If V_o was initially zero, the roles of the drain and the source are reversed in both the devices from Case III; that is, the output node of the NMOSFET acts as source whereas for the PMOSFET, with $V_i - V_G = V_{DD}$, the input node acts as source. Both devices initially operate in saturation and the output capacitor is charged via the small parallel resistance of the two conducting MOSFETs. As V_o rises to $V_{DD} - V_{TN}$, the NMOSFET is turned off while the PMOSFET is still on in ohmic mode. Thus, C_o charges until V_o equals V_{DD}, after which no current exists in the PMOSFET although it is still on. If V_o was initially at V_{DD}, it stays at V_{DD}.

It is clear from the above dc analysis that the logic function[3] of a transmission gate is described by

$$\text{Output } Y = \begin{cases} \text{Input X, if C} = 1 \\ \text{Open circuit, if C} = 0 \end{cases}$$

The propagation delay of a TG may be estimated using the channel resistances of the complementary pair. As with the inverters, output low-to-high, or charging, delay is higher than the discharging delay. For $V_i = V_{DD}$ and $V_o = 0$ (initially), both MOSFETs are in saturation. The NMOSFET offers static-on resistance R_{nd} from input to output as

$$R_{nd} = \frac{V_{DS}}{I_D} = \frac{(V_{DD} - V_o)}{k_n(V_{DD} - V_o - V_{TN})^2} \qquad \textbf{(7.86a)}$$

The PMOS resistance R_{pd} is given by

$$R_{pd} = \frac{V_{SD}}{I_D} = \frac{(V_{DD} - V_o)}{k_p(V_{DD} - |V_{TP}|)^2} \qquad \textbf{(7.86b)}$$

Hence the TG may be modeled by a single time-constant $R_P C_o$ circuit with a resistance R_P given by

$$R_p = \frac{1}{\dfrac{1}{R_{nd}} + \dfrac{1}{R_{pd}}} \qquad \textbf{(7.87)}$$

where R_{nd} and R_{pd} are the saturation resistances of the NMOS and the PMOS devices at $|V_{GS}| = V_{DD}$. The capacitance C_o arises from the two MOSFETs' interconnecting area and the load. R_P in the above equation represents the resistance of the TG only approximately. As V_o rises by the charging of C_o, the NMOSFET goes from saturation to off, and R_{nd} becomes very large for $V_o > V_{DD} - V_{TN}$. Simultaneously, the PMOSFET changes mode from saturation to ohmic, and R_{pd} decreases due to decreasing V_{SD}. In addition, the body effect in the NMOSFET complicates R_{nd}, and hence, R_p. The overall result is a nonlinear variation of R_p with increasing V_o. As a first-order approximation, we may consider R_p in Equation (7.87) constant with R_{nd} and R_{pd} evaluated for $V_o = 0$. A simpler approximation is to use only R_{pd} (at $V_o = 0$), since the PMOSFET stays on during the entire charging interval; alternatively, R_{nd} can be used in place of R_P.

| [3] Transmission gates can also be used for transmitting analog signals using appropriate gate voltages.

The above analysis may be extended for the high-to-low, or discharging, delay. Clearly, the charging and discharging delays are determined by the aspect ratios of the MOSFETs, which control the conduction parameters (and hence R_p), and the layout and interconnection areas, which contribute to capacitance. In practice, NMOSFET and PMOSFET devices of the same size are used to simplify design. Although this causes unequal conduction parameters, the resulting parallel resistance is still small.

A notable feature of the transmission gate is that when the control signal is at logic low (i.e., $C = 0$ and $\bar{C} = 1$), the gate is open-circuited from input to output. Thus, V_o is determined by other signals which are connected at the output node. This feature provides a type of wire-ANDing to realize complex switching and logic functions.

Figure 7.31 shows the use of TGs to implement a single-pole, double-throw (SPDT) switch. As shown in Figure 7.31a, the output of the switch is logic signal A if the control signal C is logic high, and it is B if C is low. This switch is implemented by connecting the outputs of two TGs together as shown in Figure 7.31b. Analog-to-digital and digital-to-analog data converter circuits (Chapter 9) use these SPDT switches widely. Observe that the same circuit in Figure 7.31b realizes a two-to-one multiplexer, $Y = A \cdot C + B \cdot \bar{C}$. Figure 7.32 shows extension of the latter for a four-to-one multiplexer. With the addition of four TGs, the configuration in Figure 7.32b eliminates the need for the four AND gates required in Figure 7.32a.

Figure 7.33 shows exclusive OR (XOR) and equivalence (XNOR) logic realizations using TGs. If true and complemented variables are available, these circuits require only four MOSFETs each.

High Current Buffers   A CMOS gate is normally designed to drive capacitive loads of 10 to 50 pF. Higher capacitive loads cannot be driven without substantially increasing the propagation delays, since the charging and discharging currents are limited by process transconductance parameters and device aspect ratios. CMOS circuits required to drive heavily capacitive loads, such as

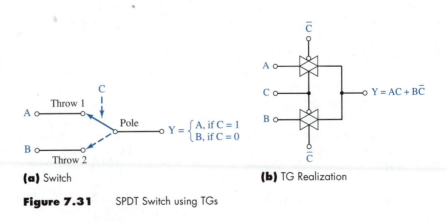

(a) Switch

(b) TG Realization

Figure 7.31   SPDT Switch using TGs

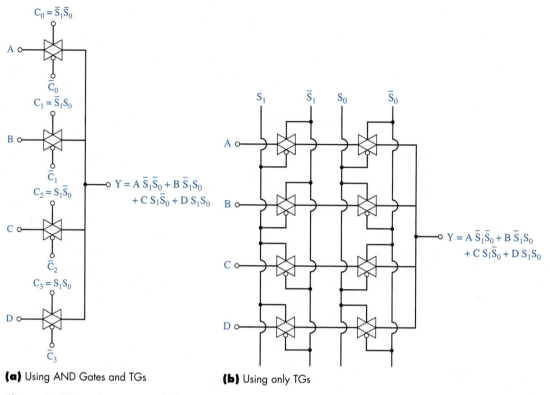

(a) Using AND Gates and TGs **(b)** Using only TGs

Figure 7.32 Four-to-one multiplexer using TGs

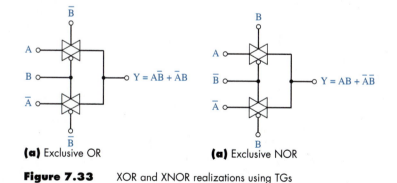

(a) Exclusive OR **(a)** Exclusive NOR

Figure 7.33 XOR and XNOR realizations using TGs

buses, or logic circuits from other families, such as the TTL, are designed to have higher conduction parameters. Such circuits, called *superbuffers,* typically have a PMOS aspect ratio twice that of the NMOS, which, as we saw earlier, enables approximately equal charging and discharging currents. An alternative is to parallel several inverters together in the same package. Within a chip, the output from an unbuffered logic circuit may be passed through a pair of inverters that are driven by the same input while their outputs are connected together. Logic in these buffered gates is formed using minimum-size standard devices while the paralleled inverters (output buffers) use large areas for high current capability. With the sharp transfer characteristics of the driving inverters, the overall gate input-output characteristics are improved; also, the VTC is independent of which of the logic (external) inputs is changing (cf. NAND and NOR gates previously discussed). This results in better noise margins ($\approx V_{DD}/2$ over the entire range of 3 to 15 V supply) than in unbuffered gates.

Tristate Buffers   CMOS technology provides tristate buffers for bus-connection of outputs from different nodes. In the circuit shown in Figure 7.34a, if the tristate control (enable) signal is at logic low (C = 0 and $\overline{C}$ = 1), the complementary pair M_{P2} and M_{N2} are both off, making Y an open circuit. For C = 1, output Y is the complement of input A. When output is at the high-impedance (open-circuited) third state corresponding to C = 0, the logic level at Y is determined by other nodes connected to it. Only one of the other nodes, however, must be active; all others must be in the high-impedance state. Figure 7.34b shows a tristate noninverting buffer where M_P and M_N are off if the control signal is at logic low.

Commercial CMOS Static Circuits   Early CMOS circuits were available in the unbuffered form as the 4000 series for SSI and MSI applications. These metal-gate circuits were designed to drive an equal current for sinking and

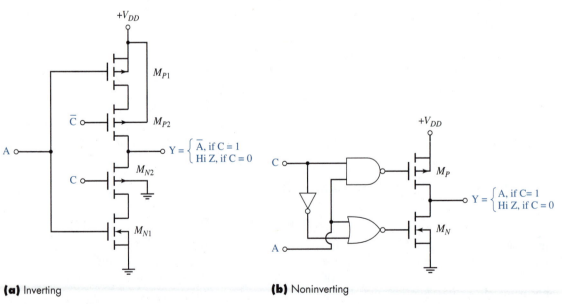

(a) Inverting   **(b)** Noninverting

Figure 7.34   Tristate buffers

sourcing. Although these circuits can be operated over a power supply range of 3 to 15 V, they are not always compatible with the popular TTL family of circuits. Also, with propagation delays in the range of 25 to 100 ns at a 5 V supply, the 4000 series does not offer a significant advantage over TTL and its many subfamilies.

Buffered versions of the 4000 series, called 4000B, are made with polysilicon gates and use a minimum feature size of 5 μm and a gate oxide thickness of 1000 Å. As seen earlier, paralleled pairs of inverters are used in buffered circuits to isolate the logic circuit from the output node. The 4000B series has higher density and noise margins, and lower propagation delay than the unbuffered series. In addition, when operating at 5 V supply, the output of each 4000 series gate can sink or source 1 mA, which is adequate to drive one TTL gate in the 74LS series. 74C/54C is another CMOS series having the same characteristics as the 4000B, but pin- and function-compatible with TTL circuits.

74HC/54HC and 74HCT/54HCT are high-speed versions of the 74C/54C. Gates in these families use a minimum feature size of 3 μm and gate oxide thickness of 600 Å. While the HC series is also pin-compatible with the LS TTL circuits, they are not voltage-compatible. This problem is overcome in the HCT series by reducing the input logic high level to 2 V at 5 V supply. Both series have reduced gate and junction capacitances and higher gains of the MOS devices, which result in the same propagation delays as for the LS TTL family. Power dissipation is 2.5 μW, compared with 2 mW of the LS TTL family. The operating supply voltage range for the 74HC/54HC series is 2 to 6 V, which makes these devices suitable for battery-powered systems. Because of its compatibility with the LS TTL series, the 74HCT/54HCT series has a supply range of 4.5 to 6.5 V.

Other high-speed CMOS technologies have progressively lowered the minimum feature size to less than 1 μm. Logic gates in these technologies achieve propagation delays approaching the ECL delays but with significantly high density and low power.

7.3.3 DYNAMIC AND DOMINO LOGIC GATES

Dynamic Gates Dynamic logic circuits use one or more *clock,* or timing signal, to synchronize inputs and outputs. Only one part of the circuit in a dynamic gate conducts during the active (high) level of clock. This reduces power dissipation and eliminates timing problems.[4] In addition, dynamic circuits use fewer devices and less silicon area to implement large-scale storage devices. Operation of these circuits, however, requires clock signals at a specified minimum frequency.

---

[4] For example, see races and hazards in C. Roth, *Fundamentals of Logic Design,* St. Paul, Minn.: West Publishing Company, 1992.

Figure 7.35 shows basic dynamic inverter implementations. Although the circuits show only E-NMOSFETs, D-NMOSFETs can be used as load devices in both. Φ is the clock in both figures, and C_i and C_o represent parasitic capacitances (on the order of 100 fF) at the indicated nodes. In Figure 7.35a, MOS-FETs M_2 and M_3 are off during the low level (off duration) of Φ. Input V_i may be present in this interval. If V_i is at logic low, M_1 is also off; otherwise, a channel is formed in M_1. In either case, output and input are isolated and voltage V_o is held constant in C_o at its previous value. (There will be a small leakage current to ground if V_o is at logic high, via the reverse-biased pn junction of the substrate of M_3.) No current is drawn from the supply. When the clock goes high, M_2 and M_3 are turned on. If V_i is now at the logic low level, M_1 is off and C_o charges to V_{DD} via M_2 and M_3. For V_i at high ($\approx V_{DD}$), M_1 is also turned on and the capacitor C_o discharges to ground. M_3, therefore, acts as a bidirectional pass gate. To discharge C_o fully at logic low output, the on resistance of M_1 must be small compared to that of M_2. Hence, the size of M_1 must be larger than that of M_2. Since the logic low output voltage depends on the conduction parameter ratio of M_1 and M_2, this circuit (as also the circuit of Figure 7.35b) is a *ratioed-logic* circuit.

In Figure 7.35b, M_1 is the bidirectional pass gate that isolates input and output when the clock is at low. If $V_i = 0$ during the logic high of the clock, the input capacitor C_i discharges to ground via the on gate M_1. M_2 is therefore turned off, and the output capacitor C_o charges to $V_o = V_{DD}$. For $V_i = V_{DD}$ during clock ($\Phi = V_{DD}$), C_i charges to $V_{DD} - V_T$ and turns M_2 on. The two conducting devices M_2 and M_3 form a voltage divider between V_{DD} and ground, and C_o discharges to logic low. When the clock changes to low, C_i holds its voltage with a leakage current through M_1. As with the previous circuit, power is dissipated only during $\Phi = V_{DD}$.

Because of junction leakage, capacitor voltages decrease during the low level of clock. With a leakage current on the order of 0.1 pA, however, the logic

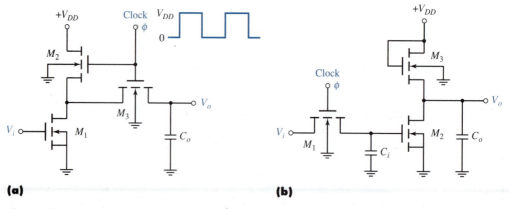

(a) **(b)**

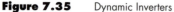

Figure 7.35 Dynamic Inverters

high value can be maintained within an acceptable voltage level for a few milliseconds. For this reason, a minimum clock frequency of a few kHz is specified. The maximum frequency of operation depends on the charge transfer rate.

Two-Phase Dynamic Circuits   To prevent unstable data being transferred from one stage to another, two or more nonoverlapping clocks are more commonly used than a single clock in dynamic circuits. In cascaded circuits such as a shift register, the use of two-phase clocks, for instance, ensures that data transfer occurs only during specified intervals. This eliminates the timing problems arising from unequal propagation delays in complex logic systems. Figure 7.36 shows cascaded two-phase ratioed inverters used as a shift register. Here again, D-NMOS devices can be used as load devices with $V_{GS} = 0$. E-NMOS loads, however, occupy less area and the circuit can be made *ratioless* using minimum-size devices for all MOSFETs. (Recall that the driver-to-load ratio K_R must be set to achieve a given V_{OL}.) Additionally, unlike a D-NMOS load, which is always on with $V_{GS} = 0$, the E-NMOS device can be switched off with a clock. Thus, power dissipation is reduced. The two clocks, Φ_1 and Φ_2, are nonoverlapping; that is, only one of them is high in each half-cycle, as shown in Figure 7.36. Although they may have different on and off times, logical AND of Φ_1 and Φ_2 at any time produces zero.

As in the dynamic circuit using a single-phase clock, all capacitances shown in Figure 7.36 are parasitic, and data transfer from input to output occurs by charge transfer between these capacitors. MOSFETs M_{11}, M_{12}, M_{13}, M_{21}, M_{22},

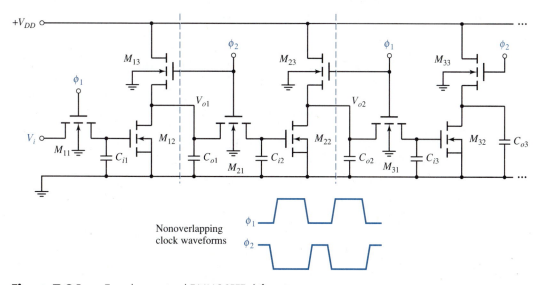

Figure 7.36   Two-phase ratioed E-NMOSFET shift register

and M_{23} constitute one stage of the shift register. That is, a logical (bit) value given at the input V_i appears after one complete period of the clock at V_{o2}. Devices $M_{11}, M_{21}, M_{31} \ldots$ are pass transistors which, when turned on, transfer charge between capacitors at the drain and the source. Each pair of devices shown vertically operates as an inverter whenever the load device is on.

We begin the analysis of the circuit with $V_i =$ high $= V_{DD}$ and $\Phi_1 =$ high $= V_{DD}$. M_{11} is turned on and the capacitor C_{i1} charges to logic high voltage V_H, which is less than V_{DD} by the threshold voltage (including body effect). If V_H is above the threshold voltage of M_{12} (as it should be in a functional circuit), M_{12} is turned on and the capacitor C_{o1} discharges to ground. With Φ_2 low, M_{21} is off; hence C_{i2} holds its charge (within leakage). The stable voltage during Φ_1 is $V_{o1} \approx$ 0 and the gate of M_{12} is at logic high (V_H).

When Φ_2 goes high (and Φ_1 low) in the next interval, the transmission device M_{11} is turned off, while M_{13} and M_{21} are turned on. With logic high voltage $(< V_H$ due to leakage, but still above threshold voltage) on C_{i1}, M_{12} is still on. Thus, capacitor C_{i2} discharges to ground via M_{21} and M_{12}. This causes M_{22} to turn off, and the output voltage V_{o2} is the same as the previous value (less leakage). Note that V_{o1} and the gate voltage of M_{22} are the complement of V_i. The value of $V_{o1} = V_{OL}$ is set by the ratio of M_{13} to M_{12}.

At the next interval of $\Phi_1 =$ high, M_{22} is still off while M_{23} and M_{31} are turned on. Capacitor C_{o2} is charged to V_H via M_{23}, and this voltage is transferred to C_{i3} via M_{31}. Hence, the stable logic value of the voltage V_{o2} during the second active interval of Φ_1 is that of V_i during the first Φ_1. That is, the data at input V_i appears at output V_{o2} after one complete period of Φ_1, which is called a 1-bit delay. Observe that while the V_{o2} output is the previous input logic value, V_{o1} holds the complement of the current input. Thus the combination of M_{11}, M_{12}, and M_{13} is acting as a master inverter, while the M_{21}, M_{22} and M_{23} combination acts as slave inverter. Input data present during Φ_1 is inverted by the master inverter and transferred to the slave inverter. During Φ_2, the slave inverter complements its input and regenerates the original data at its output. If N such master-slave combinations are cascaded, input data will appear at the output of the last slave after N clock (Φ_1) periods.

Power dissipation in each inverter stage occurs only during the logic high state of the appropriate clock. As with the single-phase clock circuits, the clock period must be above a minimum value so that the capacitor voltages after leakage can still be considered logic high. Another consideration is the loss of threshold voltage in V_H due to the saturation mode of operation. Although this loss does not propagate, V_H can be made close to V_{DD}. If the high-level voltage of the clock is raised above V_{DD}, the load devices will operate in the ohmic mode, with reduced drain-source drop.

A two-phase ratioed shift register is used as a serial access memory or recirculating register as shown in Figure 7.37. The new data bit from the input is written into the register by disabling the Recirculate signal. If Recirculate $= 1$, existing data in the register recirculates. A register for recirculating words of K-bits in parallel (K-bit word recirculating register) is implemented by duplicating the circuit of Figure 7.37 for each of the K bits. All the clocks are

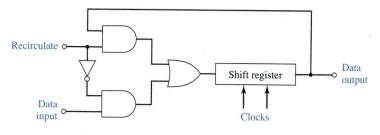

Figure 7.37   Recirculating register

connected together and also all the Recirculate controls are connected together. Such a combination of registers is used for refreshing data in display units.

A two-phase *ratioless* dynamic shift register can be designed using the same circuit configuration as in Figure 7.36 but with minimum-size devices and different clock phases. Two-phase ratioless circuits powered by clocks are also available in different configurations.

CMOS Dynamic Shift Register   CMOS transmission gates and inverters can be used in place of the E-NMOS stages in Figure 7.36. Such a CMOS dynamic shift register is shown in Figure 7.38. The transmission gates are

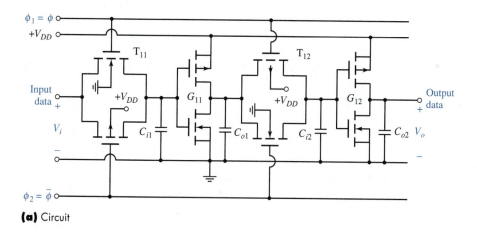

(a) Circuit

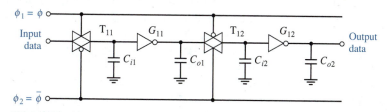

(b) Simplified Diagram

Figure 7.38   CMOS dynamic shift register

alternately controlled by $\Phi_1 = \Phi$ and $\Phi_2 = \overline{\Phi}$. Because Φ_1 and Φ_2 are the complement of each other, only alternate TGs are on during either clock (high) interval. As a result, master and slave inverter stages are isolated at any time.

The circuit operation is similar to that of the E-NMOS shift register of Figure 7.36. If V_i is logic high ($= V_{DD}$) during $\Phi_1 = \Phi$, T_{11} is closed and C_{i1} is charged to V_{DD}. This turns the NMOS device in G_{11} on and C_{o1} discharges to 0. During $\Phi_2 = \overline{\Phi}$, T_{12} closes. With G_{11} at low output (the gate of G_{11} is still high due to C_{i1}), C_{i2} discharges. The input to G_{12} is therefore low. This turns the PMOS device in G_{12} on and charges C_{o2}. The output voltage V_o is the same as the input after one clock (bit) delay. Care must be taken to ensure that Φ_1 and Φ_2 do not overlap at the logic high level. The clock frequency range is determined by the charging and discharging rates of the parasitic capacitances, which, in turn, depend on the size and layout of the devices.

Each bit of a CMOS dynamic shift register uses only four MOS devices (as opposed to six for the ratioed E-NMOS register of Figure 7.36). Although the PMOS devices take up slightly more area to have the same current capability as the NMOS devices, the circuit is ratioless. Power is required only to charge the parasitic capacitances; hence, static power dissipation is low.

Dynamic CMOS logic circuits are also available using one or more clocks. These circuits have the advantages of low device count and high clock speed. An N-input static CMOS NAND or NOR gate, for example, requires $2N$ MOS devices, with attendant high capacitance. A single-phase CMOS dynamic AND gate of the domino logic series (see below), on the other hand, uses only $N + 4$ devices. An additional disadvantage of the static CMOS circuit is that if a device in a complementary pair is open-circuited, a specific *sequence* of input combination is needed to locate the fault rather than an input combination. This may require a complex test pattern generation for easy testability. Problem 7.22 considers an example of this difficulty. Domino logic, which we consider next, uses a single-phase clock and has low device count compared to static CMOS logic. Logic faults caused by device opening or shorting are detected using a combination of inputs.

Domino Logic Gates   Domino logic utilizes a single PMOS device to charge a capacitor during the logic low of clock Φ. The charge on the capacitor is either removed to ground or retained, based on the logic formed by NMOS devices during the logic high of Φ. A static CMOS inverter provides the output drive. Figure 7.39a shows the structure of a domino gate. The charging or *precharge* interval is defined by $\Phi = 0$, and the discharge or *evaluate* phase is defined by $\Phi = 1$. Note the discharge of C_o is conditional on the logic formed by the static logic block during the evaluate phase; hence it is called *conditional discharge*.

Figure 7.39b shows a three-input NAND gate using domino logic. During $\Phi = 0$, the precharge PMOS device M_P is turned on and the capacitor C_o is charged to V_{DD}. The output of the inverter in this interval is logic low. Because M_E is off, change in inputs has no effect. Also, other domino circuits driven by Y are turned off during $\Phi = 0$. When Φ goes high, that is, during the evaluate

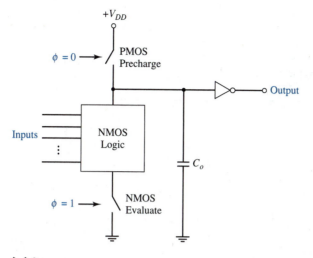

(a) Structure

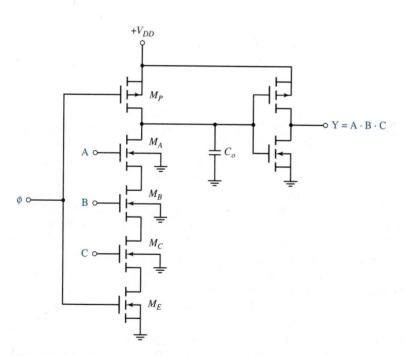

(b) NAND Gate

Figure 7.39 Domino logic

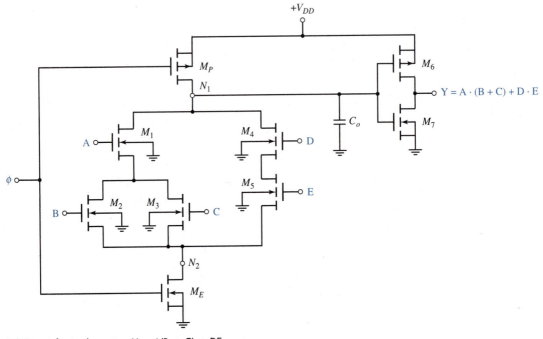

(c) Circuit for Implementing Y = A(B + C) + DE

Figure 7.39   (continued)

phase, M_P is turned off by $\Phi = 1$. C_o discharges to ground via M_A, M_B, M_C, and the evaluate NMOS device M_E if A = B = C = high. Hence the inverter output at Y goes high. If any of the inputs is low, C_o retains charge and Y becomes low. Note that M_P and M_E turn on in complementary fashion. Valid logic output is available only during $\Phi = 1$.

Figure 7.39c shows an example of realizing complex functions using domino logic. During the precharge interval ($\Phi = 0$), C_o is charged to V_{DD} and Y is low. Hence, other domino circuits driven by Y are turned off. The NMOS logic block consisting of M_1 to M_5 provides a path from node N_1 to node N_2 if X = A(B + C) + DE is high. During $\Phi = 1$, therefore, C_o discharges conditionally to ground via the evaluate device M_E. As a result, the voltage at N_1 becomes low if X = 1. Thus, output of the static CMOS inverter is Y = X = A(B + C) + DE. Note that the inputs must not vary during the evaluate phase; this prevents an indeterminate voltage at node N_1.

An advantage of the domino logic circuit is that the output of the inverter makes only one transition, if any, in either phase of clock. Just prior to the evaluate phase, Y is always low; if conditional discharge occurs, Y changes to high. Because of this single transition regardless of input, the circuit operation is free of undesirable transients and race conditions. Power dissipation from the supply occurs only during the precharge interval. Note further that there is no steady current flow from clock or supply regardless of inputs. Also, the number of devices required for an *N* variable combinational function is $N + 2$, excluding the output inverter.

When multiple circuits using domino logic are cascaded, the transition at the output of the first gate during the evaluate phase propagates through the entire circuit. If the logic block in each stage evaluates to high, the corresponding capacitance discharges to zero as the logic high output from the first gate propagates through the entire circuit. This domino effect gives rise to the name of the circuit.

As with other dynamic circuits, the frequency of the clock is limited by the maximum time required for charging and discharging of the logic node capacitor. While the precharge time is constant, the evaluate time depends on the number of variables ANDed together. Thus, ANDing of a large number of variables in the same gate slows down operation. Another limitation of the structure is that AOI output requires an additional inverter. (Note the lack of complementary outputs.) Multiple outputs, on the other hand, can be obtained from a single domino circuit. Other modifications, such as alternating logic blocks using NMOS and PMOS devices, and including latches for realizing complementary outputs [Reference 4], eliminate some of the limitations.

A low device count of domino logic results from the NMOS logic block; unlike the static CMOS, which requires a pair of NMOS and PMOS devices for each additional variable, logic is formed using only NMOS devices. With fewer PMOS devices—one for precharge and another for the buffer inverter—less silicon area is used. In addition, all devices except the pair for the buffer may be of minumum size. This results in higher chip density. Since the clock signal is used for combinational as well as sequential functions, domino logic systems require high-power clock drivers. Domino logic is used in subsystems in arithmetic and logic units, address decoders, and programmable logic arrays.

7.4 BiCMOS Logic Circuits

As we have seen in the previous sections, the dynamic response of MOSFET logic circuits deteriorates at large capacitive loads. In particular, the increase in delay time precludes the use of low-power, high-density MOSFET circuits for driving high current loads. Consequently, large-sized, on- or off-chip drivers are needed to interface MOS logic outputs to buses and commonly used bipolar logic families. *BiCMOS* digital technology is a recent development in the silicon fabrication process that combines the speed and driving capability of bipolar junction transistors with the density and low-power dissipation of CMOS devices. While it has its origin in the low-voltage analog integrated circuits developed in the late 1960s and early 1970s, high-speed, low-power digital VLSI circuits began to appear in the mid 1980s. Since then, high-performance custom and semicustom VLSI systems and *application-specific integrated circuits* (ASICs) have been developed in the BiCMOS technology. In this technology, high-gain (β) vertical npn transistors are merged with conventional CMOS processing to achieve dynamic performance close to that of the bipolar LSI systems, with little degradation in density. In addition, with low output impedance of the bipolar devices, gate delay does not increase as much with load capacitance as it does in the CMOS gates. Another advantage is the inherent compatibility of the technology with ECL and TTL levels, which enables ready interfacing with little loss of switching speed. A further advantage is that, with the development of CMOS analog circuits, it is possible to implement a complete system of all-CMOS analog and ditital circuits on the same chip with BiCMOS interface circuits. In terms of cost, power, and density, BiCMOS technology at present compares favorably with ECL. With all its advantages, BiCMOS is likely to dominate over the ECL and GaAs MESFET in ASICs and high-density (>100,000) gate arrays.

7.4.1   BASIC BiCMOS GATES

Basic BiCMOS logic circuits use CMOS logic formation and incorporate bipolar device techniques for the charging and discharging of junction capacitances of the output devices. Figure 7.40 shows the circuit of a basic BiCMOS inverter. The complementary pair of MOS transistors M_N and M_P forms the inverter, and the pair of npn transistors Q_1 and Q_2 forms the active pull-up/pull-down driver stage. We analyze the circuit operation for a capacitive load as follows.

At logic low input, the NMOSFET M_N is in cutoff while the PMOSFET M_P operates in saturation. With no channel formed in M_N, the base current for Q_2

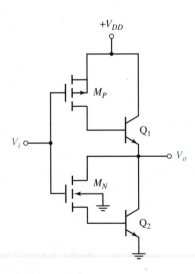

Figure 7.40   Basic BiCMOS inverter

is zero; thus, Q_2 is off. M_P provides base drive for Q_1 via its low-resistance conducting channel. If Q_1 operates in the active mode, it supplies a large emitter current—$(\beta + 1)$ times base current—to charge the load capacitance. Due to the large base drive, Q_1 may operate in saturation. In either case, the output voltage is less than the supply voltage by the base-emitter junction voltage of Q_1. Hence, $V_{OH} \approx V_{DD} - V_{BE}$. The charging time, and hence the delay t_{PLH}, are low because of the low output resistance and the high current of the on transistor Q_1.

When the input switches to logic high, M_P is turned off and M_N is turned on. The output capacitor discharges via the channel formed in M_N and supplies base current to Q_2. This turns Q_2 on, and causes a large discharge current from the capacitor. Thus, the high-to-low transition occurs rapidly. Note, however, that when the output voltage reaches the base-emitter cut-in voltage of Q_2, no further discharge takes place. Hence, the logic low voltage is $V_{OL} \approx V_{BE}$.

From this discussion, it is clear that the logic swing in a basic BiCMOS inverter is $V_{DD} - 2V_{BE}$. This less than "full-rail"[5] swing can be compensated for by using a level shifting circuit. Figure 7.41 shows the VTC for the inverter of Figure 7.40. Observe that because of the stored base charge in Q_2 (and the body effect of M_N), V_i needs to be higher than the threshold voltage of M_N in order for V_o to drop from V_{OH}. A similar effect applies for the higher end of V_i. Because of the CMOS-like VTC, the noise margins are only slightly less than $\dfrac{V_{DD}}{2}$.

The complementary pair of MOS devices in the inverter may be designed to provide equal drive current for the bases of Q_1 and Q_2. With matching transistors for Q_1 and Q_2, the circuit will then have minimum size and the propagation delays will be symmetric.

We can improve the transfer characteristics and switching speed of the basic inverter by providing paths for discharging the excess carriers from the bases of

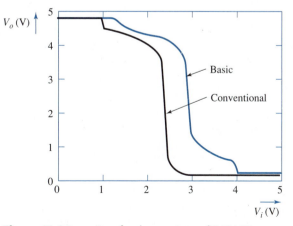

Figure 7.41 Transfer characteristics of BiCMOS inverters

[5] The bottom rail, which is ground in digital systems using only positive power supplies, is usually designated V_{SS}. In such cases, the full-rail voltage is $V_{DD} - V_{SS}$.

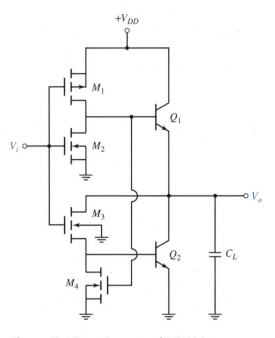

Figure 7.42   Conventional BiCMOS inverter

Q_1 and Q_2; these paths will help in turning each BJT off while the other is conducting. Figure 7.42 shows the circuit of a commonly used BiCMOS inverter with additional NMOS devices provided for base discharge. The operation of M_1 and M_3 in this circuit is the same as the NMOS devices in the basic inverter. M_2 and M_4 are turned on complement to M_1 and M_3, to discharge the base-emitter capacitances of Q_1 and Q_2. At logic low input, for example, M_1 is on and the voltage at the base of Q_1 is V_{DD}. Therefore, the gate of M_4 is at $V_{DD} > V_T$. This turns M_4 on, and the stored charge in the base of Q_2 is discharged to ground. In a similar manner, M_2 helps discharge the base of Q_1 during low-to-high transition at the input. Figure 7.41 shows the improved VTC for the conventional inverter using the same MOSFETs and BJTs as for the basic inverter.

The propagation delay of a BiCMOS gate is reduced because of the increased charging and discharging currents available from the BJTs. Due to the symmetry of the charging and discharging paths (and if the threshold voltage has the same magnitude for all the MOSFETs), the transition delay need be calculated based on the conducting MOSFET M_1 (M_3) and the BJT Q_1 (Q_2). As a first order approximation we neglect the junction capacitance of Q_1 and assume the BJT is operating in the active mode. If $V_o = V_{OL} \approx V_{BE}$ and the input switches from logic high to low ($V_{OL} \simeq V_{BE}$), then M_1 supplies the base current I_B for Q_1 given by

$$I_B = k_P(V_{DD} - V_{BE} - |V_{TP}|)^2 \qquad \textbf{(7.88)}$$

With a large β, the charging current supplied by Q_1 in the active mode is

$$I_E \approx \beta I_B \tag{7.89}$$

Thus, the time taken to charge C_L from $V_{OL} \approx V_{BE}$ to

$$V_H = \frac{(V_{OH} - V_{OL})}{2} + V_{OL} = \frac{V_{DD}}{2}$$

is

$$t_{PLH} = \frac{(V_{OH} + V_{OL})C_L}{2I_E}$$

or

$$t_{PLH} = \frac{V_{DD}C_L}{2\beta k_P(V_{DD} - V_{BE} - |V_{TP}|)^2} \tag{7.90}$$

In a similar manner it can be shown that the high-to-low propagation delay (neglecting the body effect of M_3 and the junction capacitance of Q_2) is

$$t_{PHL} = \frac{V_{DD}C_L}{2\beta k_N(V_{DD} - V_{BE} - V_{TN})^2} \tag{7.91}$$

In practice, however, Q_1 and Q_2 may operate in saturation—in which case Equations (7.90) and (7.91) must be modified. Still, the above equations show that the delays in BiCMOS gates are reduced approximately by a factor of β compared to the CMOS delays as given in Equations (7.73) and (7.74). Since a β of 100 is typical in IC BJTs, BiCMOS delays are in general smaller than CMOS delays. For a small load capacitance, however, the added capacitance from the BJTs and the additional NMOSFETs causes an increase in the BiCMOS delays over the CMOS delays. As C_L increases, it is clear from the first order approximation [Equations (7.90) and (7.91)], that the delays increase rather slowly with a slope of $1/\beta$. Hence, for capacitive loads above 500 fF, BiCMOS circuits perform better than CMOS circuits. Figure 7.43 shows the simulated switching response of the circuit of Figure 7.42 for $C_L = 1$ pF. Observe that as the load capacitance charges and V_o rises, the current from Q_1 decreases and, at very low current, the base-emitter junction voltage reduces to a small value. Therefore, the steady voltage at logic high is close to V_{DD}. In a similar manner, the steady low voltage V_{OL} is only slightly above zero. Typical inverter delays of under 1 ns have been achieved for a capacitive load of 1 pF, while delays of 200 ps have been reported for devices using 1 μm feature size.

BiCMOS NAND and NOR gates are implemented by modifying the inverter shown in Figure 7.42. Two-input NAND and NOR gates are shown in Figure 7.44.

With their higher current capability and faster response, BiCMOS circuits can be used in place of CMOS buffers. Note that the BJTs do not require as much silicon area as the MOSFETs in CMOS buffers. Additionally, with enough

```
*BiCMOS Inverter - Switching Response
M1        3        2        1   1    PENH L=5U W=25U
M2        3        2        0   0    NENH L=5U W=10U
M3        4        2        5   0    NENH L=5U W=10U
M4        5        3        0   0    NENH L=5U W=10U
Q1        1        3        4        Q2N3904
Q2        4        5        0        Q2N3904
CL        4        0        1pF
VDD       1        0        DC 5V
VI        2        0        PULSE(OV 5V 10N 0.001N 0.001N 100N 200N)
.MODEL PENH PMOS(VTO=-1V KP=40E-6)
.MODEL NENH NMOS(VTO=1V KP=100E-6 GAMMA=0.4)
.TRAN              5N 200N
.LIB "EVAL.LIB"
.PROBE
.END
```

(a) Simulation File

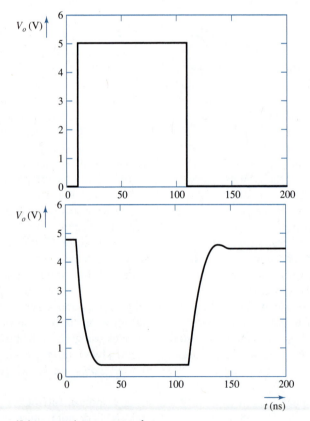

(b) Input and Output Waveforms

Figure 7.43   MicroSim PSpice-simulated switching response of a BiCMOS inverter

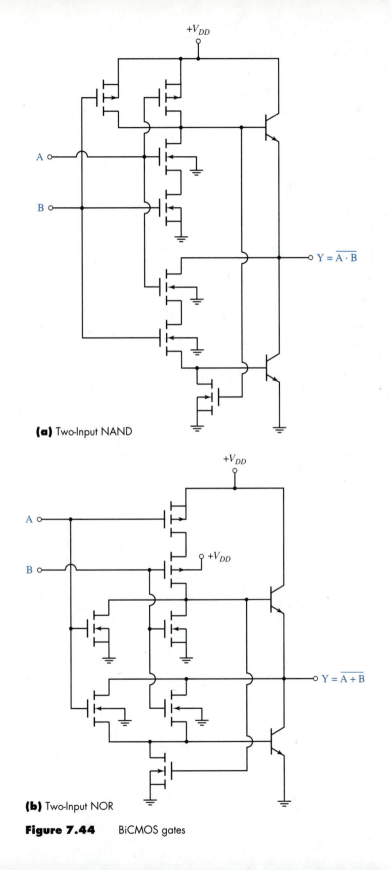

(a) Two-Input NAND

(b) Two-Input NOR

Figure 7.44   BiCMOS gates

current drive from the MOS devices to drive the BJT junction capacitances, only moderate performance is required of the BJTs.

7.4.2 Full-Swing Gates

The logic swing of less than V_{DD} in the circuits of Figure 7.44 may cause problems in driving standard CMOS gates. In addition, applications using low supply voltage will have low swing and reduced noise margins. Design engineers have developed many modifications to the conventional BiCMOS configuration for applications using supply voltages as low as 1.6 V. Figure 7.45 shows some examples of these modified "full-swing" BiCMOS circuits. These circuits provide additional low-resistance paths shunting the emitter-base junctions of the BJTs. Load capacitance C_L, therefore, continues to charge (or discharge) via these paths beyond the active junction voltage of the BJTs.

In the circuit of Figure 7.45a, resistor R_1, for example, ensures that C_L continues to charge after the emitter-base junction voltage drops below the cut-in voltage of 0.6 V. In a similar manner, R_2 carries the discharge current from C_L when V_o falls to the cut-in voltage of Q_2. Note that $R_1(R_2)$ also helps in discharging the junction capacitance of $Q_1(Q_2)$ at logic high (low) input, taking over the function of M_2 (M_4) in the conventional circuit of Figure 7.42. In the circuit of Figure 7.45b, the discharge of Q_1 is speeded up by the NMOS device M_3, which provides a lower resistance than the combination of R_1, R_2, and M_2. Observe that M_2 and M_3 act as pass transistors, whose gates are driven by the

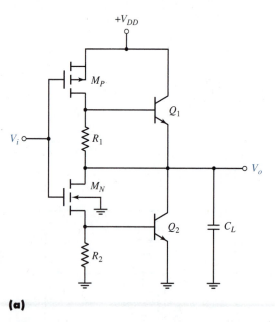

(a)

Figure 7.45   Examples of full-swing BiCMOS inverters

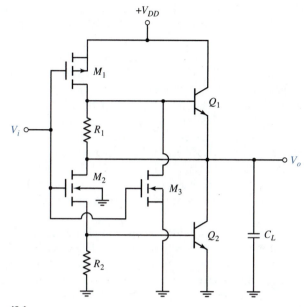

(b)

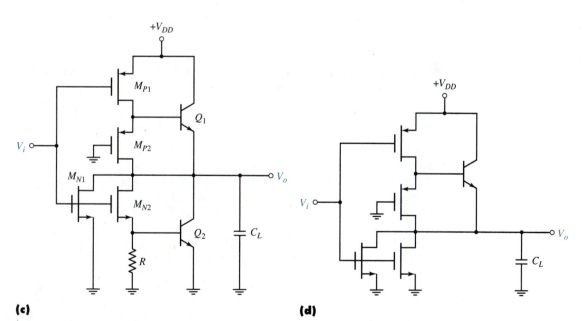

(c)

(d)

Figure 7.45 (continued)

same signal. Such devices are fabricated with common oxide and polysilicon regions covering both; thus, they are symbolized by a single gate connection through them.

Figure 7.45c uses M_{N1} for fast and complete discharge of Q_2 at logic high inputs. Additionally, PMOS device M_{P2} replaces the role of the passive resistor R_1 in Figure 7.45a and b, which occupies more area. Since M_{P2} is normally on with its gate at ground potential, its low-resistance path aids in charging of C_L to the full supply voltage V_{DD}. In addition, at logic high input, the base of Q_1 discharges to ground via M_{P2}. In applications where the load capacitance C_L is relatively low, the pull-down of V_o can be completed using only NMOS devices. Although a BJT provides a large discharge current, replacing it with MOS devices reduces the size and affords flexibility in their placement at a slightly reduced drive capability. In Figure 7.45d, C_L discharges to zero via the paralleled NMOS devices at logic high input.

Many other full-swing circuits using complementary BJTs, multiemitter BJTs, and multidrain MOS devices have evolved. These circuits strive to meet the goals of operating at low supply voltages with reduced power and increased density. Full-swing BiCMOS circuits have been used for high-speed applications involving large logic systems and static random access memory, compatible with ECL and TTL systems. In addition, BiCMOS microprocessors, gate arrays, and full- and semicustom applications with speeds rivaling ECL speeds, but at much reduced power, have been developed.

7.5   GALLIUM ARSENIDE MESFET LOGIC CIRCUITS

As we saw in Chapter 6, gallium arsenide MESFETs have current voltage characteristics similar to those of silicon NMOS devices, with the difference that gate current exists in MESFETS if the gate junction voltage exceeds the Schottky barrier voltage. Hence, the NMOS logic circuit configurations studied in section 7.1 may be used with appropriate changes in input voltage swings to prevent gate current. In this section, we study some of the commonly used GaAs MESFET logic circuits.

7.5.1   DIRECT-COUPLED FET LOGIC (DCFL) CIRCUITS

Direct-coupled FET logic circuits are formed using enhancement and depletion MESFETs, similar to the NMOS inverter of Figure 7.7a. These circuits offer high density and low power dissipation. Figure 7.46 shows the circuit of a DCFL inverter driving an identical load inverter. Unlike in the NMOS inverter, the logic high level is determined by the load circuit. Since the load inverter gate conducts when its voltage V_o exceeds the Schottky barrier voltage, V_{OH} is clamped at approximately 0.7 V regardless of the supply voltage V_{DD}. With the

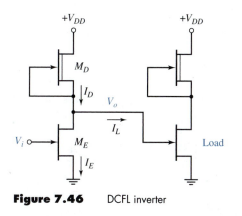

Figure 7.46   DCFL inverter

D-MESFET always conducting for $V_{GS} = 0$, V_{OL} is determined by the on resistance of the E-MESFET. Input logic levels are limited by the gate threshold and the Schottky diode conduction voltages. The following example illustrates the approximate calculation of logic levels for the DCFL inverter.

The devices for the DCFL inverter shown in Figure 7.46 have the following parameters: for the D-MESFET, $V_{TD} = -1$ V, $\beta = 200$ μA/V^2, $\lambda = 0.1$; for the E-MESFET, $V_{TE} = 0.1$ V, $\beta = 500$ μA/V^2, and $\lambda = 0.1$. The area factor for the E-MESFET is 3 while the D-MESFET has unit area. The power supply voltage is 1.5 V. Determine the logic levels and the critical voltage using the approximate MOSFET model. What is the average static power dissipation in the circuit?   **Example 7.13**

Solution

As with the MOSFET inverter of Figure 7.7, the E-MESFET M_E is off for $V_i < V_{TE} = 0.1$ V. The D-MESFET M_D, therefore, conducts via the gate of the load MESFET. Since the Schottky gate-source has a drop of about 0.7 V when conducting, the output voltage is 0.7 V for $V_i < 0.1$ V.

As V_i increases above V_{TE}, M_E begins conduction in the saturation mode and diverts some of the current from the load gate. This region in the VTC is described by the currents I_D, I_E, and I_L in M_D, M_E, and the load gate, as given by

$$I_D = I_E + I_L \qquad (7.92)$$

Using the ideal diode equation for I_L and the approximate Curtice model Equation (6.27), we can write the above equation as

$$\beta_D[-2V_{TD}(V_{DD} - V_o) - (V_{DD} - V_o)^2][1 + \lambda(V_{DD} - V_o)]$$
$$= \beta_E(V_i - V_{TE})^2(1 + \lambda V_o) + I_o\left[\exp\left(\frac{V_o}{nV_T}\right) - 1\right] \qquad (7.93)$$

where I_o is the diode saturation current, n is the ideality factor (emission coefficient) for the diode, and $V_T = kT/e$ is the thermal voltage. Note that M_D is assumed to operate in the ohmic mode.

Equation (7.93) describes the region of VTC where V_o is falling with V_i. The logic high voltage V_{OH} can be determined from this equation using the slope condition, $dV_o/dV_i = -1$. However, the exponential term in the diode current makes the equation transcendental in nature. Hence, solution by an iterative procedure or simulation may be resorted to. Instead, we can obtain an approximation to V_{OH} by considering a constant drop of 0.7 V across the Schottky barrier diode when conducting. Although the diode voltage depends on I_o and temperature, and varies slightly with current, it is reasonably constant over a range of currents (see Section 2.4). Thus, $V_{OH} \approx 0.7$ V.

V_{IL} may be approximated by the input voltage at which the load current is zero. At this input voltage, $V_o \approx 0.7$ V, which forward-biases the diode at the edge of conduction so that $I_L = 0$. Hence, setting $I_D = I_E$, and $V_o \approx 0.7$ V in Equation (7.93) results in

$$200[2(1.5 - 0.7) - (1.5 - 0.7)^2][1 + 0.1(1.5 - 0.7)]$$
$$= 1500(V_{IL} - 0.1)^2(1 + 0.07)$$

Solving this equation, we get

$$V_{IL} \approx 0.46 \text{ V}$$

For $V_i > V_{IL}$, the load MESFET is off while M_E conducts more, causing V_o to drop further. Since the drain currents are equal, we have

$$\beta_D(V_{TD})^2[1 + \lambda(V_{DD} - V_o)] = \beta_E[2(V_i - V_{TE})V_o - V_o^2](1 + \lambda V_o) \quad \textbf{(7.94)}$$

Equation (7.94) assumes that V_o has dropped sufficiently so that M_E is operating in the ohmic mode. An approximation to V_{IH} using the slope criterion and Equation (7.94) is obtained as

$$V_{IH} \approx 0.55 \text{ V}$$

V_{OL} is calculated for $V_i = V_{OH} = 0.7$ V. Using Equation (7.94) at this input, we have

$$V_{OL} \approx 0.14 \text{ V}$$

Note that we have neglected the $(1 + \lambda V_{DS})$ terms in calculating the logic levels.

The gate threshold (critical) voltage of $V_M = V_i = V_o$ occurs in the range $V_{IL} < V_i < V_{IH}$. Assuming the ohmic mode for M_D, we have

$$\beta_D[-2V_{TD}(V_{DD} - V_M) - (V_{DD} - V_M)^2][1 + \lambda(V_{DD} - V_M)]$$
$$= \beta_E(V_M - V_{TE})^2(1 + \lambda V_M) \quad \textbf{(7.95)}$$

Solving this iteratively, we get $V_M = V_i = V_o = 0.48$ V

From the logic levels, the noise margins are

$$N_{ML} = V_{IL} - V_{OL} = 0.32 \text{ V} \quad \text{and} \quad N_{MH} = V_{OH} - V_{IH} = 0.15 \text{ V}$$

and the logic swing is

$$V_{LS} = V_{OH} - V_{OL} = 0.56 \text{ V}$$

Power dissipation: At logic high output voltage of $V_o = 0.7$ V, M_D is in the ohmic mode and draws a supply current of I_{DH} given by (from the left-hand-side term in Equation (7.95))

$$I_{DH} = 0.21 \text{ mA}$$

At logic low output voltage of $V_o = 0.14$ V, M_D is in saturation and carries a current of (from the left-hand-side term in Equation (7.94))

$$I_{DL} = 0.23 \text{ mA}$$

Hence, the average supply current is

$$I_{avg} = \frac{I_{DH} + I_{DL}}{2} = 0.22 \text{ mA}$$

which gives an average static power dissipation of

$$P_d = V_{DD} I_{avg} = 0.33 \text{ mW}$$

The MicroSim PSpice simulation of the above inverter using the Curtice model (Level $= 1$) and default parameters gives the VTC shown in Figure 7.47. In the simulation input file, the load gate is represented by a Schottky diode DL. Note that in Figure 7.47, the load gate turns off at $V_i \approx 0.5$ V, which is close to the calculated value of V_{IL}. The gate threshold voltage from Figure 7.47b is $V_T = 0.51$ V, compared with the 0.48 V calculated using Equation (7.95). Similar reasonable agreement is seen for the other logic levels and currents. We must keep in mind, however, that the simplified equations used in the calculations in Example 7.13 are robust with parameters such as the diode emission coefficient, saturation currents, and resistances of the devices, while MicroSim PSpice models are sensitive to these. Consequently, simulation results may vary with the choice of these parameters. A graphical method using observed current-voltage characteristics of the devices, or an analytical method based on the equivalent circuit including resistances, may be resorted to obtain values comparable to simulation results.

DCFL characteristics vary with the geometry ratio of the depletion and the enhancement MESFETs. Unlike in MOSFET circuits, noise margins in ratioed static DCFL circuits are strongly dependent on the geometry ratio. This is because of the gate conduction of the load MESFET and the low supply voltage used. It is shown (Reference 12) that the geometry ratio of M_E to M_D must be above unity to achieve nonzero noise margins. The optimum ratio is shown as 2–3 for a noise margin of at least 0.2 V.

Power dissipation of the DCFL inverter, which is typically about 1 mW, is the lowest of all MESFET logic configurations. This feature, combined with low delay, makes the DCFL one of the most widely used MESFET logic families.

```
* GaAs DCFL INVERTER - VTC
BD        1    2    2      GDEP
BE        2    3    0      GENH 3 ; AREA FACTOR 3
DL        2    0    SBD
VI        3    0
VDD       1    0    1.5V
.DC       VI   0 1.0 0.02
.PROBE
.MODEL SBD D (N=1.2)
.MODEL GDEP GASFET (VTO =            -1 BETA=2E-4 LAMBDA=0.1 LEVEL=1)
.MODEL GENH GASFET (VTO =0.1 BETA=5E-4 LAMBDA=0.1 LEVEL=1)
.END
```

(a) Simulation File

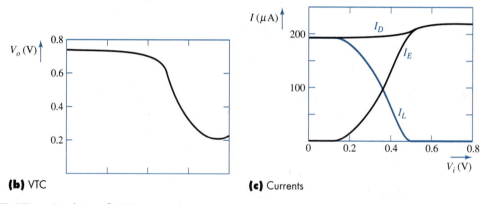

(b) VTC   **(c)** Currents

Figure 7.47   Simulation of DCFL inverter characteristics

Propagation delay: Because of the low logic swing and the low capacitance due to the semi-insulating substrate, MESFET switching delays, in general, are extremely small. We obtain an approximation to the propagation delay of a DCFL inverter using average charging and discharging currents.

Example 7.14

Estimate the propagation delays for the inverter of Example 7.13 for a lumped load capacitance of 20 fF at the output node.

Solution

Low-to-high delay: As V_i switches from V_{OH} to V_{OL}, M_E and the load diode are turned off. With the output held at $V_{OL} = 0.14$ V by the load capacitance, M_D is in saturation and provides an initial charging current, using Equation (6.27), of

$$\beta_D(V_{GS} - V_T)^2(1 + \lambda_D V_{DS}) = 0.2(-1)^2[1 + 0.1(1.5 - 0.14)] = 0.227 \text{ mA}$$

Neglecting the diode current when V_o reaches $V_{OH} = 0.7$ V, the final charging current from M_D, which is now operating in the ohmic mode, is

$$\beta_D[2(V_{GS} - V_{TD})V_{DS} - V_{DS}^2](1 + \lambda V_{DS})$$
$$= 0.2[2(1.5 - 0.7) - (1.5 - 0.7)^2][1 + 0.1(1.5 - 0.7)] = 0.207 \text{ mA}$$

Hence, the average charging current is

$$I_{ch} = \frac{0.227 + 0.207}{2} = 0.217 \text{ mA}$$

Assuming constant current charging of the load capacitance from V_{OL} to V_{OH} with I_{ch}, time taken to charge to $(V_{OL} + V_{OH})/2 = 0.42$ V is

$$t_{PLH} = \frac{(0.42 - 0.14)20}{0.217} = 25.8 \text{ ps}$$

High-to-low delay: When V_i changes from low to high, the load capacitance discharges into M_E. In addition, M_E must carry the current from M_D, which is always on. Hence the discharge current of the capacitor is given by

$$I_{dis} = I_E - I_D$$

where, again, we neglect the diode current.

Initially, at $V_o = V_{OH} = V_i = 0.7$ V, M_E is in saturation, with a drain current of

$$I_E = 1.5(0.7 - 0.1)^2(1 + 0.07) = 0.578 \text{ mA}$$

while M_D is in the ohmic mode, supplying a current of

$$I_D = 0.207 \text{ mA}$$

Thus, the initial discharge current is $0.578 - 0.207 = 0.371$ mA

At $V_o = V_{OL}$ and $V_i = V_{OH}$, the modes of M_E and M_D are reversed, and it can be verified that $I_E \approx I_D$. Hence, the average discharge current is given by

$$I_{dis} = \frac{0.371}{2} = 0.19 \text{ mA}$$

The delay time to discharge from 0.7 V to 0.42 V is, therefore, given by

$$t_{PHL} = \frac{(0.7 - 0.42)20}{0.19} = 29.5 \text{ ps}$$

The MicroSim PSpice simulation of the switching response, shown in Figure 7.48, gives comparable delays of $t_{PLH} = 22$ ps and $t_{PHL} = 21$ ps. Note that the capacitor and the MESFET currents (Figure 7.48c) agree closely with the approximate values we have calculated.

DCFL NOR and NAND gates are implemented similarly to NMOS gates, as Figure 7.49 shows. Each input device of a NOR gate draws current away from charging the load capacitance during low to high transition. Since this increases the delay t_{PHL}, fan-in is limited to three. The NAND gate has a different limita-

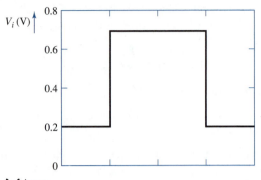

(a) Input

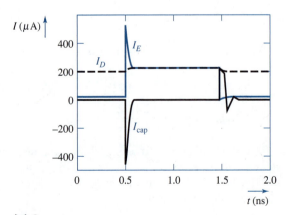

(b) Output

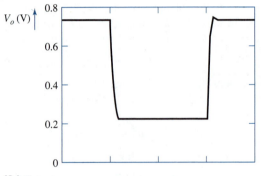

(c) Current

Figure 7.48   MicroSim PSpice simulation of a DCFL inverter switching response

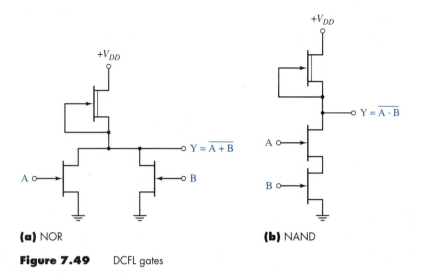

(a) NOR   **(b)** NAND

Figure 7.49   DCFL gates

tion. At logic high input, the drain-source voltage of each conducting E-MES-FET is above its threshold voltage. Thus, the output of the cascaded MESFETs exceeds the logic low voltage of the succeeding stage. To prevent logic error resulting from this output, no more than two devices are stacked in the pull-down circuit. This limits NAND fan-in to two.

With their simplicity in implementation, low power dissipation, and low delays, DCFL circuits have been used in LSI gate arrays and specialized logic functions. Relatively low noise margins, and their dependence on the geometry ratio and the precise value of threshold voltages, are the limitations of the DCFL gates. In addition, each fan-out gate contributes significant gate-source capacitance at the output node; this causes deteriorated switching performance.

7.5.2 Buffered FET Logic (BFL) Circuits

The *buffered FET logic circuit,* which is the earliest GaAs digital IC, uses only D-MESFETs with large threshold voltages. Two power supplies and level-shifting Schottky diodes are needed in BFL gates to achieve input-output compatibility with high noise margins. The current-driving capability is increased by the use of a source follower as a buffer between the output node and load gates. Figure 7.50a shows the circuit of a BFL inverter. M_1 and M_2 perform logic inversion, while the source of M_3 follows its gate voltage less a forward drop of about 0.7 V. M_4 provides constant-current biasing to M_3, which results in low output impedance. Observe that the source follower gives noninverting output. Significantly higher noise margins than with DCFL gates are obtained using negative supply ($V_{SS} < V_T$) and proper ratio for the devices.

Logic is formed at the front end of a BFL circuit, with D-MESFETs in parallel for NOR, and in series for NAND, functions. Figure 7.50b and c show examples of logic formed in BFL gates.

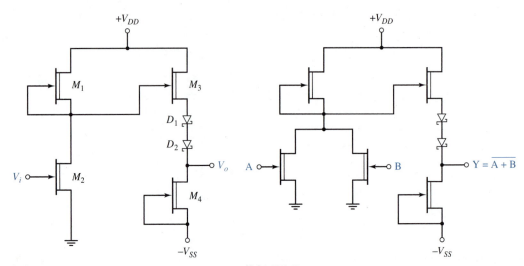

(a) Inverter

(b) NOR Gate

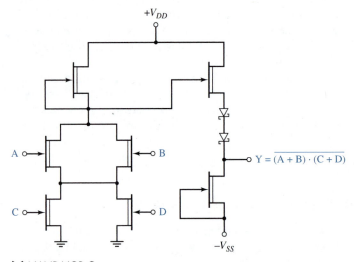

(c) NAND-NOR Gate

Figure 7.50   Buffered FET logic

The source-follower and level-shifting combination gives BFL gates high noise margins that are relatively stable at different loads. Fan-out is limited to two or three to avoid excessive delay and/or reduction in output voltage. The major limitation, however, arises from high power dissipation of 10 to 50 mW. Specialized MSI functions and SSI gates have been built using BFL with delays of the order of 50 ps.

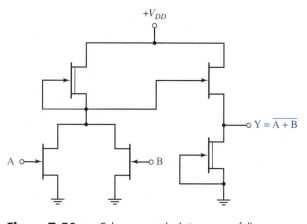

Figure 7.51   Enhancement-depletion source follower logic NOR gate

Power dissipation can be reduced by eliminating the source follower in the circuits of Figure 7.50. The resulting circuits with diode level-shifted outputs from a MESFET family known as *unbuffered FET Logic* (UFL), or, simply FET logic (FL). Both UFL and BFL circuits require a large area for the depletion devices. This precludes their application in VLSI systems. Instead, a combination of enhancement and depletion devices increases circuit density and uses only one supply voltage. Figure 7.51 shows another buffered logic circuit configuration using enhancement and depletion devices and a source follower. This circuit has larger logic swing and higher noise margins than does the DCFL. Logic swing can be increased further by adding a level-shifting diode between the source follower and the current source. Other modifications of the buffered logic circuit are possible to increase fan-out without adversely affecting switching performance. Buffered logic has been used in a variety of standard and specialized MSI functions.

7.5.3 SCHOTTKY DIODE FET LOGIC (SDFL) CIRCUITS

Schottky diode FET logic uses high-performance diodes to realize OR functions and D-MESFETs to provide load driving capability, making it similar to the DTL family (Chapter 4). The diodes occupy less area and operate at much less current compared with FETs. In addition, with low capacitance, SDFL circuits offer high switching speed and fan-out. As with DTL circuits, SDFL gates offer wire-ANDing to realize multilevel functions.

Figure 7.52a shows the circuit of an SDFL inverter. Diode D_L, which is larger in area than the switching diodes in logic realizations, is used for raising V_{IL}. M_3 serves to forward-bias the diodes and to discharge the gate capacitance of the output device M_2 during high-to-low transition at the input. For a typical threshold voltage of -1 V and supply voltages of -2.5 V and $+2.5$ V, M_3 is in saturation for the input range $0 \leq V_i \leq 2.2$ V. At $V_i = 0$, the gate of M_2 is at

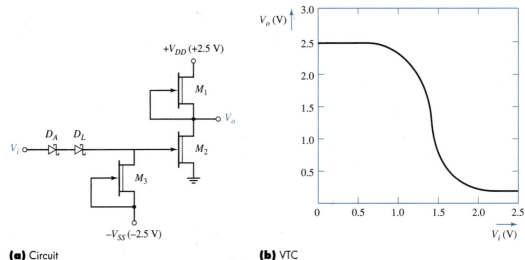

+V_{DD} (+2.5 V)

M_1

V_o

D_A D_L

V_i

M_2

M_3

−V_{SS} (−2.5 V)

(a) Circuit

(b) VTC

Figure 7.52   SDFL inverter

$V_i - 2V_D \approx -1.5$ V. Since this voltage is below V_T, M_2 is off and $V_o = V_{OH} = 2.5$ V. As V_i increases, the gate voltage of M_2 rises and M_2 begins conduction at $V_i = V_T + 2V_D \approx 0.5$ V. A further increase in V_i increases the drain current of M_2, and V_o drops. When V_i reaches about 2.2 V, gate conduction occurs in M_2. Figure 7.52b shows the VTC for the inverter with no loads connected at output. Clearly, the logic levels and noise margins are higher than in other MESFET configurations.

Figure 7.53 shows SDFL NOR and OR-AND-Inverter gates. The output stage in the OR-AND-Inverter gate uses a dual gate MESFET to complement logic at the second level. Fan-in can be increased by adding additional diodes without significantly affecting performance. Fan-out is limited to two or three because of the input current drawn by each load. Multiple gating is limited to two for high-speed performance. The SDFL circuit has been used in special-purpose MSI functions and LSI gate arrays.

7.5.4   OTHER MESFET LOGIC CONFIGURATIONS

As we have seen in the previous sections, many of the GaAs MESFET logic configurations have been derived from NMOS designs in which logic formation and output driving circuitry are arranged separately. This derivation is extended further to implement MESFET transmission, or pass transistor, logic. Unlike the NMOS transmission gate, however, gate-source conduction in MESFETs must be considered in choosing logic levels. Figure 7.54 shows basic MESFET transmission logic gates. Transmission logic gates using D-MESFETs and combined D- and E-MESFETs have been used in high-speed counters, shift registers, and multiplexers.

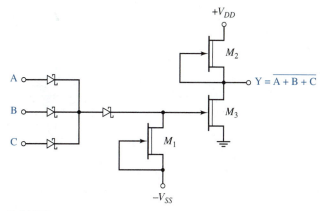

(a) NOR

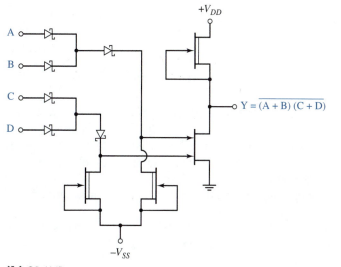

(b) OR-AND-inverter

Figure 7.53   SDFL gates

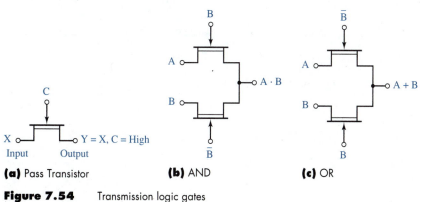

(a) Pass Transistor   **(b)** AND   **(c)** OR

Figure 7.54   Transmission logic gates

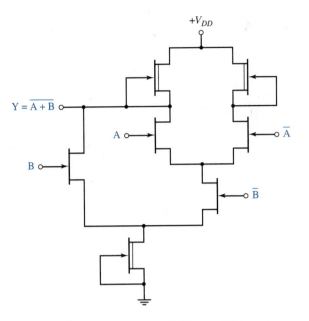

Figure 7.55   Source-coupled FET logic NOR gate

Source-coupled FET logic (SCFL) borrows from the BJT emitter-coupled logic configuration to achieve high speed and large noise margins at increased power and area. Figure 7.55 illustrates an SCFL NOR gate. Notice the similarity of this circuit with the series-gated ECL circuit of Figure 5.20. Each additional variable requires a pair of E-MESFETs, one in series with the common sources having complemented variable, and the other in parallel having true variable.

Table 7.2 provides a brief comparison of the typical characteristics of the more commonly used MESFET logic configurations.

Table 7.2   Comparison of MESFET logic families

	Type of Device	Typical Delay	Power Dissipation	Noise Margin	Fan-out	Relative Density
DCFL	E/D	30 ps	0.5 mW	0.2 V	2	Medium
BFL	D	50 ps	5 mW	0.8 V	2	Low
SDFL	D	200 ps	2 mW	0.8 V	1	High
SCFL	E/D	50 ps	2 mW	0.5 V	3	Low
Transmission gate	E or E/D	60 ps	1 mW	0.4	1	High

In addition to the logic families we have considered, various other GaAs MESFET logic configurations, including capacitor coupled, unbuffered, feedback, and dynamic domino logic have been developed. These configurations realize MSI functions with low power, high density, and/or high speed. The number of different configurations and the lack of standardization indicate the evolutionary nature of high-speed logic implementation using GaAs MESFET devices. Unlike those of the silicon CMOS devices, however, the challenges of GaAs circuits are many because of the lack of complementary devices and the lack of isolation between gate and source. Technology to meet the growing applications in high-speed signal processing and lightwave communications is overcoming these challenges with innovative circuit topologies and fabrication advances.

7.6 INTERFACING MOS AND BIPOLAR LOGIC FAMILIES

The compatibility of MOSFET—NMOS, CMOS, and BiCMOS—and MESFET logic levels with bipolar—TTL and ECL—levels requires interfacing circuits. These level-shifting circuits may form the final output stage in large-scale systems to drive the appropriate logic family. For example, in driving a standard CMOS circuit from the output of a BiCMOS gate array, logic levels must be shifted to full-rail supply voltages. A full-swing BiCMOS output circuit, or a pair of pull-up PMOS and pull-down NMOS devices, must form the driver stage. In this section, because of matching speed and power requirements, we consider only MOS–LSTTL and MOS–ECL interfaces. Input and output voltage levels of GaAs circuits compatible with ECL are obtained using buffered FET logic with appropriate MESFET sizes and supply voltages.

7.6.1 MOS–LSTTL INTERFACING

A CMOS output can be directly connected to an LSTTL input, if both circuits use 5 V supply, as shown in Figure 7.56. Device sizes for the CMOS circuit, however, must be large enough to source and sink LSTTL input currents. At logic high output, the PMOS device must be capable of supplying a maximum of 20 μA to LSTTL input. This may require a larger-than-minimum-size PMOS device. For the NMOS device, which must sink a much higher current of up to 4 mA at low output, the size becomes even larger than the PMOS device. Higher voltage operation of the CMOS circuit requires diode-clamping (shown in box) of its logic high output to limit it to the TTL level.

Going from the LSTTL output to the CMOS, the input pull-up resistor R_p connected to the supply voltage prevents voltage mismatch at the logic high level. With a value of 10 kΩ, R_p limits the sinking current for the LSTTL gate

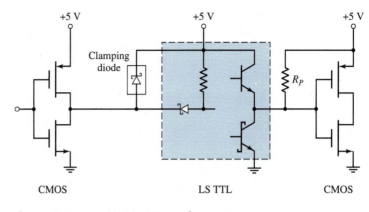

Figure 7.56   CMOS–LSTTL interfacing

to about 0.5 mA. If the CMOS circuit operates at a higher voltage, R_p must be raised to limit the sinking current and reduce wasteful power. Alternatively, an open-collector TTL output stage may be used with an appropriate pull-up resistor.

The above considerations also apply for interfacing NMOS and LSTTL outputs.

7.6.2   MOS–ECL Interfacing

Advances in CMOS logic speeds, which are approaching ECL speeds in memory applications, for example, necessitate level shifting between small-swing ECL and large-swing MOS for proper interfacing. Figure 7.57 shows a basic ECL–CMOS interfacing circuit. The two E-NMOS devices form a current mirror (similar to the BJT current mirror of Section 5.3.2), with P_1 setting the reference

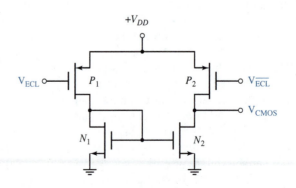

Figure 7.57   ECL–CMOS interfacing

current. Since N_1 is operating in saturation, its current, which is determined by the true ECL logic level V_{ECL} at the gate of P_1, is given by

$$I_1 = k_{N1}(V_{GS} - V_{TN})^2 \qquad \textbf{(7.96)}$$

If N_1 and N_2 have the same threshold voltage, current in N_2, which has the same gate-source voltage as N_1, is given by

$$I_2 = k_{N2}(V_{GS} - V_{TN})^2 \qquad \textbf{(7.97)}$$

Thus,

$$I_2 = \left(\frac{k_{N2}}{k_{N1}}\right)I_1 = k_R I_1, \qquad \textbf{(7.98)}$$

where $K_R = \dfrac{(W/L)_2}{(W/L)_1}$ for the same process transconductance parameter for N_1 and N_2. Since the reference current I_1 is set by the input ECL level at the gate of P_1, we can choose the geometry ratio K_R to obtain high currents at each logic level.

During low-to-high transition at V_{CMOS}, that is, at logic low ECL output (≈ -1.7 V), the pull-up PMOS device P_2 supplies the charging current to load capacitance and I_2 to N_2. Hence, for fast charging, (W/L) of P_2 must be large. When the ECL output is at logic low, P_2 conducts less current with $V_{\overline{ECL}}$ at high (≈ -0.7 V); however, N_2, which is the pull-down device, must carry both the discharge current of load and the current from P_2. Therefore, N_2 must also have a large aspect ratio (W/L). With the input ECL swing of less than 1 V, the full CMOS (output) swing of $V_{DD}-V_{SS}$ ($= V_{DD}$ in the circuit shown) can be achieved by careful design of the four MOS devices. For faster switching, ECL input levels are shifted by a diode drop before applying to the PMOS devices. Other interfacing circuits using single-ended ECL output are also possible. Figure 7.58 shows PSpice-simulated waveforms for the ECL-to-CMOS translator shown in Fig-

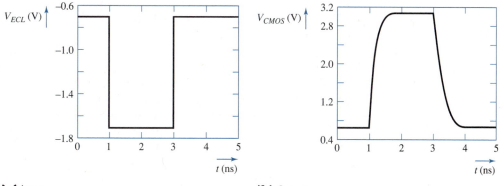

(a) Input   **(b)** Output

Figure 7.58   ECL-to-CMOS translator waveforms for Figure 7.57

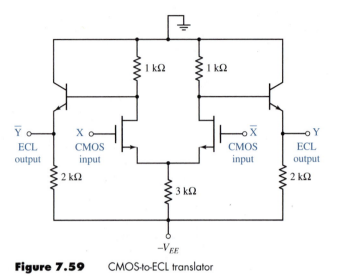

Figure 7.59   CMOS-to-ECL translator

ure 7.57 (see Problem 7.27). Higher logic swing can be obtained by proper scaling of the MOS devices.

CMOS to ECL   A basic CMOS-to-ECL translator using true and complemented CMOS input uses a source-coupled differential pair of NMOS devices as shown in Figure 7.59. As with the ECL-to-CMOS translator, the appropriate ECL level can be obtained by the choice of NMOS device sizes and resistances. Emitter-follower BJTs are used to drive ECL loads. In ICs, the source resistance is replaced by a saturated NMOS current source.

7.7   COMPARISON OF LOGIC FAMILIES

We have studied different logic families, starting with the RTL to the ECL in the bipolar families (Chapters 4 and 5), to the NMOS, CMOS, BiCMOS, and GaAs MESFET in the FET families in this chapter. Each family has different characteristics in terms of static and dynamic performance, size, and cost. Clearly, the choice of a particular logic family depends on the system specifications, the size, and the operating environment, among others. A system for real-time image processing, for example, must pack fast and high-density logic, while power dissipation is a major concern in a portable computer. Table 7.3 provides a brief comparison of typical performance characteristics of the more commonly used families.

Table 7.3 Comparison of logic families

	Supply Voltage (V)	Logic Type	Average Propagation Delay (ns)	Average Power Dissipation (mW)	Average Noise Margin (V)
TTL	+5	SSI–VLSI	10	10	0.4
LSTTL	+5	SSI–VLSI	8	2	0.5
ALSTTL	+5	SSI–VLSI	6	1	0.4
10K ECL	−5.2	SSI–VLSI	2	25	0.25
100K ECL	−4.5	SSI–VLSI	0.7	40	0.25
I^2L	+1 to +5	LSI–VLSI	0.7 to 20	0.1 to 100	0.4
NMOS	+3 to +15	LSI–VLSI	1	0.1	2*
CMOS	+1.5 to +15	SSI–VLSI	1	0.002	2.5*
BiCMOS	+1.5 to +15	MSI–VLSI	1	0.002	2.5*
GaAs MESFET		MSI–VLSI	$0.1^\dagger$	0.1 to $-10^\dagger$	0.1 to $0.2^\dagger$

* At 5 V supply voltage.
† Depends on circuit configuration.

Of all the logic families, the TTL (standard and low-power) and pin-compatible CMOS series are most widely used for low-cost, medium-speed SSI ("random logic" or "glue logic") applications, while the ECL and advanced CMOS families vie in high-speed applications. BiCMOS logic circuits are increasingly taking over from the ECL and CMOS in high-density VLSI and custom applications. Different configurations of the GaAs MESFET (as well as other GaAs devices) are used in specialized subsystem-level applications requiring high-speed operation, such as data converters, multiplexers for optical communication, and static memory. However, because of their low-yield large variations of electrical characteristics within the chip, and large power dissipation, GaAs circuits at present are limited to a very few LSI systems.

SUMMARY

The characteristics of different logic circuit configurations using MOSFETs, GaAs MESFETs, and combinations of MOSFETs and BJTs are described in this chapter.

We analyzed four configurations of E-NMOSFET inverters, depending on the type of load device. All of these inverters use an E-NMOSFET device to drive (1) a resistive load, (2) an E-NMOSFET device operating in the saturation mode, (3) an E-NMOSFET device operating in the linear mode, or (4) a D-NMOSFET device. Of these, the D-NMOSFET load configuration has better delay-power product and noise margins than the others. However, the circuit requires a few additional processing steps because of the combination of enhancement and depletion devices. The logic low level in all these circuits depends on the geometry ratio of the load MOSFET to the driver MOSFET. Hence, they are known as ratioed logic. NMOSFET logic circuits are used in LSI and VLSI systems.

CMOS logic circuits use both NMOSFETs and PMOSFETs of the enhancement type. CMOS inverters have the sharpest voltage transfer characteristics, and operate over a large range of supply voltage. They can be designed for the highest noise margins of half the supply voltage. Further, they dissipate virtually no static power. CMOS logic circuits take up more transistors, however, than NMOS circuits. The static and dynamic characteristics of CMOS NAND and NOR gates, which are ratioed, vary with different input conditions.

The switching delays in MOS logic circuits depend on lumped capacitance at the output node and the current capacity of the pull-up and pull-down devices. In general, the dynamic delay-power product is proportional to $(f/f_{max})C_L V_{DD}^2$, where f is the switching frequency, and $f_{max} = 1/(\text{propagation delay})$.

Due to their high input impedance, CMOS circuits must be protected against electrostatic discharge. Diodes at input and output clamp the voltages to safe levels. In addition, guard bands of diffusions are provided around the drain terminals to minimize the latch-up condition (in which a large current is diverted from logic circuit to ground by parasitic bipolar transistors).

Logic circuits are also implemented efficiently using CMOS transmission gates, which are formed using a pair of NMOS and PMOS devices.

Dynamic CMOS gates use one or more clock signals, and can be made ratioless. Dynamic logic circuits dissipate less power and are used in shift registers and other MSI and LSI implementations.

Domino logic is a variation of dynamic logic in which logic is formed using NMOS devices while PMOS devices charge or discharge the output capacitance. Domino circuits have high chip density.

BiCMOS logic circuits use the CMOS logic and the bipolar junction transistor output stage. These circuits have the combined advantages of low power, high density, and high drive capability. BiCMOS circuits can drive much higher capacitive loads than CMOS circuits, with delays reduced by a factor of the current gain β of the BJTs. BiCMOS circuits are used in gate arrays, microprocessors, and custom applications with speeds comparable to ECL speeds but at lower power and higher density.

GaAs MESFETs provide many different configurations for implementing logic functions. These circuits have delays in the tens of picoseconds range, but at high power dissipation, low logic swing and noise margins, and low fan-out. Approximate analyses of these circuits are carried out using simplified MOSFET equations. Direct-coupled FET logic, buffered FET logic, Schottky Diode FET logic, transmission gate logic, and source-coupled FET logic are some of the commonly used MESFET families. Low logic swing and noise margins and lack of standardization have given rise to many other configurations with varying degrees of performance characteristics. Because of high power dissipation, MESFET logic is limited to MSI and LSI functions.

Because of switching speed mismatch, interfacing between the different logic families is restricted to low-power Schottky TTL and MOS (and BiCMOS), and MOS and ECL. LSTTL–MOS interfacing requires that the sourcing and sinking current requirements are met by the driver and the load circuitry. Voltage levels are shifted to accomodate ECL–MOS interfacing.

REFERENCES

1. Uyemura, J. P. *Fundamentals of MOS Digital Integrated Circuits*. Reading, Mass.: Addison-Wesley, 1988.

2. Elmasry, M. I. (ed.). *Digital MOS Integrated Circuits*. New York: IEEE Press, 1981.

3. Uyemura, J. P. *Circuit Design for CMOS VLSI*. Boston: Kluwer Academic Publishers, 1992.

4. Weste, N. H. E., and K. Eshraghian. *Principles of CMOS VLSI Design—A Systems Perspective*. Reading, Mass.: Addison-Wesley, 1985.

5. Walsh, M. J. (ed.). *Choosing and Using CMOS*. New York: McGraw-Hill, 1985.

6. Hodges, D. A., and H. G. Jackson. *Analysis and Design of Digital Integrated Circuits*. New York: McGraw-Hill, 1988.

7. Kubo, M. et al. "Perspective on BiCMOS VLSI's." *IEEE J. Solid-State Circuits* SC-23, no. 1 (February 1988), pp. 5–11.

8. Greeneich, E. W., and K. L. McLaighlin. "Analysis and Characterization of BiCMOS for High-Speed Digital Logic." *IEEE J. Solid-State Circuits* SC-23, no. 2 (April 1988), pp. 558–65.

9. Alvarez, A. R. (ed.). *BiCMOS Technology and Applications,* Boston: Kluwer Academic Publishers, 1989.

10. Shin, H. J. "Full-swing BiCMOS Logic Circuits with Complementary Emitter-Follower Driver Configuration." *IEEE J. Solid-State Circuits* SC-26, no. 4 (April 1991), pp. 578–84.

11. Hanibuchi, T., et al. "A Bipolar-PMOS Basic Cell for 0.8-μm BiCMOS Sea of Gates." *IEEE J. Solid-State Circuits* SC-26, no. 3 (March 1991), pp. 427–31.

12. Wing, O. *Gallium Arsenide Digital Circuits*. Boston: Kluwer Academic Publishers, 1990.

13. Long, S. I., and S. E. Butner. *Gallium Arsenide Digital Integrated Circuit Design*. New York: McGraw-Hill, 1990.

Review Questions

1. What are approximate logic low and high output voltages in a resistive load NMOS inverter?

2. How must the resistor in a resistive load NMOS inverter be chosen? Why is the resistive load not preferred over an active NMOS load?

3. What is the maximum logic high voltage in an NMOS inverter with saturated NMOS load? How can this be raised?

4. How does body effect alter logic levels in a saturated NMOS load inverter?

5. How does the output vary with the input in the transition region of a saturated NMOS load inverter?

6. What are the difficulties in the implementation of a nonsaturated load NMOS inverter?

7. What is a ratioed logic?

8. How does the load-to-driver geometry ratio affect the noise margins in a D-NMOSFET load inverter?

9. Define gate threshold voltage.

10. What quantities determine static and dynamic delay-power products in an NMOS inverter?

11. What is the approximate range of supply voltage in CMOS circuits?

12. What effect does the PMOS-to-NMOS geometry ratio have on the VTC of a CMOS inverter?

13. What are the noise margins in a symmetrical CMOS inverter?

14. For a given supply voltage, how can a CMOS inverter be designed to have high noise margins?

15. What causes large capacitances at the input and output of a CMOS family relative to other logic families?

16. How does power dissipation vary in a CMOS inverter with supply voltage and frequency of switching?

17. What is the maximum frequency of operation for a CMOS inverter? What is the delay-power product at this frequency?

18. How does electrostatic discharge affect an unprotected CMOS logic circuit? How is the circuit protected against it?

19. What is latch-up in CMOS circuits? How is it minimized?

20. For an N-input NAND gate, how many active devices are needed in CMOS implementation? How many devices are needed in NMOS realization using D-NMOSFET load?

21. What is the worst-case dc design of a CMOS NAND gate? What is the worst case for switching delay?

22. Answer Question 21 above for a NOR gate.

23. How is a CMOS transmission gate implemented?

24. What limits the frequency of operation in a dynamic gate?

25. What is the logic limitation of domino logic circuits?

26. How many PMOS and NMOS devices are required to implement a four variable function using domino logic?

27. How does a BiCMOS inverter differ from a CMOS circuit?

28. What are the logic levels of a conventional BiCMOS inverter?

29. What limits the input voltage in GaAs MESFET logic circuit?

30. What causes reduced capacitance in a MESFET logic circuit compared to MOSFET capacitance?

31. Why is the fan-in of a DCFL circuit limited to two or three?

32. What is the major limitation of a BFL gate?

33. What is the basic logic function realized by an SDFL gate?

34. What are the advantages of SDFL gates over other MESFET gates?

PROBLEMS

1. Derive Equation (7.11) for the transition width of an E-NMOSFET inverter with resistive load.

2. Verify the critical values of V_{IH}, V_{IL}, V_{OL}, and V_M shown in Figure 7.2 for the resistive load inverter of Figure 7.1 with R = 5 kΩ and R = 100 kΩ. Use $V_T = 1$ V and $k = 100$ μA/V^2.

3. Calculate the threshold and critical voltages for an E-NMOSFET inverter with a saturated E-NMOSFET load (Figure 7.3a) with $V_{DD} = 5$ V, $V_{TD} = V_{TL} = 1$ V $k_D = 50$ μA/V^2, and $k_L = 10$ μA/V^2. Verify the modes of operation for the MOSFETs at the voltages calculated. What is the power dissipated in the circuit at $V_i = 0$, 2 V, and 5 V?

4. Determine the logic levels and critical voltages for the E-NMOSFET inverter with linear E-NMOSFET load (Figure 7.5a). Use $V_{DD} = 5$ V, $V_{GG} = 15$ V, $V_{TD} = V_{TL} = 2$ V, $k_D = 20$ μA/V^2, and $k_L = 5$ μA/V^2.

    What is the minimum value of V_{GG} needed to ensure linear mode of operation for the load MOSFET?

*5. Derive general expressions for the fall and rise times of a resistive load NMOS inverter in terms of V_T, V_{OL}, V_{OH}, k, and R. Show that

$$t_{PHL} = \left[\frac{C_L}{k(V_{OH} - V_T)} \right]\left[\frac{V_T}{V_{OH} - V_T} + \frac{1}{2}\ln\left(\frac{3V_{OH} - 4V_T - V_{OL}}{V_{OH} + V_{OL}} \right) \right]$$

6. Show that for a resistive load inverter $k_{dp} \approx \ln(2)$ in Equation (7.57).

*7. Show that for a D-NMOS load inverter, t_{PHL} is given by

$$t_{PHL} = \left[\frac{C_L}{2k_D(V_{OH} - V_{TD})}\right] \ln\left[\frac{3V_{OH} - 4V_{TD} - V_{OL}}{V_{OL} + V_{OH}}\right]$$

If body effect is neglected, show that t_{PLH} is given by

$$t_{PLH} = C_L\left[\frac{V_{OH} - V_{OL}}{2k_L V_{TL}^2}\right]$$

*8. Approximation to t_{PLH} may be obtained using average charging current of $I_{ch} = (I_l + I_m)/2$, where I_m is the current at $V_o = (V_{OL} + V_{OH})/2$, and I_l is the current at $V_o = V_{OL}$ with $V_i = V_{OL}$ in both cases. Estimate t_{PLH} for the D-NMOSFET load inverter of Example 7.6 using I_{ch} and linear charging. Similarly estimate t_{PHL} using linear discharging with $I_{disch} = (I_h + I_m)/2$ with $V_i = V_{OH}$.

9. Determine the constant k_{dp} in Equation (7.53) for the D-NMOS load inverter by approximating the delays given in Problem 7.

10. Using Equations (7.58) to (7.61), show that for $V_{TN} = |V_{TP}|$ and $k_P = k_N$,

$$V_{IL} = \frac{1}{4}\left[\frac{3}{2}V_{DD} + V_{TN}\right] \quad \text{and} \quad V_{IH} = \frac{1}{4}\left[\frac{5}{2}V_{DD} - V_{TN}\right].$$

What are the noise margins and the minimum supply voltage?

11. Determine the logic levels and V_M for a CMOS inverter having $V_{DD} = 10$ V, $V_{TN} = 0.8$ V, $V_{TP} = -1$ V, and $K_R = 1/4$.

12. A PMOSFET inverter is shown in Figure P7.12. Determine the logic threshold voltages and V_M. What is the output voltage V_o for $V_i = 0$? Use the following parameters: $V_{DD} = 10$ V, $R = 50$ kΩ, $V_{TP} = -2$ V, and $k_P = 50$ μA/V^2.

*13. Figure P7.13 shows another CMOS inverter configuration. Determine the VTC and indicate the modes of operation for the MOSFETs. How does

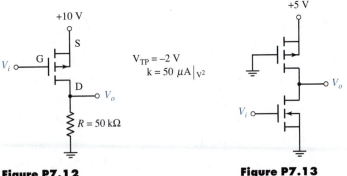

Figure P7.12   **Figure P7.13**

$K_R = k_P/k_N$ affect the VTC?

$$V_{DD} = 5 \text{ V}, \quad |V_{TP}| = V_{TN} = 1 \text{ V},$$
$$k_P = 25 \ \mu\text{A/V}^2, \quad \text{and} \quad k_N = 100 \ \mu\text{A/V}^2$$

14. Calculate the static power drawn by the CMOS inverter of Example 7.8 from the 5 V supply when $V_i = V_{DD}/2 = 2.5$ V.

*15. Calculate the gate threshold voltage for the two-input CMOS NOR gate of Example 7.12 with input B at logic low. Neglect body effect.

*16. Figure P7.16 shows the circuit of a superbuffer. Neglecting capacitance at node X, estimate t_{PLH} and t_{PHL} for $C_o = 10$ pF for the following parameters: $(W/L)_1/(W/L)_2 = (W/L)_3/(W/L)_4 = 4$, $V_{TE} = 1$ V, $V_{TD} = -2$ V, and $k = 100 \ \mu\text{A/V}^2$.

    Verify your results using a MicroSim PSpice simulation.

17. Figure P7.17 shows an implementation of NMOSFET equivalence (XNOR) function, $Y = \overline{A \oplus B} = AB + \overline{A}\,\overline{B}$. Indicate which devices are on for each of the four input combinations. Show an equivalent realization using NAND and NOR gates.

18. What is the function realized at Y in the CMOS circuit shown in Figure P7.18?

*19. Estimate the charging delay in a CMOS transmission gate for the NMOS-FET going from saturation to off. Approximate R_p by (a) the resistance of the PMOSFET at $V_o = 0$, (b) the resistance of the NMOSFET at $V_o = 0$, and (c) the parallel resistance of the two MOSFETs at $V_o = 0$. The output

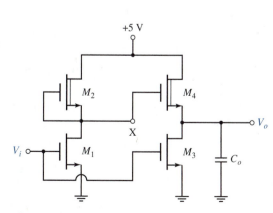

Figure P7.16

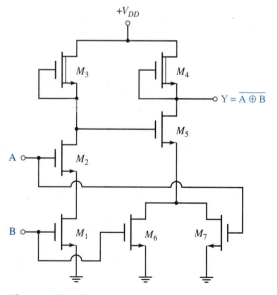

Figure P7.17

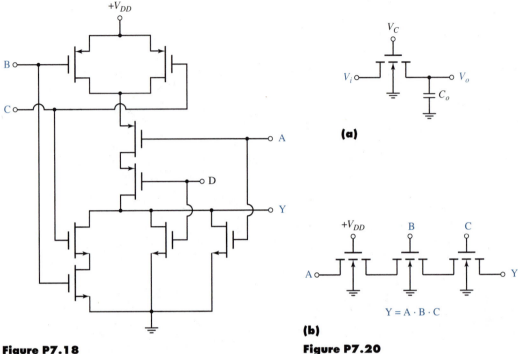

Figure P7.18

Figure P7.20

lumped capacitance is 10 pF. Other parameters are as given in Problem 7.13.

Compare the estimated delays with a MicroSim PSpice simulation result with body effect.

20. Figure P7.20a shows a "weak" transmission gate using a single NMOS device. What is the approximate logic high voltage at V_o when $V_i = V_{DD}$ and the capacitor is initially uncharged? If body effect is neglected, and V_C is constant at V_{DD}, what are the approximate time delays at V_o as V_i switches between 0 and V_{DD}? Figure P7.20b shows a three-input AND gate using NMOS pass transistors. What are the limitations of this circuit?

21. What is the function realized by the dynamic logic circuit shown in Figure P7.21?

22. Consider the two-input static CMOS NAND gate of Figure 7.25a. If the PMOS device M_{P2} is open, write down the output for the following input combinations.

A	1	0	0	1
B	1	1	0	0
Y				

If, instead of M_{P2}, M_{N2} is stuck open, what is the input sequence needed to detect the fault?

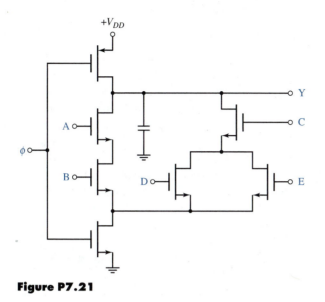

Figure P7.21

23. What is the function realized by the domino logic circuit shown in Figure P7.23?

*24. Estimate the BiCMOS propagation delays for the circuit of Figure 7.42 and compare with the MicroSim PSpice results shown in Figure 7.43 for C_L = 1 pF and 10 pF. Use the parameters given in the MicroSim PSpice input file of Figure 7.43 for the MOSFETs and a β of 100 for the BJTs.

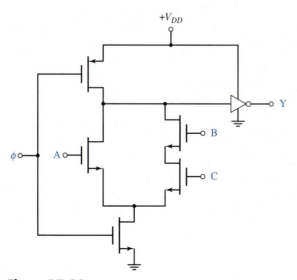

Figure P7.23

25. Determine the output voltage in the BFL inverter of Figure 7.50a for $V_i = -2$ V and $V_i = 0.5$ V. Use $V_T = -1.5$ V, $V_{DD} = 2.5$ V, and $V_{SS} = -2.5$ V, and assume a diode drop of 0.7 V. The aspect ratios are 10, 15, 5, and 10 for M_1, M_2, M_3, and M_4, respectively. Use $k = 100$ μA/V^2 and $\lambda = 0.1$ for all the devices.

26. For an SDFL inverter (Figure 53a with diode A only) using $V_{DD} = V_{SS} = 2.5$ V, determine the output voltage for $V_i = 0.5$ V and $V_i = 2$ V. The aspect ratios are 5, 15, and 10 for M_1, M_2, and M_3, respectively. Use $k = 100$ μA/V^2, $\lambda = 0.1$ and $V_T = -1.5$ V for all the devices.

27. Determine the steady voltage levels at the output of the ECL-to-CMOS translator shown in Figure 7.57 for the following parameters: $V_{TN} = 2$ V, $V_{TP} = -1.5$ V, $k_{N1} = 250$ μA/V^2, $k_{N2} = 2000$ μA/V^2, $k_{P1} = 15$ μA/V^2, and $k_{P2} = 100$ μA/V^2. The ECL input has a low level of -1.7 V and a high level of -0.7 V. •

28. For input CMOS levels of 0 and 5 V, determine the ECL levels in the circuit of Figure 7.59. The NMOS devices have $V_T = 2$ V and $k = 0.1$ mA/V^2. Neglect the base currents of BJTs.

29. *Experiments:*

A. NMOS Inverter Characteristics: In this experiment the transfer and switching characteristics of the NMOS inverter configurations are determined using a CD 4007UB MOSFET array.

(i) Wire the circuit of Figure 7.1a. Using $R = 10$ kΩ and $V_{DD} = 5$ V, vary the input voltage V_i from 0 to 5 V and measure the output voltage V_o.

(ii) Draw the voltage transfer characteristics using the data from (i). Determine the logic threshold and critical voltages. Using the parameters determined in the experiment of Problem 6.11, calculate these voltages and compare.

(iii) Apply a pulse input of 0 to 5 V amplitude and 10 μs period to the inverter. Measure the output propagation delays and rise and fall times for an external load capacitance of 10 pF. Note that at this large capacitance, the delays are primarily due to charging and discharging currents in the circuit. Calculate the delays and compare with experimental results.

(iv) Repeat (i), (ii), and (iii) for the saturated load inverter of Figure 7.3a. Note that when using two NMOSFET devices from CD 4007, the body of the load device is at ground potential while the source is at some positive voltage V_o. Hence the body effect will be significant at logic high output.

B. CMOS Inverter Characteristics: In this experiment, CMOS inverter characteristics are determined using CD 4007UB MOSFET array.

(i) Repeat (i), (ii), and (iii) of Experiment A above for the CMOS inverter of Figure 7.17.

(ii) To increase the load-driving capability of the inverter, connect all three CMOS inverters available in CD 4007UB in parallel. Measure the switching response of the paralled inverter for 10 pF load as in (iii) above. Calculate the delays and compare with experimental results.

REGENERATIVE CIRCUITS

INTRODUCTION

The logic circuits we studied in previous chapters form a class of *combinational circuits* in which the present output is determined by a combination of present inputs. Since the present output is independent of past inputs, combinational circuits are said to be *memoryless*. In this chapter, we turn our attention to the class of logic circuits in which the output is a function of both past and present inputs. These logic circuits are called *sequential circuits*. Sequential circuits have some form of *regenerative feedback* from output to input, so that the immediately preceding output, along with the present inputs, can control the present output. Sequential circuits are used as counters, shift registers, and storage (memory) and delay elements.

One subclass of sequential circuits is the family of *regenerative,* or *multivibrator, circuits* that consists of

1. The bistable multivibrator.
2. The monostable multivibrator.
3. The astable multivibrator.

Multivibrator circuits are used in digital systems as basic memory elements, time delay generators, and clock waveform generators. Basically, all three circuit types are characterized by a pair of cross-coupled logic inverters. Depending on the type of coupling (feedback) used to inject the output of one inverter into the input of the second inverter, a multivibrator has one or two stable logic levels, called *states,* at the output, or it has no output. An external signal causes a transition between the stable states, or between the stable and the unstable states.

This chapter introduces basic sequential circuits and includes the analysis and design of multivibrator circuits that use bipolar and MOS devices. We discuss the implementation of each multivibrator using discrete devices and IC gates from different logic families. We conclude this chapter with a brief description of the IC timer circuit and its operation as a monostable and an astable multivibrator circuit.

8.1   BISTABLE MULTIVIBRATOR

A basic bistable multivibrator consists of two logic inverters directly coupled from the output of one to the input of the other, as shown in Figure 8.1a. To analyze this circuit, we first remove the feedback at some convenient point, such as the input node of the first inverter G_1, as shown in Figure 8.1b.

With the voltage transfer characteristic (VTC) of each inverter as shown in Figure 8.1c, the resulting composite ("open-loop") characteristic $V_x - V_z$ for the cascaded inverters of Figure 8.1b is depicted in Figure 8.1d. Note that V_x and V_z are in phase, and the voltage gain magnitude is greater than unity in the transition region of each inverter; that is, $|dV_y/dV_x| > 1$ for $V_{IL} < V_x < V_{IH}$, and $|dV_z/dV_y| > 1$ for $V_{IL} < V_y < V_{IH}$. When the loop is closed, as in Figure 8.1a, points x and z are at the same potential. Hence, the operating point for the closed-loop circuit of Figure 8.1a is one of the three points, a, b, or c. The three possible operating points are at the intersection of the $(V_x - V_z)$ characteristic with the line $V_x = V_z$ in Figure 8.1c. Point c, however, is an unstable point, which cannot occur in a practical circuit.

Suppose, at some instant, that the circuit is operating at point c. Any minute change in voltage at x, due to ever-present noise or a power supply variation, will be amplified by G_1 and G_2, and applied back at terminal x. Since the amplified

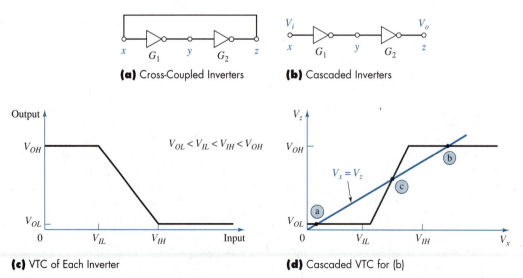

(a) Cross-Coupled Inverters   **(b)** Cascaded Inverters

(c) VTC of Each Inverter   **(d)** Cascaded VTC for (b)

Figure 8.1   General bistable multivibrator

voltage is in phase with the initial change at x, an even larger change occurs at x, and the operating point moves towards b (a) for an increase (decrease) in V_x. This process of regenerating the initial change in V_x in going through G_1 and G_2 continues, and the operating point moves quickly to b (a). The voltage gain at points a and b, however, is negligibly small. (For example, the inverter circuit is in cutoff, or in saturation, over the voltage range $V_i < V_{IL}$ and $V_i > V_{IH}$,

and $\dfrac{dV_y}{dV_x} = \dfrac{dV_z}{dV_y} \approx 0$ in that range.) Hence, no further change can occur at the

operating point a or b. As a result, with a *loop gain* $\dfrac{dV_z}{dV_x} > 1$, the circuit at any

time operates at one of the two stable points, a or b.

The bistable multivibrator thus has two stable operating points, or *states,* which are complements of each other. These states are designated Q and $\overline{Q}$ at the outputs of the two inverters, and they are given by $y = 1$, $z = 0$ and $y = 0$, $z = 1$.

A transition from one state to the other can occur only when an external voltage of appropriate amplitude is applied at x or y. The application of an external signal to cause a change in logic state is known as *triggering.* Clearly, additional input nodes are needed to inject the triggering signal without disturbing the output nodes. For example, the free input nodes in a bistable multivibrator using cross-coupled NOR gates can be used to apply the triggering signals. This is illustrated for a two-input NOR gate bistable circuit shown in Figure 8.2a. When the R and S inputs are at logic low, let the circuit be in the stable state, $Q = 1$, $\overline{Q} = 0$. When the R input goes to logic high (while the S input stays low), Q (the output of G_1) switches to logic low, after the propagation delay time t_{pd} of G_1. Following this, $\overline{Q}$ (the output of G_2) goes to logic high, after another delay t_{pd2}. After $t_{pd1} + t_{pd2}$, which is the loop delay of the circuit, the trigger input at R may be returned to its logic low level. The circuit is now in its stable state of $Q = 0$ and $\overline{Q} = 1$. This stable state is referred to as the *reset state,* and the R input is called the *reset input.* Note that another trigger at R has no effect; the circuit is already reset, or *cleared.* A trigger at the S input

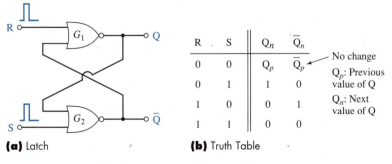

R	S	Q_n	$\overline{Q}_n$	
0	0	Q_p	$\overline{Q}_p$	No change
0	1	1	0	Q_p: Previous value of Q
1	0	0	1	Q_n: Next value of Q
1	1	0	0	

(a) Latch   **(b)** Truth Table

Figure 8.2   RS latch using NOR gates

momentarily going high and returning to low, takes the circuit to the *set state of* $Q = 1$ and $\overline{Q} = 0$; hence, the S input is referred to as the *set input*.

The functional relationship between the inputs R and S and the output Q is given in the truth table of Figure 8.2b. Observe that for $R = S = 1$, the outputs are not complements of each other. Hence, this input combination is generally not allowed in bistable circuits. Since $R = S = 0$ maintains the state of the circuit, inputs R and S are normally kept low until a state transition is required.

The bistable circuit "latches" onto its state of $Q = 1$, or $Q = 0$, until triggered by a reset or a set input. Hence, the circuit is known as a *bistable latch,* or simply a *latch.* A latch is sometimes referred to as an RS *flip-flop;* usually, however, an IC built around a latch to perform more complex functions is called a flip-flop.

A latch using two-input NAND gates is shown in Figure 8.3. The activating edge is the high-to-low transition (negative edge) at the inputs; hence, the

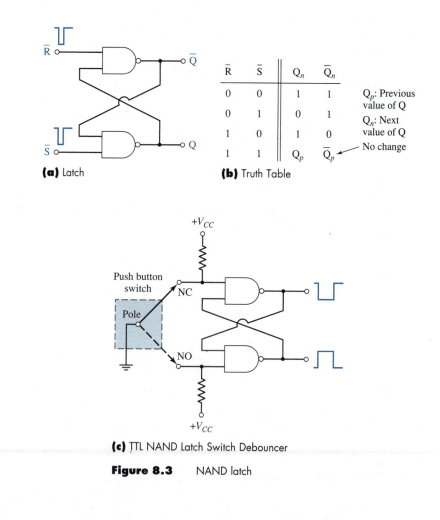

$\overline{R}$	$\overline{S}$	Q_n	$\overline{Q}_n$
0	0	1	1
0	1	0	1
1	0	1	0
1	1	Q_p	$\overline{Q}_p$

Q_p: Previous value of Q
Q_n: Next value of Q
No change

(a) Latch   **(b)** Truth Table

(c) TTL NAND Latch Switch Debouncer

Figure 8.3   NAND latch

designations $\overline{R}$ and $\overline{S}$ are used. The forbidden input combination is $\overline{R} = \overline{S} = 0$. An inverter may be added to each of the inputs to obtain the same function as a NOR latch.

NAND and NOR latches are implemented using gates from any of the logic families. A TTL NAND latch commonly used as a switch contact debouncer is shown in Figure 8.3c. As the single-pole, double-throw pushbutton switch is pressed and released, a single pulse is generated for as long as the switch is depressed.

8.1.1   CLOCKED FLIP-FLOPS

It is often necessary to trigger state changes in a latch at precisely the occurrence of a controlling *clock* signal. Rectangular pulses that occur at constant intervals are used as clock signals. The clock signal presents the activating edges of the R and S inputs to the latch only when it is active high or low. Changes at the output of the latch are therefore synchronzied with the active duration of the clock.

A clocked latch is referred to as a flip-flop. Figure 8.4a shows an RS flip-flop. The truth table for this flip-fop is the same as that for the NAND latch except that the clock must be at logic high for the state to change. Observe that changes in R and S are not transmitted to the NAND latch when the clock is at logic low; hence, Q and $\overline{Q}$ remain unchanged. The values of R and S at the positive (activating) edge of the clock affect the state according to the truth table shown in Figure 8.3b. The flip-flop may be set or reset, overriding the clock input, by providing direct set (preset) and reset (clear) inputs, as shown in Figure 8.4b. Preset and clear, however, are active low. This is indicated by the small circles in the symbol for the flip-flop shown in Figure 8.4c. For a NOR latch in the flip-flop (with AND gates replacing the NAND gates) the activating edge of the clock is the negative edge.

A clocked RS flip-flop still has the problem of a forbidden input combination. Two other flip-flops, namely, the JK and D flip-flops, incorporate additional feedback to eliminate the $Q = \overline{Q} = 1$ (or 0) state.

A basic JK flip-flop, built around a NAND latch, is shown in Figure 8.5a. As with the clocked RS flip-flop, outputs Q and $\overline{Q}$ are unchanged during the inactive (low) interval of the clock. In addition, outputs are unchanged if $J = K = 0$, regardless of the clock level. When the clock is active (high), the active low set and reset inputs of the latch may be the complements of the J and K inputs, and the latch may be triggered. For $J = K = 1$, either S or R (but not both) becomes active (low), depending on whether Q or $\overline{Q}$ is high. Hence, the output of the latch is the complement of its previous state. That is, for the input combination $J = K = 1$, this flip-flop toggles its output during the active interval of the clock.

Although the basic JK flip-flop eliminates the forbidden output for $J = K = 1$, the output is unstable for this input combination. If, during the high level

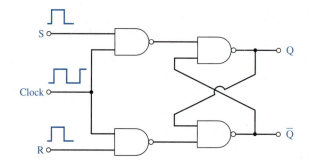

(a) Clocked RS Flip-Flop

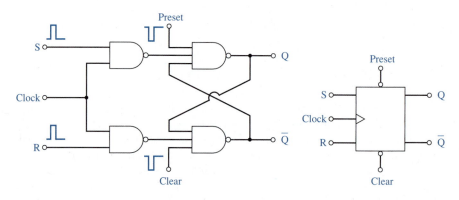

(b) Clocked RS Flip-Flop with Preset and Clear

(c) Circuit Symbol

Figure 8.4   Clocked flip-flop

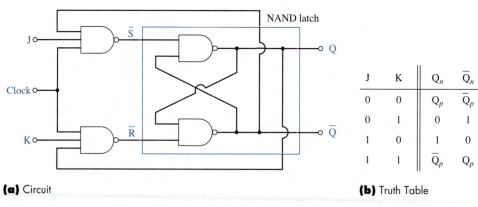

(a) Circuit

J	K	Q_n	$\overline{Q}_n$
0	0	Q_p	$\overline{Q}_p$
0	1	0	1
1	0	1	0
1	1	$\overline{Q}_p$	Q_p

(b) Truth Table

Figure 8.5   Basic JK flip-flop

of clock, both the J and K inputs remain high, the output state continues to toggle. The oscillatory outputs at Q and $\overline{Q}$ can be prevented by keeping the pulse width of the clock below the loop delay $t_{pd1} + t_{pd2}$ of the latch. Alternatively, the inputs may be restricted to remain high only for the duration of the loop delay. Since the loop delay can be extremely small (well below a nanosecond), this restriction is usually difficult to satisfy. This problem is eliminated by internally generating a narrow trigger pulse, or by using a cascade of two latches.

In an *edge-triggered flip-flop,* a narrow trigger pulse is generated from the input clock using the propagation delay of a gate. Figure 8.6a shows an edge-triggered JK flip-flop in which the clock and its complement are used to generate the trigger pulse. State transition occurs during the positive (low-to-high) edge of the clock. Figure 8.6b shows the waveforms for the circuit for $J = K = 1$ and $Q = 0$. When the clock is high, the outputs of G_1 and G_6 are low; hence, G_2 and G_5 are high and no state change occurs. Gates G_2 and G_5 stay high when the clock goes from high to low, or at the negative (falling) edge of the clock. At the positive (rising) edge of the clock, G_7 switches to low, after a delay of t_{p7}; after another delay t_{p1}, G_1 switches to low. Because of the delay $(t_{p7} + t_{p1})$ in the path from G_1 to G_2, both inputs to G_2 are high; hence, the output of G_2 is low for the interval $t_{p7} + t_{p1}{}^{1}$. (Since G_6 is low in this interval, G_5 stays high.) The narrow pulse from G_2 now triggers the latch ($G_3 - G_4$) and Q goes high, after delay t_{p3}. After delay t_{p4}, $\overline{Q}$ (G_4) switches to low, which maintains the high level of Q. Clearly, G_2 must be low until G_3 and G_4 change levels. Hence,

$$t_{p7} + t_{p1} > t_{p3} + t_{p4}$$

To ensure this operational requirement, G_1, G_2, and G_3 are designed to have longer delays than other gates. Since the state transition occurs at the low-to-high transition of the clock, the circuit is referred to as a *positive-edge-triggered flip-flop.*

Note that after the state change, which occurred after the positive edge of the clock, the output of G_7 (the complement of the clock) stays low. Hence, the J and K inputs are prevented from reaching G_2 and G_5. However, the inputs must be present just before the activating edge of the clock. This time period is known as the *setup time.* In addition, for unambiguous transmission, the inputs are required to remain stable for a period of time, called the *hold time,* just after the active clock edge. In the TTL family, setup times are typically a few nanoseconds, while hold times may be zero, depending on the circuit configuration. Any change in inputs outside of the setup and hold times is ignored by the circuit. Hence, noise spikes cannot cause state changes, unless they occur in the close vicinity of the active clock edge.

Edge-triggered *JK* flip-flops use the same logic symbol as that shown for the RS flip-flop in Figure 8.4. A negative activating edge is indicated by a circle for the clock input, in addition to the arrowhead.

| [1]This assumes equal low-to-high and high-to-low delays for all gates.

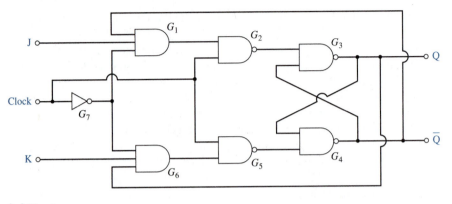

(a) Circuit

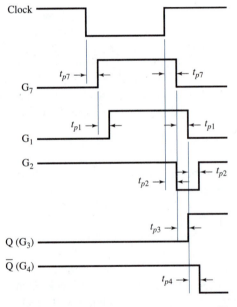

(b) Waveforms for J = K = 1   and   Q = 0, $\overline{Q}$ = 1

Figure 8.6   Edge-triggered JK flip-flop

A *master–slave JK flip-flop* prevents the toggling of its state during the clock interval by using two cascaded latches. As shown in Figure 8.7a, the master latch (G_2–G_3) is activated by the clock and the external inputs, in addition to the Q and $\overline{Q}$ outputs. The outputs of the master latch feed the slave latch (G_6–G_7), which is activated by the complement of the clock. At instant t_1 of the clock waveform (Figure 8.7b), the clock CK is rising and its complement $\overline{CK}$ is falling,

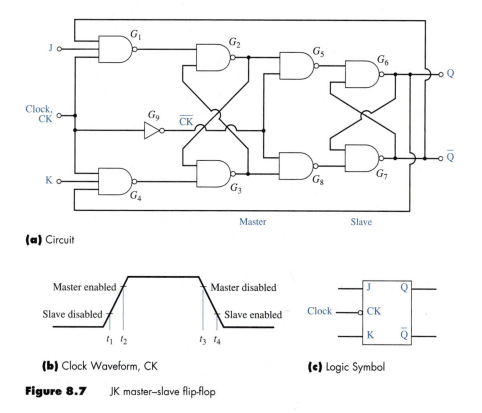

(a) Circuit

(b) Clock Waveform, CK

(c) Logic Symbol

Figure 8.7   JK master–slave flip-flop

so that the master latch outputs are prevented from reaching the slave by gates G_5 and G_8. Note that the slave is now isolated from changes in the input.

As CK reaches a sufficiently high level, at t_2, gates G_1 and G_4 respond to their J, K, Q, and $\overline{Q}$ inputs, the master latch is activated, and the slave remains inactive. At t_3 the clock CK is falling; hence, the master latch is disabled from changes in the input. As $\overline{CK}$ reaches the high level, at t_4, gates G_5 and G_8 transmit the master outputs to the slave. Hence, the slave outputs, Q and $\overline{Q}$, follow the master outputs. Note that during the transfer of the master outputs to the slave, the low level of CK inhibits the inputs to the master. The inputs can therefore exist during the entire clock interval. The minimum duration of the clock is the propagation delay of the master gates, G_1 and G_2, which is assumed to be the same as for G_3 and G_4.

A spike (from noise or an intentional signal) taking the J or K input high while the master is enabled, that is, when the clock is high, will be transferred to the master output. Once this happens, the master state cannot be changed, and the false state will subsequently be transferred to the slave during the low clock level. To prevent this, the duration of the high level of the clock is usually kept as small as possible. Figure 8.7c shows the logic symbol for the JK master–slave flip-flop. Note that the circle for the *CK* input denotes that a valid new state is available after the negative edge of clock.

As with the RS flip-flop, both edge-triggered and master–slave JK flip-flops can have direct preset and clear inputs which override the clock. For example, JK flip-flops in the TTL series are available as negative-edge-triggered (74LS73), positive-edge-triggered (7470), and master–slave (7473) types.

In a *toggle* or *T flip-flop,* the output state changes, or alternates, at every active edge or level of clock. A T flip-flop can be implemented using a JK flip-flop by making $J = K = 1$. T flip-flops are used as counters and frequency dividers.

Another commonly used flip-flop is the *D flip-flop,* which has only one input besides the clock. When clocked, a D flip-flop transfers data from its input (D) to the output; that is, the flip-flop implements the logic $Q_n = D$. In the latch shown in Figure 8.8a, the output is held at its previous value, and the clock is used to enable data transfer. This circuit is known as a *transparent data latch.* (Note that the transparent D latch can be built from an RS latch (Figure 8.4) using $S = R = D$.) Transparent latches connected in parallel are used for the

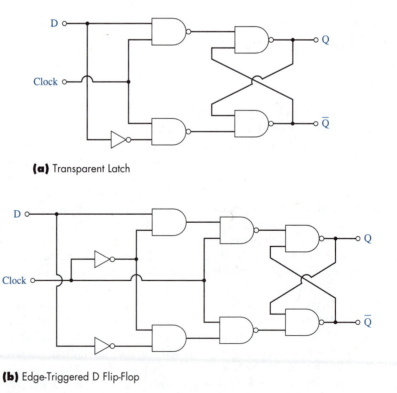

(a) Transparent Latch

(b) Edge-Triggered D Flip-Flop

Figure 8.8   *D* flip–flop

temporary storage of multibit data. Note that changes in the D input are reflected at the output when the clock is active (high) in Figure 8.8a.

As with the JK flip-flop, changes in the output when the clock is active are prevented by edge-triggering the D flip-flop. Figure 8.8b shows a positive-edge-triggered D flip-flop. In this circuit, the D input during the positive edge of the clock is transferred to the output and held until the next positive edge. Hence, there is a delay of one clock period between the input data and the output. For this reason, the D flip-flop is known as the *data* or *delay flip-flop*. As with edge-triggered JK flip-flops, setup and hold times must be observed for valid data entry. D flip-flops are also available in the master–slave configuration. Note that regardless of the circuit configuration, the logic implemented in a D flip-flop is $Q_n = D$; that is, the next state is the same as the present input.

In the TTL SSI series, for example, 7474 has two positive-edge-triggered D flip-flops with independent preset and clear inputs, while the MSI series 74LS378 has eight D flip-flops. For data latching, the MSI series 7475 contains four D latches.

Apart from data storage applications, cascaded flip-flops are used in such MSI functions as counters and shift registers. Problems 8.7 and 8.8 consider ripple and synchronous counters. An example of a four-bit shift register using D flip-flops is shown in Figure 8.9. All four flip-flops are cleared by taking the clear line low. Then, selected flip-flops are preset with 1's by presenting the data $(b_3, b_2, b_1$ and $b_0)$ and taking the parallel load line momentarily high and then bringing it back to low. Data entered in the register can be read in parallel at the Q outputs. Alternatively, data can be read serially at the output of flip-flop A, which changes at the rate of the clock. In addition, new data can be entered serially at the input and shifted at the rate of the clock. As the clock is applied the register state (the output of the flip-flops) changes as shown in Figure 8.9. Here, $s_k, k = 1, 2, \ldots$ refer to serial data entered during the kth clock interval. Note that the parallel data $(b_3, b_2, b_1,$ and $b_0)$ entered asynchronously are available sequentially at Q_A. That is, the shift register of Figure 8.9 performs parallel-to-serial data conversion.

In the second application, the shift register is used as a sequential access memory. In this application, data entered at the serial input of an N-bit register are available after N clock intervals. With the output of the last flip-flop (A in Figure 8.9) connected to the serial input, the shift register recirculates the data. Data refreshing in display units, for example, uses recirculating registers.

Shift registers are also used to implement special counters and random number generators. Reference 7, for example, discusses these applications of shift registers using TTL devices.

As indicated in our discussions, latches and flip-flops, which form the basic blocks in registers, can be implemented from basic NOR and NAND gates from any logic family. In addition, MOSFET dynamic gates can be used to implement flip-flops for reduced power dissipation. Regardless of the technology, however, the basic bistable circuit need not be built by interconnecting SSI gates. In the following sections, we consider the direct implementation of bistable circuits using bipolar and MOS devices.

	Q_D	Q_C	Q_B	Q_A
Initial state (Before clock)	b_3	b_2	b_1	b_0
After clock #1	s_1	b_3	b_2	b_1
After clock #2	s_2	s_1	b_3	b_2
After clock #3	s_3	s_2	s_1	b_3
After clock #4	s_4	s_3	s_2	s_1
After clock #5	s_5	s_4	s_3	s_2
.				
.				
.				
.				
.				

(a)

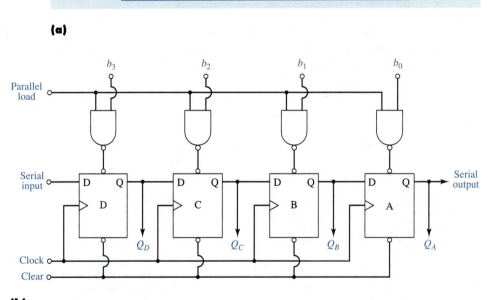

(b)

Figure 8.9   Example of a shift register

8.1.2   BJT BISTABLE MULTIVIBRATORS AND FLIP-FLOPS

A bistable latch using two BJT inverters of the RTL family is depicted in Figure 8.10. Resistors R_C and R_B are chosen such that each inverter, when disconnected from the other, can be driven to saturation with a base current of approximately $\dfrac{V_{CC}}{[R_C + R_B]}$.

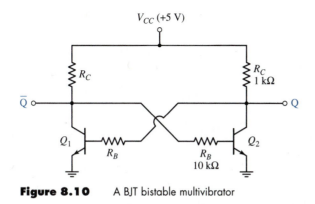

Figure 8.10 A BJT bistable multivibrator

To analyze this circuit, we assume that the bistable is in the stable state with Q_1 in saturation. Since V_{C1} is negligibly small, Q_2 is cut off. With no collector current for Q_2, the base drive for Q_1 is provided by R_C and R_B. This results in a collector voltage of less than V_{CC} for the off transistor Q_2. In the following example, we calculate stable-state voltages and currents using the nominal saturation voltages of $V_{BE(sat)} = 0.8$ V and $V_{CE(sat)} = 0.1$ V.

Calculate the stable-state collector and base voltages and currents for the BJT bistable circuit shown in Figure 8.10. Use a minimum β of 50 for both transistors. | **Example 8.1**

Solution

Assume that Q_1 is in saturation, so that

$$V_{B1} = V_{BE(sat)} = 0.8 \text{ V}, \quad \text{and} \quad V_{C1} = 0.1 \text{ V}$$

Since V_{C1} is below the cut-in voltage of Q_2, Q_2 is off. Therefore,

$$V_{B2} = V_{C1} = 0.1 \text{ V}$$

From the base circuit of Q_1, we have

$$I_{B1} = \frac{V_{CC} - V_{BE(sat)}}{R_B + R_C} = 0.38 \text{ mA}$$

and, from the collector circuit of Q_1, we have

$$I_{C1} = \frac{V_{CC} - V_{CE(sat)}}{R_C} = 4.9 \text{ mA}$$

With a minumum β of 50, the above currents ensure that Q_1 is indeed in saturation; consequently, Q_2 is cut off.

The collector voltage of Q_2 is

$$V_{C2} = V_{CC} - I_{B1} R_C = 4.62 \text{ V}$$

Hence, the stable-state voltages are: $V_{C1} = 4.62$ V, $V_{B1} = 0.8$ V, $V_{C2} = 4.62$ V, and $V_{B2} = 0.1$ V. The currents are: $I_{B1} = 0.38$ mA, $I_{C1} = 4.9$ mA, and $I_{B2} = I_{C2} \approx 0$. The logic high voltage V_{OH} is 4.62 V, and the logic low voltage V_{OL} is 0.1 V.

Because of the symmetry in the circuit, the preceding stable-state voltages and currents are the same for the second stable state in which Q_1 is off and Q_2 is in saturation. Thus, the symmetrical bistable circuit has complementary outputs in voltages, currents, and logic states. When a load is connected to each collector (such as a light-emitting diode to indicate the logic state, or another identical bistable circuit), V_{OH} drops and the base drive for the on transistor is reduced. In Example 8.2 we consider the BJT bistable outputs with loads connected.

Example 8.2

Determine the effect of loading the bistable circuit of Figure 8.10 with a 1 kΩ resistor connected at each of the collectors to ground as shown in Figure 8.11. Each of these resistors represents the approximate resistance seen at the collector of an identical bistable circuit in the off state.

Solution

In Figure 8.11, we assume the stable state is Q_1 in saturation and Q_2 cut off. With $V_{C1} = 0.1$ V, the load resistor R_L at the collector of the saturated transistor Q_1 has negligible effect on the collector current of Q_1.

Because of R_L at the collector of Q_2, current I_1 through the collector resistor of Q_2 is increased from its no-load value; hence, collector voltage V_{C2} is reduced. If the load current is I_2, the new base current for Q_1 is given by

$$I_{B1} = I_1 - I_2 = \frac{V_{CC} - V_{C2}}{R_C} - \frac{V_{C2}}{R_L}$$

or

$$I_{B1} = 5 - 2V_{C2} \qquad\qquad\qquad (8.1)$$

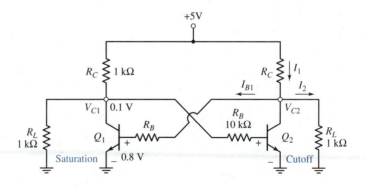

Figure 8.11   Bistable circuit with symmetrical loads connected

With Q_1 in saturation, I_{B1} is also given by

$$I_{B1} = \frac{V_{C2} - V_{BE2(sat)}}{R_B}$$

or

$$I_{B1} = \frac{V_{C2} - 0.8}{10} \tag{8.2}$$

Solving Equations (8.1) and (8.2), we have

$$V_{C2} = 2.42 \text{ V}, \quad \text{and} \quad I_{B1} = 0.16 \text{ mA}$$

We can readily verify that at this lower base drive, Q_1 is still in saturation for $\beta \geq 50$ and hence Q_2 is cut off, as assumed.

The load current is

$$I_2 = \frac{V_{C2}}{R_L} = 2.42 \text{ mA}$$

The effect of load, therefore, is to reduce the logic high voltage from a no-load value of 4.62 V to 2.42 V for a 1 $k\Omega$ load. A further increase in load current, that is, a decrease in R_L, reduces the base current for the conducting transistor. Therefore, R_L is limited by the minumum base drive needed to keep the transistor in saturation.

Other discrete bistable circuit configurations using BJTs are considered in Problems 8.5 and 8.6.

Triggering   The state of the BJT bistable circuit may be changed by applying a negative-going triggering signal to the base of the saturated transistor. Figure 8.12 shows a circuit configuration for triggering a BJT latch. In this circuit, reset (R) and set (S) triggering signals are applied at the bases of the transistors Q_R and Q_S. We can readily verify that the circuit implements the NOR latch shown in Figure 8.2. If Q_1 is on and Q_2 is off (so that $Q = 0$ and $\overline{Q} = 1$), a

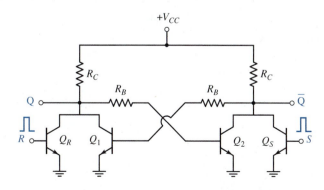

Figure 8.12   Triggering of the BJT bistable circuit

short-duration positive pulse applied to the R input has no effect. A positive pulse applied to the S input turns Q_S on and reduces the voltage at the collectors of Q_2 and Q_3 to 0.1 V. This voltage drop from V_{CC} to 0.1 V, cuts off the base drive to the on transistor Q_1. By regenerative action, then, Q_1 is quickly turned off, while Q_2 goes into saturation.

The duration of the trigger pulse at the S input must be at least equal to the loop delay consisting of the switching delays of Q_S, Q_1, and Q_2, and the pulse amplitude must be sufficient to turn Q_S on (or cause a significant drop in voltage at the collector of Q_S). Once $Q = 1$ and $\overline{Q} = 0$, only a trigger pulse applied to the R input can cause another state transition.

Observe that a positive trigger pulse is applied (via Q_S or Q_R) to the off transistor to cause a state transition. Equivalently, a large negative pulse may be applied to the base of the on transistor, Q_1 or Q_2. Narrow pulses of sufficient negative amplitude may also be applied to the collector of the off transistor to turn the on device off. However, trigger signals applied to the base can be much smaller in amplitude, because of base-to-collector voltage gain of a transistor.

If the trigger signal is a level voltage (as from another bistable circuit, for example) instead of a pulse, we may use an RC differentiator to obtain a narrow pulse.

For routing the trigger signal to the R or S input, depending on the present state, we use a symmetrical triggering scheme such as the one shown in Figure 8.13. The RC differentiators ($R_S C_S$ and $R_R C_R$) in this circuit provide sharp pulses at both the positive and negative edges of the trigger signal T. Depending on whether Q_1 or Q_2 is off, Q_R or Q_S is turned on by the narrow pulse at the positive edge of T, and the bistable circuit is reset or set. Thus, a state transition occurs at the positive edge of the trigger input T. As we saw earlier, in IC

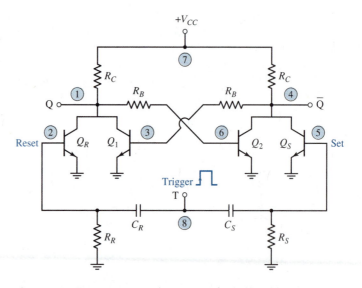

Figure 8.13   Symmetrical triggering of a BJT bistable circuit

flip-flops a narrow pulse at the positive edge of the trigger is obtained by passing the trigger and its inverted version (complement) through a gate.

In the discrete circuits shown in Figures 8.12 and 8.13, a small capacitor is usually connected across each base resistor R_B. This capacitor aids in speeding up state transitions by abruptly transferring the collector voltage drop of the currently saturated transistor to the base of the previously saturated transistor. The value of the *speed-up,* or *commutating, capacitor* (which may be determined using charge-control analysis) is on the order of 10 pF to 50 pF. In IC flip-flops, however, fast transitions are accomplished by wider pulses (of several nanoseconds) generated at the edges of the trigger input.

Figure 8.14 shows a MicroSim PSpice simulation for the symmetrical triggering circuit of Figure 8.13. The simulation starts with an initial state of Q_1 on. This starting state (or the other state, Q_2 on) must be set by an appropriate node voltage (or initial condition) to speed up the analysis of the regenerative circuit. In Figure 8.14a, the statement NODESET V(1) = 0.1V ensures that Q_1 is initially on. The waveforms in Figure 4.14b show the state transitions that occur at the positive edges of the trigger.

We may use additional transistors in parallel with Q_R and Q_S in the circuit of Figure 8.13 for direct, or manual, setting and clearing of the latch, indepen-

```
*BJT BISTABLE WITH SYMMETRICAL TRIGGER
QR    1    2    0      Q2N3904
Q1    1    3    0      Q2N3904
Q2    4    6    0      Q2N3904
QS    4    5    0      Q2N3904
RC1   7    1    1K
RC2   7    4    1K
RB1   4    3    25K
RB2   1    6    25K
CC1   4    3    22P
CC2   1    6    22P
RR    2    0    50
RS    5    0    50
CR    2    8    22P
CS    5    8    22P
VCC   7    0    DC  5V
VT    8    0    PULSE (0V  3V  100N  0.1N  0.1N  1U  2U)
.OP
.NODESET  V(1) = 0.1V
.TRAN  5N  5U
.LIB  "EVAL.LIB"
.PROBE
.END
```

(a) Input File

Figure 8.14   MicroSim PSpice simulation of the circuit in Figure 8.13

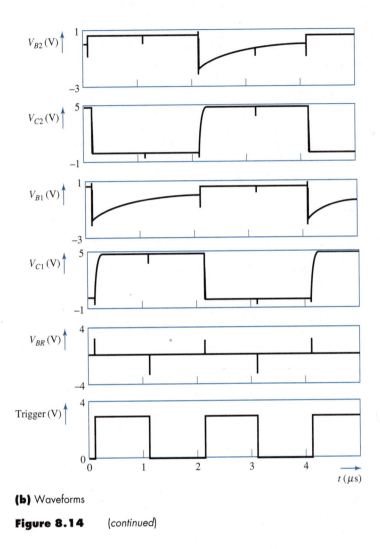

(b) Waveforms

Figure 8.14   (*continued*)

dent of the trigger signal T. Alternatively, the narrow (differentiated) trigger signal may be capacitively coupled to the collectors while Q_R and Q_S are used for direct reset and set. Other triggering schemes for the discrete BJT bistable circuit are discussed in Reference 1.

Symmetrically triggered RS latches are used for counting clock pulses. Observe that the output of the circuit in Figure 8.13 changes state, or toggles, at the occurrence of every trigger pulse; that is, the circuit implements a toggle flip-flop. Since the pulse period at the Q and $\overline{Q}$ outputs is twice that of the input trigger (or clock), the circuit is also a known as a *binary,* or *divide-by-two, counter.* A cascade of N binary counters, knows as a *ripple counter,* may be used to divide the input trigger pulse frequency (that is, multiply the period) by a factor of 2^N. Binary counters are used for pulse counting and waveform synthesis in sequential logic systems. By resetting appropriate flip-flops, based on the present state in a cascade of T flip-flops, the circuit can count in the decimal or another number base. Problem 8.7 considers this application.

In IC technologies, BJT flip-flops use gate configurations similar to those shown in Figures 8.4 and 8.8. Individual gates are realized using appropriate technology (TTL, ALS TTL, ECL, I^2L, etc.), with the circuit configurations simplified to yield high density and fast response. An example of an I^2L D latch is shown in Figure 8.15. Multicollector transistors in Figure 8.15b implement the gate configuration of $G_1 - G_6$ shown in Figure 8.15. Note that the D input, which comes from another I^2L gate, cannot be connected to both G_1 and G_5 without affecting the fan-out of the driving gate. Hence, it is derived twice from the two collectors of the driving gate.

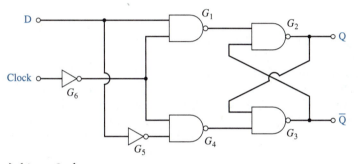

(a) Logic Configuration

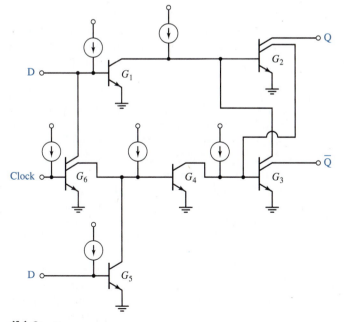

(b) Circuit

Figure 8.15 I^2L D latch

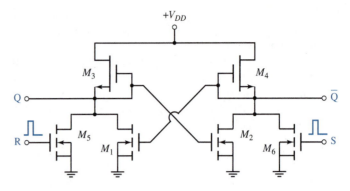

Figure 8.16   NMOSFET bistable multivibrator

8.1.3   MOSFET BISTABLE CIRCUITS

Figure 8.16 shows an implementation of a bistable multivibrator using NMOS-FET NOR gates. Regenerative feedback from the source terminal (output) of each of the DMOSFETs M_3 and M_4 to the gate terminal (input) of the other ENMOSFET ensures that one of the driver transistors (M_1 or M_2) is cut off and the other is in its ohmic mode. ENMOSFETs M_5 and M_6, which are connected in parallel with each of the driver transistors, are used for direct setting and resetting of the latch. The load device can be either a resistor or another EN-MOSFET operating in the pinch-off or ohmic mode. A depletion-type load, however, yields sharper transition characteristics (Section 7.4) than the others. As with active load NMOSFET inverters, the design parameter for the latch is the conduction parameter ratio K_R between the driver transistors (M_1, M_2, M_5, and M_6) and the load transistors (M_3 and M_4). We can readily verify the operation of the circuit.

A CMOS NOR latch is shown in Figure 8.17. As in TTL, a latch can also be implemented using CMOS NAND gates. More efficient realization of flip-flops use CMOS transmission gates (TGs).

Two TGs are used in the basic CMOS D latch shown in Figure 8.18a. The first TG, T_1, is closed when clock CK is low, and the D input is allowed to reach the input of G_1. Variations in D are reflected at the output of G_1 during the low level of CK. When CK goes high, T_1 opens and inhibits input, and T_2 closes and provides feedback to G_1. Hence, the $\overline{Q}$ and Q outputs are stable, corresponding to the previous value (during $CK = 0$) of D. To avoid data loss during switching of the clock CK from low to high, a buffered clock and its complement are used, as shown.

In the D flip-flop of Figure 8.18b, two cascaded TG latches allow the entry of input and output data. The master, or input, latch is activated during the low level of CK, while T_3 disconnects the output from the input. The Q and $\overline{Q}$ outputs correspond to the value of D during the previous low level of CK. When CK goes high, T_3 transmits the value latched by the master to the slave, or output, latch. At this time the input is disconnected by T_1. Hence, only one latch stores data at either level of the clock. Note that a new value of D is available at the Q output

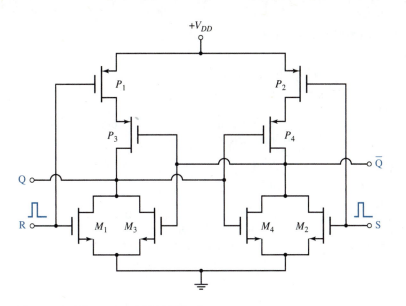

Figure 8.17   A CMOS NOR latch circuit

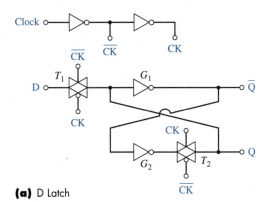

(a) D Latch

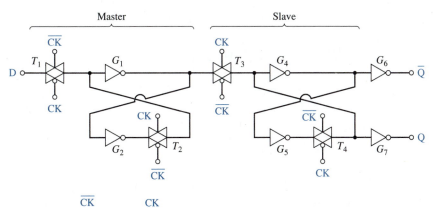

(b) D Flip-Flop

Figure 8.18   TG based CMOS flip-flops

at the positive edge of the clock; the circuit, however, is designed as a cascaded master–slave flip-flop. Data D must be stable during the setup time, which is the duration prior to the positive edge of the clock.

Direct reset and set inputs can be provided by replacing all the inverters in Figure 8.18b with two input gates (Problem 8.7).

8.1.4 SCHMITT TRIGGER

A Schmitt trigger[2] is a bistable circuit in which the logic state of the output depends on the amplitude and direction of an input signal. Figure 8.19a depicts the BJT version of a Schmitt trigger. The transfer characteristic of this Schmitt

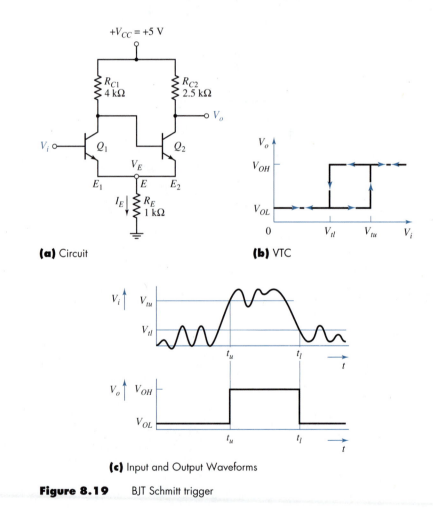

(a) Circuit   **(b)** VTC

(c) Input and Output Waveforms

Figure 8.19   BJT Schmitt trigger

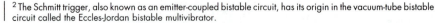

[2] The Schmitt trigger, also known as an emitter-coupled bistable circuit, has its origin in the vacuum-tube bistable circuit called the Eccles-Jordan bistable multivibrator.

trigger, shown in Figure 8.19b, exhibits *hysteresis,* or memory; that is, for a given input voltage V_i in the range $V_{tl} < V_i < V_{tu}$, the output is at either logic high or logic low, depending on whether V_i is decreasing or increasing.

Schmitt triggers are primarily used to convert slowly varying voltage waveforms to well-defined logic levels with sharp transitions. Because of hysteresis, the effect of any noise voltage within the threshold levels V_{tl} and V_{tu} is eliminated in the logic output. Figure 8.19c shows the input and output waveforms of a Schmitt trigger. With appropriate input and driver circuitry, Schmitt trigger logic gates are available in the TTL and CMOS families of integrated circuits.

We design the circuit in Figure 8.19a such that, for $V_i = 0$, Q_1 is off and Q_2 is in saturation. Note that with Q_1 off, the base current for Q_2 is supplied via R_{C1}, and R_E carries the saturation emitter current of Q_1. Our analysis of this circuit proceeds with this initial state of Q_1 off and Q_2 in saturation. The output voltage V_o is at the logic low level given by

$$V_o = V_{C2} = V_{OL} = V_{CC} - I_{C2}R_{C2}$$

Since the emitter current of Q_2 flows through R_E, the common emitter node E is at some positive voltage V_E. Therefore, Q_1 stays in cutoff for V_i below its cut-in voltage of $V_E + 0.6$ V. As we increase V_i, at $V_i = V_E + 0.6$ V, Q_1 begins to conduct.

The voltage drop in V_{C1}, due to the collector current of Q_1, reduces the base drive for Q_2. If R_{C1} is sufficiently large, the drop in V_{B2} ($= V_{C1}$) takes Q_2 out of saturation, and the emitter current I_{E2} decreases. This decrease in I_{E2} raises V_{BE1}. Consequently, I_{C1} increases and causes a further drop in V_{C1}. Eventually, the regenerative action takes Q_1 to saturation and, with $V_{BE2} = V_{CE1} \approx 0.1$ V, Q_2 is cut off. The output voltage V_o is then at $V_{OH} = V_{CC}$. The input voltage V_i at which the output V_o makes a transition from low to high level is the *upper threshold voltage,* denoted by V_{tu}, as is shown in Figure 8.19b. A further increase in V_i beyond V_{tu} drives Q_1 heavily into saturation, and the output voltage remains at V_{OH}.

If we decrease V_i from its high level, with $R_{C1} > R_{C2}$, Q_1 stays in saturation for $V_i < V_{tu}$. A further drop in V_i causes both the emitter voltage V_E and the collector voltage V_{C1} to fall, and Q_1 comes out of saturation. Device Q_2 remains cut off until V_{BE2} equals the cut-in voltage of Q_2. At $V_i = V_{tl}$, which is the *lower threshold voltage,* Q_1 is in the active mode, with $V_{CE1} = V_{BE2} \approx 0.6$ V. At this voltage, Q_2 begins to conduct. The emitter current of Q_2 causes an increase in V_E. With V_i kept constant at V_{tl}, V_{BE1} in turn decreases, and the regenerative action rapidly drives Q_1 to cutoff and Q_2 into saturation. A further decrease in V_i below V_{tl} takes Q_1 more into cutoff, with no change in Q_2, and the output stays at its logic low level. Hence, $V_o = V_{OL}$ for $V_i \leq V_{tl}$.

Observe that for $V_{tl} < V_i < V_{tu}$, the output is $V_o = V_{OL}$ if V_i is increasing, and it is $V_o = V_{OH}$ if V_i is decreasing. If V_o is at logic high for an input voltage $V_i > V_{tu}$, the transition to $V_o = V_{OL}$ occurs when V_i crosses V_{tu} and decreases below V_{tl}. Similarly, V_o switches from low to high only after V_i passes through V_{tl} and goes above V_{tu}. The difference in the input threshold voltages, ($V_{tu} - V_{tl}$), is known as the *hysteresis voltage.* As with the bistable multivibrator, regeneration and rapid state transition occur only when the loop gain $\dfrac{dV_0}{dV_i}$ is greater than unity.

The resistor values given in Figure 8.19a are typical for a TTL IC version of the Schmitt trigger. We calculate the upper and lower threshold voltages for this circuit in the following example.

Example 8.3

Calculate the upper and lower threshold voltages for the Schmitt trigger circuit shown in Figure 8.19a.

Solution

For $V_i \approx 0$ and increasing, $V_o = V_{OL}$ and Q_1 is cut off, while Q_2 is in saturation. Hence, $V_{CE2} \approx 0.1$ V   and $V_{BE2} \approx 0.8$ V.

At the emitter of Q_2, $I_E = I_{E2}$, which is given by

$$I_E = I_{B2} + I_{C2}$$

or,

$$\frac{V_E}{R_E} = \frac{V_{CC} - (V_E + 0.8)}{R_{C1}} + \frac{V_{CC} - (V_E + 0.1)}{R_{C2}}$$

Solving for the emitter voltage V_E, we have

$$V_E = 1.82 \text{ V}$$

Hence,

$$V_o = V_{OL} = V_E + V_{CE2} = 1.92 \text{ V}$$

With a base-emitter cut-in voltage of 0.6 V, Q_1 turns on, and the output switches to the logic high level when

$$V_i = V_{tu} = V_E + 0.6$$

Hence, the upper threshold voltage is

$$V_{tu} = 2.42 \text{ V}$$

At this input, Q_1 goes into saturation and Q_2 is cut off. This causes

$$V_o = V_{OH} = V_{CC} = 5 \text{ V}$$

The emitter voltage in this state depends on the input (base) current of Q_1.

As V_i is decreased, Q_1 enters the active mode, and Q_2 begins to conduct when $V_{BE2} = V_{CE1} \approx 0.6$ V. At this transition point, Q_1 is on the verge of turning off. Therefore,

$$I_{C1} \approx I_E = \frac{V_{CC} - 0.6}{R_{C1} + R_E} = 0.88 \text{ mA}$$

and

$$V_E = 0.88 \text{ V}$$

Hence, the lower threshold voltage is

$$V_i = V_{tl} = V_E + 0.7 = 1.58 \text{ V}$$

The hysteresis voltage is given by

$$V_H = 2.42 - 1.58 = 0.84 \text{ V}$$

We can reduce the hysteresis voltage to zero by adding a resistor between the emitter of Q_1 (or Q_2) and the point E in Figure 8.19a. Note that a resistor connected between E and E_1 increases V_{tl} without altering V_{tu}, while a resistor in series with the emitter of Q_2 brings V_{tu} down. Problem 8.12 considers this property of the Schmitt trigger.

TTL IC Schmitt triggers are available as inverters (the 7414, for example), and NAND gates (7413 and 74132) with totem-pole outputs. These gates typically have a hysteresis of 0.8 V.

8.1.5 CMOS SCHMITT TRIGGER INVERTER

Figure 8.20a shows the circuit of a basic CMOS Schmitt trigger inverter consisting of three NMOS and three PMOS transistors. The VTC of this circuit is shown in Figure 8.20b. The two threshold voltages in the transfer characteristic are indicated by the stylized symbol for the inverter, shown in Figure 8.20c.

We simplify the analysis of the circuit by neglecting the body effect and using device threshold voltages of the same magnitude for both the N and the P

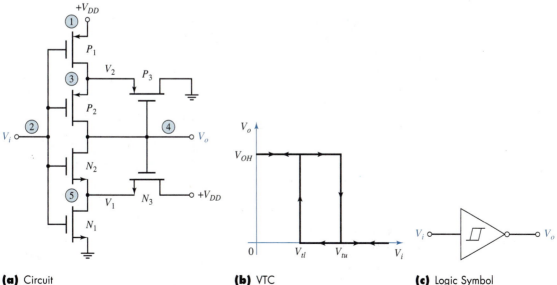

(a) Circuit   **(b)** VTC   **(c)** Logic Symbol

Figure 8.20   CMOS Schmitt trigger inverter

channel transistors. Assuming a zero body effect avoids iterative computations for the node voltages while closely approximating the actual values.

Let $V_{DD} = 5$ V and the threshold voltages be $V_{TN} = -V_{TP} = 2$ V. For $V_i = 0$, which is below V_{TN}, the NMOS devices N_1 and N_2 are off, and the PMOS devices P_1 and P_2 are on. There is negligible current from V_{DD} to ground via the chain P_1, P_2, N_2, and N_1; hence, the drain voltages V_2 and V_o are at V_{DD}. Devices N_3 and P_3 provide regenerative feedback to the input devices. Since the gate and the source of the PMOS device P_3 are at V_{DD}, P_3 is off. With $V_{G3} = V_o \approx 5$ V and $V_D = V_{DD} = 5$ V, the NMOS device N_3 is in saturation. Therefore, V_1, the source voltage of N_3, is limited to $V_{1max} = V_{DD} - V_{TN} = 3$ V. Devices N_3 and N_1 form an inverting amplifier, with V_1 dropping according to the following (see Equation 7.13):

$$V_1 = V_{DD} - V_{TN} - \sqrt{\frac{k_{n1}}{k_{n3}}}(V_i - V_{TN}) \tag{8.3}$$

As V_i increases above $V_{TN} = 2$ V, N_1 turns on and N_2 stays off. While V_o stays at logic high (V_{DD}), V_1 now falls according to Equation 8.3.

Device N_2 turns on when

$$V_{GS2} = V_i - V_1 = V_{TN} \tag{8.4}$$

Hence, at $V_i = V_1 + V_{TN}$, both N_1 and N_2 conduct. Let us look at the mode of operation for N_1 at this input. Since N_2 is beginning to conduct, from Equation 8.4 we have

$$V_i - V_{DS1} = V_{TN}$$

or $$V_{DS1} = V_i - V_{TN}$$

This puts N_1 at the edge of the pinch-off region.[3] With both N_1 and N_2 conducting, the output voltage V_o drops to zero. Therefore, the upper threshold voltage $V_i = V_{tu}$ at which the output switches from high to low (because of inverter action) corresponds to N_2 beginning to conduct while N_1 and N_3 are in saturation. The value of V_{tu} is calculated by equating the saturation drain currents in N_1 and N_3 just prior to the output transition.

The current in N_3 is

$$I_{D3} = k_{n3}(V_{DD} - V_1 - V_{TN})^2$$

which, from Equation 8.4, becomes

$$I_{D3} = k_{n3}(V_{DD} - V_i)^2 \tag{8.5}$$

The current in N_1 is

$$I_{D1} = k_{n1}(V_i - V_{TN})^2 \tag{8.6}$$

---

[3] Because of the body effect, the threshold voltage of N_2 is slightly higher than the zero bias voltage V_{T0}. With source and body connected together, N_1 has no body effect and its threshold voltage is constant at V_{T0}. Hence, N_1 is actually in the ohmic mode.

Setting $I_{D3} = I_{D1}$, we have

$$V_i = V_{tu} = \frac{V_{DD} + V_{TN}\sqrt{\dfrac{k_{n1}}{k_{n3}}}}{1 + \sqrt{\dfrac{k_{n1}}{k_{n3}}}} \qquad (8.7)$$

For $V_{TN} = 2$ V, $V_{DD} = 5$ V and the same conduction parameter $k_{n1} = k_{n3}$, Equation 8.7 gives $V_{tu} = 3.5$ V.

This value of upper threshold voltage is slightly less than the actual voltage, since the body effect in N_2 requires a higher input voltage V_i to turn on.

For V_i above V_{tu}, the output voltage remains low, with all the NMOS devices conducting and all the PMOS devices off.

When V_i decreases from V_{DD}, a similar process results in the reversal of the operation of the PMOS and the NMOS devices. Device P_1 turns on first, and V_2 rises linearly with decreasing V_i, since P_1 and P_3 form an inverting amplifier. However, P_2 does not conduct until V_i is reduced to V_{tl}, as given by

$$V_i = V_{tl} = V_2 - |V_{TP}| \qquad (8.8)$$

At this input P_1, P_2, and P_3 conduct, and V_o rises to V_{DD}. Hence, the lower threshold voltage V_{tl} is determined by equating the currents in P_1 and P_3 at the onset of conduction of P_2. Assuming constant threshold voltages and $V_o \approx 0$, we can show that P_1 is at the edge of pinchoff and P_3 is in saturation. Equating the currents in P_1 and P_3 and using Equation (8.8), we obtain

$$V_{tl} = \frac{(V_{DD} - |V_{TP}|)\sqrt{\dfrac{k_{p1}}{k_{p3}}}}{1 + \sqrt{\dfrac{k_{p1}}{k_{p3}}}} \qquad (8.9)$$

For $k_{p1} = k_{p3}$ and $V_{TP} = -2$ V, we have $V_{tl} = 1.5$ V. As with V_{tu}, Equation 8.9 underestimates the actual value of V_{tl}.

Equations (8.7) and (8.9) are the design equations for the CMOS Schmitt trigger. Since the conduction parameter k of a MOSFET depends on the width-to-length ratio W/L [Equation (6.9)], the above equations may be rearranged as follows:

$$\frac{(W/L)_{n1}}{(W/L)_{n3}} = \left[\frac{V_{DD} - V_{tu}}{V_{tu} - V_{TN}}\right]^2 \qquad (8.10)$$

for the NMOS transistors, and

$$\frac{(W/L)_{p1}}{(W/L)_{p3}} = \left[\frac{V_{tl}}{V_{DD} - V_{tl} - |V_{TP}|}\right]^2 \qquad (8.11)$$

for the PMOS transistors.

Equations (8.10) and (8.11) specify the device dimensions for given upper and lower threshold voltages. Example 8.4 illustrates the design of a CMOS Schmitt trigger.

Example 8.4 | **A** CMOS Schmitt trigger with an upper threshold voltage of 3 V and a lower threshold voltage of 1.5 V is to be designed for use with $V_{DD} = 5$ V and $V_{TN} = -V_{TP} = 1$ V. Specify the aspect ratios for the MOSFET devices. Verify the design using a MicroSim PSpice simulation.

Solution

For the NMOS transistors, we have, from Equation 8.10,

$$\frac{(W/L)_{n1}}{(W/L)_{n3}} = \left[\frac{5-3}{3-1}\right]^2 = 1$$

```
* CMOS SCHMITT TRIGGER      VTC
MP1    3    2    1    1     PCH
MP2    4    2    3    1     PCH
MP3    0    4    3    1     PCH L = 5U  W = 15U
MN3    1    4    5    0     NCH
MN2    4    2    5    0     NCH
MN1    5    2    0    0     NCH
VDD    1    0              DC 5
VI     2    0              PWL (0  0  10U  5V  20U  0)
.PROBE
.MODEL PCH PMOS (VTO = -1   KP = 8E-6   GAMMA = 0.4)
.MODEL NCH NMOS (VTO = 1    KP = 20E-6  GAMMA = 0.4)
.TRAN/OP 0.25U 20U
.OPTIONS DEFL = 5U DEFW = 5U
.END
```

(a) Input File

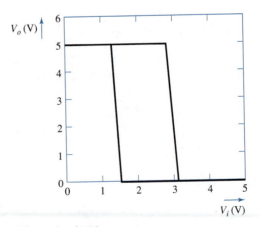

(b) Simulated VTC

Figure 8.21   MicroSim PSpice simulation of the CMOS Schmitt trigger of Example 8.4

For the PMOS transistors, we have, from Equation (8.11),

$$\frac{(W/L)_{p1}}{(W/L)_{p3}} = \left[\frac{1.5}{5 - 1.5 - 1}\right]^2 = 0.36$$

Hence, an aspect ratio of $W/L = 3$ may be used for P_3, and a ratio of unity may be used for all other transistors.

Figure 8.21 shows the MicroSim PSpice simulation of the designed CMOS Schmitt trigger, using typical process transconductance parameters and body effect coefficient. Recall that the parameter KP in the simulation file corresponds to $2\mu C_{ox}$, so that the device transconductance parameter is $KP(W/L)$. (The graph of the voltage transfer characteristics is obtained from the results of the transient analysis for a piecewise linear input, changing the x-axis to V_i instead of time t.) The simulated voltage transfer characteristic shows $V_{tl} \approx 1.4$ V and $V_{tu} \approx 3.1$ V.

In an IC Schmitt trigger circuit, the output V_o shown in Figure 8.20 is passed through an inverting buffer and a second inverter driver stage, to increase the fan-out. Also, the input V_i is applied via a diode protection circuit, which changes the threshold voltages slightly. Typical CMOS Schmitt trigger gate threshold voltages are 2.5 V for V_{tu} and 1.6 V for V_{tl}.

8.2   MONOSTABLE MULTIVIBRATOR

A monostable multivibrator, commonly called a *one-shot,* produces a fixed-duration pulse at the activating edge of an input trigger. As the name implies, this multivibrator has a single stable state, from which it makes a transition to a quasistable state with the application of a trigger. After a specific interval, the circuit returns to its initial stable state. The duration of the quasistable state, which is the pulse width of the one-shot, is determined by the circuit parameters. Monostable multivibrators are used to generate single pulses of a fixed width from a much larger or smaller pulse or level signals. We consider the implementation of the one-shot using BJTs and MOSFETs in the next section.

8.2.1   MONOSTABLE MULTIVIBRATOR BJT

A discrete monostable multivibrator using BJTs is shown in Figure 8.22a. Observe that this circuit configuration is similar to that of the discrete BJT bistable multivibrator of Figure 8.10, except that the output (collector) of the first inverter is capacitively coupled to the input (base) of the second inverter. Transistor Q_T is used to amplify the trigger signal.

We design the circuit so that for $V_i = 0$ V, it is in its stable state with Q_1 cut off and Q_2 in saturation. When a positive-going trigger signal of sufficient amplitude is applied at $t = 0$, Q_T is turned on, causing V_{C1} to fall to $V_{CE(sat)} \approx 0.1$ V. The drop in V_{C1} is transferred by the timing capacitor C to the base

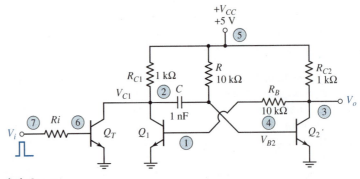

(a) Circuit

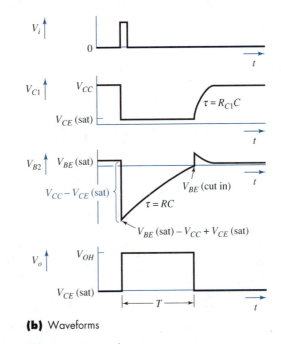

(b) Waveforms

Figure 8.22   A BJT one-shot circuit

of Q_2. Device Q_2, therefore, turns off rapidly by the regenerative action, and the output voltage V_o rises from logic low ($V_{CE(sat)}$) to logic high ($< V_{CC}$). The circuit is now in its quasistable state, with Q_1 on and Q_2 off. In this state, the capacitor C charges toward V_{CC} via R and Q_1. The base voltage of Q_2 rises from an initial value of $V_{BE(sat)} - (V_{CC} - V_{CE(sat)})$ to a final value of V_{CC} with a time constant of RC. Neglecting the small on-resistance of Q_1, this rise in base voltage V_{B2} is given by

$$V_{B2} = V_{CC} + [V_{BE(sat)} + V_{CE(sat)} - 2V_{CC}]e^{-t/RC} \qquad \textbf{(8.12)}$$

At $t = T$, V_{B2} reaches the cut-in voltage of Q_2. The regenerative process then turns Q_2 on and Q_1 is cut off; the circuit therefore returns to its stable state.

From Equation (8.12), the duration of the quasistable state T, during which the output is at logic high, is given by

$$T = RC \ln\left(\frac{2V_{CC} - V_{BE(sat)} - V_{CE(sat)}}{V_{CC} - V_{BE(cut-in)}}\right)$$

For V_{CC} above a few volts, this timing equation may be approximated by

$$T \approx RC \ln 2 \qquad\qquad \textbf{(8.13)}$$

The design of a BJT one-shot for a given pulse width is governed by Equation (8.13). From the RC product obtained in Equation (8.13), we choose R such that $R < \beta R_{C2}$ to keep Q_2 in saturation in the stable state. Also, capacitor C must be a low-leakage device.

The collector and base voltage of the BJT one-shot are calculated as illustrated in the following example.

For the given circuit parameters of the one-shot shown in Figure 8.22, calculate and sketch the stable and quasistable state voltages at the collector and base of the two transistors. Use a β of 100 for both transistors.
What is the pulse width of the one-shot output?

Example 8.5

Solution

In the stable state, the saturation base current of Q_2 is supplied via R, as given by

$$I_{B2} = \frac{V_{CC} - V_{BE2(sat)}}{R} = 0.42 \text{ mA}$$

With a collector current of

$$I_{C2} = \frac{V_{CC} - V_{CE2(sat)}}{R_{C2}} = 4.9 \text{ mA},$$

Q_2 is in saturation for $\beta = 100$. Therefore Q_1 is cut off, with

$$V_{BE1} = V_{CE2(sat)} = 0.1 \text{ V} \qquad \text{and} \qquad V_{CE1} = V_{CC} = 5 \text{ V}$$

Hence, the stable-state voltages are:

$$V_{B1} = 0.1 \text{ V}, \quad V_{C1} = 5 \text{ V}, \quad V_{B2} = 0.8 \text{ V}, \quad \text{and} \quad V_{C2} = V_o = 0.1 \text{ V}$$

In the quasistable state, Q_1 is in saturation, with a base current of

$$I_{B1} = \frac{V_{CC} - V_{BE1(sat)}}{R_{C2} + R_B} = 0.38 \text{ mA}$$

and a collector current of

$$I_{C1} = \frac{V_{CC} - V_{CE1(sat)}}{R_{C1}} = 4.9 \text{ mA}$$

Note that this current neglects the charging current of the capacitor.

The collector voltage of Q_2 is

$$V_{C2} = V_{CC} - I_{B1}R_{C2} = 4.62 \text{ V}$$

The base voltage of Q_2 rises exponentially with a time constant of $\tau = RC = 10 \ \mu s$ from $V_{B2}(0^+) = 0.8 - 4.9 = -4.1$ V to $V_{B2}(\infty) = 5$ V as

$$V_{B2} = 5 - 9.1e^{-t/\tau}$$

The voltages in the quasistable state are:

$$V_{B1} = 0.8 \text{ V}, \quad V_{C1} = 0.1 \text{ V}, \quad V_{C2} = V_o = 4.62 \text{ V}$$

and V_{B2} rises exponentially as indicated.

The time T at which V_{B2} reaches the cut-in voltage of $V_{BE2(\text{cut-in})} = 0.65$ V is given by

$$T = 10 \ \ln(9.1/4.35)\mu s = 7.38 \ \mu s$$

Hence, the pulse width is 7.38 μs at the output.

As Q_1 switches from saturation to cutoff (via the active mode) at $t = T^+$, its collector voltage initially jumps from 0.1 V to a slightly more positive value in the active mode. This increase in V_{C1} is transferred by the timing capacitor to the base of Q_2; hence, Q_2 is driven into heavy saturation, thereby causing a slight overshoot in V_{B2} at $t = T^+$. As the capacitor C charges via Q_2, with a time constant of approximately $R_{C1}C$, the overshoot decays to $V_{BE(\text{sat})}$; V_{C1} returns to V_{CC} with the same time constant. Therefore, the pulse at the collector of Q_1 has the same duration as V_o, but with an exponentially rising positive-going edge. Note further that the negative transition in V_{B2} at the beginning of the quasistable state is approximately equal to $-V_{CC}$. The emitter–base junction of Q_2 must withstand this large reverse bias. Figure 8.22b shows the waveforms for the one-shot.

The transition from the stable to the quasistable state in Example 8.5 above assumes that the input trigger exists at least up to the time Q_2 is turned off, but no later than when Q_1 is turned back on. Figure 8.23 shows a MicroSim PSpice simulation for the one-shot shown in Figure 8.22, with a trigger of sufficiently long duration. Here again, we begin the simulation with the stable initial state of Q_2 in saturation. This initial state is ensured by the NODESET statement. As with the bistable circuit, an RC differentiator may be used to turn Q_T on momentarily, if the trigger pulse width is longer than the output pulse width.

The BJT one-shot can be triggered at the negative edge of a pulse by applying the trigger to the collector of Q_1. The drop in V_{C1} is then coupled via the capacitor C to the base of Q_2 to turn Q_2 off. This triggering method eliminates the trigger amplifier Q_T. However, since the resistance to ground of the collector is very small when Q_1 is on, the trigger source must be buffered. Figure 8.24 shows a simple diode-buffered trigger circuit. The diode conducts during the negative edge of the input, thus triggering the one-shot at the negative edge of V_i.

```
* BJT MONOSTABLE   MULTIVIBRATOR
Q1        2      1    0    Q2N3904
Q2        3      4    0    Q2N3904
QT        2      6    0    Q2N3904
RC1       5      2         1K
RC2       5      3         1K
RTIME     5      4         10K
CTIME     4      2         1000P
RB        3      1         10K
RI        7      6         1K
VCC       5      0         5V
VTRIG     7      0         PULSE (0V  3V  2U  0.5N  0.5N  3U  20U)
.NODESET  V(3) = 0.1V
.LIB "EVAL.LIB"
.TRAN/OP  1U  20U
.PROBE
.END
```

(a) Input File

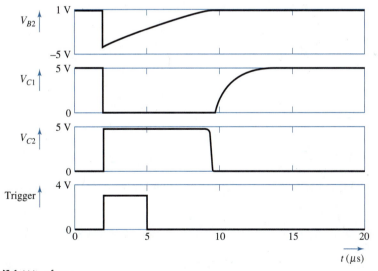

(b) Waveforms

Figure 8.23   MicroSim PSpice simulation of positive-edge triggered BJT one-shot

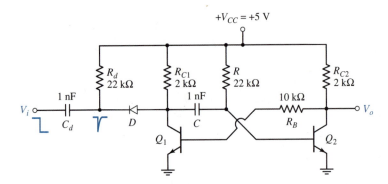

Figure 8.24   A negative-edge triggered BJT one-shot

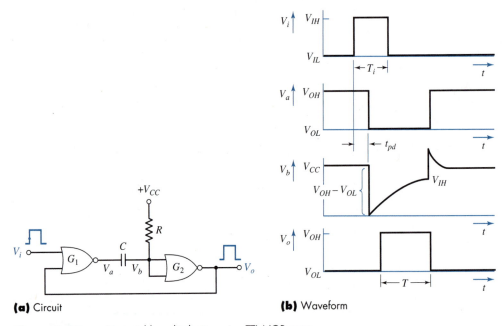

(a) Circuit   **(b)** Waveform

Figure 8.25   Monostable multivibrator using TTL NOR gates

8.2.2 TTL NOR GATE MONOSTABLE CIRCUIT

One-shots can be built using NAND or NOR gates from any of the logic families. Figure 8.25a shows a TTL NOR gate one-shot. When the trigger voltage V_i is at logic low, the circuit is in its stable state, with V_a and V_b at logic high and V_o at logic low. As V_i goes from low to high, which is the activating edge of this one-shot, the output of gate G_1 switches to logic low, after a propagation delay of t_{pd}. This high-to-low transition at V_a is transferred by the capacitor C to the input of gate G_2. After another delay of t_{pd}, V_o switches to logic high. At this point, the input trigger may be returned to logic low. The circuit is now in its quasistable state. Capacitor C charges via R and the low output resistance of G_1, and V_b rises toward V_{CC}. At $t = T$, V_b reaches the logic high threshold V_{IH}, and V_o switches to logic low (after t_{pd}). Voltage V_a therefore returns to logic high (after another delay of t_{pd}), and the circuit is back in its stable state. Figure 8.25b shows the waveforms at V_a, V_b, and V_o after the application of the trigger signal V_i.

Pulse width T is calculated using the charging equation for C, as illustrated in Example 8.6.

Using typical TTL threshold voltages of $V_{OH} = 3.8$ V, $V_{OL} = 0.1$ V, and $V_{IH} = 1.6$ V, determine the pulse width of the NOR gate one-shot shown in Figure 8.25, for $R = 15$ kΩ and $C = 1$ nF. Neglect gate propagation delays. | **Example 8.6**

Solution

After the application of the trigger input at $t = 0$, V_a switches from $V_{OH} = 3.8$ V to $V_{OL} = 0.1$ V. Hence, at $t = 0^+$, V_b jumps from $V_{CC} = 5$ V to

$$V_b(0^+) = V_{CC} - V_{OH} + V_{OL} = 1.3 \text{ V}$$

The capacitor now charges toward $V_b(\infty) = V_{CC} = 5$ V with a time constant of $\tau \approx RC = 15 \ \mu$s.

The charging voltage V_b is given by

$$V_b = V_{CC} + (V_{OL} - V_{OH})e^{-\frac{t}{RC}} \qquad \textbf{(8.14)}$$

or

$$V_b = 5 - 3.7e^{\frac{-t}{\tau}}$$

At $t = T$, V_b reaches V_{IH}. Hence, the pulse width T is given by

$$T = RC \ln \left[\frac{V_{OH} - V_{OL}}{V_{CC} - V_{IH}} \right] \qquad \textbf{(8.15)}$$

or

$$T = 15 \ln \left(\frac{3.7}{3.4} \right) = 1.27 \ \mu\text{s}$$

For proper timing operation, the input trigger width T_i must satisfy $2t_{pd} \leq T_i \leq T + 2t_{pd}$. Otherwise V_a will only be the complement of V_i. Additionally, to ensure that the circuit starts from its stable state, a succeeding trigger signal must not be applied until at least five time constants ($5RC$) after the pulse width T. This delay allows the capacitor to charge to $V_b = V_{CC}$.

From Equation (8.15), it is clear that a slight change in V_{IH}, V_{OH}, or V_{OL} may cause an incorrect pulse width or improper operation of the circuit. Because of this sensitivity to logic levels, the basic TTL gate one-shot is used in pulse stretcher applications where timing is not critical.

A TTL NAND gate one-shot for generating a short pulse at the negative edge of a trigger is shown in Figure 8.26. Gate G_2 shapes the differentiated (positive-going) signal to the TTL signal, and G_1 acts as a buffer for the trigger source. Resistor R is no larger than a few hundred ohms to prevent the input (sourcing) current of G_2 from raising V_b above V_{IL}. Figure 8.26b shows the waveforms for this one-shot.

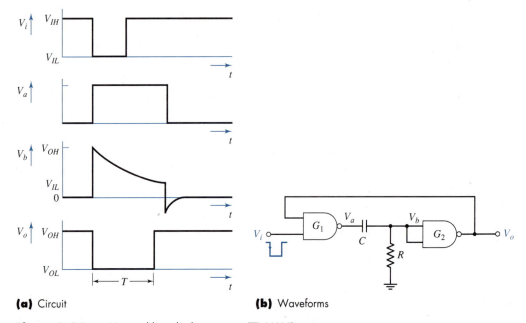

(a) Circuit

(b) Waveforms

Figure 8.26 Monostable multivibrator using TTL NAND gates

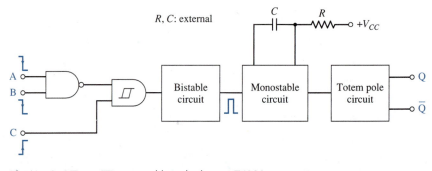

Figure 8.27 TTL monostable multivibrator, 74121

Both NAND and NOR gate TTL one-shots suffer from timing instability due to variations in the logic threshold voltages with loading. Integrated circuit monostable multivibrators of the TTL family, such as the 74121, use the BJT one-shot circuit of Figure 8.22, with compensating circuitry for temperature and supply variations. These multivibrators may be triggered using any one of three inputs. A Schmitt trigger buffer is provided for one of the inputs, as shown in Figure 8.27, to allow triggering with a slowly varying or noisy signal. A narrow trigger pulse is obtained from an input by setting and resetting an internal bistable circuit. The IC provides complementary totem-pole outputs. The pulse width is given by $T \approx 0.7RC$. Widths ranging from tens of ns to a few tens of seconds can be obtained using an external resistor R (from 2 kΩ to 30 kΩ) and a capacitor C (from 10 pF to 10 μF).

The TTL family of monostable multivibrators also includes a pair of retriggerable one-shots, the 74122 and 74123. These circuits may be triggered while they are in the quasistable state, so that their pulse widths can be extended. Additionally, these circuits provide a reset input, which can be used to terminate the output pulse prematurely.

8.2.3 CMOS NOR GATE MONOSTABLE CIRCUIT

A CMOS monostable multivibrator using two NOR gates is shown in Figure 8.28a. While the circuit configuration is identical to that of the TTL NOR gate one-shot, the CMOS version has a more stable pulse width, because of its higher noise margin and logic swing.

Operation of the CMOS one-shot is similar to that of the TTL NOR gate one-shot shown in Figure 8.25. For $V_i = 0$, the circuit is in its stable state, with V_a and V_b at logic high ($\approx V_{DD}$) and the output V_o at logic low (≈ 0). When a positive trigger signal is applied, at $t = 0$, the output of G_1 goes low (≈ 0), after a delay t_{pd}, and V_b drops abruptly from V_{DD} to 0. Therefore, V_o jumps to logic high, after the propagation delay of G_2, and the circuit enters its quasistable state. Capacitor C is now charging toward V_{DD}, via R and the on-gate G_1. With a time constant of approximately RC, the charging voltage V_b is given by

$$V_b \approx V_{DD}(1 - e^{-t/RC}) \qquad \textbf{(8.16)}$$

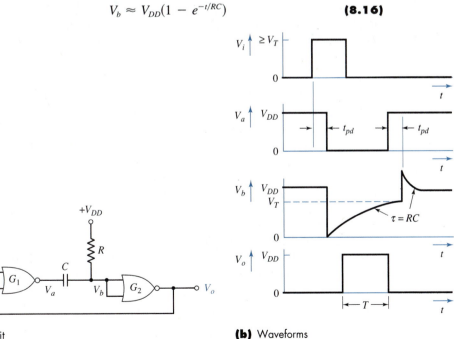

(a) Circuit **(b)** Waveforms

Figure 8.28 A CMOS NOR gate one-shot

At $t = T$, V_b reaches the input logic high threshold voltage of $V_{IH} \approx V_T$ for G_2, and V_o falls to logic low. Hence, the output (high) pulse width T is given by

$$T \approx RC \ln \frac{V_{DD}}{(V_{DD} - V_T)} \qquad \text{(8.17)}$$

If $V_{IH} = V_T = \dfrac{V_{DD}}{2}$, Equation (8.17) reduces to

$$T \approx RC \ln 2$$

After another delay of t_{pd}, V_a rises to logic high, causing V_b to rise by approximately V_{DD}. (Because of the internal protection diode connected between each input terminal and V_{DD}, V_b jumps to $V_{DD} + V_{Diode}$.) Since $V_b > V_{DD}$, the capacitor discharges and brings the voltage V_b down to the steady value of V_{DD}. The circuit is now back in its stable state. The waveforms at various points are shown in Figure 8.28b.

As with the TTL NOR gate one-shot, the trigger pulse width T_i must satisfy $2t_{pd} \leq T_i \leq T + 2t_{pd}$, and the next trigger pulse can only be applied after a delay of about five time constants.

With large V_{OH} and V_{IH}, the pulse width of the CMOS one-shot is less sensitive to temperature variations than a TTL one-shot. In general, however, the stability of the pulse width with supply voltage, temperature, and passive component variations cannot be relied upon in systems requiring precise timing signals. For stable and accurate timing signals, a high-frequency clock is generated within the system, and different periods are obtained by appropriate division of the clock, using counters. In the following sections we consider the generation of clock waveforms.

8.3　ASTABLE MULTIVIBRATOR

An astable multivibrator is a cross-coupled inverter circuit without stable states. The circuit switches autonomously between two quasistable states and thus produces a square wave at the inverter outputs. The duration of each quasistable state is determined by external components. Astable circuits are used as clock generators to synchronize events in digital systems.

In the simplest configuration of an astable circuit, two cascaded one-shots are connected as shown in Figure 8.29. Each one-shot is triggered by the negative edge of the Q output of the other. The period of the square wave is given by $T = T_1 + T_2$, where T_1 and T_2 are the pulse widths of the one-shots. The duty cycle, which is the ratio of the logic high output pulse duration to its period, is given by

$$\% \text{ duty cycle} = \left(\frac{T_1}{T}\right) 100$$

Square waves with a duty cycle of up to 50 percent are used as clock signals.

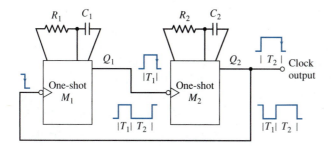

Figure 8.29   Astable multivibrator using one-shots

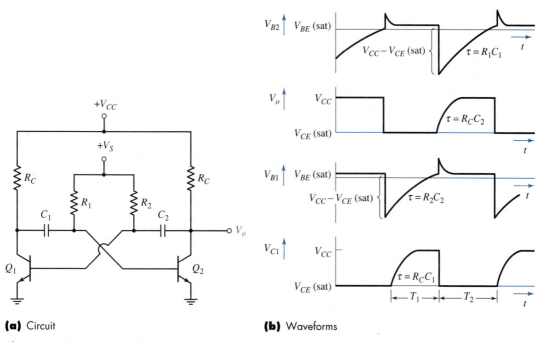

(a) Circuit   **(b)** Waveforms

Figure 8.30   BJT astable multivibrator

Astable multivibrators can be implemented without the use of one-shots. We consider astable circuits using discrete BJTs and logic gates in the following sections.

8.3.1   BJT Astable Circuit

A discrete BJT astable circuit uses two capacitively-coupled inverters, as depicted in Figure 8.30. We analyze the circuit for the case in which the auxiliary power supply V_S is larger than the forward junction voltages of the transistor. For

$t < 0$, if Q_1 is in saturation and Q_2 is cut off, capacitor C_1 charges toward V_S via R_1 and Q_1. This charging exponentially raises the base–emitter junction voltage V_{BE2}. At $t = 0$, V_{BE2} reaches the cut-in voltage, and Q_2 is rapidly turned on by regeneration. With a drop in V_{C2} from V_{CC} to $V_{CE(sat)}$, Q_1 is turned off, and V_{C1} rises to V_{CC}. The base voltage of Q_1 therefore falls to $V_{BE(sat)} + V_{CE(sat)} - V_{CC}$.

The circuit is now in one of its quasistable states, and the capacitor C_2 charges toward V_S via R_2 and Q_2; hence, V_{BE1} increases exponentially, as given by

$$V_{BE1} = V_S + (V_{BE(sat)} + V_{CE(sat)} - V_{CC} - V_S)e^{-\frac{t}{R_2 C_2}} \qquad \textbf{(8.18)}$$

where the on-resistance of the transistor is neglected.

At $t = T_1$, C_2 is charged to the cut-in voltage of Q_1, and the circuit quickly switches to its second quasistable state. In this quasistable state, Q_1 is on and Q_2 is off, with the base of Q_2 having dropped to $(V_{BE(sat)} + V_{CE(sat)} - V_{CC})$. Time T_1, when V_{BE1} reaches $V_{BE(cut-in)}$, is given by

$$T_1 = R_2 C_2 \ln \left[\frac{V_{CC} + V_S - V_{BE(sat)} - V_{CE(sat)}}{V_S - V_{BE(cut-in)}} \right] \qquad \textbf{(8.19)}$$

Capacitor C_1 is now charging with a time constant of $R_1 C_1$, via the on-transistor Q_1; hence V_{BE2} rises according to

$$V_{BE2} = V_S + (V_{BE(sat)} + V_{CE(sat)} - V_{CC} - V_S)e^{-t/R_1 C_1} \qquad \textbf{(8.20)}$$

At $t = T_2$, V_{BE2} reaches the cut-in voltage, $V_{BE(cut-in)}$. The duration T_2 is given by

$$T_2 = R_1 C_1 \ln \left[\frac{V_{CC} + V_S - V_{BE(sat)} - V_{CE(sat)}}{V_S - V_{BE(cut-in)}} \right] \qquad \textbf{(8.21)}$$

Again, we neglect the on-resistance of Q_1, compared to the resistance of the timing resistor R_1. Note that during T_2, the turn-off collector voltage of Q_2 rises rather slowly after an initial jump, as seen with the BJT one-shot. When Q_1 is turned on at the end of T_1, the drop in V_{C1} is coupled to V_{B2}, and Q_2 comes out of saturation. This abruptly raises V_{C2} and causes a jump in V_{B1}. Device Q_1 is therefore driven heavily into saturation, resulting in an overshoot of the base voltage V_{B1}. The capacitor C_2 begins to charge via the collector resistor of Q_2 and the base–emitter junction of Q_1. Hence, the overshoot in V_{B1} decays and the collector voltage V_{C2} rises, both with the time constant of approximately $R_C C_2$. In a similar manner, the collector voltage V_{C1} rises with a time constant of $R_C C_1$ when Q_1 is turned off. Thus, the rising (positive) edges of $V_o(=V_{C2})$ and V_{C1} are rounded, as shown in Figure 8.30.

From Equations (8.19) and (8.21), the frequency of the square wave $f = 1/T = 1/(T_1 + T_2)$, for a symmetrical output, that is, for $R_1 C_1 = R_2 C_2 = RC$, can be approximated as

$$f \approx \frac{1}{RC} \ln \left[1 + \frac{V_{CC}}{V_S} \right] \qquad \textbf{(8.22)}$$

Equation (8.22) shows that for a variable voltage V_S much larger than the saturation junction voltages of the BJT, the frequency is inversely proportional to V_S. Thus, the astable circuit can be used as a variable frequency square-wave generator, or a voltage-to-frequency converter. In this application, the RC product is kept constant so that a square wave of frequency varying with V_S, as given in Equation (8.22), can be obtained. Such a square wave is useful in systems requiring inexpensive analog-to-digital converters (see Chapter 9).

For generating a square wave at a fixed frequency of $f \approx \dfrac{1}{RC \ln 2}$, voltages V_S and V_{CC} may be the same. The durations of saturation for Q_1 and Q_2 are then

$$T_1 \approx R_2 C_2 \ln 2 \approx 0.69 R_2 C_2 \quad \text{and} \quad T_2 \approx R_1 C_1 \ln 2 \approx 0.69 R_1 C_1$$

Note that T_1 and T_2 can be independently controlled. This feature enables the generation of a fixed-period square wave with different duty cycles.

Design a BJT astable multivibrator to generate a clock of 50 μs period and 60 percent duty cycle. The saturation base current must not be more than 20 percent of the collector current, and the collector current is limited to 2 mA. Use V_{CC} = 5 V.

Example 8.7

Solution

With V_{CC} = 5 V, a maximum collector current of 2 mA is obtained using

$$R_C \approx \frac{5}{2} = 2.5 \text{ k}\Omega$$

At a 60 percent duty cycle with a 50 μs period, the saturation intervals for the transistors are

$$T_1 = 50 \times 0.6 = 30 \ \mu s \quad \text{and} \quad T_2 = 20 \ \mu s$$

Hence

$$0.69 R_2 C_2 = T_1 = 30 \ \mu s \quad \text{or} \quad R_2 C_2 = 43.478 \ \mu s$$

and

$$0.69 R_1 C_1 = T_2 = 20 \ \mu s \quad \text{or} \quad R_1 C_1 = 28.986 \ \mu s$$

Since the saturation base current is limited to $0.2 \times 2 = 0.4$ mA, the timing resistance for each transistor is

$$R_1 = R_2 \geq \frac{5 - 0.8}{0.4} = 10.5 \text{ k}\Omega$$

Hence, $R_1 = R_2 = 12$ kΩ can be used.

From the RC time constants, the capacitances are given by

$$C_1 = 2.415 \text{ nF} \quad \text{and} \quad C_2 = 3.623 \text{ nF}$$

Note that with $R_1 = R_2 = 12$ kΩ, each transistor must have a minimum β of $\dfrac{I_{C(\text{sat})}}{I_B} = 6$ for astable operation.

The rounded (positive) edge of the square-wave output can be improved (sharpened) by isolating the collector of the off transistor from the timing capacitor, as shown in Figure 8.31. When Q_1 in this circuit is turned on, the voltage at the collector of Q_2 rises to V_{CC}, independent of the charging of C_2. The time constant for V_{B2} is still $R_1 C_1$ while C_2 is charging via R_4. The on and the off times of the output are therefore unaffected (see Problem 8.17).

A problem with the BJT astable circuit of Figure 8.30 is that it may be in the blocked state with both transistors in saturation. The astable circuit shown in Figure 8.32, with the emitter feedback resistor, prevents the blocked state.

8.3.2   TTL ASTABLE CIRCUIT

Figure 8.33 shows a TTL NAND astable circuit, which is a cascade of two one-shots shown in Figure 8.29. Each RC pair determines the on and the off times of the square-wave output. Although both the gates, G_1 and G_2, are used as inverters, the NAND (or NOR) gate causes the circuit to function as a gated clock generator. In the NAND implementation, for example, keeping one of the unused inputs at logic low disables the clock circuit. The on and off intervals of

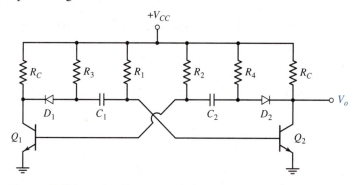

Figure 8.31   Astable circuit with sharper output waveform.

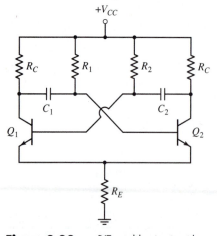

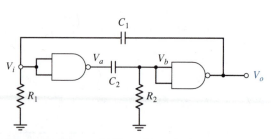

Figure 8.32   BJT astable circuit with emitter feedback

Figure 8.33   A TTL astable multivibrator

the square wave can be controlled independently by the two RC time constants. As with the TTL one-shots, however, this circuit suffers from timing instability. In addition, the circuit has a blocked state similar to the BJT astable circuit. When power is turned on, the circuit may stay in the stable but unwanted state of V_i and V_b at logic low and V_a and V_o at logic high. A positive transient voltage injected at one of the gate inputs will take the circuit out of this blocked state.

8.3.3   CMOS ASTABLE CIRCUIT

Unlike the TTL circuit, the CMOS astable circuit shown in Figure 8.34a has a well-defined period and a sharp transition between logic levels. Suppose for $t < 0$, V_a is at logic high. Since V_o is at approximately ground potential, capacitor C is charging and V_i is increasing exponentially, with a time constant of RC.

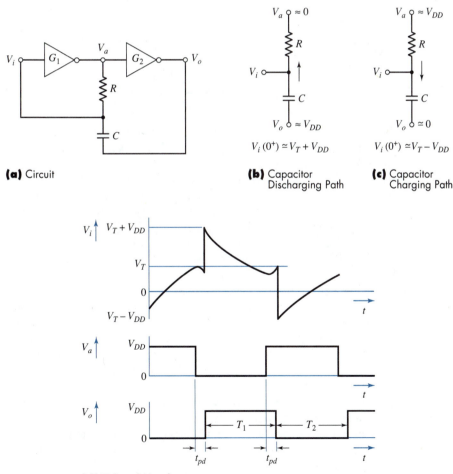

(a) Circuit

(b) Capacitor Discharging Path

(c) Capacitor Charging Path

(d) Voltage Waveforms

Figure 8.34   A CMOS astable multivibrator

At $t = 0$, V_i reaches the gate threshold voltage of approximately V_T, and V_a switches (after t_{pd}) to logic low (≈ 0). Therefore, V_o rises (after another delay of t_{pd}) to logic high ($\approx V_{DD}$). Capacitor C couples this rise in V_o to the input of gate G_1. Therefore, V_i jumps from V_T to approximately $V_T + V_{DD}$. Referring to the discharge path in Figure 8.34b, voltage V_i decreases according to

$$V_i = (V_{DD} + V_T)e^{-t/RC} \tag{8.23}$$

At $t = T_1$, V_i decreases to V_T again, and V_a switches to V_{DD}, after a delay of t_{pd}. This causes V_o to fall from V_{DD} to zero, after a further delay of t_{pd}, and V_i falls by V_{DD}. The capacitor now charges towards V_{DD}. From Figure 8.34c, V_i increases as

$$V_i = V_{DD} + (V_T - 2V_{DD})e^{-t/RC} \tag{8.24}$$

As V_i reaches V_T, at $t = T_2$, V_a switches and the cycle repeats.

From Equations (8.23) and (8.24), the time intervals T_1 and T_2 are given by

$$T_1 = RC \ln\left[\frac{V_{DD} + V_T}{V_T}\right] \tag{8.25}$$

$$T_2 = RC \ln\left[\frac{2V_{DD} - V_T}{V_{DD} - V_T}\right] \tag{8.26}$$

Period T of the square wave is $T = T_1 + T_2$. A symmetrical square wave (that is, $T_1 = T_2$) results only when $V_{DD} = 2V_T$ and the period then becomes $T = 2RC \ln 3 \approx 2.2RC$.

The voltages in Figure 8.34d and the derivations in Equations (8.25) and (8.26) are valid if the gate output voltage is zero when the gate is on and is at full supply voltage when the gate is off. Additionally, for more accurate timing intervals, we must consider the effect of the input protection diode in gate G_1. If the diode is included in the analysis, we modify the charging equation so that $V_i(0^+) \approx -0.7$ V, and the discharging equation must have $V_i(0^+) \approx V_{DD} + 0.7$ V.

Also note the behavior of the capacitor voltage V_i when it reaches V_T. If the capacitor is charging and V_i reaches V_T, V_a does not fall instantaneously. Voltage V_i continues to rise during the propagation interval t_{pd} of G_1. Voltage V_a then goes to logic low, but V_o does not change from low to high until after the propagation delay of G_2. Hence, in the interval between t_{pd} and $2t_{pd}$, the capacitor reverses its charging direction, as shown in Figure 8.34d, before responding to the change in V_o. A similar direction reversal occurs for V_i when C is discharging.

As with other CMOS circuits, the CMOS astable circuit is susceptible to variations in temperature, power supply, and threshold voltages. In addition, changing either R or C changes both T_1 and T_2 in the same proportion. Therefore, the period of the square wave can be varied by varying either R or C while keeping the same duty cycle.

Example 8.8

Determine the pulse intervals and period for a CMOS clock circuit, using $R = 10$ kΩ and $C = 1$ nF. The supply voltage is 5 V and the threshold voltage is 2 V.

If R is increased to 12 kΩ, how much does the period of the clock change?

Solution

From Equation (8.25), we have

$$T_1 = 10 \ln\left(\frac{5 + 2}{2}\right) = 10 \ln(3.5) = 12.53 \ \mu s$$

and

$$T_2 = 10 \ln\left(\frac{10 - 2}{5 - 2}\right) = 10 \ln\frac{8}{3} = 9.81 \ \mu s$$

Hence, the period is $T = 22.34 \ \mu s$ and the duty cycle is $\frac{T_1}{T} = 56$ percent.

If R is increased to 12 kΩ, the time constant for both charging and discharging increases. Hence, T_1 and T_2 increase to

$$T_1 = 12 \ln(3.5) = 15.03 \ \mu s \quad \text{and} \quad T_2 = 12 \ln\left(\frac{8}{3}\right) = 11.77 \ \mu s$$

Consequently, the period changes to $T = 26.8 \ \mu s$, while the duty cycle remains the same at $\frac{T_1}{T} = 56$ percent.

Figure 8.35 shows modified CMOS astable circuits. The large resistance R_S ($\geq 10R$) in Figure 8.35a reduces the effect of variations in V_T. In Figure 8.35b, the off and on times of the square wave can be independently controlled using R_1 and R_2.

If a quartz crystal is used to replace part of the timing circuit, the square-wave output of an astable circuit can be made quite stable with temperature and

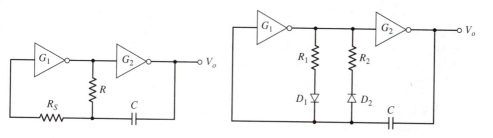

(a) Large Resistance **(b)** Independent Control of Square Valve

Figure 8.35 Modified CMOS astable circuits

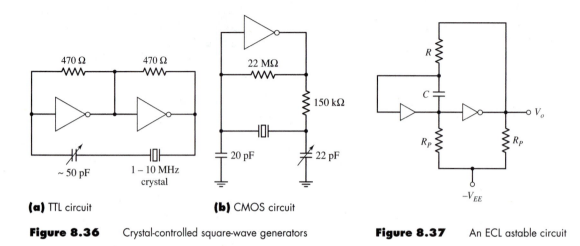

(a) TTL circuit   **(b)** CMOS circuit

Figure 8.36   Crystal-controlled square-wave generators   **Figure 8.37**   An ECL astable circuit

power supply variations. Figure 8.36 shows TTL and CMOS crystal-controlled square-wave generators. For generating high-frequency square waves (above 50 MHz), astable circuits employing ECL gates may be used. Figure 8.37 shows an ECL astable circuit (see Problem 8.22).

8.4   THE 555 INTEGRATED CIRCUIT TIMER

Monostable and astable multivibrators may be implemented with precise time durations using the commercially available IC popularly known as the 555 timer. The 555 timer is available in both bipolar and CMOS technologies. Although it does not belong to any of the logic families we studied so far, the timer is capable of operating over a wide range of supply voltages and can drive high source and sink current loads. In this section, we briefly consider the operation of the timer as an externally triggered (monostable) circuit and as a free running (astable) circuit.

Figure 8.38 shows the block diagram of the 555 timer. The circuit consists of two voltage comparators C_1 and C_2, an RS latch F, a BJT (or CMOS) switch Q_1, and an output driver (inverting amplifier) A. The comparator is the interface between the analog and digital circuitry. The comparator output is at logic high if its noninverting input V_+ is greater than its inverting input V_-, and at logic low if $V_+ < V_-$. The two voltages, V_{tu} and V_{tl}, at which C_1 and C_2 switch output states are derived internally by the voltage divider chain consisting of the three equal resistors each labeled R_T. Thus, $V_{tu} = (2/3)V_{CC}$, and $V_{tl} = V_{CC}/3$. The RS latch is set (or reset) by the low-to-high transition of the comparators. The latch can also be reset independently by a high-to-low trigger at the reset input.

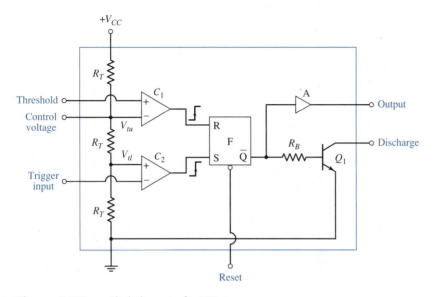

Figure 8.38   Block diagram of a 555 timer

8.4.1   MONOSTABLE MULTIVIBRATOR OPERATION

The timer connected for monostable operation is shown in Figure 8.39a. The external resistor R and the capacitor C determine the pulse width. The circuit is in the stable state with S and R at logic low and $\overline{Q}$ at logic high when the trigger input voltage is above $V_{CC}/3$. With $\overline{Q}$ at logic high, the timing capacitor C is held discharged by transistor Q_1. When the trigger voltage goes low (that is, below $V_{tl} = V_{CC}/3$), the output of comparator C_2 goes high, which sets the latch F. The circuit is now in its quasistable state with the output V_o at logic high. Since $\overline{Q}$ is at logic low, the BJT switch across capacitor C is open. Hence, C charges toward V_{CC} with a time constant of RC. At $t = T$, the capacitor voltage V_{th} reaches $V_{tu} = 2V_{CC}/3$, which causes the reset input of the latch to switch to logic high. The latch is therefore reset, and the $\overline{Q}$ output discharges the capacitor. The circuit is now back in its stable state. Waveforms for the one-shot are shown in Figure 8.39b. We can show that the pulse width of the logic high output is given by

$$T = RC \ln 3 \qquad\qquad \textbf{(8.27)}$$

Clearly, T is independent of the supply voltage. Any variation in V_{CC} affects both the threshold and input voltages for both comparators. The pulse width is also independent of the absolute values of the resistors R_T. The threshold voltages, and hence the pulse width, depend only on the ratio of the resistors.

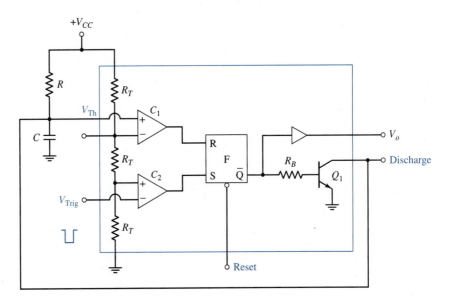

(a) Circuit

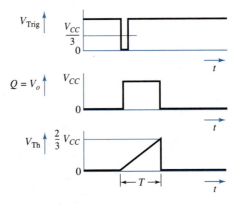

(b) Waveforms

Figure 8.39   Timer as a monostable multivibrator

Additional trigger pulses applied during the quasistable state have no effect on the output, since the latch is already set. The latch can be prematurely reset (and the pulse width shortened) by applying an independent negative trigger to the reset input. When not used, the reset input is normally tied to the supply voltage, to avoid false resetting of the latch by extraneous (noise) signals. Timer-based monostable multivibrators are used for generating time delays of 10 μs to nearly 100 s.

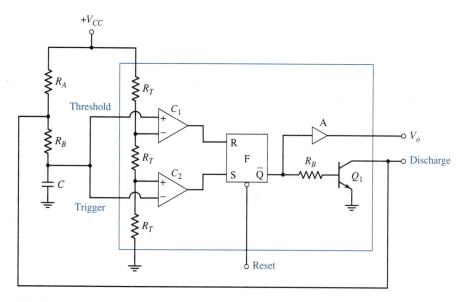

(a) Circuit

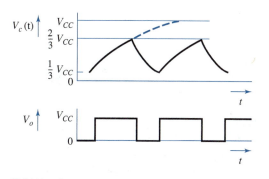

(b) Waveforms

Figure 8.40 Timer as an astable multivibrator

8.4.2 ASTABLE MULTIVIBRATOR OPERATION

Figure 8.40a shows the timer connected for astable operation. The two external resistors R_A and R_B and the capacitor C determine the pulse width and the duty cycle of the square wave output. With the BJT switch Q_1 connected to the junction of R_A and R_B, the capacitor discharges via R_B when $\overline{Q}$ is at logic high, and charges via the series combination of R_A and R_B when $\overline{Q}$ is at logic low. Note that the trigger input and the threshold input are both connected to the capacitor.

As a result either both comparators C_1 and C_2 are at logic low, or only one of them is at logic high.

To study the operation of this circuit, we assume that the latch is initially set. With $\overline{Q}$ at logic low, the capacitor begins to charge toward V_{CC}, with a time constant of $(R_A + R_B)C$. As the capacitor voltage V_C reaches the lower threshold voltage of $V_{tl} = V_{CC}/3$, the output of comparator C_2 switches to logic low, which has no effect on the latch, and the capacitor voltage continues to increase. When V_C reaches the upper threshold voltage of $V_{tu} = (2/3)V_{CC}$, the output of C_1 goes high, which resets the latch. The BJT switch Q_1 is now turned on, and the capacitor begins to discharge toward zero from its initial value of $(2/3)V_{CC}$, with a time constant of $R_B C$. When V_C drops to $V_{CC}/3$, the output of C_2 changes state to logic high, and the latch is set. Therefore, the $\overline{Q}$ output goes low and the cycle repeats. Figure 8.40b shows the waveforms of the astable circuit.

Since the capacitor voltage varies between $V_{tl} = V_{CC}/3$ and $V_{tu} = (2/3)V_{CC}$, the on time (C charging) and the off time (C discharging) of the output are given by

$$T_{on} = (R_A + R_B)C \ln 2 \qquad\qquad \textbf{(8.28)}$$

$$T_{off} = R_B C \ln 2 \qquad\qquad \textbf{(8.29)}$$

Period T of the square wave becomes

$$T = T_{on} + T_{off}(R_A + 2R_B)C \ln 2 \qquad\qquad \textbf{(8.30)}$$

and the duty cycle is given by

$$\text{Duty cycle} = \frac{T_{on}}{T} = \frac{R_A + R_B}{R_A + 2R_B} \qquad\qquad \textbf{(8.31)}$$

From Equation (8.31), we see that the duty cycle is always greater than 50 percent for nonzero values of R_A and R_B.

Timers are used to generate fairly stable square wave outputs with periods of from tens of microseconds to a few seconds. It is also possible to obtain square wave outputs with a 50 percent duty cycle (equal on and off times) (Problem 23).

SUMMARY

This chapter presented the operation and implementation of the three classes of regenerative, or multivibrator, circuits:

1. The bistable multivibrator.

2. The monostable multivibrator.

3. The astable multivibrator.

A bistable multivibrator, also known as a latch, has two stable states, and the circuit can be switched from one state to the other by means of an external trigger signal. Bistable multivibrators are used in flip flops, registers, counters, and other sequential logic functions. NOR or NAND gates from any logic family are used for implementing bistable multivibrators.

A Schmitt trigger is another bistable multivibrator, which switches states when an input signal crosses one of two threshold voltages. Schmitt triggers are used to convert slowly varying or noisy voltage waveforms into logic signals with fast rise and fall times.

A monostable multivibrator, also known as a one-shot, is normally in a stable state from which it can be triggered by an external signal to go to an unstable state. After a fixed interval, determined by a resistor-capacitor combination, the circuit returns to its stable state. Monostable circuits are used to generate time delays at the positive or negative edges of trigger signals. Cross-coupled NAND and NOR gates of any logic family are used to implement monostable multivibrators.

An astable multivibrator switches autonomously between two unstable states with fixed intervals. Clock waveform generators, or square wave oscillators for synchronous systems, are implemented using astable circuits. Two inverters coupled back-to-back by timing capacitors generate square waveforms. Other implementations of the astable circuits are possible with different RC configurations. Gated and precision clock generators are also based on astable circuits.

The popular IC timer known as the 555 timer, though not a digital circuit belonging to any logic family, is used as an astable or a monostable multivibrator.

REFERENCES

1. Millman, J., and H. Taub. *Pulse, Digital, and Switching Waveforms.* New York: McGraw-Hill, 1965.

2. Taub, H., and D. Schilling. *Digital Integrated Electronics.* New York: McGraw-Hill, 1977.

3. Hodges, D. A., and H. G. Jackson. *Analysis and Design of Digital Integrated Circuits.* New York: McGraw-Hill, 1988.

4. Uyemura, J. P. *Fundamentals of MOS Digital Integrated Circuits.* Reading, MA: Addison-Wesley Publishing Company, 1988.

5. Roth, C. H. Jr. *Fundamentals of Logic Design.* St. Paul, MN: West Publishing Co., 1992.

6. *Linear Databook,* Santa Clara, CA: National Semiconductor Corp., 1982.

7. *Designing with TTL Integrated Circuits.* ed. R. L. Morris, and J. R. Miller. New York: McGraw-Hill, 1971.

REVIEW QUESTIONS

1. What are the basic logic elements needed to form a bistable latch?

2. What happens to the outputs of an RS NOR latch when: (a) $R = S = 1$, and (b) $R = S = 0$?

3. Why are the resistors to supply needed in the TTL switch debouncer of Figure 8.3c?

4. How does a clocked RS flip-flop differ from an RS latch in operation?

5. Show the logic diagram of divide-by-two counter with manual set and reset inputs, using TTL NOR gates.

6. How does an edge-triggered flip-flop differ from a master-slave flip-flop in operation?

7. What is the noise restriction of a master-slave flip-flop?

8. What is the output of a JK flip-flop for the forbidden input combination of an RS flip-flop?

9. What problem may occur in a D latch during a high clock level? How is this prevented in a D flip-flop?

10. A sinusoidal voltage of $4 \sin(2000\pi t)$ is applied to a BJT Schmitt trigger with threshold voltages of 1.5 V and 2.2 V. Sketch the input and output voltage waveforms.

11. What is hysteresis in a Schmitt trigger?

12. What are the design parameters for a CMOS Schmitt trigger?

13. Calculate the approximate pulse width of a TTL NOR gate monostable with $R = 1$ kΩ and $C = 1000$ pF.

14. What is the restriction on the trigger pulse width in a CMOS monostable NOR gate?

15. What is the pulse width of a CMOS NOR gate monostable with $R = 1$ kΩ and $C = 1000$ pF?

16. For the same R and C as in Question 15, what is the pulse width of a discrete BJT one-shot?

17. The Q output of a one-shot with a pulse width of 5 ms is connected to the input of a second one-shot with a pulse width of 10 ms. Sketch the waveforms at the Q outputs of the one-shots for a positive trigger applied to the first one-shot if: (a) both are triggered at the positive edge, and (b) the second one-shot is triggered at the negative edge.

18. In a TTL NAND gate astable or monostable circuit, why must the resistance from input to ground be kept low?

19. What is the duty cycle and the period of the square wave obtained from the CMOS astable circuit of Figure 8.34a, with $R = 1$ kΩ and $C = 1000$ pF?

20. Calculate the period and the duty cycle of a discrete BJT astable circuit using $R_1 = R_2 = 15$ kΩ, $C_1 = 0.01$ μF, and $C_2 = 0.02$ μF.

21. To obtain the same square-wave period as in Question 20, can 150 Ω resistors and 10 μF and 20 μF capacitors be used? Explain.

22. What causes emitter–base junction breakdown in BJT astable and mono-stable circuits?

23. In the astable circuit of Figure 8.31, how must R_3 and R_4 be chosen?

PROBLEMS

1. Analyze the JK flip-flop shown in Figure P8.1 for different combinations of inputs and output. State if the flip-flop is edge-triggered or master–slave.

2. Analyze the D flip-flop shown in Figure P8.2 for different combinations of D and Q. State the type of D flip-flop.

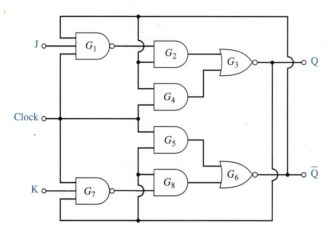

Figure P8.1

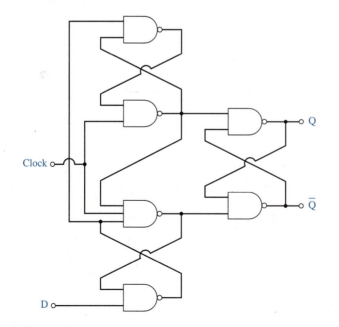

Figure P8.2

3. Determine the stable-state collector and base voltages of the bistable circuit in Figure 8.10 when a green LED is connected to the collector of Q_1 and a red LED is connected to the collector of Q_2. A 500 Ω resistor is in series with each LED to limit the diode on current. Assume a voltage drop of 1.7 V for each diode when turned on, and a β of at least 50 for the BJTs.

   If the diode series resistance can be varied, what is the maximum current possible through the diode under bistable conditions?

4. A symmetrical BJT bistable circuit is to be designed for the loads shown in Figure 8.11. The power supply voltage is 5 V and the transistors have a minimum β of 50. The collector current of each transistor is limited to 4 mA and the base drive cannot exceed 1.25 times the minimum required. If the loads draw a maximum current of 2 mA each, determine the values of R_B and R_C and the logic high output voltage.

5. To ensure cutoff of the off transistor, a negative bias supply is used in a discrete BJT bistable circuit, as shown in Figure P8.5. Determine the stable state collector and base voltages and currents in this circuit.

6. A self-biased bistable circuit is shown in Figure P8.6. Determine the stable state collector, emitter, and base voltages and currents.

   If a 5 kΩ load is connected from the collector of each transistor to ground, what are the stable state voltages?

7. Three T flip-flops, each built using a JK flip-flop with $J = K = 1$, are cascaded as shown in Figure P8.7, forming a ripple counter. If clock signal CK of period T_c is applied to the clock input of flip-flop A, sketch the wave forms at Q_A, Q_B, and Q_C as the clock changes state. Show the sequence of logic values for Q_A, Q_B, and Q_C in a table. How is the frequency of the waveform

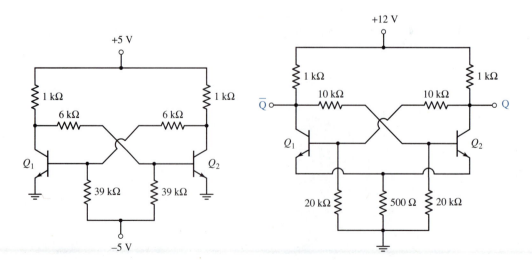

Figure P8.5   **Figure P8.6**

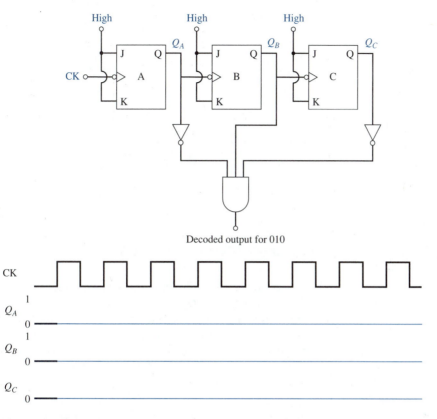

Decoded output for 010

Figure P8.7

at Q_C related to the input frequency at T_A? If each flip-flop has a propagation delay of t_f, what is the minimum period T_c required for unambiguous counting? If the output $Q_A Q_B Q_C = 010$ is decoded using the decoder circuit shown in the figure, sketch the decoder output along with Q_A, Q_B, and Q_C. Assume that the decoder gate delays are small compared with t_f. What problem is evident in the decoder output?

If each flip-flop has an additional reset input available for resetting independently of the clock, design the feedback circuitry needed so that the ripple counter operates as a divide-by-six counter; that is, show how all three flip-flops can be reset after every six clock pulses from CK. (Hint: use the state that results *after* the application of the sixth pulse to determine the reset condition.)

8. A three-bit synchronous counter is shown Figure P8.8. Sketch the waveforms at Q_A, Q_B, and Q_C as the clock changes state. If the decoder shown in Figure P8.7 is used, determine if there is a decoder glitch.

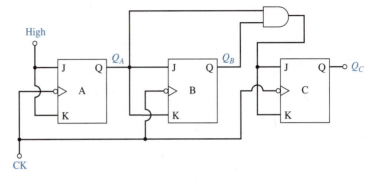

Figure P8.8

* 9. Show the logic diagram of a TG-based D flip-flop (Figure 8.18b) with active low direct set and reset inputs.

*10. Consider the BJT bistable circuit of Figure 8.13 without the differentiating circuit but with the bases of Q_R and Q_S tied together via a 15 kΩ resistor in series with each base. Rectangular trigger pulses are applied at T to obtain a divide-by-two output. If all transistors are 2N3904, determine the allowable pulse width and the maximum repetition rate of T by running MicroSim PSpice simulations. Use the manufacturer's data for the parameters of 2N3904, or the device library file in MicroSim PSpice.

*11. Obtain MicroSim PSpice simulation resuts for the transient analysis of the CMOS NOR latch shown in Figure 8.17. Use a CD4000UB MOSFET array and assume the initial conditions are $Q \approx V_{DD} = 5$ V and $\overline{Q} \approx 0$. Include the parameters measured for the MOSFETs in Experiment 6.11 in Chapter 6. Let the R input be a pulse that changes from 0 to 5 V. Determine the propagation delay from R to Q.

*12. Calculate the upper and lower threshold voltages for the Schmitt trigger in Figure 8.19a with an 800 Ω resistor inserted (a) between E and E_1, and (b) between E and E_2. Note that in the first case, V_{tu} is unchanged, and in the second case, V_{tl} is unchanged.

Determine the resistance values required to eliminate hysteresis in each case; that is, calculate R_{e1} (with $R_{e2} = 0$) such that V_{tl} is raised to 2.42 V, and calculate R_{e2} (with $R_{e1} = 0$) such that V_{tu} is lowered to 1.58 V.

*13. Determine the threshold voltages for the BJT Schmitt trigger shown in Figure P8.13. Obtain the VTC and compare the results using MicroSim PSpice simulation with 2N3904 transistors.

14. Determine the upper and lower threshold voltages in terms of R_I, R_F, and V_{DD}, for the CMOS Schmitt trigger shown in Figure P8.14. Use $R_I < R_F$, $V_T = V_{DD}/2$, $V_{OL} = 0$, and $V_{OH} = V_{DD}$.

Can the circuit be implemented using TTL inverters? Explain.

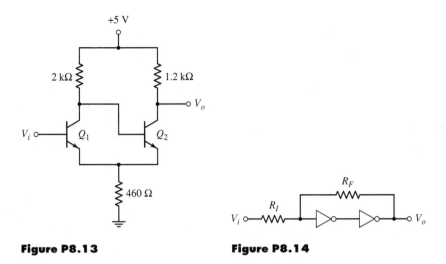

Figure P8.13   **Figure P8.14**

15. Calculate the stable state voltages at the base and collector of Q_1 and Q_2 for the BJT one-shot shown in Figure 8.24. Sketch the waveforms at these terminals after the application of a trigger at V_i.

16. Design a one-shot using BJTs to generate a pulse of 25 μs duration at the negative-going edge of a square-wave input with a 1 ms period. The saturation collector current of each transistor is limited to 5 mA, and is no more than two times the base current. Use $V_{CC} = 5$ V.

17. Assuming ideal diodes, show that the on and off times of the square wave output from the astable circuit of Figure 8.31 are the same as those given in Equations (8.19) and (8.21).

*18. Calculate the period of the square wave for the astable circuit of Figure 8.32, for $R_C = 1$ kΩ, $R_1 = 10$ kΩ, $R_2 = 15$ kΩ, $R_E = 500$ Ω, $C_1 = C_2 = 0.01$ μF, and $V_{CC} = 5$ V. Sketch the waveforms at V_o and V_E.

*19. Verify the operation and the time period of the BJT astable circuit of Problem 18 using a MicroSim PSpice simulation.

20. Derive the expressions for the period of the square wave at V_o in Figures 8.35a and b. Assume a zero voltage drop for the diodes and zero propagation delays for the gates.

21. A simple square wave oscillator using a single Schmitt trigger inverter is shown in Figure P8.21. Show that the period of the square wave is given by

$$T = T_{on} + T_{off}$$
$$= RC \left[\ln \frac{V_{tu} - V_{OL}}{V_{tl} - V_{OL}} + \ln \frac{V_{OH} - V_{tl}}{V_{OH} - V_{tu}} \right]$$

Assume that the inverter draws negligible input current.

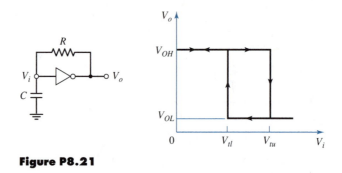

Figure P8.21

22. Refer to the ECL astable circuit shown in Figure 8.37. Derive expressions for the on and off times of the square wave in terms of V_{OL}, V_{OH}, V_{IL}, V_{IH}, and V_R, which is the reference voltage.

23. An astable circuit using a 555 timer is connected as shown in Figure P8.23. Show that

$$T_{on} \text{ (high output)} = R_A C \ln 2$$

$$T_{off} \text{ (low output)} = \left[\frac{R_A R_B}{R_A + R_B}\right] C \ln\left[\frac{R_B - 2R_A}{2R_B - R_A}\right]$$

What is the restriction, if any, on the range of values for R_A and R_B? Calculate the period and the duty cycle for $R_A = 51$ kΩ and $R_B = 22$ kΩ.

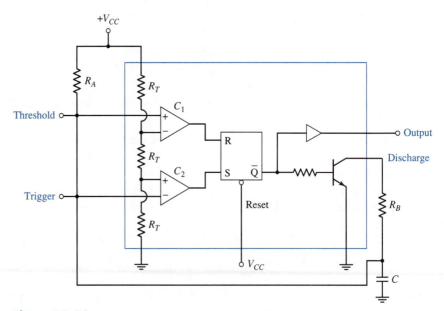

Figure P8.23

24. *Experiment*

    A. *BJT Schmitt trigger characteristics*

      a. Connect the Schmitt trigger circuit shown in Figure 8.19a, using 2N3904 transistors and $R_{C1} = 5$ kΩ, $R_{C2} = 2.5$ kΩ, and $R_E = 1$ kΩ. Apply a triangular wave input that increases from 0 to 5 V and decreases to 0 with a period of 1 ms. Display the input and output on an oscilloscope in the x–y mode and measure V_{OL}, V_{OH}, V_{tu}, and V_{tl}. Alternatively, for better measurement accuracy, apply a variable dc voltage at the input. Slowly increase the dc source voltage from 0 to 5 V and then decrease it to 0. Note the input and output voltages in both directions to determine V_{OL}, V_{OH}, V_{tu}, and V_{tl}.

      b. Connect a 1 kΩ potentiometer R_1 between E and E_1 and vary R_1 until $V_{tu} = V_{tl}$. Record the value of R_1 in circuit.

      c. Connect the 1 kΩ potentiometer between E and E_2 and adjust it until $V_{tu} = V_{tl}$. Record the value of resistance used.

      d. Calculate the voltages V_{OL}, V_{OH}, V_{tu}, and V_{tl}. Determine the new threshold voltages with the potentiometer in circuit for (b) and (c). Compare calculated values with experimental results.

    B. *BJT and CMOS monostable multivibrators timing characteristics*

      a. Connect the negative edge-triggered BJT monostable circuit shown in Figure 8.24a, using 2N3904 transistors and a 1N914 diode. Note the stable state voltages at the bases and collectors of the transistors. Apply a square wave input of 0 to 5 V, with on and off times of 500 μs. Record the duration and amplitude of the output pulse at the collector of Q_2.

      b. Trigger the monostable circuit for $RC = 10$ μs, 50 μs, 100 μs, and 500 μs. Record the output pulse width in each case.

      c. Calculate the stable state voltages at the bases and collectors of the transistors and the output pulse amplitude when triggered. Determine the pulse width for each value of RC. Compare the calculated values with the experimental results. Using the model parameters for 2N3904 and 1N914, perform a MicroSim PSpice simulation for any value of RC used in the experiment. Compare the voltage amplitudes and pulse width with the experimental and calculated values.

      d. Connect the CMOS monostable circuit shown in Figure 8.28a, using CD 4007UB. Note that both G_1 and G_2 (which is an inverter) can be built using one CD 4007UB chip (see Figure P6.11 for pin numbers). Keep $V_{DD} = 5$ V and $R = 10$ kΩ, use $C = 1$ nF, 5 nF, 100 nF, and 500 nF. Trigger the circuit with a square wave of 0 to 5 V, with on and off times of 500 μs. Record the output pulse amplitude and width for each case.

      e. Calculate the pulse width for each RC used and compare with the experimental results.

    C. *BJT and CMOS astable multivibrators timing characteristics*

      a. Connect the BJT astable circuit shown in Figure 8.30a, using 2N3904 transistors, with R_1 and R_2 returned to $V_{CC} = 5$ V. For $R_c = 2.2$ kΩ, $R_1 = 25$ kΩ, $R_2 = 10$ kΩ, and $C_1 = C_2 = 0.001$ μF, observe the voltage waveforms at the collectors and bases of the transistors. Record the period and the on and the off times of the square wave at the collector of Q_2.

b. Change R_2 to 15 kΩ, 25 kΩ, and 33 kΩ and record the period in each case.

c. Calculate the period for each case in (a) and (b). Compare the calculated periods with the experimental results.

d. For $R_1 = R_2 = 25$ kΩ, return R_1 and R_2 to a variable supply V_S for voltage-to-frequency conversion. Vary V_S from 1 V to 5 V in steps of 1 V and record the period in each case.

e. Calculate the frequency of the square wave, using Equation (8.22). Plot the calculated and the observed frequencies as a function of V_S, and compare the linear regions for voltage-to-frequency conversion.

f. Connect the CMOS astable circuit shown in Figure 8.34a, using CD 4007UB. For $V_{DD} = 5$ V and $R = 10$ kΩ, use $C = 1$ nF, 5 nF, 100 nF, and 500 nF and record the period of the square wave in each case.

g. Calculate the period for each case using Equations (8.25) and (8.26). Plot the calculated and the observed periods as a function of the RC time constant.

ANALOG–DIGITAL DATA CONVERTERS

INTRODUCTION

Topics in the previous chapters on combinational and sequential logic circuits dealt with processing data in the form of digital signals. There are many advantages to processing digital signals, which are discrete in time (occurring, for example, at the clock rate) and discrete in amplitude (with only two levels, logic low and logic high). Most physical signals, however, are continuous in both time and amplitude. Transducers such as a thermocouple or microphone produce continuous electrical signals (voltages or currents) that are analogous to physical variables. To process these *analog signals* using digital circuits, we must convert the signals so that they are discrete in time and amplitude. On the other hand, many systems require continuously varying signals for their operation. Therefore, processed digital data must be converted to continuous signals before being applied to such continuous systems. The stored digital data read from a compact disc (CD) in a CD player, or from memory in a speech synthesizer, for example, are converted to analog signals for use with a speaker. The interfacing between continuous, analog signals and discontinuous, digital signals requires special analog–digital data converters. In this chapter, we study the common techniques for analog–digital conversion of data and the systems used to implement these techniques.

9.1 DIGITAL-TO-ANALOG (D/A) CONVERSION

Digital-to-analog converters (DACs) convert discrete-time digital signals to continuous-time analog signals. DACs are also used in some of the popular, low-cost analog-to-digital converters.

A D/A conversion process must satisfy two requirements: (a) the voltage (or current) output of the process must be proportional to the equivalent decimal value of the digital (binary) data, and (b) the converted voltage (or current) must be continuous as the digital input changes from one binary value to another. An

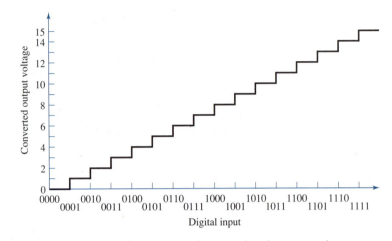

Figure 9.1   Digital input data and converted analog output voltage

obvious solution to make the converted signal continuous is to keep the voltage (or current) constant between the input digital values. Figure 9.1 shows an example of the transfer characteristic between digital input data and converted analog output voltage. For a 4-bit straight binary input in this figure, the output voltage is equal to the decimal value of the digital data. As the input switches from one value to the next consecutive value, the output voltage jumps by 1 V. Hence, a staircase waveform results, with each stair step lasting for the duration of the input data.

Typically, a low-pass (smoothing) filter is used after the D/A conversion process to round out the staircase waveform. Filtering is especially helpful in attenuating large voltage spikes that arise when more than one bit changes from 1 to 0. For example, if the input changes from 0111 to1000, the actual data applied to the DAC may momentarily go to 0000 due to progagation delays in the driving digital circuitry. This change will cause a 'glitch' of 7 V to 0 V at the output before reaching the final value of 8 V, according to Figure 9.1. Low-pass filtering reduces the amplitudes of such unwanted negative spikes of voltage.

In general, a DAC produces a current output proportional to the natural (straight) binary value of the digital input.[1] For most applications requiring analog voltage, an operational amplifier (op amp) may be used as a current-to-voltage converter (see appendix). Although the binary values for N-bit digital data are in the range 0 to $(2^N - 1)$, the data origins and the required range of converted voltage may be different. Data from a set of flip-flops (or a register), for example, may need to be converted to generate an audible tone in a micro-computer speaker; or, the logic levels generated by a set of thumbwheel switches may need to be converted before being applied to control the speed of a dc motor. The analog voltage range is different for each case. Therefore, the con-

---

[1] Modifications for other weights, such as sign–magnitude and one's complement, can be made in the converter, as will be shown later.

version process can choose any output voltage as a maximum corresponding to the largest binary weight of $(2^N - 1)$, and can then proportionally scale down the voltage for lower digital values. In practice, the maximum voltage is determined by the application requirements and the converter technology.

9.1.1 BINARY-WEIGHTED RESISTOR NETWORK

The simplest circuit for obtaining a voltage proportional to the straight binary value of digital data is shown in Figure 9.2. The reference voltage source V_R sets the maximum voltage corresponding to $(2^N - 1)$. Each single-pole double-throw (SPDT) switch connects the corresponding resistor to V_R if the bit b_k ($k = 0, 1, \ldots, (N - 1)$) is 1 (logic high), or to ground if b_k is 0 (logic low). CMOS transmission gates (see Section 7.3.2), for example, can be used to implement these switches. The least significant bit (lsb) controls switch S_0, the next higher bit (bit 1) controls switch S_1, and so on to the most significant bit (msb). The pole of switch S_{N-1} is connected to resistor $R_{N-1} = R$.

To obtain a current proportional to the bit position, each resistor is binary-weighted according to the bit to which it is connected. For a 4-bit ($N = 4$) converter, shown in Figure 9.2, the binary value ranges from 0000 (decimal 0) to 1111 (decimal 15). Therefore, resistor R_2 connecting bit 2 has twice the value of R_3 (which connects the msb), R_1 has four times R_3, and R_0 has eight times R_3.

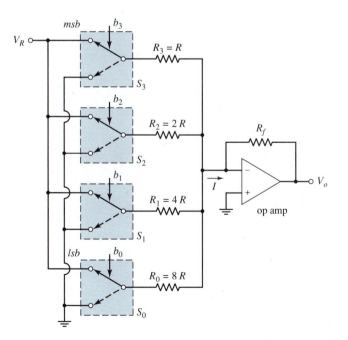

Figure 9.2   D/A converter using binary-weighted resistor network

Because of these binary-weighted values of resistors, this circuit is known as a binary-weighted resistor DAC. The op amp serves as a buffer between the reference source and the output.

Using ideal amplifier characteristics (see appendix), current I at the summing (inverting) node (or virtual ground) can be calculated for any of the 16 different input bit combinations. When the lsb is 1 and the rest of the bits are 0, that is, for data 0001, current I_1 into the virtual ground node is given by

$$I_1 = \frac{V_R}{R_0} = \frac{V_R}{8R}$$

For data 1000, in which only the msb is 1, current I_8 is

$$I_8 = \frac{V_R}{R_3} = \frac{V_R}{R}$$

Since the network is linear, current for the bit combination 1001 (decimal 9) = 1000 (decimal 8) + 0001 (decimal 1) is obtained using superposition as

$$I = I_9 = I_8 + I_1 = \frac{V_R}{R} + \frac{V_R}{8R} = \frac{V_R}{R}\left(1 + \frac{1}{8}\right)$$

Hence, the output voltage V_o for 1001 is

$$V_o = -IR_f = -V_R\frac{R_f}{R}\left(1 + \frac{1}{8}\right)$$

In general, if $b_k = \{0, 1\}$, for $k = 0, 1, 2, 3$ represents the input data, with b_0 as lsb and b_3 as msb, the current at the summing node is given by

$$I = V_R\left(\frac{b_3}{R_3} + \frac{b_2}{R_2} + \frac{b_1}{R_1} + \frac{b_0}{R_0}\right)$$
$$= \frac{V_R}{R}\left(b_3 + \frac{b_2}{2} + \frac{b_1}{4} + \frac{b_0}{8}\right)$$

Hence, the output voltage V_o for the four-bit converter is given by

$$V_o = -V_R\frac{R_f}{R}\left(b_3 + \frac{b_2}{2} + \frac{b_1}{4} + \frac{b_0}{8}\right)$$

This equation demonstrates that the magnitude of output voltage V_o is proportional to the straight binary value of the input data. For an N-bit converter employing N binary-weighted resistors $R, 2R, \ldots 2^k R \ldots 2^{(N-1)} R$ for bits b_{N-1}, $b_{N-2}, \ldots b_{N-1-k}, \ldots b_0$, respectively, in the scheme shown in Figure 9.2, the current at the summing node is given by

$$I = \frac{V_R}{R}\left(b_{N-1} + \frac{b_{N-2}}{2} + \cdots \frac{b_k}{2^{(N-k-1)}} \cdots \frac{b_0}{2^{(N-1)}}\right)$$
$$= \frac{V_R}{R}\sum_{k=0}^{N-1}\frac{b_k}{2^{(N-k-1)}} \tag{9.1}$$

and the corresponding output voltage is given by

$$V_o = -V_R \frac{R_f}{R} \left(b_{N-1} + \frac{b_{N-2}}{2} + \cdot \cdot \frac{b_k}{2^{(N-k-1)}} \cdot \cdot + \frac{b_o}{2^{(N-1)}} \right)$$

$$= -V_R \frac{R_f}{R} \sum_{k=0}^{N-1} \frac{b_k}{2^{(N-k-1)}}$$

$$= -\frac{V_R}{2^{(N-1)}} \frac{R_f}{R} \sum_{k=0}^{N-1} b_k 2^k \tag{9.2}$$

By a suitable choice of the reference voltage V_R and the resistances R_f and R, V_o can be varied over any range with each output voltage proportional to the binary input data.

The current through each resistor is either zero or a value proportional to the binary weight of the input bit. This causes different currents to be switched from the reference source as inputs change; consequently, transients occur in the output waveform. Current transients from V_R can be eliminated by connecting the digital SPDT switches at the summing node, as shown in Figure 9.3. Each switch diverts a constant current from V_R to the summing node or to ground.

To compensate for the bias current effect when all bits are at zero, a resistance may be connected from the noninverting terminal to ground. (Note

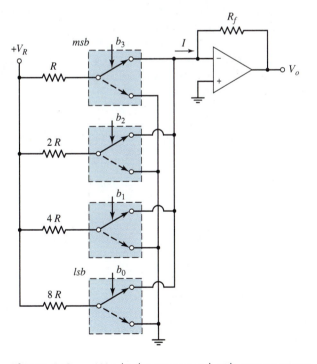

Figure 9.3 Weighted resistor network with constant current

that if $i_+ \approx i_- \neq 0$, equal resistances from the inverting and noninverting terminals to ground enable v_+ and v_- to track each other. Hence, $V_o \approx 0$ for all bits at 0.) In bipolar technology, constant currents proportional to the binary weights are generated and switched to the summing node, or to ground, using diodes or differential amplifiers.

Although the binary-weighted resistor network is conceptually simple to implement, in practice it requires a wide range of precision and tracking resistors or current sources. For an 8-bit converter, for example, if a 1 kΩ resistor is chosen for the msb (R_7), the other resistors are 2 kΩ, 4 kΩ, 8 kΩ, 16 kΩ, 32 kΩ, 64 kΩ, and 128 kΩ. For proper digital data tracking, all these resistors must have extremely low tolerances and temperature variations. In integrated circuits, it is difficult to fabricate a wide range of resistors with high precision. In addition, unequal amplifier bias (input) currents, though small, can cause significant errors in the output voltage when coupled with a large resistance. On the other hand, choosing a small value for the msb will drain a large current from the reference supply.

Example 9.1

An 8-bit DAC uses $V_R = 5$ V in the circuit shown in Figure 9.2.

(a) What is the value of the smallest resistance used such that the maximum current from the reference supply is 10 mA? (b) If the feedback resistance R_f is the same as the resistance connected to the msb, what is the converter resolution, that is, by what amount does the output voltage change for a change in the lsb of the input data? (c) What is the maximum nominal output voltage? (d) What is the voltage corresponding to the input 10101101? (e) If each resistor used has ±10 percent tolerance, what is the range of maximum and minimum output voltages due to resistance variation?

Solution

a. The maximum current i_{max} from the reference supply occurs for the input 11111111. From Equation 9.1, we have

$$i_{max} = \frac{5}{R}\left(\frac{255}{128}\right) = 10 \text{ mA}$$

Hence, $R \geq 996 \ \Omega$

b. With $R_f = R_7$, the smallest change in output voltage occurs for a change in the input lsb. With $b_0 = 1$ and all other bits 0 in Equation 9.2, the converter resolution is

$$V_o = \frac{5}{128} = 0.0391 \text{ V}$$

c. With all bits at 1, the nominal maximum, or full-scale, output voltage is, from Equation 9.2,

$$V_{omax} = 5\left(\frac{255}{128}\right) = 9.961 \text{ V}$$

d. V_o for 10101101 = 173 (decimal) which is

$$V_o = 173\left(\frac{5}{128}\right) = 6.758 \text{ V}$$

e.   Since $V_{o\text{max}} = 5\left(\frac{255}{128}\right)\frac{R_f}{R}$, the worst-case maximum and minimum voltages are

$$\text{Maximum: } V_{o\text{max}} = 5\left(\frac{255}{128}\right)\left(\frac{1.01}{0.99}\right) = 10.162 \text{ V}$$

$$\text{Minimum: } V_{o\text{max}} = 5\left(\frac{255}{128}\right)\left(\frac{0.99}{1.01}\right) = 9.764 \text{ V}$$

Thus, the resistance variation causes a maximum change of approximately 400 mV at the output.

For the minimum voltage with lsb = 1, the range is

$$\frac{5}{128}\left(\frac{1.01}{0.99}\right) \geq V_{o\text{min}} \geq \frac{5}{128}\left(\frac{0.99}{1.01}\right)$$

or

$$40 \text{ mV} \geq V_{o\text{min}} \geq 38 \text{ mV}$$

Therefore, the change in voltage for the lsb is 2 mV.

   This example shows that the resistance variation has a greater effect on the higher end of the output voltage. More importantly, a variation in V_o due to a resistance variation is much higher than that due to changing inputs at the lower end. This means that because of imprecise resistor values, it is impossible to determine if an output voltage is caused by a particular combination of input bits.

   In addition, we see that the tolerances of resistance ratios, rather than the absolute resistance values, must be kept small to achieve accurate conversion. In practice, a full-scale voltage variation due to no more than $\pm 1/2$ lsb is specified for D/A converters. For the previous example, this corresponds to 19.5 mV. To achieve this accuracy, the ratio R_f/R must have a tolerance of less than 0.2 percent. This tolerance is not difficult to obtain in integrated circuits. However, the wide range of resistances that must track with temperature can be avoided by modifying the circuit. For this reason, D/A converters employing binary-weighted networks are not used for more than 4-bit conversion.

9.1.2   R-2R RESISTOR LADDER NETWORK

If another resistor is added for each bit, as shown in Figure 9.4, the range can be reduced from $R: 2^{(N-1)}R$ to $R: 2R$. We analyze the circuit by examining two cases and using the principle of superposition.

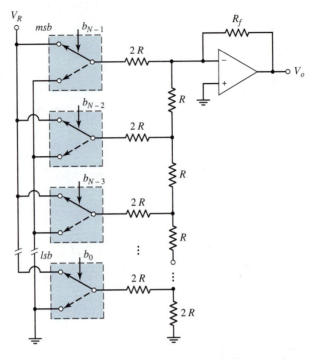

Figure 9.4   D/A converter using R-$2R$ ladder network

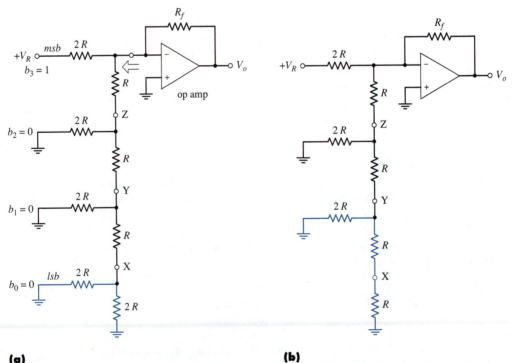

(a)

(b)

Figure 9.5   Thevenin equivalent voltage and resistance for a 4-bit converter for input 1000

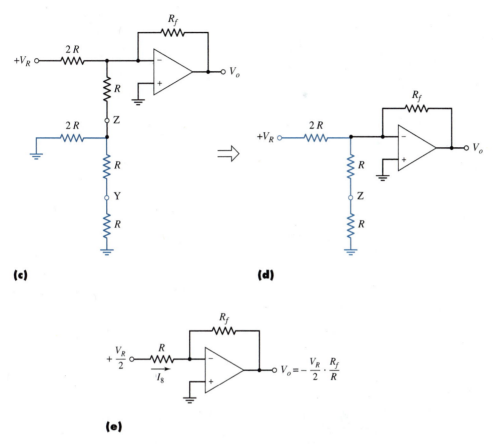

(c)   **(d)**

(e)

Figure 9.5   (continued)

Figure 9.5a shows a 4-bit converter with an input data of 1000 (decimal 8). Looking to the left from the summing node of the op amp, the Thevenin equivalent circuit is derived as shown in Figures 9.5a-e. With the Thevenin resistance $R_T = R$ and the Thevenin voltage $V_T = V_R/2$ at the inverting terminal, the input current into the summing node is

$$I_8 = \frac{V_R/2}{R} = \frac{V_R}{2R}$$

Hence, the output voltage for input 1000 is

$$V_o = -\frac{V_R}{2}\left(\frac{R_f}{R}\right)$$

For input data 0001 (decimal 1) from Figure 9.6, $R_T = R$ and $V_T = \dfrac{V_R}{8}$, which yields

$$I_1 = \frac{V_R}{16R}$$

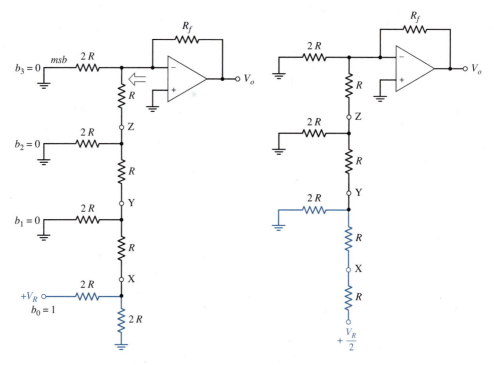

(a)

(b)

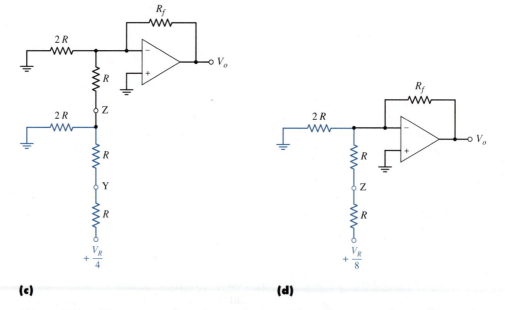

(c)

(d)

Figure 9.6   Thevenin equivalent voltage and resistance for a 4-bit converter for input 0001

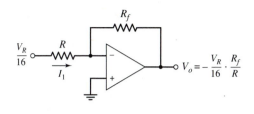

(e)

Figure 9.6   (continued)

and

$$V_o = -\frac{V_R}{16}\left(\frac{R_f}{R}\right)$$

This discussion shows that, in general, for the 4-bit input $b_3 b_2 b_1 b_0$, the current into the summing node is given by

$$I = \frac{V_R}{R}\left(\frac{b_3}{2} + \frac{b_2}{4} + \frac{b_1}{8} + \frac{b_0}{16}\right)$$

and the output voltage V_o is

$$V_o = -\frac{V_R}{16}\left(\frac{R_f}{R}\right)[8b_3 + 4b_2 + 2b_1 + b_0]$$

If we extend this equation for an N-bit converter, we get an output voltage of

$$V_o = -\frac{V_R}{2^N}\left(\frac{R_f}{R}\right)[2^{(N-1)}b_{n-1} + \ldots 2^1 b_1 + b_0]$$

or

$$V_o = -\frac{V_R}{2^N}\left(\frac{R_f}{R}\right)\sum_{k=0}^{N-1} 2^k b_k \qquad \textbf{(9.3)}$$

Note that the output resistance of the ladder network, that is, the resistance seen by the summing junction, is the same ($= R$) regardless of the input bit combination.

Similar to the modified binary-weighted network of Figure 9.3, we can rearrange the ladder network as shown in Figure 9.7a so that the current from V_R is constant. In addition, the currents through the resistors do not change when the input data values are switched. Hence, the difference in the propagation delays, associated with the longer path and higher parasitic capacitance between the lower bits and the summing node, is eliminated. Also, with the currents switched between ground and the summing node, the SPDT switches need not be designed for switching V_R. Commercial DACs use a pair of NMOS transistors, as shown in Figure 9.7b, for each bit, to switch the binary-weighted

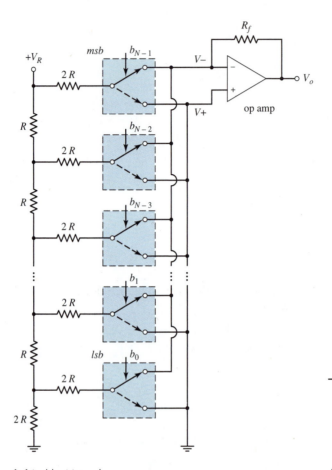

(a) Ladder Network

(b) NMOS SPDT Switch

Figure 9.7   Modified ladder network DAC

current to the op amp or ground. The width of each switch transistor, and hence its on resistance, is adjusted proportionally to its current. This keeps the voltage at the left of each switch at a constant value, regardless of which transistor in the switch is conducting. Exact current switching to the inverting or noninverting terminal of the op amp occurs with the complementary, nonoverlapping bit signals b_k and $\bar{b}_k$. Identical current sources, using matched bipolar transistors and resistors, are also used to drive the resistor ladder.

DACs using the $R\text{-}2R$ ladder network are commercially available as monolithic circuits for 6 to 20-bit conversion. These ICs commonly derive a stable reference voltage from supply voltages and generate an output current proportional to the binary weight of the input data. An op amp with a feedback resistor is used to convert this current to the required output voltage range.

9.1.3 CHARGE REDISTRIBUTION DAC

A DAC can also use charge redistribution in a bank of binary-weighted capacitors, similar to switching currents in a resistor network. In this scheme, known as switched capacitor DAC, MOS capacitors redistribute a steady-state charge to obtain the analog output voltage.

Figure 9.8a shows a simplified circuit for a 4-bit converter using a weighted capacitor bank. Circuit operation requires two steps, which take place during two nonoverlapping clocks, ϕ_1 and ϕ_2. In the first step, which is the reset phase, all capacitors are connected to ground during the logic high of ϕ_1. Hence, at the end of ϕ_1, the capacitors are discharged and $V_o = V_+ = 0$. In the second step,

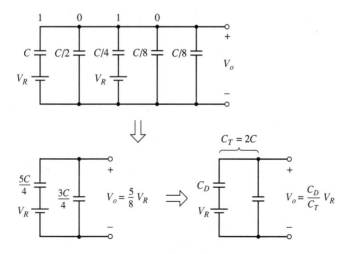

(a) Circuit

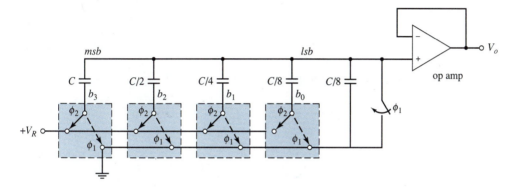

(b) Equivalent Capacitive Divider for Input 1010

Figure 9.8 Four-bit DAC using charge redistribution

which occurs when ϕ_2 goes high, each double-throw switch is connected to V_R if b_k is logic high, or to ground if b_k is logic low. The rightmost capacitor (with value $C/8$) is now between the top plate of the other capacitors and ground. The op amp buffers the output V_o from the capacitor bank.

The equivalent circuit at the end of ϕ_2 for the data 1010 is shown in Figure 9.8b. From this figure, we see that V_o for 1010 is given by

$$V_o = \frac{V_R \frac{5}{4} C}{\frac{5}{4} C + \frac{3}{4} C} = \frac{5}{8} V_R$$

In general, the output voltage V_o can be written as

$$V_o = (C_D/C_T) \, V_R \tag{9.4}$$

where C_D is a variable capacitance connected to V_R during ϕ_2, and C_T is the total capacitance of the bank.

Since C_D depends on the digital data $(b_k, k = 0, N - 1)$ as

$$C_D = C \sum_{k=0}^{N-1} 2^{-(N-1-k)} b_k \tag{9.5}$$

and

$$C_T = C \sum_{k=0}^{N-1} 2^{-k} + \frac{C}{2^{(N-1)}} = 2C,$$

we obtain

$$V_o = \left[\sum_{k=0}^{N-1} 2^{-(N-1-k)} b_k \right] \frac{V_R}{2}$$

or

$$V_o = \frac{V_R}{2^N} \sum_{k=0}^{N-1} b_k 2^k \tag{9.6}$$

As with the weighted-resistor network, DAC resolution increases with a large number of capacitors. However, large capacitors (with the largest capacitance of 2^{N-1} times the smallest capacitance) are not practical in MOS technology because of their increased physical size. Also, higher capacitances cause large transient currents when discharging during ϕ_1. However, Equation (9.6) shows that V_o depends only on the ratio of the capacitances. Therefore, the tolerance and variation of each capacitor can be high as long as the ratio of capacitance values is within acceptable limits.

A converter for higher than 8 bits can be implemented using two stages of smaller ratios of capacitors. Figure 9.9 shows a two-stage 8-bit DAC. During the ϕ_1 phase, the feedback capacitor is discharged and the converter capacitors are connected to V_R. The capacitor corresponding to data bit b_k,

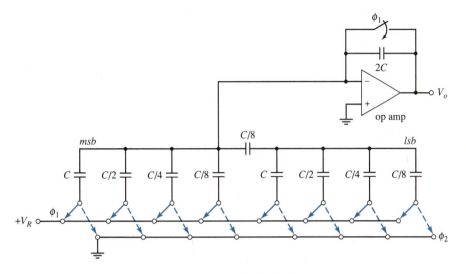

Figure 9.9 Four-bit DAC using two stages of weighted capacitors

$k = 0 \ldots (N - 1)$, is connected to ground during ϕ_2. We can verify (Problem 4) that the output V_o is similar to that given in Equation (9.6).

Single chip C-$2C$ capacitor ladder networks, similar to resistor networks, are used to implement high resolution DACs. Capacitor network DACs employing NMOS technology are used in medium-speed analog-to-digital converters.

9.1.4 CONVERSION FROM OTHER THAN STRAIGHT BINARY DATA

Conversion of data that is not in straight binary form can be performed by suitably modifying the reference voltage source or adding an appropriate voltage at the summing junction of the op amp. For sign–magnitude data, for example, a positive (or negative) reference voltage is selected by the sign bit (msb), while the other bits determine the magnitude of the current, based on their weights. Alternatively, the converted output voltage of the magnitude bits may either be inverted or passed without change, depending on the sign bit. The bipolar output corresponding to the sign–magnitude input data is used in bipolar voltmeters.

Figure 9.10 shows a basic scheme for bipolar operation, using a second op amp. The SPDT switch connects the noninverting input of the output op amp to voltage V_m, which corresponds to the magnitude bits $\{b_k, k = 0 \ldots (N - 2)\}$, if the sign bit b_{N-1} is 0 (positive), or to ground if b_{N-1} is 1 (negative). Note that if the reference voltage V_R is negative, using the resistive circuit of Figure 9.7, for example, then V_m is positive, with a magnitude proportional to bits $\{b_k, k = 0 \ldots (N - 2)\}$. Therefore, for $b_{N-1} = 1$, the output op amp inverts the voltage and V_o becomes the negative of V_m. Resistor with value $R/2$ is used

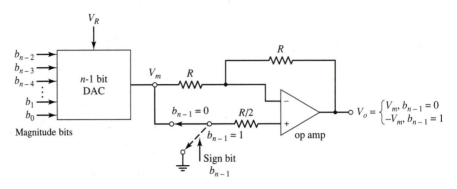

Figure 9.10   Bipolar D/A conversion for sign–magnitude data

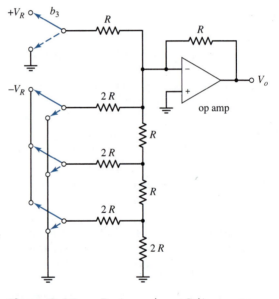

Figure 9.11   Two's complement D/A converter

for bias current compensation. If $b_{N-1} = 0$, V_m is applied to both the inverting and noninverting inputs of the op amp. Under ideal op amp characteristics, this results in $V_+ = V_m = V_-$. Hence, $V_o = V_m$. In practice, small but finite bias currents into the op amp terminals cause a deviation in the output voltage from V_m.

The weighted-capacitor network of Figure 9.8 can be modified for bipolar operation by connecting the capacitor C for the msb to V_R during Φ_1 (high) and to ground during Φ_2 (high).

For a 4-bit two's complement code, whose weights are given in Table 9.1, the implementation shown in Figure 9.11 may be used. We can show (Problem 5) that the output voltage ranges from $-\dfrac{7}{8}V_R$ to $+\dfrac{7}{8}V_R$ for input codes

Table 9.1	Four-bit two's complement code

Input	Weight of Analog Voltage
0000	+0
0001	+1
0010	+2
0011	+3
0100	+4
0101	+5
0110	+6
0111	+7
1000	−8
1001	−7
1010	−6
1011	−5
1100	−4
1101	−3
1110	−2
1111	−1

from 1001 to 0111. Note that the input 1000 gives the largest magnitude output $|V_o|$ of $-V_R$, and $V_o = 0$ corresponds to the input 0000.

Data conversion from the binary-coded decimal (BCD) form to analog voltage is frequently required in analog displays. Because each digit differs from its neighbors by a weight of 10, a weighted resistor network is impractical for more than one digit. A ladder network, which combines binary-weighted resistors for each digit and a series resistor between digits, achieves the required resolution for two decades of input. (Although the scheme can be extended for any number of digits, the conversion of more than two decades is usually not useful in visual displays.) Figure 9.12 shows a two-digit BCD-to-analog converter, where the series resistance R_S between the units and the tens decades is chosen based on the weight of 10. In this figure, d_{03}, d_{02}, d_{01}, and d_{00}, for example, represent the four bits for the units digit. Example 9.2 illustrates the calculation of R_S.

Calculate the value of R_S in Figure 9.12 such that the output voltage for the decimal input $d0$ is 10 times that for $0d$, where $0 \le d \le 9$. What is the maximum output voltage if the op amp feedback resistance is R?

Example 9.2

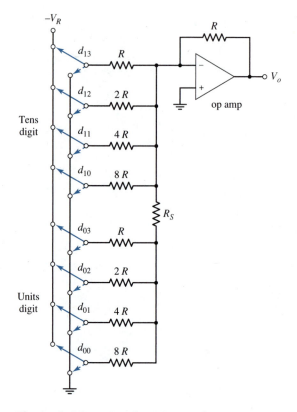

Figure 9.12   Two-digit BCD-to-analog converter

Solution

Consider the case $d = 8$. For the decimal input of 80, the BCD bit combination is 1000 0000. Reducing the circuit using the Thevenin equivalent at the inverting input of the op amp, as shown in Figure 9.13a, results in the Thevenin voltage given by

$$V_T = -\frac{8V_R(8R + 15R_S)}{240R + 225R_S}$$

and the Thevenin resistance given by

$$R_T = \frac{8R(8R + 15R_S)}{225R_S + 240R}$$

Therefore, the output voltage for decimal 80 is

$$V_o = V_{80} = V_R$$

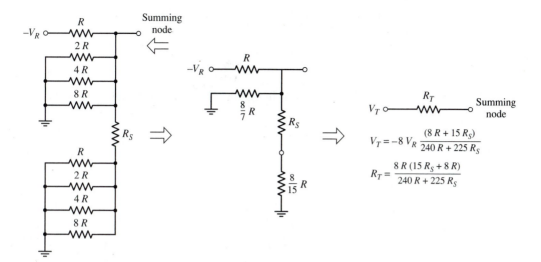

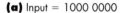

(a) Input = 1000 0000

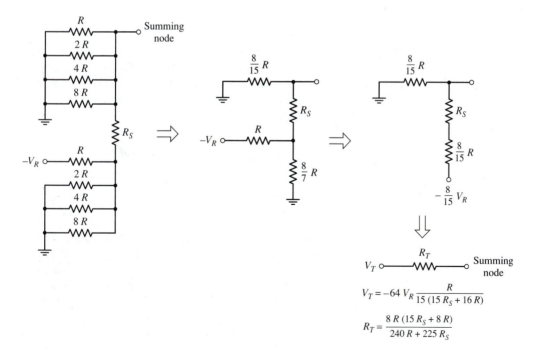

(b) Input = 0000 1000

Figure 9.13   Thevenin equivalent at the summing node of op amp

For the decimal input of 08, which is 0000 1000, the Thevenin voltage is given by (Figure 9.13b)

$$V_T = -\frac{64 V_R R}{15(16R + 15R_S)}$$

while the Thevenin resistance is the same as before. Hence, the output voltage is

$$V_o = V_{08} = \frac{8RV_R}{8R + 15R_S}$$

For $V_{80} = 10V_{08}$, then, we have

$$R_S = 4.8\ R$$

Note that the resistance seen by the summing node of the op amp is the same, regardless of the input combinaion. This resistance is given by

$$\frac{8R(8R + 15R_S)}{225R_S + 240R} = \frac{16R}{33}$$

For the op amp feedback resistance of R, the maximum output voltage, corresponding to decimal 99, is $\dfrac{9.9}{8}\ V_R$.

The above configuration, which combines a 4-bit binary-weighted resistor network with the ladder network, can be used for straight binary input, as well. In that case, we can show that the series resistance is $R_S = 8R$ (Problem 7). Similarly, we can modify the two-stage capacitor network shown in Figure 9.9 for 8-bit straight binary input to accept a BCD input.

9.1.5   MULTIPLYING DACS

As we have seen in this section, the output of a DAC, in general, is expressed as

$$V_o = kV_R$$

where the value of k depends on the input and V_R is a constant voltage. Therefore, output V_o may be considered a product of the digital input and the analog voltage V_R. DACs used in applications where V_R is a variable are known as multiplying D/A converters. For example, each of the circuits shown in Figures 9.2 to 9.4 may be used as an amplifier with digitally controlled gain. For an N-bit DAC with an analog (continuous) signal applied for V_R, we can vary the gain of the amplifier in discrete steps from 0 to $(2^N - 1)$. (The SPDT switches must be designed for transmitting bipolar analog signals.) With a bipolar reference voltage and two's complement input data, we can obtain four-quadrant multiplication of V_R by the input data. An application of a multiplying DAC is the generation of a power series waveform by cascading two or more stages, as

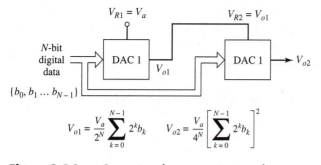

$$V_{o1} = \frac{V_a}{2^N} \sum_{k=0}^{N-1} 2^k b_k \qquad V_{o2} = \frac{V_a}{4^N} \left[\sum_{k=0}^{N-1} 2^k b_k \right]^2$$

Figure 9.14   Generation of a power series waveform

shown in Figure 9.14. Other applications include analog voltage division by an input binary number, and the generation of sine waves in a modem.

9.1.6 D/A CONVERSION ERRORS

As we saw in Example 9.1, the primary sources of error in a D/A converter are component mismatches, variations of resistances with temperature, and the accuracy and stability of the reference voltage with temperature. Other errors arise from the nonideal behavior of the op amp (typically, from the bias current and dc offset voltage causing $v_- \neq v_+$), and the unequal, nonzero resistances contributed by the SPDT switches.

The conversion speed depends on the response time of the switches and the settling time of the op amp. In discrete resistor networks, careful layout and wiring reduce stray capacitances and the attendant settling time delays for producing a stable output.

Monolithic DACs are currently available in the bipolar, CMOS and BiCMOS technologies with output settling times as low as a few nanoseconds for 8-bit resolution. DACs from 6- to 20-bit resolution are used in such diverse applications as digital audio, imaging, waveform synthesis, and color graphics computer systems.

9.2 ANALOG-TO-DIGITAL (A/D) CONVERSION

The process of converting a continuous analog voltage to a discrete number inherently involves error. The error arises from *quantizing* a continuous variable within a finite number of intervals. Quantizing an analog voltage assigns each voltage to a digital 'bin.' For an N-bit conversion, the number of bins, or distinct voltage levels, is 2^N. Quantization errors occur because an infinite number of possible analog voltage amplitudes are assigned to only 2^N discrete values. Clearly, the larger the number of levels, that is, number of bits in the conversion, the smaller the quantization error. Consider, for example, the conversion of a

0-to-8 V analog signal into 3-bit digital data. With eight levels available for the assignment of each voltage amplitude, two possible voltage ranges for each of the 3-bit converted data are shown in Table 9.2.

Table 9.2	Range of analog voltage for a 3-bit A/D conversion	
	Analog Voltage	
Digital Data	**Range 1**	**Range 2**
000	0–1	0.0–0.5
001	1–2	0.5–1.5
010	2–3	1.5–2.5
011	3–4	2.5–3.5
100	4–5	3.5–4.5
101	5–6	4.5–5.5
110	6–7	5.5–6.5
111	7–8	6.5–7.5

In Range 1, the quantization error for any voltage is in the range -0.5 V to 0.5 V. This is also the case for most of the values in Range 2. If most of the values are in the vicinity of zero and below the maximum, as is the case in most practical situations, Range 2 is preferred. If $N = 4$, each bin has a range of 0.5 V, instead of 1 V, and the resolution is improved.

Note that in Range 1, each of the quantization levels is the same, regardless of the input analog voltage; in Range 2, most of the levels are also the same, except the first. Hence, these two quantizers are referred to as *linear* or *uniform quantizers*. A *nonuniform quantizer* emphasizes low voltage analog signals by assigning smaller quantization levels; as the analog signal level increases, the quantization level also increases. Nonuniform quantizers are employed in speech transmission, where the dynamic range is high with low level signals occuring more frequently than large-amplitude signals.

A *1-bit quantizer* is a basic element in the analog-to-digital conversion process. With only 1 bit, this quantizer assigns logic 0 or logic 1 to the input analog voltage V_i, depending on whether V_i is below or above a given reference voltage. An analog voltage comparator may be used as a 1-bit quantizer. Figure 9.15 shows an analog voltage comparator and its characteristics. The ideal input–output characteristic, shown in Figure 9.15b, indicates that the device is a hard limiter. Since V_o assumes only one of two values,

$$V_o = \begin{cases} V_L \text{ (low)}, & \text{if } V_+ < V_- \\ V_H \text{ (high)}, & \text{if } V_+ > V_- \end{cases}$$

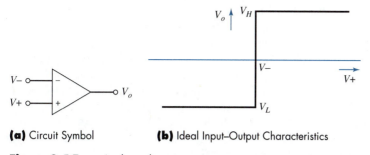

(a) Circuit Symbol   **(b)** Ideal Input–Output Characteristics

Figure 9.15   Analog voltage comparator

the comparator is a 1-bit quantizer, with the quantization level set by one of the inputs. With appropriate level shifting, the output levels can be made compatible with TTL or other logic levels. Thus, the quantizer performs one-bit analog-to-digital conversion. Any delay in the conversion process is due only to the response time of the comparator. Therefore, conversion can be very fast.

Quantization, and hence conversion, may be extended to more than two levels by using more than one comparator. This is the technique used in flash, or parallel, analog-to-digital converters.

9.2.1   PARALLEL/FLASH ADC

The extension of 1-bit quantization to 2-bit analog-to-digital conversion is shown in Figure 9.16. As we see in the conversion table in Figure 9.16b, the range of analog voltage is $0 < V_a < V_R$, which is converted to 2 bits, $b_1 b_0$. Three

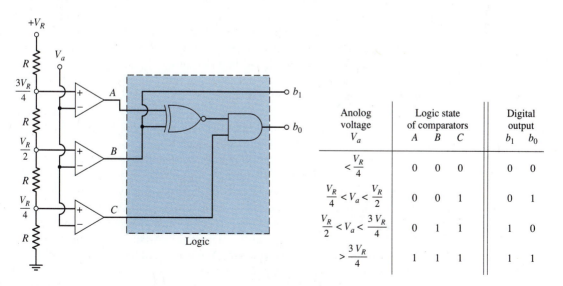

Analog voltage V_a	Logic state of comparators			Digital output	
	A	B	C	b_1	b_0
$< \dfrac{V_R}{4}$	0	0	0	0	0
$\dfrac{V_R}{4} < V_a < \dfrac{V_R}{2}$	0	0	1	0	1
$\dfrac{V_R}{2} < V_a < \dfrac{3 V_R}{4}$	0	1	1	1	0
$> \dfrac{3 V_R}{4}$	1	1	1	1	1

(a) Basic Circuit   **(b)** Conversion Table

Figure 9.16   Two-bit analog-to-digital convertor

comparators are needed to quantize the analog signal into four levels, $0-V_R/4$, $V_R/4 - V_R/2$, $V_R/2 - 3V_R/4$, and above $3V_R/4$.

The two gates in the logic block in Figure 9.16a encode the comparator outputs to 2-bit digital data. Also, each 2-bit combination covers a range of $V_R/4$. Hence, when the output is converted back to analog voltage by a DAC, quantization errors result between the converted and the original analog input voltages. For example, if the analog input voltage is slightly below $V_R/4$, the digital output is $b_1 b_0 = 00$, while the corresponding DAC voltage is 0. The maximum quantization error is, therefore, $V_R/4$. The maximum error, however, is the same regardless of the digital output.

A problem with this scheme is that erroneous outputs may occur, due to unequal delays of the comparators. If the subsequent circuitry captures the digital value before it becomes stable, errors due to changes in V_a are propagated. To prevent this, all comparator outputs are sampled and stored in latches at the occurrence of a clock signal. Encoding logic is used at the output of the latches to determine the digital value. Figure 9.17 shows a 3-bit clocked converter. The analog voltage below $V_R/14$, rather than $V_R/8$, is assigned a digital value of 000, and $V_a > (13/14) V_R$ is assigned a value of 111. Because of these smaller ranges at the extremes, any quantization error is limited to $V_R/14$ for outputs 000 and 111. At other outputs, the error lies between $-V_R/14$ and $+V_R/14$ for any output. Therefore, the quantization error has a maximum magnitude of $V_R/14$. Note that the 8-to-3 encoding logic is simplified because of the comparator outputs: if C_k is high, then all C_i, $i = 0, 1, \ldots (k - 1)$, are also high.

Converters that use comparators and encoding logic are parallel converters, since all bits are available in parallel rather than in sequence. These converters are quite fast with the delay of only the response time of the comparators. Hence, they are called *flash converters*. (Encoding of comparator outputs can be performed later, once a representation for the analog voltage has been obtained.)

To speed up the entire operation, ECL comparator–latch combinations are used. ECL flash converters with conversion rates of up to 500 million analog samples per second are available. A data size of more than 8 bits, however, is very complex because of the large number of comparators ($2^N - 1$ for an N-bit converter) and the large amount of encoding logic required. In addition, 2^N precision resistors are needed at the inputs of the comparators. Conversion errors are due principally to comparator resolution and uncertainty ("jitter") at comparator outputs when both inputs are equal.

High-resolution flash conversion may be accomplished at reduced conversion rates. Figure 9.18 shows a half-flash, or multirange flash converter for 8 bits using two 4-bit flash converters. This 8-bit half-flash converter uses 30 comparators instead of the 255 required by an 8-bit flash converter. The first converter, ADC1 in Figure 9.18, performs a coarse conversion to obtain the four most significant bits, b_7, b_6, b_5, and b_4. The error between V_a and the converted analog voltage at the output of the 4-bit DAC is amplified by the differential amplifier. Since this error is due to 4-bit quantization by ADC1, an amplifier gain of 16 enables the second identical ADC (the fine converter) to use the same reference voltage and perform the lower 4-bit conversion.

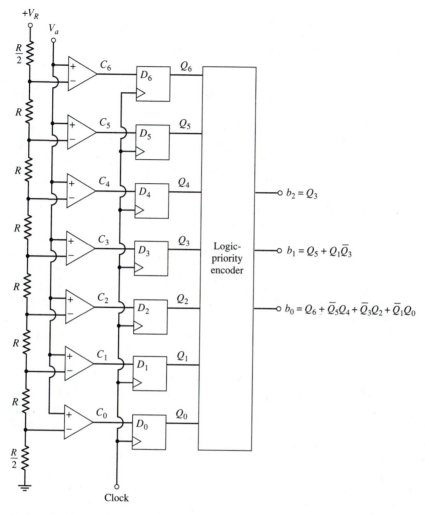

Figure 9.17   Three-bit flash converter

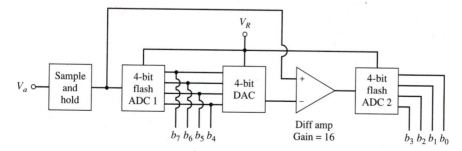

Figure 9.18   Eight-bit half-flash converter

Assume, for example, that the reference voltage V_R is 32 V, so that the resolution in analog voltage is 2 V for ADC1. With an amplifier gain of 16, the quantization error of ADC1 (which is a maximum of 2 V) is converted to the lower four bits by ADC2. However, ADC1 and the DAC must have a high resolution to limit the error to one bit out of eight. To avoid this stringent requirement on ADC1, the differential amplifier uses a gain of 8 and ADC2 employs 5-bit quantization.

The msb of ADC2 is combined with the lsb of ADC1 to form the fourth most significant bit at the final output. For example, if the msb of ADC2 is 1, this indicates an error of more than one bit (that is, more than 2 V of quantization error due to nonlinearity) in the coarse converter; hence, the output of ADC1 is increased by 1.

Note that the DAC must still have 8-bit resolution. Additionally, response times of the amplifier, the registers (holding the outputs of both ADCs), and the output correction logic circuitry must all be small. Typically, ECL logic circuitry is combined with voltage comparators to achieve fast response times.

Because of the increased time delay between the inputs of the first and second converters, the input voltage must remain constant. A sample-and-hold circuit (see below), which holds the sampled input voltage constant, reduces the conversion rate while allowing the resolution to increase.

9.2.2   SAMPLE-AND-HOLD IN A/D CONVERSION

A sample-and-hold circuit maintains a constant-amplitude analog signal during the conversion process. A continuous-time analog signal must be sampled at periodic intervals, with a sampling rate of at least twice the highest frequency (*bandwidth*) of the signal. This is known as the *sampling theorem,* which ensures that the digitized version (sampled signal) bears all the information of the analog signal. The minimum sampling rate at twice the highest frequency of the signal is called the *Nyquist rate*. While the time interval between samples is small enough to satisfy the sampling theorem, the signal amplitude may vary during conversion, especially if the ADC is slow. Therefore, the amplitude at each sample must be held constant while the signal is being converted.

A basic sample-and-hold circuit consists of an analog switch with a capacitor, as shown in Figure 9.19a. The switch is operated at the sampling rate by the sampling clock. When the switch is closed during the nth sampling interval nT_s, capacitor C charges to

$$V_s = V_a(t) \mid t = nT_s$$

In Figure 9.19a, the capacitor is shown as charging linearly from its previously held voltage to its new voltage. The rate of decay of voltage V_s during conversion (that is, during the hold time T_h) depends on the leakage and dielectric absorption of the capacitor, the off resistance of the switch, and the input current of the converter. Figure 9.19b shows a buffered sample-and-hold circuit. Using a low-leakage capacitor, such as Teflon or polystyrene, a drop in voltage V_s with time

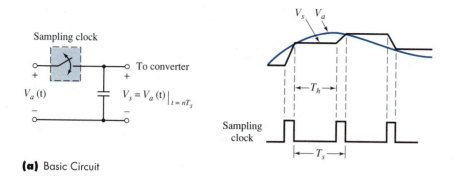

(a) Basic Circuit

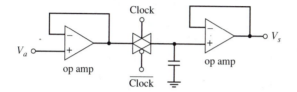

(b) Buffered Circuit

Figure 9.19   Sample-and-hold

is reduced to tens of mV/s. The response time and on resistance of the switch and the slew rate of the input op amp determine the minimum on time for the sampling clock. Sample-and-hold circuits for sampling rates of higher than 1 MHz and droop rates of less than 20 mV/s are available. These circuits are used as front-end circuits for medium speed successive approximation ADCs (see next section).

A *pulse-code modulation (PCM)* communication system converts the constant-output analog voltage V_s of the sample-and-hold circuit during T_h to a digital representation, using an ADC, and transmits the digital value. At the PCM receiver, the received digital data are converted back to analog voltages, using a DAC and a reconstruction filter. Such a system is employed in the telephone transmission of speech.

The rest of this section examines commonly used types of analog-to-digital converters.

9.2.3   SUCCESSIVE APPROXIMATION ADC

Successive approximation ADC is a feedback-type converter that is widely used in medium-speed applications such as PCM speech transmission. The conversion process is analogous to determining the weight X of an unknown mass using a chemist's weight balance and a given number of known weights. Assume, for example, the known weights are 1 kg, 2 kg, 4 kg, and 8 kg, and the unknown

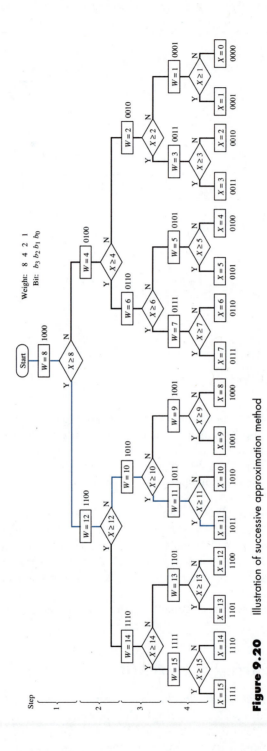

Weight: 8 4 2 1
Bit: $b_3\, b_2\, b_1\, b_0$

Figure 9.20 Illustration of successive approximation method

mass is 11.5 kg. Instead of starting at an arbitrary weight, the systematic procedure begins at half the maximum measurable weight of 8 kg. Since $X > 8$ kg,
we retain the 8-kg weight at the beginning of the second step and add the next
lower weight of 4 kg. We now see that $X < 12$ kg; hence, the 4 kg weight is
removed. In step 3, we add the next lower weight of 2 kg. Since $X > 10$ kg, we
retain the 2-kg weight in step 4, and add the last available weight of 1 kg. With
all weights used, we determine that X is between 11 kg and 12 kg. Figure 9.20
shows this process in a flowchart. Although the flowchart looks lengthy, we note
that there are only four decision (comparison) blocks in each of the 16 different
paths from the beginning to the end. At each new weight setting, the corresponding binary value $b_3 b_2 b_1 b_0$ is indicated in the flowchart.

Figure 9.21a shows the implementation of successive approximation ADC.
The process starts with the digital value 100 . . 00, which is half the digital
range of the converter, in an N-bit register. This register is referred to as a
successive approximation register (SAR). Each approximation step takes place
during a clock interval. The DAC converts the SAR contents to analog voltage
V_d, which is compared against the input voltage V_a. The comparator output C is
used to (a) retain the current bit value of 1 in the SAR if C = high, that is,
$V_a > V_d$, or, (b) change (complement) the current bit value to 0 if C is low. At
the start of the next clock pulse, the next lower-order bit is set to 1. The process
of setting a bit to 1 and comparing V_d with V_a is carried out in sequence for all
N bits. At each step, a voltage increment (or decrement) of half the magnitude
of the previous step results at the output of the DAC. Figure 9.21b shows the
waveforms at the DAC output for a given value of V_a. (An offset voltage corre-

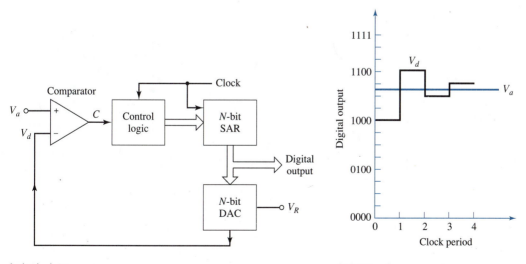

(a) Block Diagram

(b) Waveforms

Figure 9.21   Successive approximation ADC

sponding to $-(1/2)$ lsb is usually added to V_d before applying V_d to the comparator. This keeps the conversion error in the range of $\pm(1/2)$ lsb.)

By starting at the midpoint 100 . . 00, the circuit completes the conversion in exactly N clock pulses regardless of the analog signal amplitude. This is an advantage in applications in which the time required for conversion must be a constant such as in uniform sampling. (Recall that in uniform sampling, analog voltage samples are obtained at fixed intervals of a sampling clock [see Figure 9.19a]).

Another implementation of successive approximation ADC, shown in Figure 9.22, uses a string of resistors and an analog switching matrix. The control logic in this figure activates the appropriate switches and causes a binary fraction of V_R to be applied as V_d at each step. For the 3-bit converter shown, the msb is initially 1 and the two lsb's are 0, that is, $A = 1$, $B = C = 0$. This data combination closes switches s_{21}, s_{10}, and s_{00} in the path from $(7/16)V_R$ to the inverting input of the comparator. If the analog input voltage V_a is above $V_d = (7/16)V_R$,

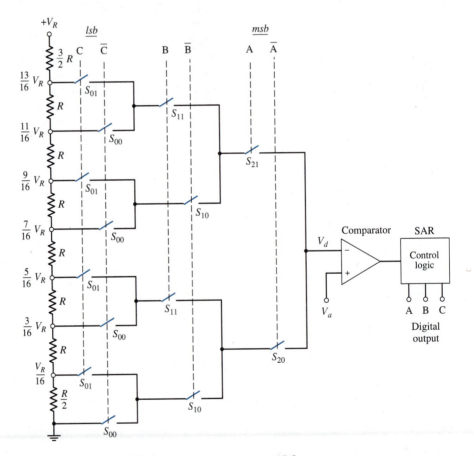

Figure 9.22   Modified successive approximation ADC

the comparator output goes high; at the next step, then, the control logic applies an input of $ABC = 110$ to the switches. This results in $_d = (11/16)V_R$. If the comparator output now goes low, the next approximation for the data is 101, which corresponds to $V_d = (9/16)V_R$. At the third and last step, the converted result is either 100 if $(7/16)V_R < V_a < (9/16)V_R$, or 101 if $(9/16)V_R < V_a < (11/16)V_{rR}$. Note that the polarity of the reference voltage is the same as that of the analog voltage.

Bipolar analog signals are converted by connecting the top end of the resistor string to $+V_R$ and the bottom end to $-V_R$ (instead of ground) (see Problem 11). Other binary codes can be obtained by suitably modifying the resistor string and reference voltages.

For an N-bit converter, this technique requires $(2^{(N+1)} - 2)$ switches and 2^N resistors. Thus, this technique uses a large number of switches and resistors compared to the previous scheme. The advantage, however, lies in the ability to implement all elements of the entire converter in a single chip using MOS technology.

Successive approximation ADCs with built-in or external sample-and-hold circuits are available for 8- to 16-bit resolution with conversion times in the range of a few μs to tens of μs. The conversion speed is essentially dependent on the comparator switching speed, the clock rate, and the bit resolution. Both R-$2R$ ladder and weighted-capacitor (NMOS) networks are used for the DACs.

9.2.4 CHARGE REDISTRIBUTION ADC

Charge redistribution converters implement the successive approximation algorithm by using a bank of capacitors. Instead of a string of resistors, binary-weighted capacitors redistribute the charge (as in a redistribution DAC) to obtain a voltage V_d proportional to the assumed digital data.

The 4-bit charge redistribution ADC shown in Figure 9.23 operates in three phases: sample, hold, and redistribute. During the sample phase ($\phi_s = 1$), the switch at the upper plates of all capacitors is connected to ground, while the bottom plates are connected to the input signal V_i. In the next phase, which is the hold phase ($\phi_h = 1$), the bottom plates of the capacitors are connected to ground, while the ground connection at the top plates is removed. The voltage at the top plates of the capacitors is therefore $V_- = -V_a = -V_i|_{\text{(sampled during } \phi_s)}$, relative to the bottom plates. The charge stored in the capacitor bank is $2V_aC$. The redistribute phase ($\phi_r = 1$) begins with the digital value of 1000. In this phase, the rightmost capacitor in the bank is connected to ground. The switches for the other capacitors connect to V_R or ground, depending on the bit value of 1 or 0. This phase takes N clock pulses during ϕ_r for an N-bit converter. The convergence of the digital value close to V_a is obtained by successively incrementing V_- to zero.

Figure 9.23b shows the waveforms for converting V_a in the range $(11/16)V_R > V_a > (10/16)V_R$. During the first clock in phase ϕ_r, the leftmost capacitor (corresponding to the msb) is connected to V_R and all other capacitors are at ground. We now have a capacitive voltage divider with an external voltage of V_R

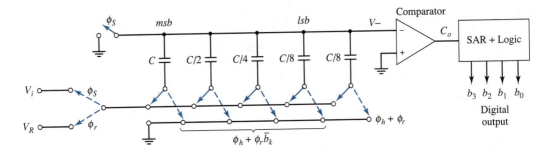

(a) Circuit

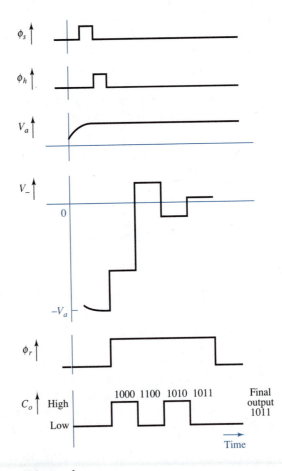

(b) Waveforms

Figure 9.23   Charge redistribution ADC

applied to two series capacitances of C each. Since the initial charge in the bank is $2V_aC$, a net voltage of $V_- = (V_R/2) - V_a$ appears at the inverting input of the comparator. With $V_a > (V_R/2)$, the comparator output C_o goes to logic high. At the beginning of the next clock, the control logic keeps the msb at 1, and the digital value of 1100 is applied. At this value, V_- becomes $(3/4)V_R - V_a$, which is positive. This causes C_o to go low; hence, at the third clock, 1010 is tried. Now, $V_- = [(5/8)V_R - V_a] < 0$, which makes $C_o =$ high. Hence, the final value tried is 1011. At the end of the fourth clock, the digital output is 1011 or 1010, depending on whether C_o is high or low.

As with the charge redistribution DAC, a higher resolution ADC can be obtained by using a cascade of capacitor banks. These ADCs are available in the CMOS technology with 8- to 16-bit resolution.

9.2.5 TRACKING AND DUAL SLOPE ADCs

A simpler alternative to successive approximation ADC uses a counter and a DAC, as shown in Figure 9.24. The reference voltage V_R of the DAC is such that V_d at full scale (that is, for the counter output of 11 .. 11) is equal to the maximum analog input voltage (V_a) to be converted. Conversion begins by resetting the counter to 0 at the arrival of a new sample of the analog voltage V_a. With the DAC output $V_d < V_a$, the comparator output is at logic high, and the clock is applied to the counter. As the stair step voltage V_d increases due to the increasing counter output, the comparator output switches to logic low when V_d reaches V_a. The counter output now corresponds to the digital valve of V_a within ($\frac{1}{2}$) lsb. Conversion resolution is improved by using more bits for the counter and the DAC. The disadvantage of this simple scheme is that the DAC output V_d always starts at 0 before tracking V_a. Hence, the conversion time is variable and depends on the amplitude of V_a.

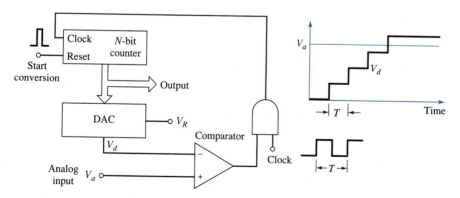

Figure 9.24   Counter ramp ADC

Conversion can be speeded up by using an up/down counter. Instead of resetting the counter at the beginning of each conversion, the comparator output is used to count up or down from the previous counter value. The counter output and the DAC voltage therefore track the analog input voltage V_a. Tracking ADCs with up/down counters are useful in low-cost, low-speed applications such as the measurement of slowly varying signals from temperature transducers and strain gauges.

Another type of ADC that uses a counter is the dual slope converter shown in Figure 9.25. Instead of using a DAC in the feedback loop, this converter employs an integrator to integrate the analog voltage V_a over a fixed time interval. Charge accumulated in the integrator capacitor is discharged at a constant rate

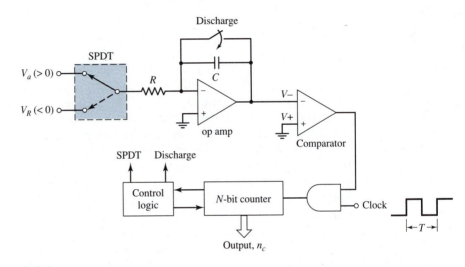

(a) Circuit

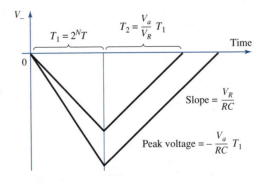

(b) Integrator Output Waveform

Figure 9.25   Dual slope converter

to zero. The ratio of the charging and discharging time intervals is proportional to the digital value of V_a.

Conversion in the circuit of Figure 9.25 begins by discharging the capacitor C and connecting the SPDT switch at the input of the op amp integrator to the input voltage V_a (> 0). As the capacitor charges, the N-bit counter is simultaneously enabled to increment from zero. Since the charging current derived from V_a is (V_a/R), the output of the integrator (which is connected to the inverting input V_- of the comparator) ramps down with a slope of (V_a/RC) (Figure 9.25b). With the reference (noninverting) voltage V_+ at zero, the comparator output is at logic high for positive V_a; hence, the clock signal passes to the counter. When the counter reaches the full count and resets to zero after 2^N clock pulses, the integrator input is switched to a reference voltage $V_R < 0$. Since V_R is negative, the accumulated charge in C is discharged at the rate of (V_R/RC). The comparator output is still high, so the counter continues to count from its reset state. When V_- reaches zero, the comparator output switches to low, and the counter is disabled. The output of the counter n_c is now proportional to the digital value of V_a.

The time T_1 during which C is charging is $2^N T$, where T is the period of the clock. The output of the integrator, V_-, reaches its peak value of $V_p = -(V_a/RC)T_1$ at the end of T_1. During T_2, C discharges from its initial voltage V_p with a slope of (V_R/RC). Time interval T_2, when the comparator output switches from high to low, is therefore given by

$$\left(\frac{V_R}{RC}\right)T_2 = \left(\frac{V_a}{RC}\right)T_1$$

or

$$T_2 = \left(\frac{V_a}{V_R}\right)T_1 \tag{9.7}$$

Since the counter output n_c at the end of T_2 is proportional to the time interval T_2 as $T_2 = n_c T$, and $T_1 = 2^N T$, the above equation becomes

$$n_c = \left(\frac{V_a}{V_R}\right)2^N \tag{9.8}$$

if $|V_R| = V_a|_{\max}$, a counter output of $n_c = 2^N$ corresponds to the full-scale digital value for $V_a = V_a|_{\max}$, and $n_c < 2^N$ is proportional to $V_a < V_a|_{\max}$, within the least significant count.

The resolution of this converter can be improved by increasing the size of the counter. Although the converter features a variable conversion time, it has other advantages. Equation (9.8) shows that conversion is independent of the tolerances of both R and C. Only the clock must be stable with time for repeatability of the converted data. Because of the inherent low-pass filtering in integration, high-frequency noise at the input is eliminated. In addition, if $T_1 = 2^N T$ is chosen as a multiple of the power line period, the effects of low-

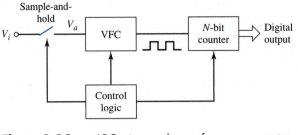

Figure 9.26 ADC using a voltage-to-frequency converter

frequency power supply harmonic noise is minimized. To take advantage of the latter, however, a long integration time must be allowed. For this reason, dual slope converters are usually used for converting slowly varying or noisy signals. High-accuracy digital voltmeters, for example, are built using dual slope converters. IC converters using the dual slope technique are available in the MOS and bipolar technologies with an external capacitor for integration.

9.2.6 ADC USING VOLTAGE-TO-FREQUENCY CONVERSION

A voltage-to-frequency converter (VFC) provides a simple method of converting slowly varying analog signals to digital data. Figure 9.26 shows the scheme for a VFC-based ADC. Any free-running square-wave generator (astable circuit, described in Section 8.3) with provision for controlling the frequency with an external voltage can be used as a VFC. In addition, voltage-controlled oscillators are available as single-chip integrated circuits. By counting the number of clock pulses over a fixed time interval, we obtain a digital value proportional to the input analog voltage.

The conversion resolution depends on the sensitivity of the VFC with voltage, the counter resolution, and the counting duration. VFCs are available for high-frequency (in the MHz range) and high-precision (0.1 percent nonlinearity) clock generation. VFC-based converters are useful in low-cost applications for slowly varying signals.

SUMMARY

This chapter presents the techniques and circuits for digital-to-analog and analog-to-digital conversion. D/A conversion requires generating a voltage proportional to the digital input data. Using a reference voltage source V_R, an N-bit DAC converts straight binary data to a voltage kV_R, with a resolution of $V_R/2^N$ where k is proportional to the binary input.

DACs are implemented using (a) binary-weighted resistors, (b) R-$2R$ resistor ladders, or (c) binary-weighted capacitors. In resistive networks, a voltage proportional to the digital data is obtained by resistive voltage division and the closing of analog voltage switches. Alternatively, binary-weighted currents can be switched into a buffer amplifier and converted to voltage. Charge redistribution yields the needed output voltage in a weighted-capacitor bank.

Conversion of other than straight binary data can be performed using bipolar reference voltages and modifying the converter circuit.

A DAC can be used as a programmable-gain amplifier by applying the time-varying signal as the reference voltage V_R. A DAC used in such an application is called a multiplying DAC. Other uses of multiplying DACs include power-series and sine waveform generation and voltage division.

The voltage comparator is a basic analog–digital interface device. Its output is at logic high or low, depending on the relative voltages at its inverting and noninverting inputs. By nature, a comparator quantizes an analog voltage as being above or below a reference voltage. Hence, a comparator is a 1-bit analog-to-digital converter.

Paralleling many comparators, each with a different reference voltage, and encoding the digital outputs of the comparators, we obtain an N-bit digital conversion of analog voltage. Such an ADC, called a flash or parallel ADC, is very fast, with only the propagation delay of the comparators. However, a large number of comparators is required.

For converters requiring long time intervals, analog signals must be sampled and held constant during conversion. A sample-and-hold circuit uses a capacitor to charge to the analog voltage at the time of sampling; with low-leakage capacitors and buffered circuitry, this charge can be held virtually constant during the conversion interval.

Successive approximation ADCs use N step digital approximation to obtain N-bit binary data. A DAC in a feedback loop converts the digital data and a comparator controls the approximation. This ADC takes N clock periods to complete a conversion.

A variation on the successive approximation algorithm uses a switch matrix or a capacitor bank in place of a DAC in MOS implementations.

Several other methods of analog-to-digital conversion use a counter for slowly varying signals. In the counter ramp method, a counter is activated at the beginning of a new analog voltage sample V_a. A DAC converts the counter output and applies the converted voltage V_d to a comparator. When V_d equals V_a, the counter is disabled. The conversion time depends on the analog voltage. An up/down counter is used to track the input voltage and reduce the conversion time.

A dual slope ADC uses a capacitor to integrate the analog voltage over a fixed time period. The capacitor is discharged at a constant rate, using a reference voltage, and a counter is run simultaneously. The output of the counter at the end of the discharge interval is proportional to the analog voltage.

Output pulses from a voltage-to-frequency converter are counted over a fixed interval to obtain the digital equivalent in a VFC-based ADC.

REFERENCES

1. Hoeschele, D. F., Jr. *Analog-to-Digital/Digital-to-Analog Conversion Techniques.* New York: Wiley, 1968.

2. Hnatek, E. R. *A User's Handbook of D/A and A/D Converters.* New York: Wiley, 1976.

3. *Data Converter Reference Manual* I and II, Analog Devices, Norwood, MA: 1992.

4. *Analog-Digital Conversion Handbook,* ed. D. H. Sheingold. Englewood Cliffs, NJ: Prentice Hall, 1986.

5. McCreary, J. L., and P. R. Gray. "All-MOS Charge Redistribution Analog-to-Digital Conversion Techniques—Part I." *IEEE J. Solid-State Circuits* SC-10, no. 6 (December 1975), pp. 371–79.

6. Singh, S. P.; A. Prabhakar; and A. B. Bhattacharyya. "C-2C Ladder-Based D/A Converters for PCM Codes." *IEEE J. Solid-State Circuits* SC-22, no. 6 (December 1987), pp. 1197–1200.

7. Allen P. E., and E. Sanchez-Sinencio. *Switched Capacitor Circuits.* New York: Van Nostrand Reinhold Company, 1984.

8. Lee Y. S.; L. M. Terman; and L. G. Heller. "A Two-Stage Weighted Capacitor Network for D/A-A/D Conversion." *IEEE J. Solid-State Circuits* SC-14, no. 4 (August 1979), pp. 778–81.

9. Hamade, A. R. "A Single Chip all-MOS 8-bit A/D Converter." *IEEE J. Solid-State Circuits* SC-13, no. 6 (December 1978), pp. 785–91.

10. Haykin, S. *Communication Systems.* New York: Wiley, 1994.

REVIEW QUESTIONS

1. What is the ratio of largest-to-smallest resistance in a weighted resistor network DAC for 8-bit conversion?

2. How many capacitors are needed for an 8-bit charge redistribution DAC? What is the ratio of largest-to-smallest capacitance?

3. What is a multiplying DAC?

4. What is a 1-bit quantizer?

5. How many comparators are needed for an 8-bit flash converter?

6. What is the conversion time for an 8-bit ADC, using the successive approximation method?

7. How many resistors and switches are needed in an 8-bit switch matrix ADC?

8. How many capacitors are needed for a single-stage 8-bit successive approximation ADC? What is the ratio of largest-to-smallest capacitance?

9. If the clock period is 1 ms and the current digital output is 1000 0000, what is the time needed to convert the next sample, which corresponds to 1000 1000, using an 8-bit tracking ADC with up counting only?

10. What is the time needed in the previous problem if the converter uses an 8-bit up/down counter?

11. If the clock period is 1 ms, what is the worst-case (maximum) conversion time for an 8-bit ADC using the dual slope method?

12. What happens if $V_a > V_R$ in a dual slope ADC?

PROBLEMS

1. A 4-bit DAC using bipolar current sources is shown in Figure P9.1. Derive the expression for V_o.

2. Calculate the output voltage V_o for the 4-bit charge redistribution DAC shown in Figure 9.8a for inputs 0001, 1000, and 1111, and verify Equation 9.4.

3. Calculate the output voltage V_o for the two-stage DAC circuit of Figure 9.9.

4. Show that by connecting C in Figure 9.8a to V_R during ϕ_1 and to ground during ϕ_2, a bipolar output is obtained. Calculate V_o.

5. Calculate the output voltage V_o for the two's complement DAC circuit of Figure 9.11, and verify the weights shown in Table 9.1.

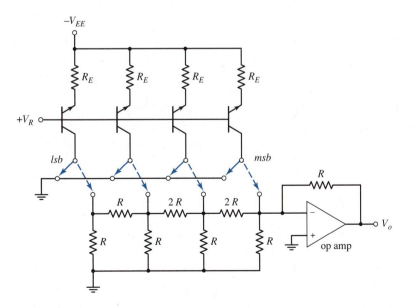

Figure P9.1

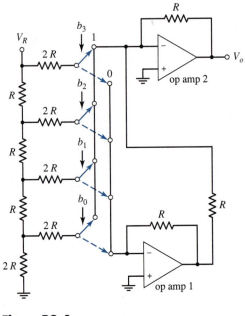

Figure P9.6

6. A 4-bit offset binary DAC is shown in Figure P9.6. Calculate and sketch the transfer characteristic between the digital input $b_3b_2b_1b_0$ and converted voltage V_o. Modify the circuit to obtain $V_o = 0$ for $b_3b_2b_1b_0 = 1000$.

7. Show that $R_S = 8R$ for an 8-bit straight binary D/A converter using the scheme shown in Figure 9.12. What is the resistance seen by the summing node of the op amp?

* 8. A BCD-to-analog converter may be constructed using the ladder network shown in Figure P9.8. If each decade box has the $R\text{-}2R$ ladder network shown for the units digit, determine the reference voltage applied to each box. What is the output voltage V_o?

9. How is V_o related to V_i in the multiplying DAC circuit shown in Figure P9.9?

10. *a.* Verify the encoding logic for the 3-bit flash ADC shown in Figure 9.17.
    b. Determine the encoding logic required to obtain the following gray code output. Note that if comparator C_k is high, then all C_i, $i = 0, 1, .. (k - 1)$, are also high.

	None	C_0	C_1	C_2	C_3	C_4	C_5	C_6
b_2	0	0	0	0	1	1	1	1
b_1	0	0	1	1	1	1	0	0
b_0	0	1	1	0	0	1	1	0

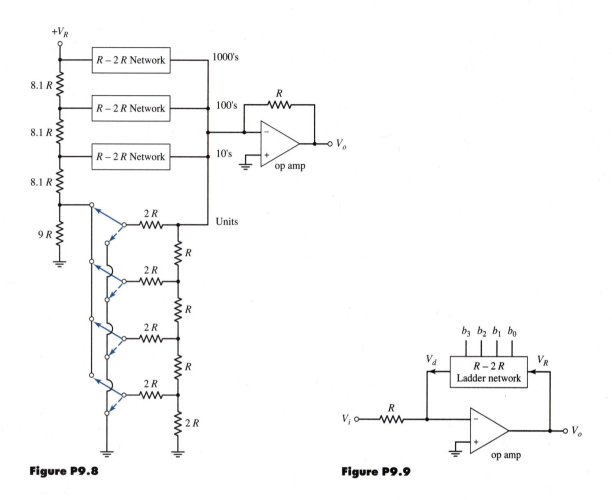

Figure P9.8

Figure P9.9

11. The 3-bit ADC shown in Figure 9.22 has +5 V connected to the upper end of the resistor string and −5 V connected to the lower end (instead of ground). Show which switches are successively closed for an analog voltage of (a) −2 V, and (b) +4 V.

What are the digital values in each case?

12. A flash DAC can be implemented similar to the fiash ADC shown in Figure P9.12. In this figure, each switch is closed by the appropriate decoded value as indicated for 2-bit conversion. Determine the output voltage for each input combination. Compare the circuit complexity (including the number of gates in the decoder and the number of switches) and the speed of this converter with the R-$2R$ ladder and the charge redistribution converters.

*13. A commonly used VFC for ADC is shown in Figure P9.13. Assume that V_a is held constant for a long time. Sketch the waveforms at the monostable and

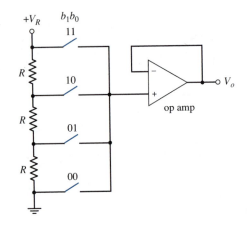

Figure P9.12

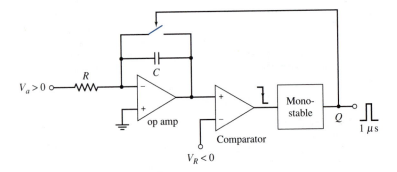

Figure P9.13

the op amp outputs. Derive the frequency of the square wave in terms of V_a, R, C, and V_R. If the pulse width of the monostable circuit is neglected compared to the period, what is the frequency for $V_a = 5$ V, given $R = 10$ kΩ, $C = 1000$ pF, and $V_R = -10$ V? What is the frequency resolution $\Delta f/\Delta V$? If an 8-bit counter is used, what is the counting duration (i.e., the hold time of V_a) for the full-scale output of counter at $V_a = 10$ V? If the same time interval is used for converting V_a in the range of $0 < V_a < 10$ V, how much must V_a change for a change in the lsb of count?

BASIC OP AMP CHARACTERISTICS

The simplified model of an op amp is shown in Figure A9.1. In practice, R_i is very large and R_o is very small compared to other circuit resistances, and the open-loop gain A is large. Hence, in most applications, an op amp may be considered ideal with the following characteristics: (a) input bias current are zero, that is, $i_+ = i_- = 0$, or, $R_i = \infty$, and (b) input voltages track each other, that is, $v_+ = v_-$ for finite v_o. Note that if v_+ is different from v_-, v_o is limited by the power supply, $+V_S$ or $-V_S$. With IC amplifiers that have open loop gains above 100,000, input resistances in the tens of MΩ range, and output resistances less than 100 ohms, the ideal amplifier assumptions introduce negligible errors in data converter applications. The effect of nonzero bias currents i_+ and i_- (on the order of 10^{-15} A) makes $v_+ \neq v_-$. This effect may be reduced by connecting resistances from the noninverting and inverting terminals to ground. Voltage drops in these resistances cause $v_+ \approx v_-$.

In the inverter amplifier configuration shown in Figure A9.2, the noninverting terminal is grounded, and the input is applied to the inverting terminal via R_1. With ideal amplifier property (b), we have $v_+ = v_- = 0$. Since $v_- = 0$ by virtue of grounding the noninverting input, the inverting input terminal is commonly referred to as virtual ground. This terminal is also known as the summing junction, because of the way currents from different input sources add at this node.

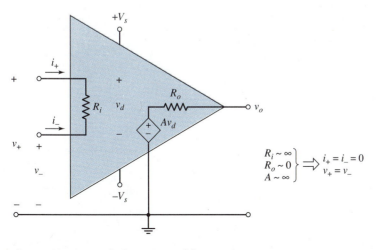

Figure A9.1   Ideal op amp model

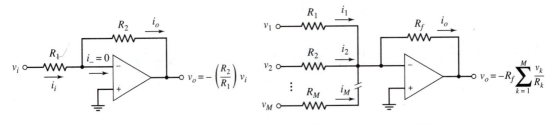

Figure A9.2   Inverting amplifier   **Figure A9.3**   Summing amplifier

With $v_- = 0$, the current from the input source is $i_i = v_i/R_1$, which is proportional to the input voltage v_i. Hence, the circuit performs voltage-to-current conversion. Since current $i_- = 0$, current i_o through resistor R_2 is $i_o = i_i = v_i/R_1$, and the output voltage $v_o = v_- - i_o R_2 = -v_i (R_2/R_1)$. The voltage gain of the circuit is therefore given by $v_o/v_i = - R_2/R_1$.

When M input sources are connected as shown in Figure A9.3, the current from each source is summed at the virtual ground node, and the output current is given by $i_o = i_1 + i_2 + .. + i_M = \dfrac{v_1}{R_1} + \dfrac{v_2}{R_2} + .. + \dfrac{v_M}{R_M}$. For this reason, the noinverting terminal of the op amp is known as the summing junction. The output voltage v_o is given by $v_o = -R_f\left(\dfrac{v_1}{R_1} + \dfrac{v_2}{R_2} + .. + \dfrac{v_M}{R_M}\right)$.

10

SEMICONDUCTOR MEMORIES AND VLSI SYSTEMS

INTRODUCTION

In this chapter, we study the structure and organization of semiconductor memories used for storing large quantities of data. As we dicussed in Chapter 8, a large-size semiconductor memory can be built using a collection of flip-flops or bistable multivibrators, each of which stores one data bit. While bipolar and *static* MOS memories use flip-flops, *dynamic* MOS memories employ charge storage in parasitic capacitances to represent data. Because of their regular structure and the complexity of the associated circuitry for data transfer in both types of memories, their designs are suitable for VLSI design methodologies. With the ever increasing need for large memory capacity, the designs of the memory circuits and devices for fast data access, high density, and low power dissipation are important. The first part of this chapter discusses the basic circuits used in static and dynamic *read/write memories* and nonvolatile *read-only memories*.

Combinational logic functions may be realized by storing truth tables in read-only memories. By modifying the memories as arrays of AND–OR or quivalent gates, whose inputs can be programmed, we can implement many desired functions. Additionally, we can fabricate large arrays of gates, or blocks of multiple levels of combinational circuits, leaving user-programmable connections for input and output. With the addition of flip-flops, these circuits can be programmed to realize desired combinational and sequential functions. Such *programmable logic devices,* with thousands of gates and flip-flops available, are a cost-effective alternative to designing a complex system as a custom-made VLSI system. These devices use modified circuit configurations discussed in Chapters 4, 5, and 7 to obtain low voltage operation, high density, or low power dissipation with adequate noise margins. Because of their increasing applications in ASICs, the size and the importance of available programmable logic devices are growing. In the second part of this chapter we discuss the structure and organization of some of these programmable logic devices.

10.1   BASIC RANDOM ACCESS MEMORY ORGANIZATION

Semiconductor memory organization may be viewed as a three-dimensional array of memory cells, each of which stores one bit of data. If N bits of data are accessed in parallel or, simultaneously, for reading from or writing to the cells, the *data width* or *word size* of the memory is N bits. The *size,* or *capacity,* of the memory refers to M words, so that the memory storage capacity is MN bits. To access each of the M words uniquely, an address size of K bits, where $2^K = M$, is required. Memory size M is usually expressed in powers of 2, with $2^{10} = 1,024$ denoted by K (Kilo), and $2^{20} = 1,048,576$ denoted by M (Mega), etc. A group of eight bits of data is a *byte,* abbreviated B. Memory size is specified in terms of KB, MB, etc. (Note that K and M refer to sizes that are slightly more than 1,000 and 1,000,000, respectively.)

Figure 10.1 shows the signals and buses in a typical semiconductor memory chip. The capacity of the chip is MN bits, which are organized as M words $\times$ N bits. The read or write operation is activated by the $R/\overline{W}$, or write enable (WE), line that is connected to all M words. In general, the input and output are separated in a *Read/Write Memory* (RWM); hence, there are two data buses, as shown in Figure 10.1. Any word of N bits in the chip can be accessed independently of any other word. Data from the selected word are available on the output data bus when the $R/\overline{W}$ signal is at logic high. Writing to a selected word is accomplished by placing the data on the input data bus and taking the $R/\overline{W}$ signal low.

The outputs from each word are buffered by tristate gates, so that only the selected word activates the output data bus. The inputs to each word are connected together and data are written to a selected word from the input data bus. The inputs may also be buffered by tristate gates. Additionally, an external data bus may be connected to a bidirectional buffer before being connected to/from the memory cells. Further, a memory chip typically has a Chip Enable line which, when deactivated, puts all input and output lines in the high-impedance state, so that the buses are relinquished for use by other devices.

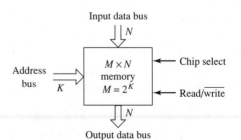

Figure 10.1   Typical signals and buses in an RWM chip

Because any word can be accessed randomly, the memory system described above is commonly known as *Random Access Memory (RAM)*. The *access time* is the time required for valid data to become available from, or stored to, a word; this time is independent of the physical location of the word. Since data can be retrieved from or stored to any word, a better description of the memory is *Read/Write Memory* (RWM), as opposed to *Read-Only Memory (ROM)* (see Section 10.5). On the other hand, a shift register (Chapter 8) is a *serial access memory*.

Since the data are lost when power is shut down, semiconductor RWMs are called *volatile* memories. ROMs are *nonvolatile* under normal operating conditions. In addition, a RAM (RWM) that retains data indefinitely with the power on, but does not require external *refreshing*, is called *Static RAM (SRAM)*. *Dynamic RAM (DRAM)* stores data in a leaky capacitor whose charge must be restored periodically to retain the data. Data from dynamic RAM are read by transferring charge to a transistor; hence, DRAM readout may be destructive.

A conceptually simple RWM organization scheme is shown in Figure 10.2. The random access of any word is possible with the *address decoder* which

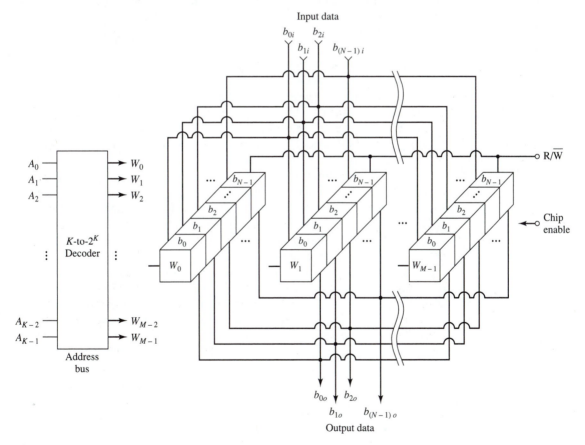

Figure 10.2   Basic memory organization

selects one out of $2^K(= M$, usually) lines, where K specifies the number of *address bits.* A typical memory size of 1 K words, for example, needs a $\log_2(1024) = 10$-bit address bus to access any of the 1024 words; hence, a 10-bit-to-1024-line decoder is required. This type of addressing a word by selecting one of M words using a single decoder is known as *one-dimensional* or *linear addressing.* The data width, or word size, can be increased from N to $N + P$ by increasing the number of cells in each of the M words by P; however, the address width remains the same. The disadvantage of linear addressing is the large size of the decoder required. To decode a 2-bit address in order to access one of four words, for example, four two-input AND gates and four inverters are required. As the size M of the memory grows, the total number of gates in the decoder increases linearly with M and the number of inputs per gate increases as $K = \log_2 M$. To reduce the size of the decoder, a two-dimensional decoding scheme is commonly used.

The two-dimensional address decoding scheme splits the K ($= \log_2 M$)-bit address lines into two groups of K_1 and K_2 bits each such that $K_1 + K_2 = K$. Usually $K_1 = K_2 = K/2$, so that two identical decoders of $(K/2)$-to-$(M/2)$ are needed, instead of a single K-to-M decoder. The two decoders, referred to as row (X-) and column (Y-) decoders, select a row and a column. The word located at the intersection of the selected row and column is activated for reading or writing. The implementation of these decoders is considered in Section 10.4.

The timing requirements for RWM are shown in Figure 10.3. *Read access time* is the delay between the time a stable address on the address bus is presented (with the chip selected) and the time the data stored at that address is available on the data bus. This delay arises from the propagation delays of (a) the address decoder, (b) the sense amplifier, which 'senses' the data stored in the memory cells (as discussed in the following section), and (c) the output data buffer. If the chip select line is not active at the same time that the address lines are valid, a slightly smaller chip-select-to-output-time delay occurs. The *read cycle time* is the minimum time required between successive address changes for the read operation. This time is usually determined by the external system clock. As seen in Figure 10.3, the read cycle time is slightly longer than the read access time, to allow for a margin of safety, as well as for such operations as error detection and parity checking.

During the write cycle, a valid address and chip select must be active before the write pulse becomes active (and the data to be stored are ready). This is the *write setup time.* The data must be valid for the *data setup time* during the active write pulse, and must stay valid for the *data hold time* after the write pulse. The *write time* or *write pulsewidth* is the minimum duration of the active write pulse. The *write cycle time* is the time required between successive write operations; this is also the minimum time required for the address to be valid during write operations. The write cycle time is slightly longer than the sum of the write time and the write setup time.

The *memory cycle time* is the minimum time needed for data transfer, and is the longer of the read and the write cycle times.

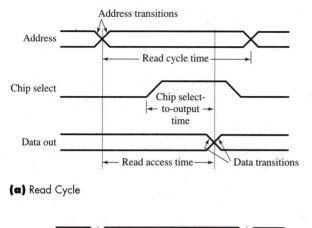

(a) Read Cycle

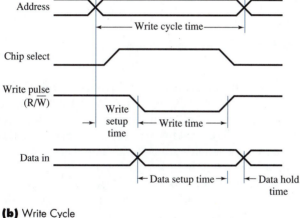

(b) Write Cycle

Figure 10.3   Timing requirements for RWM access

10.2 STATIC READ/WRITE MEMORY CELLS

Each bit, or cell, in a static RAM (SRAM) is a flip-flop with capability for addressing, reading, writing, and combining with other cells to create larger storage capacity. RAM cells are implemented with bipolar (TTL, ECL, I²L), MOS (NMOS, CMOS) and BiCMOS technologies. Access time, power dissipation, and chip density depend on the technology used.

10.2.1 BIPOLAR STATIC RAM CELLS

Figure 10.4 shows a TTL static RAM cell. The flip-flop, consisting of multi-emitter transistors Q_1 and Q_2, uses the additional emitters for the coincident row and column selections. True data (DATA) and complement data ($\overline{\text{DATA}}$) bit-

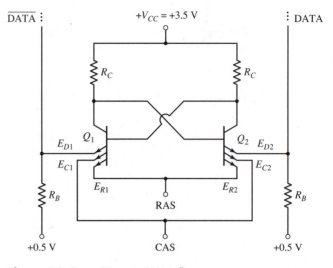

Figure 10.4 TTL static RAM cell

lines from each word of N bits are connected to the corresponding bit positions in all words. Hence, there are N pairs of DATA ad $\overline{\text{DATA}}$ lines, also known as *one-sense* and *zero-sense lines,* each running through all M cells. The same DATA and $\overline{\text{DATA}}$ lines are used to carry cell input and output data. Auxiliary circuits, known as sense amplifiers and write amplifiers, are used to control the direction of data flow.

Cell operation has three modes: standby, read, and write. With circuit supply voltages at 3.5 V and 0.5 V, the logic low and logic high levels may be at less than 0.5 V and greater than 3 V, respectively. Further, the *Column Address Select (CAS) lines* and *Row Address Select (RAS) lines* are at active high when a word (and hence all the cells in that word) is selected. In the static condition, let Q_1 in Figure 10.4 be in saturation and Q_2 be off, so that the stored data bit is 1. If the cell is not selected, that is, either CAS or RAS is low (or both are low) at about 0.3 V, the saturation base and collector currents of Q_1 are shared by the low address select line. The $\overline{\text{DATA}}$ line connected to the emitter E_{D1} of Q_1 is at a voltage higher than 0.3 V, so E_{D1} is off. With a saturation collector–emitter voltage of 0.2 V and base-emitter voltage of 0.8 V, the collector of Q_1 is at 0.5 V and the base is at 1.1 V. At these voltages, Q_2 is off and the voltage at the DATA line is controlled by other cells.

The cell is addressed for reading by raising both CAS and RAS to 3 V. Emitters E_{C1} and E_{R1} are now turned off, while E_{D1} conducts. Current from E_{D1} is sensed by the sense amplifier (Figure 10.5a) and the output of the amplifier goes to logic low. Valid data from the cell is therefore available at the DATA outut, after the response time of the flip-flop and the sense amplifier. The sense amplifier may also have complementary outputs, as shown in Figure 10.5a.

When data are being written to the cell, the DATA and $\overline{\text{DATA}}$ lines connected to the complementary outputs of the write amplifier (Figure 10.5b) raise

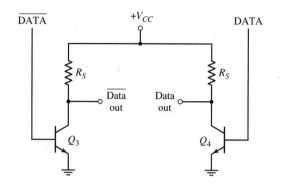

(a) Sense Amplifier

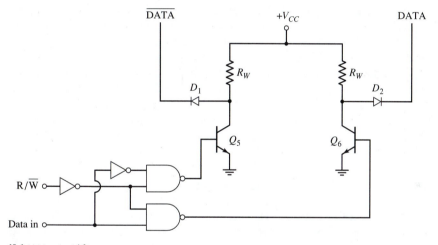

(b) Write Amplifier

Figure 10.5   Amplifiers for the RAM cell of Figure 10.4

the voltage of E_{D1} (or E_{D2}) to high, while lowering the other. If, in the present example, a 0 is to be stored in Q_2, the $\overline{\text{DATA}}$ line goes high to 3 V. Since the cell is selected with CAS and RAS also high, Q_1 is turned off. Device Q_2 now conducts via R_C, E_{D2}, and R_B. The collector of Q_2 becomes low (at about 0.8 V), so that Q_1 stays off, and the stored data remain until changed.

If CAS (or RAS) alone is high, the on transistor continues to conduct via the emitter connected to RAS (CAS). Hence, even when the DATA (or $\overline{\text{DATA}}$) line is low (while reading from or writing to other cells), unaddressed cells are virtually disconnected from both the DATA and $\overline{\text{DATA}}$ lines.

Examples of sense amplifier and write amplifier circuits are shown in Figures 10.5a and b. Both diodes in the write amplifier are reverse-biased during the read operation ($R/\overline{W} = 1$), so that the DATA (or $\overline{\text{DATA}}$) line carries the on transistor current. The sense amplifier output (Data out) therefore corresponds to the stored bit. During writing ($R/\overline{W} = 0$), $Q_6(Q_5)$ is off for the data input of 1 (0). Hence, the voltage at $E_{D2}(E_{D1})$ is raised, regardless of the state of $Q_2(Q_1)$, and the new data are stored. In standby mode, $R/\overline{W} = 1$, and both Q_5 and Q_6 are on. Hence, the DATA and $\overline{\text{DATA}}$ lines are unaffected by changes in data input, and the flip-flop remains in its previous state. Since CAS or RAS is low, Q_3 and Q_4 are driven by other cells. Note that both a sense amplifier and a write amplifier are used for each bit in an N-bit word with all the M words connected together.

The access time of the TTL cell depends on the switching speed of the transistors and the charging time for the long capacitive lines, DATA and $\overline{\text{DATA}}$.

Power dissipation is a major consideration in a large-sized SRAM. For the TTL cell, excluding the amplifiers, standby power is nearly 1 mW. Hence, the maximum number of cells per chip, without external cooling, is limited to between 500 and 1000.

Variations of the TTL SRAM cell have been developed for reduced power dissipation and increased density per chip. Faster access time is available using nonsaturating ECL-type cells with multiple emitters for address selection. Bipolar SRAMs using Schottky-diode-coupled RAM cells provide low power dissipation and a better delay–power product. Still, because of the their high power dissipation compared to MOS memories, bipolar SRAMs are limited in size to below 32 Kbits. Bipolar (TTL and ECL) SRAMs have typical access times in the range of 5 to 10 ns. Applications of these memories are limited to small systems and those requiring high-speed access such as buffer and cache memories.

10.2.2 MOS STATIC RAM CELLS

NMOS and CMOS static RAMs have much lower power dissipation and higher density compared to bipolar RAMs of the same size. MOS RAMs have the generic structure shown in Figure 10.6. The load devices L_1 and L_2 for the flip-flop may be NMOS depletion- or enhancement-type transistors, or PMOS-FETs in the case of a CMOS cell. Each can also be an undoped polysilicon resistor. In an NMOS flip-flop, large-valued polysilicon resistors reduce the standby power of the on transistor. Although the complexity of forming undoped polysilicon sheet resistors increases the chip complexity, both the chip area and the power dissipation are decreased. The width-to-length aspect ratios (W/L) for NMOS load devices are kept small to reduce the conduction parameters and hence the standby current drain. Access between external data and the cell is provided via the pass transistors (transmission gates) P_1, P_2, P_3, and P_4, which are connected to row and column address select lines. When RAS is at logic low, the cell is isolated from the DATA and $\overline{\text{DATA}}$ lines. To read the cell, both RAS and CAS are raised to logic high, and sense amplifiers connected to the DATA and $\overline{\text{DATA}}$ lines detect the stored bit in the cell.

In Figure 10.6, assume, for example, that Q_1 is on and Q_2 is off, so that the data bit stored in the cell is 0. If $R/\overline{W} = 0$, then Q_3 in the Read/Write amplifier is off. With the gates of P_2 and P_4 (RAS and CAS) at logic high, $\overline{\text{DATA}}$ line goes

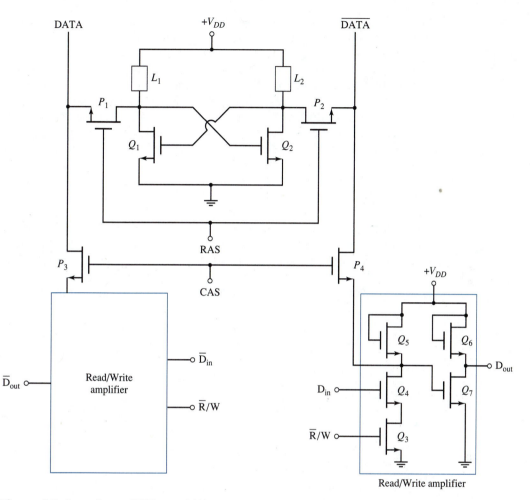

Figure 10.6 Generic MOS static RAM structure

high and Q_7 is turned on. Hence, D_{out} is at logic low. Since Q_1 is on, it carries the current of Q_5 in the Read/Write amplifier (bottom left block) connected to the DATA line. Hence, the gate of Q_7 in the DATA line is at low voltage and $\overline{D}_{out}$ is at logic high. Note that reading the cell value is nondestructive; that is, the value stored in the cell is unaltered by the reading process. This is ensured by designing Q_1 (as well as Q_2) such that its resistance is small compared to that of the pass transistors. With a large conduction parameter for Q_1, its drain will be at a voltage below the threshold voltage of Q_2, so that Q_2 will stay off during the reading process.

The cell value is altered in the flip-flop by asserting the DATA or $\overline{DATA}$ line. For example, to write a 1 in the cell, D_{in} and $\overline{R}/W$ are raised to V_{DD}. Devices Q_3, Q_4, and Q_5, which form a NAND gate with inputs D_{in} and $\overline{R}/W$ (bottom right amplifier), conduct and bring the voltage at the $\overline{DATA}$ line to logic low. Hence, the drain of Q_2 is at low voltage, which turns Q_1 off and raises the gate voltage of Q_2. The cell now stores a 1, with Q_1 off and Q_2 on. Writing to the cell is effected by forcing the gate of the on transistor to logic low. To ensure that the gate of Q_1 (which is connected to the drain of Q_2) is brought below the threshold voltage, the resistance of the load device L_2 must be large relative to the resistances of the on transistors (P_2, P_4, Q_3 and Q_4). Similarly, for writing a 0, L_1 must have a high resistance relative to the pass transistors and the NAND gate transistors. L_1 and L_2 are also required to have high resistances to reduce the standby current drain. Hence, the load devices have low conduction parameters.

Input and output data are connected via tristate buffers, which are controlled by chip select, output enable, and write enable signals. The use of tristate buffers enables the input and output data lines to be connected together and reduces the pin count. Figure 10.7 shows the block diagram of a typical 2 K × 8 bit memory chip. The bidirectional data bus, $D_0 - D_7$, is controlled by the pair of 8-bit buffers for reading and writing data. In general, the chip select signal, when deactivated, disables the data control and input/output (I/O) circuits, as well as the row and column address decoders. This reduces the standby power. The chip operates on a single 5 V supply, with signals compatible with

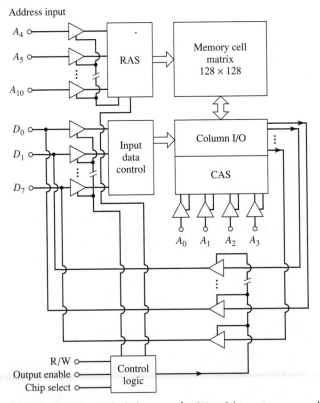

Figure 10.7   Block diagram of a 2K × 8 bit static memory chip

TTL levels. Chip select is also used to extend the size of the memory (Problem 10.2).

The memory access speed is limited by parasitic capacitances. The combined capacitance of the address select lines and the parallel pass transistors contribute to a large lumped capacitance. The layout must be carefully considered to minimize the distributed resistance and capacitance. To reduce parasitic capacitances, the substrate is usually biased at a negative voltage. (Note that when the source-to-substrate voltage is negative, the number of injected carriers from the source (n^+) to the substrate (p) is reduced). This negative bias voltage is generated on the chip, using a ring oscillator or an RC oscillator; the memory chip therefore operates on a single external positive supply. In addition, if the gates of the cell (Q_1 and Q_2 in Figure 10.6) are operating near their threshold voltages, only a small change in voltage (from $V_T + \Delta V$ to $V_T - \Delta V$) at the DATA and $\overline{\text{DATA}}$ lines is needed for writing data. Similarly, the sense amplifier may be designed to detect and amplify a small change in voltage between the 1 and 0 values and the valid output logic levels. These low voltage changes reduce the overall memory access time.

Figure 10.8 shows a CMOS static RAM cell. The flip-flop and the pass transistors for the DATA and $\overline{\text{DATA}}$ lines form a cell of six transistors. As with other CMOS circuits, the use of complementary devices reduces the power

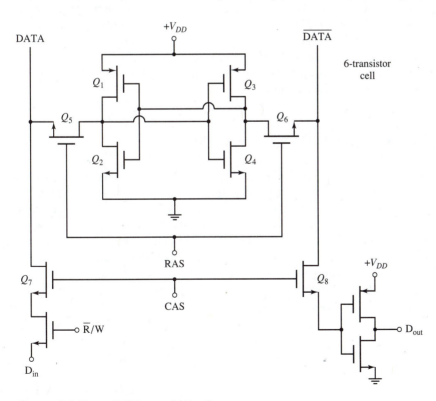

Figure 10.8   CMOS static RAM cell

consumption. Although a single-ended scheme for reading and writing is shown in Figure 10.8, more elaborate sense amplifiers and write buffers are commonly used to reduce the access time and improve performance. The amplifier–buffer combination senses 0 for reading and forces 0 for writing at the source–drain junction of the P and N MOSFETs. Design considerations for high-speed operation are similar to those of the NMOS cell.

Variations of the six-transistor CMOS cell provide increased density and reduced power consumption. The PMOS devices in Figure 10.8, for example, can be replaced by undoped polysilicon resistor loads. However, the design must consider data loss with temperature as well as transient currents due to unstable values of the load resistances.

In a BiCMOS memory cell, the NMOS pass transistor Q_6 in Figure 10.8 is replaced by an npn transistor (in the common-collector configuration), for high drive capability. Over the past decade CMOS and BiCMOS static RAMs have been increasing in density and speed. The Motorola MCM6206, for example, is a 32 K $\times$ 8 bit CMOS SRAM with a read cycle time of 30 ns. In addition to TTL compatible signals and tristate outputs, this chip has a chip enable input. When the chip enable is deactivated, the chip is put in low-power standby mode with the power supply current reduced from 140 mA to below 40 mA. The Motorola MCM6706 is a pin compatible BiCMOS version of the same size SRAM as MCM6206. While the CMOS chip has a maximum rating of 1 W of power dissipation, MCM6706 has 2 W rating with a high output drive capability and a 10 ns read cycle time. Specialty SRAMs with less than 10 ns access times are currently available in sizes ranging from 64 K $\times$ 18 to 64 K $\times$ 72 bits. These RAMs, which operate at 3.3 V supply, are used as cache memories in faster personal computers and workstations.

10.3   MOS DYNAMIC RAM

As in dynamic logic circuits, charge storage in MOSFET parasitic capacitances is used for temporary storage of data in dynamic RAMs (DRAMs). However, because of gate leakage currents, the charge must be restored (or refreshed) at periodic intervals for long time storage. Data *refreshing* is usually carried out once every 2 ms, using support circuitry. In addition, the readout may be destructive, in which case every read operation must be followed by a write operation. These drawbacks are more than overcome by the advantages of high density, low power dissipation, and reduced complexity. Therefore, for large-sized memories (typically above 1 MB) DRAMs are preferred over SRAMs.

10.3.1   FOUR TRANSISTOR NMOS DYNAMIC RAM

Figure 10.9a shows a four-transistor NMOS dynamic RAM cell. This circuit is the same as the generic static RAM cell shown in Figure 10.6, with LX_1 and LX_2 operating as row-selection devices, as well as load devices. With clock ϕ, the

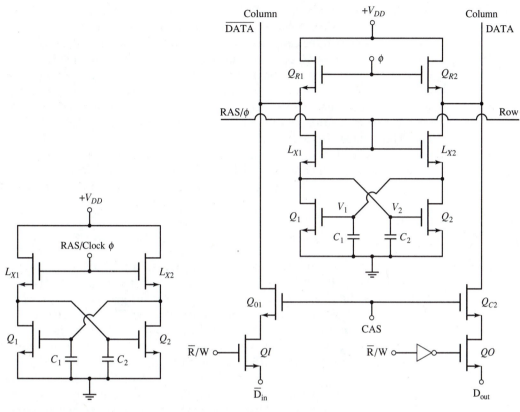

(a) RAM Cell   **(b)** DRAM Cell as Part of Memory

Figure 10.9   Four-transistor NMOS dynamic RAM

cross-coupled inverters that form the latch have dynamic loads. Charge is stored in the lumped parasitic capacitance at the gate node of Q_1 or Q_2 (whichever device is on) during the high level of the clock. When the clock is at logic low, this charge leaks through the reverse-biased n^+p (drain–substrate) junction. Since the leakage rate is low (a few pA), logic state of the latch is held for a short time. Activating the clock for approximately a microsecond at 2 ms intervals restores the lost charge.

The four-transistor cell forms part of a large memory, as shown in Figure 10.9b. To understand the operation of this memory, assume that Q_1 is on with its gate voltage V_1 (across capacitor C_1) above the threshold voltage, and Q_2 is off with its gate voltage $V_2 \approx 0$. As C_1 loses charge, refresh clock ϕ is applied. When ϕ is high, QR_1, LX_1, QR_2, and LX_2 are turned on. C_1, therefore, charges toward the supply V_{DD} via QR_2 and LX_2, while Q_1 conducts via QR_1 and LX_1. During this time, V_2 is held low by the conducting device Q_1. Therefore, the charge lost in C_1 is restored, while C_2 is virtually unchanged. Note that the on transistor gate voltage V_1 need only be raised slightly. During the refresh cycle (ϕ = high), the inverter device Q_1 (Q_2) has the on devices QR_1 and LX_1 (QR_2 and LX_2) as an active load.

The refresh devices QR_1 and QR_2 are shared by an entire column of memory, as are the devices connected to the DATA and $\overline{\text{DATA}}$ lines.

For writing 0 to the cell, the $\overline{\text{DATA}}$ line is taken high, while RAS and CAS are high. The gate of Q_2 goes high and charges C_2, while C_1 discharges via the drain of Q_2. With its gate voltage at zero, Q_1 is now off. Hence, the stable state of the latch is Q_1 off and Q_2 on. A 1 is written to the cell by taking the $\overline{\text{DATA}}$ line low, which discharges C_2 via Q_1 and charges C_1 toward V_{DD} via QR_2 and LX_2.

To read data from the cell, RAS and CAS are activated and $\overline{R}/W$ is taken low. The voltage across C_1 is then available at D_{out} via LX_2, QC_2, and Q_0. To avoid excessive discharge of C_1 (and the loss of data) during the read operation, C_1 must be large compared to the DATA line capacitance; otherwise, each read operation must be followed by a write operation to replenish the charge in C_1.

10.3.2   THREE-TRANSISTOR NMOS DYNAMIC RAM

If the devices for read and write operations are separated, data can be read nondestructively. Figure 10.10a shows the circuit of the three-transistor early NMOS DRAM cell. In this cell, the capacitance at the gate node of Q_2, rather than a cross-coupled inverter latch, stores the bit value. To read the cell, $\overline{D}_{\text{out}}$ is made high by *precharge* (see below), and the read line is taken high. If C is initially charged, the gate of Q_2 will be above the threshold voltage and Q_2 will be turned on. With Q_3 also on, the precharged $\overline{D}_{\text{out}}$ line conducts via Q_2 and Q_3, and $\overline{D}_{\text{out}}$ goes to zero. If no charge is initially stored in C (bit 0), the precharged $\overline{D}_{\text{out}}$ line stays high. The sense amplifier must detect this change in the $\overline{D}_{\text{out}}$ line. Note that capacitor C is not discharged during the read operation; hence, the cell may be read repeatedly, within 2 ms intervals, before being refreshed.

To write a 1 to the cell, the write line is taken high, which turns Q_1 on and charges C to the logic high voltage in the D_{in} line. A 0 in the D_{in} line discharges Q_1 when the write line is high. In practice, the D_{in} line may be precharged first, regardless of the value of input bit, and then C and D_{in} are discharged when a 0 is written.

To refresh the cell, the complement of the stored data ($\overline{D}_{\text{out}}$) is first read. This value is inverted and then rewritten in the cell. Again, in practice, a cell may be refreshed only if the stored value is 1.

One possible memory scheme that includes common refresh and precharge circuitry for a column of cells is shown in Figure 10.10b. For reading or writing, the D_{in} and D_{out} lines are first *precharged* to logic high via Q_4 and Q_5 by taking the Precharge line high. During this phase, the RX, WX, Read, Write, and Refresh lines are all low. After precharge, both the RX and Read lines go high, while the Precharge line stays low for the read operation. If charge is stored in C, Q_2 turns on; with Q_3 also on, the D_{out} line discharges to ground. If no charge is stored in C, Q_2 is off and the D_{out} line stays high. Hence, the complement of the stored bit value appears at the $\overline{D}_{\text{out}}$ (Q_9) output. This output may be inverted before being connected to an external pin. Note that the stored charge in C is used only to turn Q_2 on without being discharged during the read process.

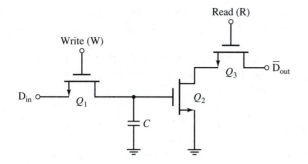

(a) Basic Cell

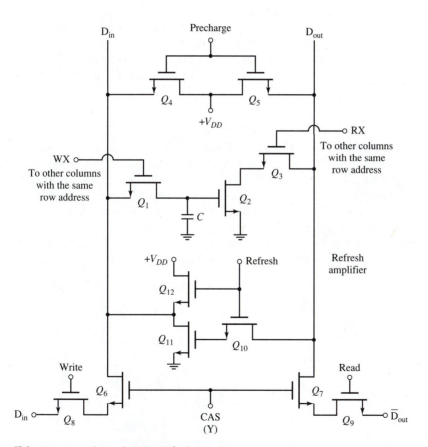

(b) Memory with Read–Write–Refresh Circuitry

Figure 10.10   Three-transistor NMOS dynamic RAM

For writing 0 ($D_{in} = 0$) to memory, the precharged D_{in} line is discharged to ground via Q_6 and Q_8, while the stored charge in C, if any, discharges through Q_1. If $D_{in} = 1$, C charges via Q_1.

To restore the charge in C, the RX and Refresh lines are taken high, while WX is low. If sufficient charge is stored in C, Q_2 and Q_3 are on. The gate of Q_{11} is then low, so the D_{in} line goes high. Thus, in the first phase of the refresh cycle, the refresh amplifier outputs the complement of the stored bit value on the $\overline{D}_{in}$ line. The WX signal then goes high to write the stored data back to the cell.

Each column of memory cells needs a refresh amplifier, and the cells in a column are refreshed sequentially. In Figure 10.10a, note that in a row of cells the WX and RX lines connect the Write and Read gate inputs. All the NY cells in a row are refreshed simultaneously by activating the appropriate WX and RX lines. If each row takes a time of T_R to refresh and there are NX rows, a total time of $NX \cdot T_R$ is needed for the complete refresh of $NX \cdot NY$ cells. During this time, no external data transfer can occur.

Although the stored charge in the capacitor is not discharged during the read process, precharging enables faster detection of a logic high at the D_{out} line than with a fixed (static) pull-up device. However, the operation of three-transistor dynamic memory requires complex timing and support circuitry.

10.3.3   ONE-TRANSISTOR CELL

Further reduction in the number of active devices is achieved in the one-transistor DRAM cell. As with the three-transistor cell, this cell also utilizes parasitic capacitance to store charge. The one-transistor cell is the most widely used cell in large capacity DRAMs available commercially. Although many variations are available, the basic circuit for the cell is shown in Figure 10.11a.

Parasitic capacitance C is formed of polysilicon-oxide-silicon and is accessed via the pass, or access, transistor Q for reading and writing. Note that the channel-to-body capacitance of the E-MOSFET is included in the storage ca-

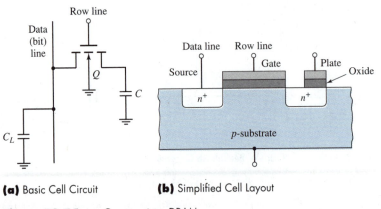

(a) Basic Cell Circuit   **(b)** Simplified Cell Layout

Figure 10.11   One-transistor DRAM

pacitance C. To write to the cell, the Row line (gate) is raised while the Data (bit) line is at a low (0) or high (1) level. Capacitance C is then charged or discharged via Q to the value in the bit line. The stored cell value is transferred to the Data line by raising the Row line to high. Because of the charge transfer between the bit line stray capacitance C_L and the storage capacitance C, we have $V_C = V_{CL}$ = voltage in the Data line in equilibrium. Since C_L is much larger than C ($C_L \approx 10C$) due to the large number of cells in a column, the direct transfer of logic high cell value during reading may correspond to an extremely low voltage. For a reasonable difference in the voltages read for the low and high logic levels, the Data line is precharged, as in the three-transistor cell. Note, however, that precharging the bit line causes the 0 bit stored in the cell to be lost, since the equilibrium voltage is higher than that of the storage capacitance C prior to the read operation.

The following example illustrates the typical voltages available.

Example 10.1

The storage capacitance C in Figure 10.11a has a value of 40 fF and the voltages are approximately 4.5 V and 0 at logic high and logic low, respectively. If the data line has a total capacitance of 300 fF and it is precharged to 5 V before reading, what are the line voltages corresponding to the logic high and logic low levels?

Solution

At $V_C = 4.5$ V and the bit line voltage of 5 V, the total stored charge is

$$Q_s = V_C C + V_{CL} C_L = 4 \cdot 5 \times 40 + 5 \times 300 = 1680 \text{ fC}$$

This charge is distributed between C and C_L so that the equilibrium voltage read at logic high level is

$$V_H = \frac{Q_s}{C + C_L} = \frac{1680}{340} = 4.94 \text{ V}$$

Thus, the data line voltage changes very little from its precharged value of 5 V. At $V_C = 0$ and the data line voltage of 5 V, the total stored charge is

$$Q_s = 5 \times 300 = 1500 \text{ fC}$$

Hence, the logic low equilibrium voltage read is

$$V_L = \frac{1500}{340} = 4.44 \text{ V}$$

Clearly, V_C changed after the cell was read, thereby destroying the stored cell value of 0.

To read the logic state of the cell, a sense amplifier connected to the Data line must detect and amplify the difference between the high and low equilibrium voltages at the Data line, given by $\Delta V = 4.94 - 4.44 = 0.5$ V.

As with other dynamic RAMs, the one-transistor cell loses its data due to leakage of the storage capacitor. Hence, circuitry for periodic refresh must be included, in addition to read-followed-by-refresh circuitry. The cell can be compact, as shown in Figure 10.11b; the source terminal of the MOSFET doubles as the column or bit line, and the gate is the row line. The diffused junction forms the drain of the MOSFET cell. This compact structure, in which charge storage occurs in the n^+ doped region beneath the polysilicon plate, permits a high chip density. Many other structures, including those without the source terminal of the MOSFET, have been developed for the one-transistor cell (see References 6 and 7).

Figure 10.12 shows a column of cells in a simplified organization of a 16 Kbit memory with a sense amplifier. The memory is organized as 128×128 cells, that is, 128 rows with 128 cells in each row. The storage capacitor plates of all cells are connected to V_{DD}. Each column of 128 cells is divided into two groups of 64, separated by the sense amplifier. At each end of a column, a dummy cell is used to raise the differential voltage for data sensing. The sense amplifier is a regenerative amplifier consisting of a cross-coupled flip-flop (Q_{S1}, Q_{S2}, Q_{S3}, and Q_{S4}), and a gating device (Q_{SG}). When there is a voltage difference between Q_{S3} and Q_{S4}, the difference is amplified by Q_{S1} and Q_{S2}, and the output lines of the amplifier (half-bit lines) go to low and high voltages.

The sequence of operation for reading the cell begins with precharging. When ϕ_P goes high, the precharged devices Q_{P1} and Q_{P2} conduct, and the half-bit lines are set to V_{DD}. Hence, the drain ends of both dummy cells Q_{D1} and Q_{D2} are charged to zero. The precharge clock ϕ_P is taken low after about 100 ns.

The selected row is then activated by the 7-to-128 row address decoder output. If the selected row is in the top half, for example, the charge stored in each cell in the selected row is redistributed with its corresponding half-bit line charge. For a stored cell value of logic high, the equilibrium voltage at the top half-bit line is about 4.9 V for a supply voltage of $V_{DD} = 5$ V; if a zero is stored in the cell, the top half-bit line voltage drops by a larger amount, from V_{DD} to about 4.5 V. While none of the rows in the bottom half-bit line is selected, the bottom dummy row is activated (ϕ_{DB} = high) during activation of the selected top row. (The process is reversed when the selected row is in the bottom half.) Since the dummy cell is at zero voltage, the bottom half-bit line has a reduced equilibrium voltage. The size of Q_{D2} (as well as Q_{D1}) is such that when it is activated, the charge redistribution causes a drop of about 250 mV at the bit line, to about 4.75 V—midway between the drops corresponding to the low and high values stored in the selected cell. Therefore, when the differential sense amplifier is activated by the isolation signal (ϕ_{SI} = high) and the gating signal (ϕ_{SG} = high), the difference in the bit line voltage is amplified.

If the selected top cell value is high, the top half-bit line goes to V_{DD} by boosting the amplified signal. The bottom half-bit line, which carries the dummy cell value of zero, goes to zero voltage. For a high stored cell value, the Data and $\overline{\text{Data}}$ lines carry high and low bit values respectively, when the column address select line goes high.

If the selected cell is in the bottom half of the column, the complement of the stored cell value appears in the Data line. Therefore, the value must be

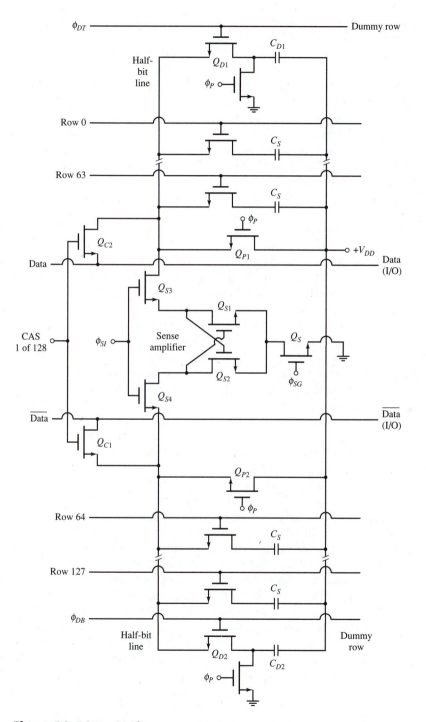

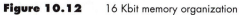

Figure 10.12   16 Kbit memory organization

inverted before being presented to the external output when the row address selector activates the bottom rows, 64 to 127.

The use of dummy cells causes a higher difference in voltage for the differential sense amplifier. Additionally, the true and complemented half-bit (column sense) lines provide a balanced input to the amplifier. Since the difference between logic high and low levels sensed is small, a balanced input increases the signal-to-noise ratio of the sense amplifier.

Writing to the selected cell is accomplished in a manner similar to reading. The true value of the input on the Data line (upper rows 0 to 63) and the complemented value on the $\overline{\text{Data}}$ line (lower rows 64 to 127) are obtained via Q_{C1} and Q_{C2} when the column address selection is active. If the cell to be written to is in the upper half, the sense amplifier takes the top half-bit line high (low), and the cell is charged (discharged) when the row is selected. To write in the bottom half-cells, the data are inverted by the input data buffer before being applied to the Data and $\overline{\text{Data}}$ lines.

The cells are refreshed sequentially in rows. Precharging and sensing of the cells in a row recharges the half-bit lines to V_{DD} and 0. Hence, all the cells in the selected row are refreshed, while external data transfer is inhibited by an inactive column address select signal. Early designs of the 16 Kbit memory use about 100 ns of precharge time and 150 ns of address activation time for each row. Hence, the entire memory requires 128×250 ns $= 32 \ \mu$s for refresh. Data transfer cannot occur during the refresh interval of 32 μs and refresh is performed once every 2 ms. Complex timing and control circuitry is therefore required to coordinate external data transfer with the refresh cycle. Refresh in processor-based memory systems, such as personal computers, is carried out by periodically interrupting the normal execution cycle, or is done during part of the instruction cycle when memory is not accessed for execution. Integrated DRAMs include circuitry for the refresh operation, as well as for address decoding, data buffering, and control and timing generation.

One-transistor cells have achieved the highest chip density over the last two decades. Innovations in cell structure and advances in circuit design using multiplexed, as well as nonmultiplexed, column address lines and on-chip bias voltage generation have rapidly increased the size and productivity of dynamic RAMs. In addition, error correcting codes and redundant circuitry are incorporated to overcome sensitivity to manufacturing defects. The defects in DRAMs are due to alpha particles emitted by traces of radioactive materials present in the packaging of the chip; the particles discharge the low charge stored in the cells. This type of random loss of stored data is known as *soft error*. Noise due to the simultaneous switching of a large number of bits in a word at the output buffer is minimized by careful layout and circuit design.

Currently 64-Mbit DRAM chips are available in organizations of 16 M $\times$ 4 bits, 8 M $\times$ 8 bits, and 4 M $\times$ 16 bits, with access times of below 50 ns. These operate at 3.3 V internal supply and feature low-voltage TTL compatibility. Because of the lower supply voltage operation, steps are taken to minimize the effects of increased switching noise and an extended refresh time of 15 ms.

CMOS DRAMs offer a wide operating voltage and a higher tolerance to process variations with comparable speeds. DRAMs with access times below 10 ns have been reported, with BiCMOS drivers and address decoders.

10.4 ADDRESS DECODERS

Since memory cells occupy small areas and consume low power, address decoders may determine the overall size, power dissipation, and, above all, access time of a memory system. The decoders must therefore be efficiently designed for optimum complexity, speed, and size.

An N-to-2^N decoder can be directly implemented using NOR or NAND gates. Bipolar (TTL) RAMs, for example, use TTL NAND decoders. MOS RAMs usually employ NOR decoders because of their fast access time. As an example, Figure 10.13a shows a 2-to-4 NOR decoder. Figure 10.13b shows the decoder with NMOS enhancement drivers and depletion loads. Buffers are added at the input of $A_0, A_1, \overline{A}_0$, and $\overline{A}_1$ to drive all the devices. This type of NOR (or NAND) static decoder may be used for small values of N, so that power dissipation is not excessive. For reduced static power consumption, CMOS implementation is preferred. Alternatively, an all-NMOS realization such as that shown in Figure 10.14 can be used. Using both true and complemented variables, the circuit in Figure 10.14 requires a pair of NMOS devices for each input (address) variable. There are two devices for each decoded output, with the pull-up device driven by the variable and the pull-down device driven by the complement. Paralleling the pull-down devices and cascading the pull-up devices realizes the NOR function. Although the device count is higher than in the circuit shown in Figure 10.13b, the power dissipation is much lower.

NOR gate decoding shown in Figure 10.13b can be extended for any number of variables by paralleling the driver devices; however, the capacitance at the output node increases, which results in an increased delay time. Power dissipation also increases with increasing N. For these reasons, two-stage decoders are used for decoding a large number of bits, typically above five. Examples of two-stage NOR-NAND and NAND-NOR decoders are shown in Figure 10.15. A two-input NOR gate at the output stage (Figure 10.15c) is preferred in MOS realizations for low output capacitance and, hence, faster rise time.

The use of complementary devices in two-stage decoding reduces the static power dissipation. Figure 10.16 is an example of complementary two-stage decoding in which the NAND and NOR gates decode half of the address bits, while a complementary (pseudo-CMOS) inverter passes the high or low level to the output. For the address $A_5 A_4 A_3 \ A_2 A_1 A_0 = 110\ 101$, for example, the bottom NOR gate output is 1 and the rightmost NAND gate output is 0. Since the gates of the PMOS and NMOS devices are low, the PMOS device transmits a 1 to the decoded output (on row a_{53} for the present example), while the NMOS device is off. Other NMOS devices connected to the same NAND output (column) are also off; however, since the NOR outputs of each of those rows are

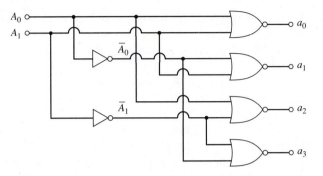

(a) Gate Realization

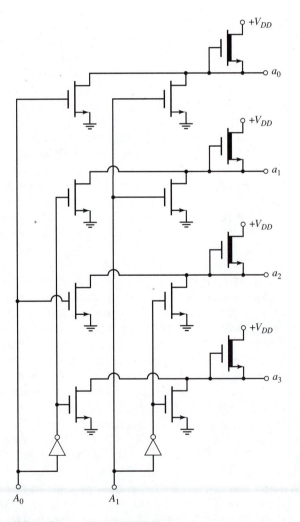

(b) Realization Using NMOS Devices

Figure 10.13 A 2-to-4 address decoder using NOR gates

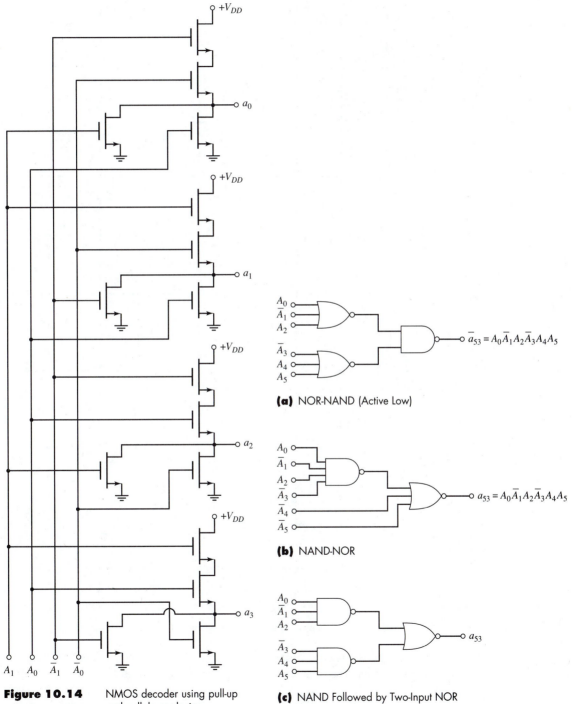

Figure 10.14 NMOS decoder using pull-up and pull-down devices

(a) NOR-NAND (Active Low)

$$\bar{a}_{53} = A_0 \bar{A}_1 A_2 \bar{A}_3 A_4 A_5$$

(b) NAND-NOR

$$a_{53} = A_0 \bar{A}_1 A_2 \bar{A}_3 A_4 A_5$$

(c) NAND Followed by Two-Input NOR

$$a_{53}$$

Figure 10.15 Two-stage decoders

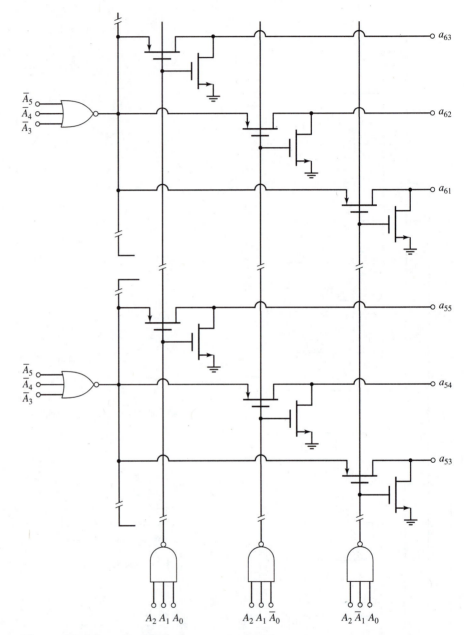

Figure 10.16   Complementary two-stage decoder

low, only a 0 is passed. In columns other than $A_2 A_1 A_0$ (101), the NAND outputs are high; hence, the NMOS devices are turned on, resulting in a 0 for all output lines except a_{53}.

Dynamic decoders offer reduced static power dissipation and have increased density. Examples of cascaded dynamic NOR and CMOS domino AND decoders are shown in Figure 10.17. During $\phi = 0$ (precharge phase), the

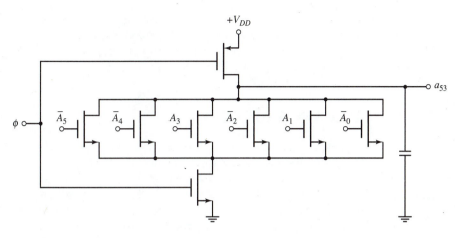

(a) NOR Decoder

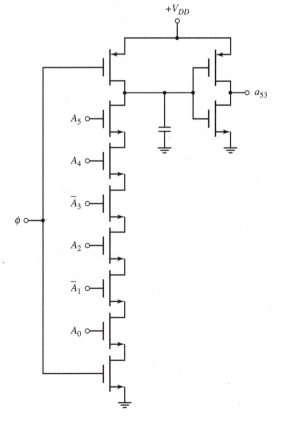

(b) CMOS Domino AND Decoder

Figure 10.17 Dynamic decoders

PMOS devices in both circuits charge the parasitic output capacitance to logic high. The output logic is evaluated when ϕ is high and the address bits are applied. Multiphase dynamic decoders can also be implemented. Domino decoders, as seen in Chapter 7, are faster, smaller in size, and free of glitches.

10.5   READ-ONLY MEMORY

As the name implies, *a read-only memory (ROM)* stores information that can only be read. The information is stored during manufacture or later; it cannot be altered as easily as in a read-write memory. ROMs are also random access memories, with fixed data stored in all addresses. ROMs are typically used in computer systems for storing monitor programs (initialization routines and boot-strap loaders) and routines to run during interrupts. Microinstructions and constants required to run scientific programs in dedicated digital systems are also stored in ROMs. In these applications, a ROM is used as a look-up table. ROM look-up tables are also used for generating arbitrary waveforms and in character displays. Another type of application includes the usage of ROMs for implementing combinational logic functions, such as code converters. Also, fast arithmetic operations are realized as combinational functions with the operands as address inputs. For example, a combinational multiplier uses multiplier and multiplicand data as inputs and obtains stored product at the output. Such applications are fast with only gate propagation delays. Additionally, the encoding and decoding of information in digital communication systems employ ROMs. In all of these applications, the stored information is constant. Hence, nonvolatile ROMs are advantageous over volatile RAMs.

The structure of a ROM is shown in Figure 10.18. The capacity of the ROM in this figure is $2^K \times N$ bits, where K denotes the number of address bits. Decoding the K bit address may be accomplished using a single K-to-2^K decoder for smaller sized ROMs. For large sizes, two-dimensional decoding is employed, as in RAMs. In addition, because of the read-only operation, column decoding is usually combined with stored data in special tree decoders (see Section 10.5.2). The stored data may be considered as N-bit encoded values of the 2^K possible input combinations. Alternatively, each output bit represents a unique combination of K-bit input. Hence, the output corresponds to realizing N combinational functions of K variables each. Note that for each K-bit address, only one set of N-bit data is available.

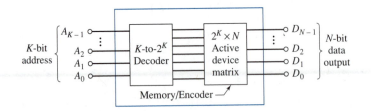

Figure 10.18   Basic ROM structure

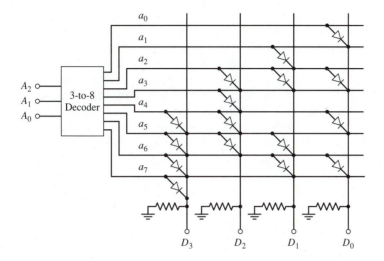

(a) Using Diodes

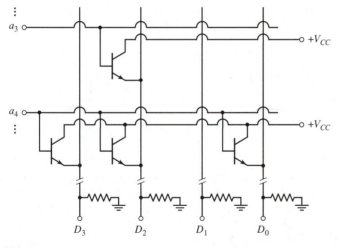

(b) Using BJTs

Figure 10.19 Example of an 8 × 4 ROM

10.5.1 MASK-PROGRAMMABLE ROM

Figure 10.19 depicts an earlier implementation of an 8 × 4 mask-programmable ROM (see below) that uses a diode matrix. Although this size is too small to warrant a ROM, it illustrates the application of ROM as a code converter. The truth table for converting a 3-bit binary code to the corresponding 3-bit gray code with odd parity[1] is shown in Table 10.1. This table is

───────────

[1] An odd parity bit makes the total number of 1's in the output an odd number.

implemented in the ROM with eight rows of word lines and four columns of data or bit lines. For the 3-bit external address $(A_2 A_1 A_0)$, one of the eight rows (A_0 through A_7) goes high via the internal address decoder. The presence of a diode (or other nonlinear element) at the intersection of the addressed row and a column connects the asserted row input to the column output $(D_3 D_2 D_1 D_0)$. Note that the conducting diodes draw power from the decoder output. This loading of decoder output is eliminated with the use of bipolar transistors (Figure 10.19b), which buffer the decoder output. Since all transistors connected to the same row have the same base and the same collector potentials, a single transistor with multiemitters is usually employed. Bipolar ROMs use junction diodes, Schottky diodes, or BJTs with fast access times.

Table 10.1			Truth table for a 3-bit code converter				
A_2	A_1	A_0	**Row**	D_3	D_2	D_1	D_0
0	0	0	a_0	0	0	0	1
0	0	1	a_1	0	0	1	0
0	1	0	a_2	0	1	1	1
0	1	1	a_3	0	1	0	0
1	0	0	a_4	1	1	0	1
1	0	1	a_5	1	1	1	0
1	1	0	a_6	1	0	1	1
1	1	1	a_7	1	0	0	0

As the number of inputs and outputs increases, the number of active devices, and hence the ROM size, also increases. To reduce the chip area and power consumption, ROMs are presently built using the MOSFET technology. Typically, an enhancement-mode NMOSFET is fabricated at each of the $2^K \times N$ crosspoints, or cells. By selectively omitting the drain, source, or gate terminals of the MOSFETs at desired locations, a logic 1 is stored. Alternatively, the circuit can be configured such that the absence of an active MOSFET indicates a stored 0. Figure 10.20 shows part of the code converter ROM of Figure 10.19 using NMOSFETs. Gate electrodes are masked off at selected locations so that the induced channel in these (inactive) devices cannot be enhanced. Consider, for example, output D_0. When the decoder output a_2 (or a_4) is logic high, inactive MOSFET Q_{20} (or Q_{40}) cannot turn on; hence, D_0 is pulled high to a voltage of approximately V_{DD}. If a_3 is selected, active device Q_{30} conducts and pulls down the voltage at D_0 to 0. For a higher drive capability, depletion pull-up devices are used.

Each column output in Figure 10.20 goes low when any of the rows connected to it goes high (assuming an active device is at the column–row intersection). Hence, the cell array in this ROM may be considered a NOR cell array. Alternatively, pull-down devices may be arranged in series and driven by active-low row decoder outputs. Each column output in such cells implements the

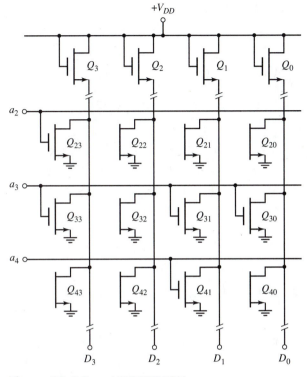

Figure 10.20   NMOSFET ROM

NAND function of row decoder outputs. The presence of an active MOSFET in the NAND cell ROM array indicates a stored logic 1. As with logic gates, NAND arrays are slower compared to NOR arrays.

Source and drain n diffusions are present at each crosspoint of rows (decoded address lines) and columns (output data lines), regardless of the bit stored. The presence of gates to activate the MOSFET distinguishes stored 1's from 0's. In bipolar ROMs (using diodes or BJTs, as shown in Figure 10.19), a contact may be omitted at selected crosspoints. In both bipolar and MOSFET ROMs, stored data are therefore, "programmed" during manufacture by the formation of active devices at selected locations and the masking off at other locations. Hence, these ROMs are known as *mask-programmed ROMs*. The nonvolatility of stored data arises from the absence of active devices (as in bipolar ROMs), or the presence of inactive nonlinear devices (as in MOSFET ROMs).

Regular patterns of the source and drain diffusions in the NMOSFET ROM result in a simple structure and a high chip density. Figure 10.21 shows the basic structure of a NOR cell ROM. As seen in the figure, two columns of n diffusions, one for drains and the other for sources, are formed over the p substrate for each output bit. Polysilicon rows connecting the address decoder outputs serve as gate terminals. For storing logic 0, a thin oxide is formed underneath the polysilicon row at the intersection with the columns of diffusion. With a normal thickness of less than 0.1 μm, the oxide layer creates an active MOSFET with a threshold

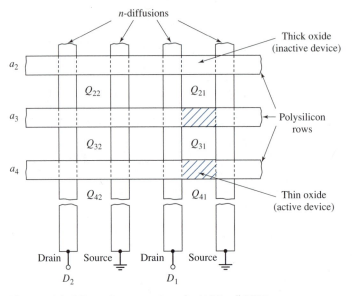

Figure 10.21   Basic structure of a NOR cell ROM

voltage of about 1 V. At the intersection where no active device is desired, that is, where a logic 1 is to be stored, a thick oxide or ion implantation is used so that the threshold voltage is higher than the normal row decoder voltage. When a row is selected (by logic high), gates with low threshold voltages enhance the induced channel and pull the drain (column output) voltage low. However, thick oxide gates are unable to form the channel below them; hence, the column outputs are pulled high by the driver devices.

NAND cell ROMs have a similar structure with active devices formed for storing 1's. For mask-programming, NOR cells can be customized to the desired data using almost-completed ROMs. After forming the columns of diffusions, mask programming may be used for selective implanting to store 0's. Mask programming may also use contact programming in which active-device contacts are made only where 0's are to be stored. Because of the series connection of active MOSFETs, NAND cells must be completely programmed during manufacture. However, NAND cells offer higher chip density than NOR cells. Mask-programmed MOS and CMOS ROMs with densities in the hundreds of Mbits have been developed as VLSI systems.

Unless a large number of devices with the same data patterns is needed, mask-programmed ROMs are expensive to fabricate. In addition, even a single error in the stored data may render a ROM useless. Hence, designers must be absolutely certain about the bit patterns required before committing them to a ROM. It is also uneconomical to use ROMs for implementing K variable combinational logic functions with M outputs if the outputs are zero for most of the 2^K input combinations. User-programmable ROMs offer the flexibility to alter stored data so that they can be used in the developmental stages of a system. Before discussing programmable ROMs we will consider several ROM address decoding schemes.

10.5.2   ROM ADDRESS DECODERS

As with RAMs, it is more efficient to use two-dimensional decoding to address ROM cells. The NOR decoder shown in Figure 10.13, for example, can be used for row (N_R bits) and column ($K - N_R$ bits) decoding. However, because the cells are only to be read, column decoding and output data may be combined. Figure 10.22 shows an implementation of the 8×4 code converter illustrated in Figure 10.19 (with the truth table given in Table 10.1). This implementation uses a 2-bit row decoder and a 1-bit column decoder. With a 4×8 cell organization, each of the four row decoders selects a pair of rows in the truth table.

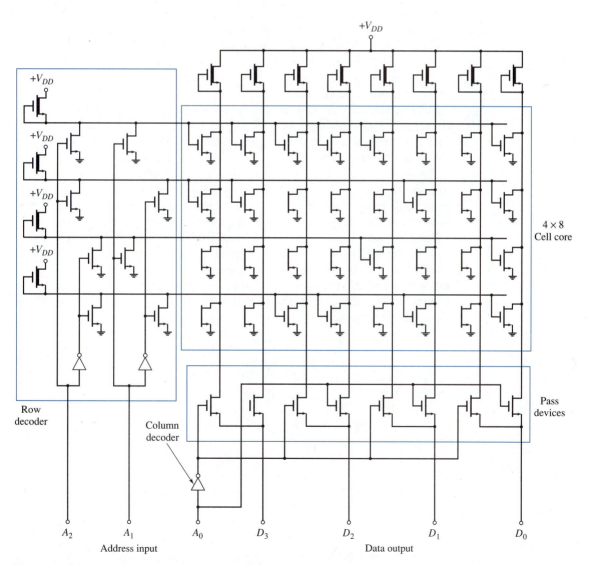

Figure 10.22.   ROM address decoding combined with data output

Address bit A_0 selects one of the two columns for each output bit. Note that the cells in the rows are arranged such that for any combination of $A_2 A_1$, the cell values alternate for $A_0 = 0$ and $A_0 = 1$. Hence, the pass devices transmit the corresponding values for the given combination of $A_2 A_1 A_0$.

CMOS transmission gates may be used in place of single MOSFETs to achieve a larger difference between the logic low and high voltages. Alternatively, sense amplifiers may be employed for each output bit. In addition, the outputs are usually gated with an external chip select/output enable input. As with RAMs, chip select input enables memory extension, both in the number of words and in the number of output bits.

In general, for $N_C = K - N_R$ bits, there are 2^{N_C} columns of cells for each output bit. Hence, NOR column decoders are employed to select one out of 2^{N_C} columns; for each output bit, a pass device connects the selected column to the output. Figure 10.23 shows an organization for a 16 Kbit ROM. This memory is organized as $2 K$ words of 8 bits each. With an 11-bit address, $N_R = 7$ and $N_C = 4$ are chosen. This results in 128 rows and a core of 128×128 cells. Each of the 16 column decoder outputs activates a distinct group of eight transmission devices, which are connected to the 128 columns of cells. Each output bit from the selected row is obtained as a 16-to-1 data selector output. In addition, chip select and/or output enable lines are usually provided to extend the number of words and the word size.

A more efficient decoder that reduces the number of active devices is the tree decoder shown in Figure 10.24. With 3-bit column decoding, each row has eight cells. When a row decoder is active (high), the column decoder provides only one of the eight cell values at the output via binary tree decoding. This column decoder uses $8 + 4 + 2 = 14$ MOSFETs for each output bit, with no pass devices. In NOR decoding with a pass transistor, a total of $(8 \times 4) + 1 = 33$ MOSFETs are required. Hence, there are fewer devices in the tree decoder. Tree decoding can readily be extended to any number of column address and output bits. However, the speed is slower due to delays in the series pass transistors.

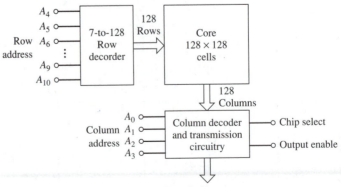

Figure 10.23   Organization of a 2 K × 8 bit ROM

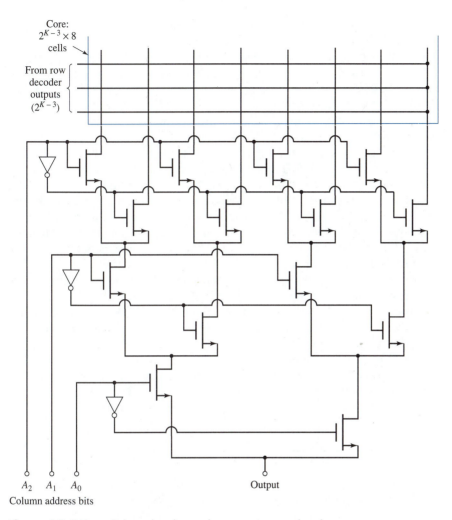

Figure 10.24   Column decoding and output, using tree decoder

Row address decoding for bipolar ROMs employ TTL-type multiemitter gating circuitry. For column decoding and data routing, a TTL NAND decoder is used to drive an output transistor. The emitter current of the transistor whose collector is connected to the column output (Figure 10.19b) defines the logic state of the selected cell.

10.5.3   PROGRAMMABLE ROM (PROM)

For flexibility in defining stored data patterns after final packaging by the manufacturer, both bipolar and MOS ROMs can be made user-programmable. In *field-programmable ROMs*, a fusible link is included in each of the nonlinear devices to connect the rows to the columns. The link is burned out, or open-circuited, where no active device is needed, by passing a sufficiently large current during programming. Once burned, the patterns stay permanent. Although the links can be included in both bipolar and MOS cells during manufacture, in practice only bipolar cells are currently available as programmable ROMs. Figure 10.25 shows an array of bipolar cells with fusible links.

Links were initially made of a thin film of nichrome connecting the cell emitters with the column lines. Problems with nichrome links included unreliable contacts, corrosion, and the reversal of broken links making contact. In later designs, polysilicon fuse technology has taken the place of nichrome. In this technology, a layer of about 0.3μm of polysilicon is deposited as a notched strip fusible link at each crosspoint. Due to the low resistance of polysilicon, there is very little voltage drop at low operating currents. During programming, or "burning of the PROM," current pulses in the range of 20 mA to 30 mA are passed through selected cell links. (Since such high currents are not suitable with the high impedances of MOSFET devices, fusible links are limited to bipolar ROMs.) The high temperature (about 1400°C) caused by heating of the fuse oxidizes the silicon and forms an insulator. Once oxidized, the silicon does not reverse to form an electrical contact. Hence, the PROM stays programmed permanently. In Schottky bipolar PROMs, titanium-tungsten fuses are used instead of polysilicon fuses.

Field-programmable ROMs offer user programmability at a lower cost than mask-programmed ROMs. PROMs are available in sizes of 32×8 bits to $2\,K \times 8$ bits, with access times in the tens of ns range. They also provide TTL compatibility with open-collector or tristate output. Large-capacity PROMs

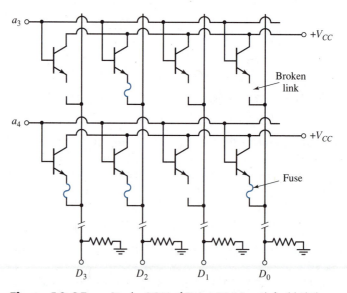

Figure 10.25   Bipolar ROM of Figure 10.19, with fusible links

with electrically fusible polysilicon links have also been recently developed. These PROMs are used in a variety of applications involving look-up tables and fast arithmetic operations. However, PROMs are only programmable once and consume a large silicon area and power. For more flexibility in altering the data patterns as design changes occur, for example, several types of reprogrammable ROMs are available, as described in the following sections.

10.5.4   ERASABLE PROGRAMMABLE ROM (EPROM)

EPROMs are MOSFET devices that have found wide usage in software development applications and prototype systems, as well as in finished products. Each storage cell in an EPROM is a special MOSFET consisting of two gates: the *control* or *select gate,* and the *floating gate.* The simplified structure of a floating gate MOSFET is shown in Figure 10.26a. In this n-channel enhancement-type MOSFET, the first polysilicon gate above the channel region is the floating gate, which is left unconnected. The second gate is the normal control gate of the MOSFET, which is connected to the row decoder, or word selector, output.

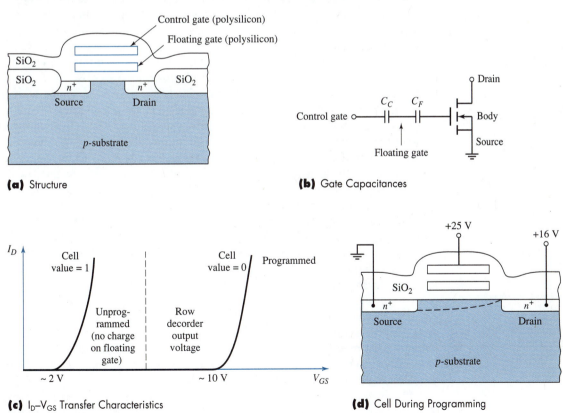

(a) Structure

(b) Gate Capacitances

(c) I_D–V_{GS} Transfer Characteristics

(d) Cell During Programming

Figure 10.26   Floating gate EPROM cell

Capacitances between the channel and the gates (Figure 10.26b) cause the applied control gate voltage to divide between the two gates. When the cell is not programmed, that is, when the floating gate has no charge on it, the control gate behaves as a regular gate of an *n*-channel enhancement-type MOSFET, with a slightly higher threshold voltage of about 2 V (Figure 10.26c). This higher threshold voltage is a result of the capacitive voltage division. When such a cell is used in the NOR ROM array of Figure 10.20, for example, a logic 0 output appears at the selected column if the selected row decoder output exceeds the threshold voltage. In this state, the cell is storing a logic 1.

To program, or store, a 0 in the cell, the floating gate must be charged so that a threshold voltage of much higher than 2 V would be needed at the control gate. During programming, a high voltage of about 16 to 20 V (depending on the technology), is applied between the source and the drain. Simultaneously, a voltage of about 25 V (or the same as the drain voltage, depending on the technology) is applied to the control gate (Figure 10.26d). Since the cell was initially unprogrammed, the control gate voltage establishes the channel. The MOSFET then operates in the pinch-off region, with the channel pinched off at the drain end, due to the large drain voltage. Electrons are accelerated through the channel, acquiring a high energy when they reach the drain. In addition, the high electric field in the drain–substrate region causes avalanche breakdown, which also contributes to high-energy, or *hot, electrons.* (Note that at the high electric field, the drift velocity of the electrons becomes comparable to the thermal velocity, and the total energy of the electrons increases. Because of the high kinetic energy of these electrons, which is related to temperature, the lattice becomes hot.) The scattered hot electrons are attracted by the high electric field established by the positive control gate voltage. Due to their energy, some of the electrons penetrate the thin insulating oxide and reach the floating polysilicon gate. The floating gate thus acquires charge. As more negative charge accumulates on the floating gate, the field strength is reduced and further accumulation is inhibited. This self-limiting process of charging the floating gate is referred to as the *floating gate avalanche injection.*

Once the floating gate is charged, the charge remains trapped, even after the large programming voltages are removed. Note that there is no discharge path available, since the gate is surrounded by insulating oxide. Because the stored charge on the floating gate causes a negative voltage, electrons from the surface of the substrate are repelled. Hence, a much larger voltage is required at the control gate to reestablish the channel than when the floating gate was uncharged. The cell is now programmed and is storing a logic 0. The new threshold voltage of the device is higher than the row decoder output (Figure 10.26c). The column value read for the selected row is therefore logic high (equivalent to no active device at the row–column intersection) when the programmed cell is used in the NOR array of Figure 10.20.

The charge can remain stored on the isolated floating gate for many years. Hence, a programmed EPROM is sometimes known as *read-mostly memory.* Note that for a nominal row decoder output of 5 V, a cell value of logic high is read if the cell was programmed, and a low is read if the cell was not programmed.

To store different data patterns, all cells must first be returned to their unprogrammed state; that is, the electrons on the floating gate must be transported back to the substrate. This is accomplished by illuminating the cells with a strong ultraviolet light at a wavelength of below 400 nm (typically, a short wavelength of 253.7 nm) for about 20 minutes. Electrons trapped on the floating gate acquire sufficient photon energy from the ultraviolet light to overcome the inherent potential energy barrier; therefore, they escape through the oxide layer to the substrate. To permit erasure using ultraviolet radiation, an EPROM package is equipped with a transparent quartz window. Once erased, the cells can be selectively reprogrammed. Note that the radiation erases all the cells at once. Since sunlight and fluorescent lighting contain low-level ultraviolet radiation (in the 300 to 400 nm), the exposure of EPROM packages to these sources must be prevented to avoid the loss of data over a long period. An opaque cover or label over the quartz window is normally used as a precaution.

EPROMs can be erased and reprogrammed many times. For permanent data storage, as in the final design of a system, for example, one-time programmable EPROMs are available without the quartz window. These MOS ROMs replace the bipolar mask-programmed and field-programmable ROMs. Also, these EPROMs have higher reliability and density, lower power dissipation, and lower cost, but they operate at a lower speed.

EPROMs are available with TTL compatible voltage levels. The sizes range from 1 K to 1 M bits, organized for 4, 8, or 16 bit outputs. The Intel 2764A, for example, is a 64 Kbit (8 K $\times$ 8 bit) EPROM. EPROM access times have steadily improved to under 100 ns. After erasure, the external address and data are supplied in the programming mode with a specified programming supply voltage. For each address, a TTL pulse of 5 ms to 50 ms duration is used to program the cells to store 0's.

10.5.5 ELECTRICALLY ERASABLE PROM (EEPROM)

A disadvantage of EPROMs is that they must be removed from circuit for the erase and reprogram processes. Depending on the intensity of the ultraviolet light source, the erasure may be slow, taking up to an hour. This is particularly undesirable if new data patterns call for changing only a few of the previously stored cell values. A variation of the floating gate MOSFET offers the versatility of in-circuit programming and fast erasure, for single byte modification of the data. Figure 10.27a shows the structure of a commonly used floating gate MOSFET for the electrically erasable programmable ROM (EEPROM) cell.

As seen in this figure, the MOSFET is similar to the floating gate device used in an EPROM except that the floating gate is placed very close to the drain. The thickness of the insulating oxide between the drain and the floating gate is extremely small (below 20 nm). At this thickness, a high electric field, of the

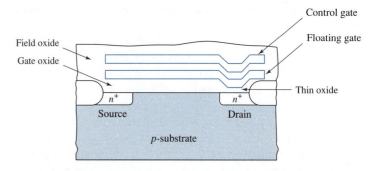

(a) MOSFET Structure

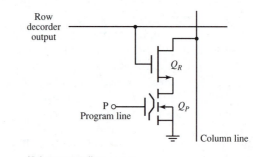

(b) ROM Cell Connection

Figure 10.27　　EEPROM cell

order of 10^9 V/m, causes electrons to tunnel through the oxide. The thin oxide above the drain is therefore called the tunnel oxide, and the MOSFET is known as a *floating gate tunnel oxide (flotox)* cell. The reversible tunneling process is known is *Fowler-Nordheim tunneling.* As with the EPROM cell, the control gate has a low threshold voltage if the floating gate is uncharged.

To program the cell with a 1, that is, to charge the floating gate, a voltage of about 20 V is applied to the control gate, with the drain at ground. This voltage creates a sufficiently high electric field in the oxide region to cause current flow. Electrons flowing from the n^+ drain, by tunneling, accumulate on the floating gate. The presence of this negative charge, which remains on the floating gate even after the programming voltage is removed, effectively raises the threshold voltage at the control gate. Hence, the cell is programmed to store a logic 1. To discharge the floating gate, the supply is reversed; the drain voltage of 20 V relative to the control gate now removes electrons from the floating gate, again by tunneling. The discharge may be such that a net positive charge is left on the floating gate. The threshold voltage of the MOSFET is therefore reduced to a negative value of about -2 V. The stored value of the cell is now a logic 0. With a nominal positive control gate voltage of 5 V (which is below the voltage corresponding to the charged floating gate), the stored value can be read by detecting the conduction of the device (0 for on, 1 for off).

Figure 10.27b shows the circuit configuration for an EEPROM cell. The additional enhancement-type MOSFET Q_R is needed for proper reading of the cell value in Q_P. The gates of both Q_R and Q_P are at logic high ($\approx$ 5 V) during the read process. If Q_P has a stored 1 (floating gate charged), the device will be off at 5 V of control gate voltage; hence, the column line will be at logic high in the NOR cell array (Figure 10.20). For a stored 0 (floating gate discharged), Q_P conducts and brings the column line down to logic low. When the row decoder output goes low after the reading, Q_R turns off and breaks contact with the column line. In the absence of Q_R, Q_P will still be on with its negative threshold voltage.

For writing to the cell, the row decoder output is brought to 20 V, which turns Q_R on. If the gate of Q_P is taken to 20 V while the column line voltage is held at zero, the floating gate becomes charged and a 1 is stored in the cell. For writing a 0, the column line voltage is brought to 20 V with the gate of Q_P held at zero. This discharges Q_P and puts it in the depletion mode. Control circuitry must be used to ensure proper voltage levels at the row and column lines and at the gate of Q_P, for both reading and writing. Note that the floating gate is charged and discharged by the high electric field in the tunnel oxide region. Degradation of the oxide insulation occurs after repeated read and write operations. Typically, 10,000 to 100,000 write cycles are guaranteed for commercial EEPROM devices.

The need for 5 V and 20 V supplies is a disadvantage for writing to the EEPROM cells. The external high voltage supply is eliminated in modern EEPROMs by on-chip voltage generation from a single 5 V source. Still, the size of an EEPROM is large due to the addition of conventional MOSFETs for each storage cell MOSFET. The size is also increased because of the larger gate areas needed for tunneling in the storage cell MOSFET. The Intel 2864A is an example of a 64 Kbit (8 K $\times$ 8 bit) EEPROM. This chip operates with an external 5 V supply; it has a read access time of 200 ns and write cycle time of 200 ns. Presently, NAND-structured EEPROMs for sizes 1 Mbit and above have been reported.

Flash memory is a variation of EEPROM that uses only one transistor per cell. All the cells in a flash memory chip can be erased simultaneously while in the circuit, like in EEPROMs. Although flash memory chips lack byte-wise erasure and programming, they offer high density at a lower cost than EEPROMs.

In one type of flash memory technology, each cell is formed similar to an EPROM cell (Figure 10.26), but with thinner gate oxides above the channel and between the control and floating gates. The thin oxide (about 10 nm) enables Fowler-Nordheim tunneling of electrons for cell erasure. Cells in a flash memory may be arranged similar to those in a ROM (Figure 10.20), but with the source terminal available for erasing.

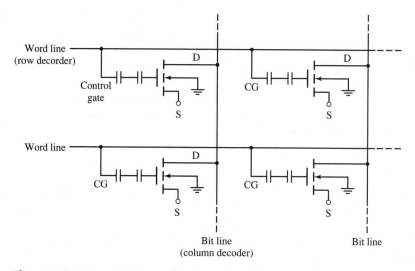

Figure 10.28   NOR array of flash memory cells

Figure 10.28 shows a NOR array of flash memory cells. The bit lines in this figure correspond to column outputs, similar to those shown in Figure 10.22; pass devices select one of many bit lines to each output line.

All cells are simultaneously erased, that is, the floating gates (storing 1's) are discharged by taking the control gates (word lines) to ground and the source terminals to 12 V (with the drains floating). At this high electric field between the source and the gate, the electrons previously stored in the floating gates tunnel through the thin oxide layers to the sources. The cells are programmed to store a 0; that is, their floating gates are charged, similar to EPROM, by the hot electron injection technique. For programming, or writing a 0, to selected bits of a word, the control gates of the selected word are driven to 12 V by the word line, with the source terminals at ground. With the selected bit lines (the drains) taken to about 7 V, the electrons in the channels acquire a high energy and hot electrons are generated. These hot electrons accumulate on the floating gates, thereby raising the threshold voltage of the selected cells in the word. The selected bits of the word thus store a 0. During the reading of a word, the sources are at ground, while the selected word line takes the control gates to 5 V. With a drain bias voltage, the programmed cells (bits of the selected word) storing a 0 in the NOR array of Figure 10.28 are off. Since the corresponding bit lines carry no current, the bit values of 0 are read. The word line voltage turns on those cells with stored 1's (no charge on the floating gates), and the currents in the bit lines are sensed to read 1's.

The tunneling mechanism of simultaneously erasing an entire array takes from less than a second to a few seconds, while programming each word takes about 10 μs. With single transistor cells, erasure is monitored carefully to avoid depletion-mode operation of the cells. To ensure operation in the enhancement mode, the entire array is usually programmed to 0's first and then the biases are switched for erasure.

The Intel 28F256 is the flash memory version of the Intel 2764A 8 K $\times$ 8 bit EPROM, with a comparable access time and number of reprogramming cycles. Currently, flash memories with densities above 1 M bits are available.

Both EEPROMs and flash memories are sometimes referred to as *non-volatile random-access memories (NVRAMs)*. Although these memories are used in read-mostly applications, their quick programming (writing) ability is used to advantage as RWMs. For example, saving RAM data during a power failure is an application of NVRAMs. In this application, only a short interval of battery backup is required for writing data from the RAMs to the NVRAMs. Once power is restored, the data are copied from the NVRAMs to the RAMs. In another application, the fast parallel access (below 100 ns) of words of data (as opposed to serial access) and the reliability and low-power operation of NVRAMs target hard disk drives in computers. With their increasing density and lower cost, flash memories, in particular, are expected to replace disk drives.

For reduced standby and operating power, dynamic ROMs can be used. These ROMs store data in static cell but use dynamic circuitry for address decoding and buffering. One or more clock signals are therefore required for operation.

10.6 SEQUENTIAL ACCESS MEMORY

Unlike in RWMs and ROMs, the access time for data transfer in serial, or sequential, access memory depends on the address location. Recirculating registers, which are implemented using shift registers (Section 7.3.3) fall in this class of memories. Large shift registers form *first-in/first-out (FIFO) memories*. FIFOs are used as interfaces between fast and slow data transfer devices, such as printer interfaces. Charge-coupled devices are another type of serial memory.

Serial access memories offer high density with a small amount of external wiring. (A 1 K $\times$ 1 bit serial memory, for example, requires only the clock signal, instead of 10 address lines.) However, serial access memories are slow in terms of access time. Because of limited applications, serial access memories are not widely used for large digital storage requirements.

Presently, dual-port memories are available for both serial and random accessing. In computers that interface with slower devices, such as network routers, data may enter the serial port of the memory in packets; the processor can read and process the data in a random sequence at the random port. Unlike in a FIFO memory, data read via the random port are not lost. The dual-port memory thus combines FIFO and SRAM capabilities. As an example, the IDT70825 from Integrated Device Technology is an 8 K $\times$ 16 bit serial access/ random access memory (SARAM), with control circuitry for accessing 16 bits of data on the sequential port or the standard RAM port. A pointer/buffer enables splitting the memory into two buffers with programmable sizes.

10.7 PROGRAMMABLE LOGIC DEVICES

The evolution of *programmable logic devices (PLDs)* began with PROMs. Although we considered ROMs as memories in the previous sections, an alternative view of a ROM is a matrix consisting of an AND array and an OR array. For example, each row in a $2^3 \times 4$ bit ROM (Figure 10.19) produces a *minterm*[2]: $\bar{A}_2 \bar{A}_1 \bar{A}_0$ in row a_0, $\bar{A}_2 \bar{A}_1 A_0$ in row a_1, etc. Each column represents the logical OR of the minterms of those rows that are connected to the column with active devices (or links). The result is that each column realizes a two-level AND–OR logic function.

In general, a $2^K \times N$ bit ROM has all the 2^K rows of minterms available, with each row implementing a logical AND (minterm) of K variables; hence, there are 2^K AND gates, each with K inputs. For each column, there is an OR gate with a maximum of 2^K inputs; hence, there are N OR gates, each with different inputs. Thus, a ROM has fixed AND and mask-programmed OR arrays, while a programmable ROM has a fixed AND array and a programmable OR array. Inputs to the OR array are programmed by blowing fuses, for example, in PROMs. To simplify the logical representation of such an arrangement, the notation shown in Figure 10.29 is used. In this figure, which shows the code converter of Figure 10.19, the gate inputs are shown by single lines; each '×' at a row–column intersection represents contact (an intact fuse, for example). Such a PROM can be programmed to realize N (4 in Figure 10.29) combinational functions in K (3 in Figure 10.29) variables each.

A *programmable logic device* (Figure 10.30) is a generalization of the PROM in which one or both of the AND and OR arrays can be programmed by the user to implement a set of N functions. Programming is accomplished using fusible links, as in field-programmable ROMs. The normally closed electrical contact of a fuse at a crosspoint location is opened by burning the fuse with current heating.

Alternatively, PLDs are fabricated with *antifuses* at the crosspoints. A dielectric of multilayer oxygen-nitrogen-oxygen (ONO) formed between n^+ diffusion and polysilicon, or a layer of amorphous silicon between metal layers, forms an antifuse. The antifuse normally has a very high resistance between its terminals. Passing sufficient current through an antifuse with a high voltage (10 to 20 V) applied across its terminals melts the dielectric material and creates a permanent low-resistance link. At 16 V and 5 mA programming current, the melted ONO dielectric, or *programmable low-impedance circuit element (PLICE),* has a resistance of about 600 Ω. The PLICE resistance drops to about 100 Ω at 15 mA. Although the antifuse itself is small in size (about 1 μm in length), the high-voltage transistors needed for programming take up considerable chip area. For reprogrammability, PLDs use EPROM- or EEPROM-like (floating gate) cells.

Newer PLDs, such as some field-programmable *gate arrays* (see Section 10.7.3), use SRAM cells to determine the connections at crosspoints. A

---

[2] A K variable minterm is a product (AND) term of K variables, with each variable appearing once in true or complemented form.

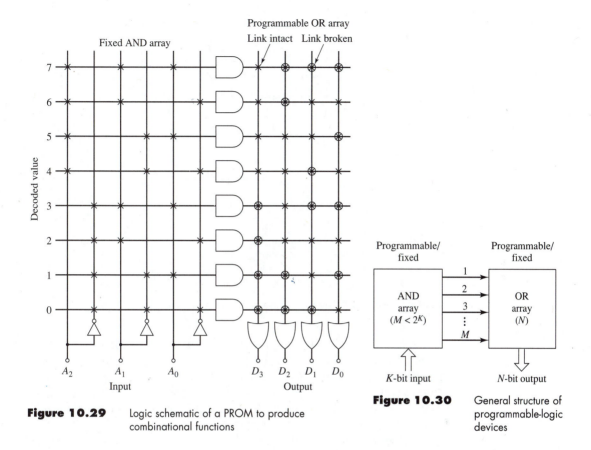

Figure 10.29 Logic schematic of a PROM to produce combinational functions

Figure 10.30 General structure of programmable-logic devices

nonvolatile memory, such as an EPROM or a magnetic disk, contains programming data for the PLD. When the PLD is ready to be programmed, the data are loaded ("written") in the SRAM cells, and the output of the cells is used to control the pass gates at the crosspoints. Reprogramming the PLD for different output functions is facilitated by reloading the SRAM. However, since the SRAM is volatile, this type of PLD must be programmed after every power-up. Also, the large size of SRAM cells contributes to the overall size of the SRAM-based *electrically erasable PLDs (EEPLDs)*. While EPROM-based PLDs (EPLDs) are nonvolatile, they require additional processing steps over those needed for fabrication of the cells.

A major difference between PROMs and PLDs is that PLDs need not fully decode all of the 2^K minterms. This is justified by the fact that most practical K variable logic functions do not require all of the product terms. In such cases, a ROM, which may be considered a mask-programmed PLD, has a large wasted area. By including a clock and providing feedback from the output, PLDs can also implement sequential logic functions.

PLDs fill the need to design a large, sophisticated system involving a large number of inputs and outputs, with the flexibility to incorporate design changes.

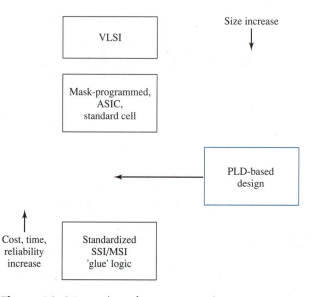

Figure 10.31   Place of PLDs in system design

At one end of this design task (Figure 10.31) is the use of standardized low-cost SSI and MSI devices. In this approach, the overall cost, size, power and cooling requirements, and inventory of the standard or 'glue logic' all increase with the size of the system. Also, with a large number of many different devices, reliability decreases. In addition, design changes cannot be made without the time-consuming rewiring of individual devices.

The other design approach is the LSI/VLSI design that specifically meets the given system requirements. Yet another approach is the *application-specific integrated circuit (ASIC)*, which is implemented using mask- or factory-programmed gate arrays. Both VLSI (full custom) and ASIC (semicustom) designs result in small size, low power, and high performance and reliability. However, a VLSI or an ASIC cannot be justified in terms of developmental and manufacturing costs and time unless a large volume is needed. It is estimated that the overhead costs for producing VLSI or ASIC chips are in the range of $20,000 to $200,000, with a production time of one year or more. In addition, design modifications and error corrections require redesign and refabrication. As shown in Figure 10.31, a design using PLDs results in reduced cost and provides reprogramming capability for design changes. For high-volume chips, after the correctness of the design is proven using PLDs, the size, power, and reliability can be improved by implementing the system as a VLSI circuit.

In general, PLDs may be classified according to the programmability of the AND and OR arrays. PLDs with programmable AND and fixed OR arrays, for example, are known as programmable array logic (PAL) devices. In programmable logic arrays (PLA), both the AND and OR arrays are programmable. Because of their growing size and functionality, PLDs are themselves implemented as VLSI circuits. This section introduces the structure of some of the basic PLDs.

10.7.1   PROGRAMMABLE LOGIC ARRAYS

A *programmable logic array (PLA)* has field-programmable AND and OR arrays, much like a ROM. As with ROMs, the arrangement of gates is fixed in
PLAs; hence, they are known as *fixed architecture PLDs* (as oppesed to field-
programmable gate arrays, which are *flexible architecture PLDs*). Instead of a
full K-to-2^K address decoder (full decoding), as in a ROM, a PLA has much
fewer than 2^K AND gates for K input variables. For N outputs from the PLA,
there are N OR gates, each with programmable inputs from all of the AND gate
outputs. Figure 10.32 shows the PLA realization of the code converter of Figure 10.19a. Each x indicates a connection at the intersection. For this three-input PLA, each AND gate has six possible (true and complemented) inputs
available, although not all six can be connected. (Some AND gates have only
one input connected; see the minimized logic equations in Figure 10.32.) With a
total of six AND gates, each of the four OR gates has a maximum of six possible
inputs. Hence, any output can produce the sum of up to six different terms. The
advantage of a PLA arises if two or more output functions have a common
product term that can be shared by different OR gates; that is, for a reduction
in PLA area, different output functions may be jointly minimized. In addition,
since "don't care" minterms need not be included in the logic function, less area

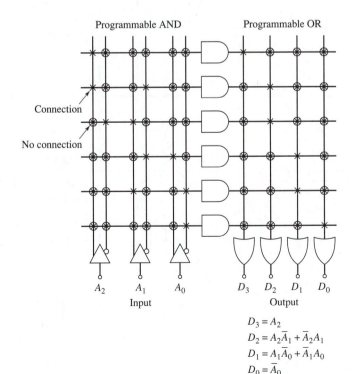

$$D_3 = A_2$$
$$D_2 = A_2\overline{A}_1 + \overline{A}_2 A_1$$
$$D_1 = A_1\overline{A}_0 + \overline{A}_1 A_0$$
$$D_0 = \overline{A}_0$$

Figure 10.32   Example of a PLA

is required.[3] Also, with an inverter in the path of each output, the true or complemented output can be obtained. This flexibility allows a function Y, with many terms, to be realized more efficiently using $\overline{Y}$ and inverting the result at the output.

The regular structure of PLAs with programmable AND and OR arrays is used in complex systems involving many variables. When the number of logic functions or the number of minimized terms is large, several PLAs are combined, with shared inputs. Additionally, PLAs with feedback are used as finite state machines in the control logic of sophisticated systems. The regular, compact structure of PLAs is so useful that a design using PLAs may be retained in semicustom systems independent of the technology.

PLAs in the bipolar technology are implemented using junction or Schottky diode AND gates and emitter follower OR gates. The bipolar AND–OR implementation of a PLA is shown in Figure 10.33. As with PROMs, titanium-tungsten fusible links are used at the intersections of the rows and columns. Fuse links in bipolar PLAs occupy a large area and require special circuitry to blow them by passing a high current.

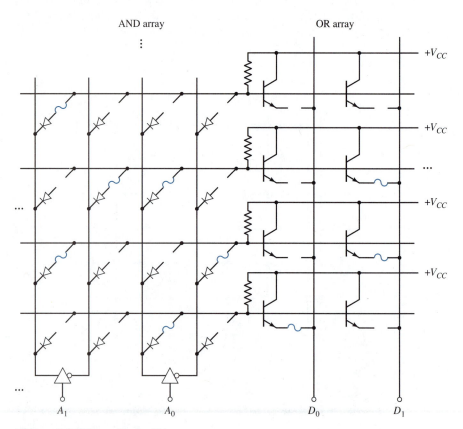

Figure 10.33   Bipolar PLA

[3]"Don't care" minterms can assume either logic low or high values since they do not affect the output.

Commercial field-programmable logic arrays have true and complemented input buffers and totem-pole drivers. Also, true or complemented outputs can be obtained. These PLAs are specified by the number of inputs, AND gates, and outputs. PLAs are available for inputs and outputs from 10 to 30. Because of excessive capacitance and load, product terms are limited to 200. For example, the Texas Instruments TIFPLA839 has 14 inputs, 32 AND gates, and six outputs. Note that for such a large number of inputs, computer-aided tools are necessary to minimize the logic functions before they are implemented on the PLAs.

Both NMOS and CMOS PLAs are available as mask-programmed logic arrays. Although these PLAs commonly have a NOR–NOR structure, the arrays are still referred to as AND and OR arrays. Figure 10.34 shows an NMOS NOR–NOR PLA. Only MOSFETs with connections to either a column or a row are shown in the circuit. Because of the NOR–NOR canonical realization, each row is considered to be carrying a product term, and each output (column) provides the sum-of-product result.

Note that the horizontal (AND) and vertical (OR) arrays in Figure 10.34 are identical in form except for the 90 degree rotation. This regularity in structure,

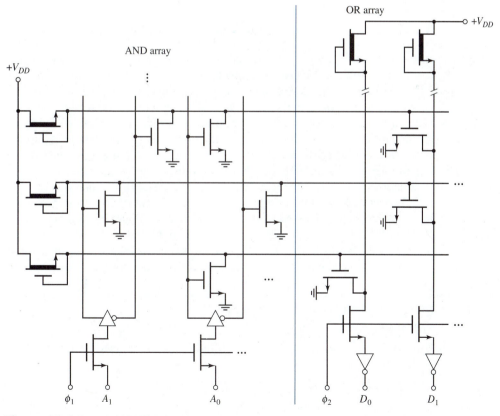

Figure 10.34   NMOS PLA

and the ability to realize any two-level logic function, make the MOS PLAs popular in semicustom devices. The input and output are passed via pass transistors or transmission gates, which are controlled by clocks ϕ_1 and ϕ_2. (To reduce static power dissipation and speed up operation, the depletion pull-up devices for the two arrays may also be clocked.) During ϕ_1, the inputs are applied to the AND array, and the NOR combination of inputs results on the rows. The columns are activated by the NOR combination of rows during ϕ_2. Inverters at the outputs give the final sum-of-product results.

CMOS PLAs use PMOS enhancement devices to pull up row and column voltages. Dynamic NMOS and CMOS PLAs offer high density and low power at a reduced speed compared to bipolar PLAs.

PLAs offer increased flexibility in the design of complex systems. They are efficient for implementing functions with a large number of variables in the product terms. For simple functions, however, they waste chip area and affect speed. Logic OR of M variables, for example, requires M input columns, M AND gates (rows in the AND plane), and one output column. With registers at the input and output, PLAs are used to perform programmable logic sequencing and finite-state machine operations. As the level of complexity increases, however, sophisticated computer-aided design tools are required for error-free design of systems up to the mask level layout.

10.7.2 PROGRAMMABLE ARRAY LOGIC DEVICES

A *programmable array logic (PAL)* device has the same structure as a ROM, but has a programmable AND array and a fixed OR (or NOR) array. Because of the fixed OR array, a PAL device is less expensive and easier to program. Also, the absence of fusible links in the OR array and the circuitry required to blow them yields a higher density and faster response time than a comparable field-programmable logic array. However, the lack of shared rows (AND gate outputs) with the columns (OR gate inputs) (Figure 10.34) requires that each output function be simplified, with no common product term with others.

Figure 10.35 shows the logic schematic of a PAL14L4, which is a 14-input and 4-output (AND–OR Inverter) PAL available from Monolithic Memories Inc. This device has 16 AND gates, each with a maximum of 28 inputs, and 4 NOR gates. All of the AND gate outputs are connected to the NOR gates. (In some devices, however, the OR or NOR gates may not all have the same number of inputs.) True and complemented buffers are provided for each input variable.

Example 10.2   Design a one-bit full-adder using a PAL14L4.

Solution

Using A, B, and C_i as the inputs to the full-adder, we have the following functions for the sum S and the output carry C_o.

$$S = \overline{A}\,\overline{B}C_i + \overline{A}B\overline{C}_i + A\overline{B}\,\overline{C}_i + ABC_i$$
$$C_o = AB\overline{C}_i + A\overline{B}C_i + \overline{A}BC_i + ABC_i$$

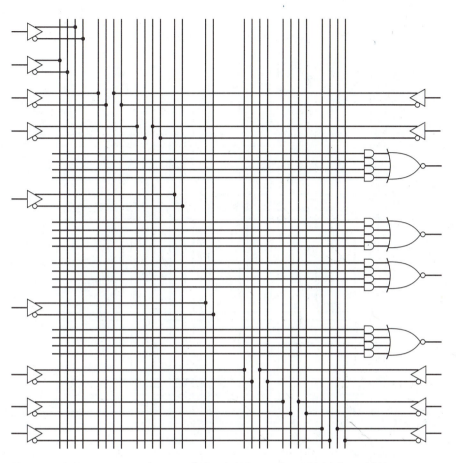

Figure 10.35   Logic schematic of PAL14L4 (Copyright ©1995 Advanced Micro Devices, Inc. Reprinted with permission of copyright owner. All rights reserved.)

The implementation using these equations is shown in Figure 10.36. If fuses are used at the intersections of the rows and columns, dots represent fuses left intact to make contact. Unused AND gates may be indicated by an asterisk (*) inside the gates. Although ABC_i is a common term in both S and C_o, it cannot be used from a single AND gate output. Also, the sum of the products must be limited to four terms to use the given PAL.

PAL devices are available for inputs of from 10 to 20 or more and outputs of from 1 to 10. Outputs may be available from OR or NOR gates, depending on the device. Some devices have flip-flops, so that "registered" outputs are available. PAL devices are used in programmable logic sequencers and other sequential systems where the number of devices is such that a standardized SSI and MSI implementation may take a large number of gates and a custom LSI would

$$S = \overline{\overline{A}\,\overline{B}\,\overline{C}_i} + \overline{A}B\overline{C}_i + A\overline{B}\overline{C}_i + AB\overline{C}_i$$
$$C_o = \overline{\overline{A}\,\overline{B}\,C_i} + \overline{A}B\overline{C}_i + \overline{A}\,\overline{B}\,\overline{C}_i + A\overline{B}\,\overline{C}_i$$

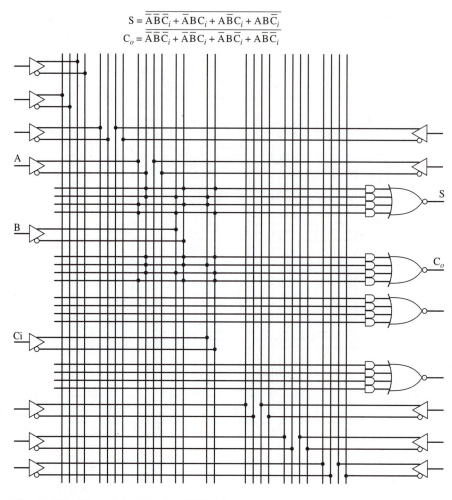

Figure 10.36   Full-adder using PAL14L4

be expensive and time-consuming. PALASM,[4] ABEL,[5] and proLogic[6] are some of the PLD programming languages commercially available.

Generic Array Logic (GAL)[7]   A GAL device is a versatile implementation of a PAL device with EEPROM cells for reprogrammability. A multifunction output stage, referred to as an *output logic macro cell (OLMC),* has D flip-flops and combinational circuitry which can be programmed for combinational or sequential bidirectional input/output. With active high (true) or low (complemented) output programmability and the flexibility of OLMC, GAL devices can be more efficient than PAL devices in many applications. The Lattice Semicon-

---

[4]PALASM is a registered trademark of Advanced Micro Devices.
[5]ABEL is a registered trademark of Data I/O Corp.
[6]proLogic is a registered trademark of proLogic Systems Inc.
[7]GAL is a registered trademark of Lattice Semiconductor.

ductor Corporation GAL16V8A, for example, has 10 inputs and 8 outputs; it can be programmed to have up to 18 inputs using the 8 bidirectional outputs.

10.7.3 PROGRAMMABLE GATE ARRAYS

A gate array, also known as an *uncommitted logic array,* has hundreds to thousands of NAND or NOR gates ("sea-of-gates") arranged in regular arrays without connections. Identical gate patterns and processing steps are used in manufacturing the gate arrays. The interconnections for a specified application are customized using wiring channels that separate the arrays. Because of the flexibility in the interconnections, gate arrays are not limited to implementing only two-level functions, as in PLAs and PAL devices. For example, many semicustom designs in computers use mask-programmed gate arrays; the PLDs for these designs would be large and have impractically complex interconnection structure.

Mask-programmed gate arrays consist of rows (or columns) of transistors that can be interconnected both within and between the rows to implement logic functions. Additional circuitry provides interfaces between external input and output signals and the internal signals. Standard bipolar (TTL, ECL I^2L, ISL, and STL) gate devices are modified as discussed in Chapters 4 and 5 (Sections 4.4.14, 5.4, and 5.6) to obtain high-density gate arrays. MOS (both NMOS and CMOS) and BiCMOS arrays offer a higher density at reduced power dissipation, while GaAs arrays are preferred for high speed.

Figure 10.37 shows part of a CMOS array and the interconnections for two-input NAND and NOR functions. The top rectangle in Figure 10.37a, with p^+ diffusion, represents a row of four PMOSFETs, and the bottom rectangle, with n^+ diffusion on a p well, represents four NMOSFETs. Each small square in the p^+ and n^+ diffusion regions serves as a source or drain connection, and the vertical blocks of polysilicon strips (tunnels) running through the rectangles at the next level act as gates. The two horizontal metal strips are used to carry power (V_{DD}) and ground (V_{SS}). There are two contact points for each source or drain which are separated by the V_{DD} (V_{SS}) line in the PMOS (NMOS) devices. Transistor channels are formed in the diffusion area underneath the polysilicon strips and within the left and right contact points. The area of PMOSFETs is twice that of NMOSFETs to obtain approximately equal rise and fall times in switching. Figure 10.37b shows the equivalent transistors. Examples of the interconnections required for two-input NAND and NOR functions are shown in Figures 10.37c and d.

The gate array in Figure 10.37 shows common gates for each pair of complementary devices. A number of variations of the basic CMOS gate array are possible to implement different functions with minimum interconnections and high density. Currently CMOS and BiCMOS gate arrays with hundreds of thousands of gates and with delays of less than a nanosecond have been developed for use in communications systems, fast minicomputers, and graphics systems.

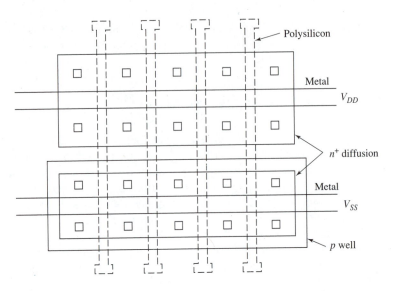

(a) Layout

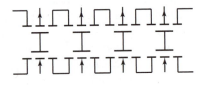

(b) Equivalent Transistors

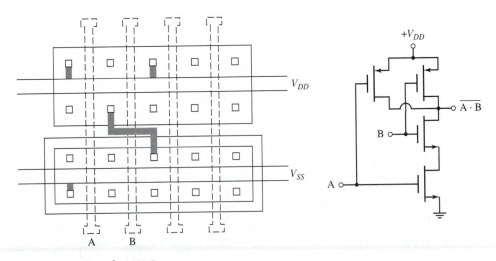

(c) Interconnections for NAND

Figure 10.37   CMOS gate array

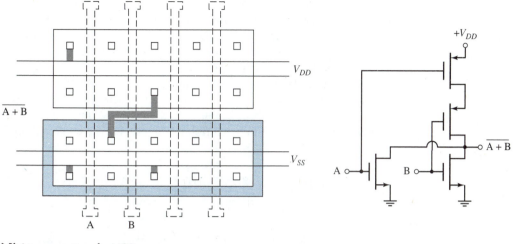

$\overline{A + B}$

(d) Interconnections for NOR

Figure 10.37 *(continued)*

Although gate arrays offer high density and speed with the flexibility to realize multilevel logic functions, they are not efficient in implementing memory functions. This limitation is due to the matrix of unconnected transistors, rather than logic cells. Also, a large chip area is needed to provide connections between the rows of these devices. In addition, depending on the function realized and the number of connections and external signals needed, very few gates may be used while all the interconnection space is taken up. Another disadvantage is the nonuniformity in gate delays. Depending on the location of the gates and the length of the interconnection paths, delays may vary if the devices are not carefully placed. Because of the different layout configurations in devices from different manufacturers, and the different placement of gates and interconnection paths, the size and cost of an array for a given function may also vary.

A *field-programmable gate array (FPGA)* is a reprogrammable gate array that uses antifuse, SRAM, EPROM, or EEPROM technology for repeated programming of user-defined functions. A major difference between FPGAs and other PLDs is the organization. FPGAs incorporate logic blocks instead of fixed AND–OR (or other two-level) gates or unconnected transistors. Each logic block may consist of complementary transistor pairs ("fine-grain" programmable blocks), basic gates, flip-flops, multiplexers, or RAMs. (Because of the large number of programmable logic blocks, FPGAs are sometimes referred to as *complex PLDs (CPLDs)*.) Drivers, or I/O blocks, are included at the periphery for interfacing with external signals. Instead of being programmed by modifying the fixed interconnections, as in PLDs, FPGAs are programmed by modifying the routing of the interconnections in the logic blocks. The functionality of different logic blocks and the electrical (user) programmability of the routing are utilized efficiently in designing prototype ASICs and semicustom systems, using

FPGAs. In general, FPGA-based designs are faster, with lower power dissipation, than comparable designs using many PLDs. The complexity of the FPGAs, however, requires careful design using CAD tools. Table 10.2 shows some of the characteristics of a few commercially available FPGAs and CPLDs.

Figure 10.38 shows an example of a "coarse-grain" logic block used in the Actel Act-2 FPGA. With the three 2-to-1 multiplexers, two of which are driven by $s_1 s_2$ and the third driven by $s_3 + s_4$, this block can produce any of a variety of two-, three-, and four-variable combinational functions. These functions must be expressed in terms of 2-to-1 multiplexer functions. The combinational

Table 10.2   Characteristics of FPGAs and CPLDs

No. of Equivalent Gates	Programming Method	Basic Cell	Architecture	Manufacturer
1200–20,000	Antifuse (PLICE)	Multiplexer	Gate array (rows)	Actel
1500–12,00	Antifuse	Multiplexer	Matrix	QuickLogic
1200–20,000	SRAM	Look-up table	Matrix	Xilinx
2000–20,000	EPROM	Extended PLA	PLA	Altera
900–3600	EEPROM	Extended PLA	PLA	AMD

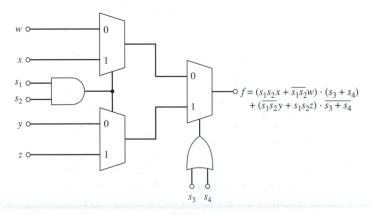

$$f = (s_1 s_2 x + \overline{s_1 s_2} w) \cdot (s_3 + s_4) + (\overline{s_1 s_2} y + s_1 s_2 z) \cdot \overline{s_3 + s_4}$$

Figure 10.38   Logic block in Actel Act-2 FPGA
(Courtesy of Actel—Reprinted with permission)

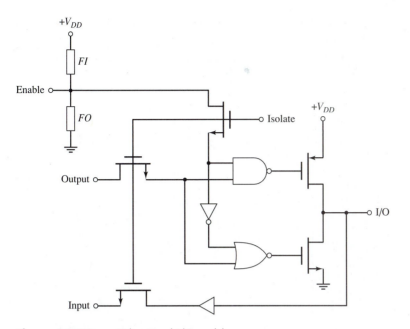

Figure 10.39   Bidirectional I/O module

logic block shown is followed by a sequential block built around two latches, which may be configured as D or toggle flip-flops. (In turn, a flip-flop can also be built using two logic blocks.)

The input/output module is shown in Figure 10.39. This bidirectional block can be programmed for input by blowing the antifuse *FI*, and for output by blowing *FO*. (Recall that an antifuse offers a high resistance when intact and a low resistance when burned.) If both *FI* and *FO* are left in the circuit, the input Enable can be used to control bidirectional data flow.

A variation of the multiplexer-based logic block with 14 inputs and a flip-flop is employed in the QuickLogic FPGA (Figure 10.40). In this device, the logic function is carried out by a 4-to-1 multiplexer, which is controlled by the outputs of gates *A* and *F*. Both types of logic blocks, shown in Figures 10.38 and 10.39, can realize a large number of functions with a relatively small number of devices. The price paid for this flexibility is the large routing, and hence the large size, of the interconnections when all (or most) of the inputs are present. Two-terminal antifuses at the crosspoints can reduce the overall size of a multiplexer-based FPGA.

Another type of logic block is based on a function generator or look-up table (LUT), and is used in the Xilinx FPGA. By programming a *K*-bit address static memory, the block can realize any function with up to *K* inputs. An SRAM stores the complete truth table in the form of an LUT. In this respect, the LUT architecture of the Xilinx FPGA is similar to that of an EEPROM, but is volatile.

As with ROMs, the LUT is largely underutilized for functions with only a

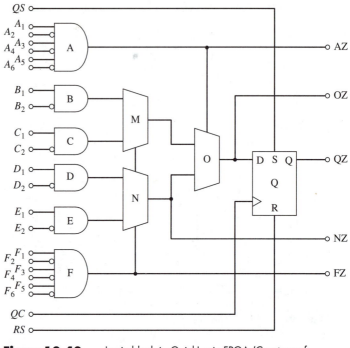

Figure 10.40   Logic block in QuickLogic FPGA (Courtesy of QuickLogic—Reprinted with permission)

few variables. The logic block of the Xilinx XC3000 series, shown in Figure 10.41, has a five-variable LUT. The SRAM-controlled interconnect points can be programmed such that this LUT can be reconfigured as two four-variable LUTs, with two outputs, or a single-output five-variable LUT. Connections to the two flip-flops and the multiplexers are also controlled by SRAM cells. More LUTs and multiplexers and two fixed nonprogrammable connections in the Xilinx XC4000 series FPGAs enable fast addition and the implementation of a small amount of memory.

PLA-based logic blocks are also used in CPLDs. These EPLDs or EEPLDs with floating gate interconnection cells have a large number of inputs and can realize multilevel logic functions. By programming the input AND gates, as in PLAs, true or complemented two-level sum-of-products output is obtained from each logic block. The output may also be registered using a programmable flipflop. Altera MAX 7000 series CPLD, for example, has a maximum of 10,000 gates with 256 logic blocks called macrocells.

Standard cell design is yet another approach to ASICs using PLDs. Unlike gate arrays with the same size and large routing channels, or FPGAs with logic blocks, a library of standard cells is available for use in a design. These call modules consist of basic gates, flip-flops, and commonly used functions such as decoders, multiplexers, and full adders, as well as memories. Each cell in the family is individually designed and tested, by simulation in the technology of use, for speed, driving capability, or other performance criterion. Therefore, the transistors in the cells may not all have the same size; however, the devices and

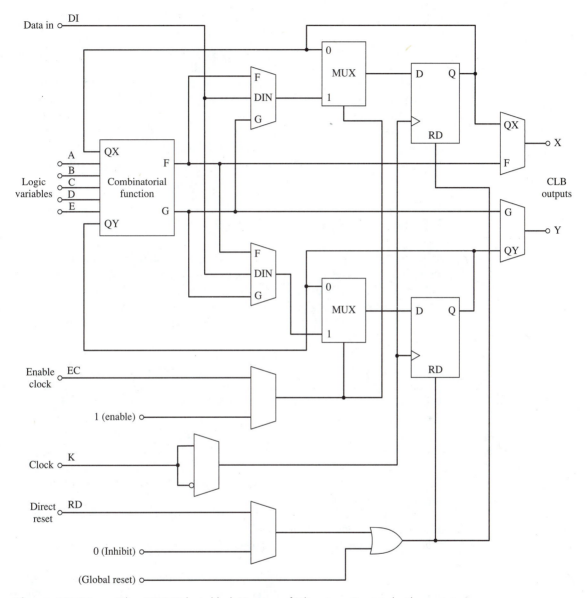

Figure 10.41 Xilinx XC3000 logic block (Courtesy of Xilinx, Inc.—Reprinted with permission)

connections in a cell have the same layout. Each individually designed cell is treated as a building block (fine-grain block), so that the final ASIC design with many blocks can be reduced in size.

NMOS and CMOS standard cell libraries are available with hundreds of cells. For example, the SCMOS standard cell library[8] includes SSI gates, flip-flops, clock generators, MSI counters, shift registers and multiplexers, with more cells added in the updates. Unlike in fixed arrays, these cells can be placed

---

| [8]The library is available from the Institute for Technology Development, Starkville, MS.

anywhere in the layout, which affords flexibility and efficiency in the wiring. With individual cell placement, each ASIC designed using standard cells requires a complete and unique set of masks. As a result, the cost and mask-making time for placement and routing of the cell interconnections are increased. Still, a design using standard cells has lower development time and cost than a comparable custom IC design. In terms of size and performance, however, standard cell designs in general are less efficient.

Designing with standard cells proceeds with CAD tools that use the database of the cell library. Simulation of the final design is performed using individual cell dimensions, time delay equations, and power dissipation, with temperature dependency.

10.8   VLSI DESIGN ISSUES

As discussed in Chapter 1 the design of a digital system from specifications to final product involves functional, logic, circuit, and chip level design. Simulation and/or verification at every design level is required for an error-free system, regardless of random logic, ASIC, or VLSI design. For ASIC and VLSI design, however, CAD tools are needed at all levels, because of the complexity of the system.

As discussed earlier, field-programmable logic devices with different architectures and programming capabilities offer a less expensive alternative to full-custom design. In addition, considerations such as nonrecurring engineering (NRE) cost and manufacturing time (time-to-market) may override the design choice in many cases.

Among the possible design methodologies of full-custom, mask-programmed, and field-programmable devices, EPLDs and EEPLDs are presently used overwhelmingly in applications with medium circuit complexity (up to 10,000 gates) and low volume. FPGAs are particularly useful in early engineering prototypes.

Although these applications have clock speeds of below 50 MHz, the flexibility in making design changes at low cost and in a short time outweighs other factors in systems that are not easily accessible and whose specifications may change over time.

For higher speeds and especially newer architectures in computers under development, one-time programmable gate arrays and PALs offer a cost advantage over ASIC. Other considerations, such as the number of pins in the final chip, power, and path-dependent delays lead to different programming technologies (EPROM, EEPROM, antifuse, and SRAM).

Regardless of the choice used in the development of prototypes, the production of a large volume of tested designs is invariably carried out in ASICs. ASICs are also built using standard cell libraries for high density in terms of number of functions at reduced design time. While ASICs take up considerable NRE cost, the conversion of tested designs from many PLDs or gate arrays to ASICs reduces the overall size and cost and increases the reliability. However, sophisticated CAD tools are required to convert the designs.

CAD tools can also be used to design as ASIC and convert the design to PLD for prototyping. Therefore, CAD tools must be capable of using the description of each architecture, design, and technology in a hardware description language (HDL), such as VHDL or Verilog HDL, and must be able to generate the design in the target technology.

As the complexity of the system increases, the design using any architecture or technology requires design automation tools to carry out all phases of the design process. In this regard, *silicon compiler*s are gaining widespread applications along with hardware description languages and simulation packages. A silicon compiler is an "expert" system that accepts the behavioral description of a proposed system and produces a design capable of being implemented directly in a custom or standard VLSI system. In addition, a silicon compiler synthesizes the IC layout, functional simulation model, and timing model in terms of a library of structures for PLDs, memories, and others.

For an error-free design with an efficient use of the silicon area, thousands of design rules are stored in the library for different processes. Using these rules and the process selected for the system, the compiler produces the layout and interconnections, along with plots and other documentation. The compilation process thus eliminates the individual design and simulation at the logic and circuit levels, and the attendant modeling and parameter extraction and layout verification. However, a designer must be knowledgeable in all levels of design to be able to refine the resulting design at each level. As with programming PLDs using specific symbolic languages, designing from a high level of abstraction requires a detailed understanding of the process technologies, circuit configurations, and logic structures, so that efficient and cost-effective refinements can be made in the final design.

As the minimum feature size of silicon devices decreases toward $0.1~\mu\text{m}$ and the number of active devices approaches tens of millions, there are physical limitations and challenges of VLSI design. It has been shown that the product of the maximum applied voltage and the frequency is limited to 200 V GHz for silicon. Current technology has not yet reached this physical limitation of dynamic range. The operating temperature, however, limits the number of gates per chip at any switching delay. Typically, allowing a 100°C rise in the temperature of the device from its environment, the packing density N_G of VLSI circuits is related to the delay t_p, as (see Reference 27):

$$\frac{N_G}{t_p} \leq 4 \times 10^5 \text{ gates/ns}$$

This tradeoff between gate density and delay must be considered as chip density increases. In addition, it has been shown that the switching energy, or delay-power product, must be high enough (> 5 MeV) to operate well in a noisy environment. This energy also limits the power supply voltage at the low end since in most cases the delay-power product can be related to supply voltage and stray capacitance.

Low-voltage operation also requires the device threshold voltages to be scaled below the supply voltage in MOS and CMOS devices. Conventional BiCMOS circuits, on the other hand, require a base–emitter junction voltage of

0.7 V. To operate at low supply voltages, this junction voltage must be circumvented using innovative circuit configurations. For MOS devices to achieve a channel length of below 0.1 μm and operate at a supply of 2 V or less, the threshold voltage must be below 0.5 V at room temperature. Several problems face the process of attaining such a low threshold voltage using the classical MOS scaling technique.

The scaling of voltages due to the built-in junction potential and the charge in the channel does not occur in the same way as the scaling of dimensions and doping. In addition, the low thickness of the gate oxide, which is required to scale the voltage, causes tunneling and reliability suffers due to the low punch-through voltage. Advanced scaling processes and new device structures are essential to overcoming the current limitations in low-voltage operation.

Further limitations occur when the device size nears the depletion widths. Short-channel and parasitic effects become dominant, and the interaction of adjacent devices causes severe problems. Additionally, the size and performance of the devices are affected by leakage currents (at less than the threshold voltage), threshold voltage variations, and hot-carrier effects (due to high electric fields). Furthermore, increased source-drain resistance, decreased mobility, and increased capacitances all require accurate device model parameters for the simulation process. It is estimated that over 1000 parameters are required to model a device with a 0.5 μm feature size. In addition, interconnect delays account for 50 percent or more of the total delay. Computer-aided design and simulation tools must therefore be sophisticated enough to take into account all of the effects due to high density and low voltage operation. The breakthroughs in substrate engineering, device technology and design, and circuit design and modeling are gradually overcoming some of the limitations in the current VLSI design process.

SUMMARY

This chapter presented basic circuit configurations for semiconductor read/write and read-only memories. Random access refers to accessing any word in memory with the same delay independent of its location. Although both RWM and ROM have random access capability, RWM is usually referred to as random access memory (RAM). Static memory retains data without clocking. Dynamic memory needs periodic data refreshing.

Memory containing M words of N bits may be organized in one dimension with K bits of address such that $2^K \geq M$. Two-dimensional organization reduces the size of address decoding (2^K-to-1) by using row and column decoding.

Memory access time refers to the delay from the time an address is presented to the time data transfer to/from that address is completed. Read or write cycle time corresponds to the minimum time necessary to read or write valid data at any address. The minimum time needed for data exchange is the memory cycle time.

Bipolar, MOS, CMOS, and BiCMOS static RAMs are built using static flip-flop cells for each memory bit. For high density and low power, MOS dynamic RAMs are available using four-, three-, or one-transistor cells. These DRAMs store data as charge in stray capacitances and hence require periodic refreshing to restore any leaking charge. One-transistor DRAMs have reached densities nearing 1 Gbit.

Apart from memory cells, address decoding occupies significant chip area and contributes to increased access time. Various bipolar and MOS decoder configurations are possible to reduce the delay and increase the density.

A read-only memory stores fixed data at each address location by connecting active devices to supply or ground during manufacture. Because of the fixed data at each location, efficient decoders are implemented in ROMs by combining the decoding with output lines. ROMs with fusible links can be programmed by the user. For repeated programmability, floating gate MOSFETs are used. These EPROM devices lose their charge, or become erased, by exposure to ultraviolet light; the devices are programmed by selectively charging the MOSFETs.

Electrically erasable programmable ROMs use an electron tunneling mechanism through a thin gate oxide. Although they have a lower density than EPROMs, EEPROMs can be programmed in-circuit, with the flexibility to alter data at any location without affecting other addresses. Modified EEPROM cells are used in nonvolatile RAMs (NVRAMs) or flash memories, which are read-mostly memory.

Programmable logic devices (PLDs) use the programmability of various types of ROM to implement combinational and sequential functions. In addition to fuses, EPROM-, and EEPROM-like cells, PLDs also use antifuses, which are normally open links. Instead of full decoding, PLDs may incorporate a hard-wired subset of a truth table, or an array of two-level gates, along with flip-flops. With selective connections of the inputs and outputs to the gates and flip-flops, a large number of complex combinational and sequential functions can be produced. The programmability and high density of PLDs make them useful in designing ASICs where design changes can be made rapidly and inexpensively. With the increasing complexity of PLDs, sophisticated computer-aided design tools are needed to design, simulate, and verify the ASICs before going to the final products.

The design of VLSI circuits and ASICs must consider nonrecurring engineering costs and the flexibility required in making design changes. As the size of active devices decreases and the density increases, the limitations due to power dissipation and propagation delay must be considered. In addition, the scaling of circuit parameters for low-voltage operation, while maintaining high density, is a major design constraint.

REFERENCES

1. Luecke, G.; J. P. Mize.; and W. N. Carr. *Semiconductor Memory Design and Application.* New York: McGraw-Hill, 1973.

2. Hodges, D. A.; and H. G. Jackson. *Analysis and Design of Digital Integrated Circuits.* New York: McGraw-Hill, 1988.

3. Taub, H.; and D. Schilling. *Digital Integrated Electronics*. New York: Mc-Graw-Hill, 1977.

4. Uyemura, J. P. *Circuit Design for CMOS VLSI*. Boston: Kluwer Academic Publishers, 1992.

5. Child, J. "Faster processors ignite SRAM revolution," *Computer Design* 33, no. 8 (July 1994), pp. 97–103.

6. Lu, N. C. C. "Advanced Cell Structures for Dynamic RAMs," *IEEE Circuits and Devices Magazine*, January 1989, pp. 27–36.

7. Rideout, V. L. "One-Device Cells for Dynamic Random-Access Memories: A Tutorial," *IEEE TED* ED-26, June 1979 pp. 839–852.

8. Mead, C.; and L. Conway. *Introduction to VLSI Systems*. Reading, MA: Addison-Wesley, 1980.

9. Pashley, R. D.; and S. K. Lai. "Flash memories: The best of two worlds," *IEEE Spectrum*, December 1989, pp. 30–33.

10. *Flash Memory* I and II, Mt. Prospect, IL: Intel Corporation, 1994.

11. Child, J. "Specialty SRAM combines best of dual-port SRAMs and FIFOs," *Computer Design* 32, no. 6 (June 1993), pp. 34–36.

12. El-Ayat, K. A., et al. "A CMOS Electrically Configurable Gate Array," *IEEE JSSC* 24, no. 3 (June 1989), pp. 752–762.

13. Burns, S.G.; and P.R. Bond. *Principles of Electronic Circuits*. St. Paul, Minn: West Publishing, 1987.

14. Rose, J., et al., "Architecture of Field-Programmable Gate Arrays: The Effect of Logic Block Functionality on Area Efficiency," *IEEE JSSC* 25, no. 5 (October 1990); pp. 1217–1225.

15. Rose, J., A. E. Gamal; and A. Sangiovanni-Vincentelli. "Architecture of Field Programmable Gate Arrays," *Proc. IEEE* 81, no. 7 (July 1993), pp. 1013–1029.

16. Trimberger, S. "A Reprogrammable Gate Array and Applications," ibid, pp. 1030–1041.

17. Greene, J.; E. Hamdy; and S. Beal. "Antifuse Field Programmable Gate Arrays," ibid, pp. 1042–1056.

18. Brown, S. D.; R. J. Francis; J. Rose; and Z. G. Vranesic. *Field Programmable Gate Arrays*. Boston: Kluwer Academic Publishers, 1992.

19. Tuck, B. "FPGAs race for the gold in product development," *Computer Design* 31, no. 4 (April 1992), pp. 88–104.

20. Jenkins, J. H. *Designing with FPGAs and CPLDs*. Englewood Cliffs, NJ: Prentice-Hall, 1994.

21. Chan, P. K.; and S. Mourad. *Digital Design Using Field Programmable Gate Arrays*. Englewood Cliffs, NJ: PTR Prentice-Hall, 1994.

22. *FPGA Databook and Design Guide*. Sunnyvale, CA: Actel Corporation, 1994.

23. *QuickLogic*. Santa Clara, CA: QuickLogic Corporation, 1995.

24. *The Programmable Logic Data Book*. San Jose, CA: Xilinx, 1994.

25. *Altera Data Book*. San Jose, CA: Altera Corporation, 1995.

26. Johnson, S. C. "VLSI circuit design reaches the level of architectural description," *Electronics* 57, (May 3, 1984), pp. 121–128.

27. Nagata, M. "Limitations, Innovations, and Challenges of Circuits and Devices into a Half Micrometer and Beyond," *IEEE JSSC* 27, no. 4 (April 1992), pp. 465–472.

28. Tiwary, G. "Below the half-micron mark," *IEEE Spectrum* 31, no. 11 (November 1994), pp. 84–87.

29. Fawcett, B. K. "Tools to speed FPGA development," ibid, pp. 88–94.

REVIEW QUESTIONS

1. For a 2048 word x 16 bit memory, what is the address bus width? What is the data bus width?

2. What is the basic circuit in SRAM cells?

3. What is two-dimensional decoding? Why is it preferred?

4. Compare the number of gates required in the decoding circuitry for a 10 bit address if
   a. linear decoding is used, or
   b. two-dimensional decoding with 6 bits for the rows and 4 bits for the columns is used.

5. What is the function of a sense amplifier in an SRAM?

6. How is data stored in a DRAM cell?

7. What is memory refreshing?

8. Why is NOR address decoding preferred over NAND decoding?

9. Compare SRAMs and DRAMs in terms of density, speed, and overall circuit complexity.

10. Define mask-programmed and user programmable ROMs.

11. If L bits are used in column decoding of a MOS ROM using a tree decoder, how many active devices are needed?

12. In terms of data erasure, how do EPROM and EEPROM differ?

13. How are floating gate and tunnel oxide gate ROM devices programmed?

14. What is a flash memory? How does it differ from SRAM?

15. Compare sequential access memories with SRAMs in terms of speed and density.

16. How does an antifuse differ from a fuse?

17. What are the disadvantages of using fuses and antifuses in terms of chip size, circuit complexity, and programmability of PLDs?

18. How does a K-variable EEPROM differ from a K-variable PLA?

19. What is the difference between a PLA and a PAL device?

20. When is a PLD preferred over a VLSI for an ASIC design?

21. What is a standard cell design?

22. What are the constraints in increasing the density of VLSI circuits? State the constraints in designing circuits for low-voltage operation?

PROBLEMS

1. A memory of 16 words with a word size of 8 bits uses a linear addressing scheme.
   a. Determine the address size and the number of AND gates and inverters required for the address decoder.
   b. Repeat (a) if the number of words is increased to 64.
   c. Repeat (a) if the work size is increased to 32 bits while the number of words is kept at 64.

2. Memory chips are available as $1 \text{ K} \times 1$ bit with a 10-bit address and one bit each for data input and output. Each chip has an active low chip enable and R/W inputs. Show an organization for building a $4 \text{ K} \times 8$ bit memory using these chips. Indicate how each chip is selected using an appropriate number of external address bits and address decoding circuitry.

3. Calculate the standby power per cell for the TTL SRAM cell shown in Figure 10.3. Assume $R_C = 20 \text{ k}\Omega$, the logic low levels at RAS and CAS are 0.3 V. $V_{CE(\text{sat})} = 0.2$ V, and $V_{BE(\text{sat})} = 0.8$ V. If the maximum power dissipation per chip is 1 mW, what is the memory capacity possible per chip?

4. A dynamic memory cell has a capacitance of 100 fF. If the initial voltage is 4 V and the leakage current is 5 pA, what is the voltage after 2 ms?

5. A one-transistor dynamic memory cell has a capacitance of $C = 50$ fF, and the bit line capacitance is $C_L = 240$ fF. The bit line is precharged to a voltage of 5 V for reading. If C has a voltage of 4 V for high and 0 for low, determine the bit line voltages corresponding to the low and high levels. What is the voltage difference that a sense amplifier must detect between the 0 and 1 bit values?

6. Calculate the number of NOR and NAND gates in Figure 10.15a, b, and c for full decoding of a 6-bit address.

7. Show a diode ROM realization of a 2×2 multiplier. Compare the ROM realization with NOR and NAND cells of MOSFETs. If an 8×8 multiplier is implemented directly using a single ROM, what is the size of the ROM in the total number of cells?

8. A 1 K PROM is available as 256 words $\times$ 4 bits. The chip has two active low enable inputs. If the row address decoder has five bits, show the memory organization including the size of the data selector. If an expanded memory of $1 \text{ K} \times 8$ is needed, how many 256×4 chips are required? Show an organization for the expanded memory.

9. A 2-bit comparator has four inputs, $A = a_1 a_0$ and $B = b_1 b_0$, and three outputs, $G = 1$ if $A > B$, $E = 1$ if $A = B$, and $L = 1$ if $A < B$. Show a

mask-programmed PLA implementation for the comparator. If the function is realized as a 4-bit ROM, compare its size to the PLA. If G and E are the only outputs from the PLA array (note the L can be obtained from a single NOR gate as $L = \overline{G + E}$), compare the number of active devices and the size with a ROM.

10. Implement the full-adder functions using the PAL10L8 shown in Figure P10.10.

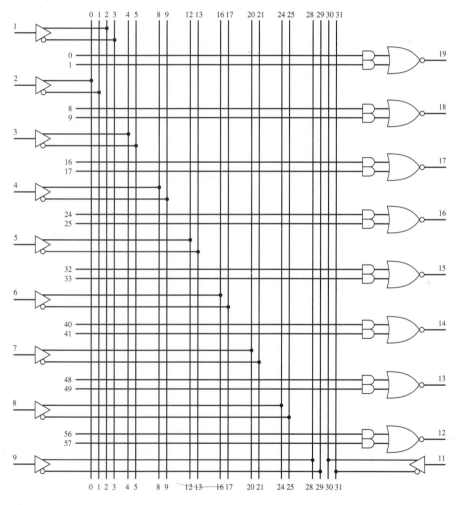

Figure P10.10

11. Implement a 2-bit combinational multiplier with a 4-bit output using a PAL14L4.

12. On the CMOS gate array layout of Figure 10.37a, show how to implement the functions: (a) $\overline{AB + CD}$, and (b) two-input exclusive OR.

13. Show the inputs to realize $Y = AB + C\overline{D} + \overline{E}$ in (a) Figure 10.38, and (b) Figure 10.40.

14. *Experiment: Access time measurement for an SRAM*
    Wire an SRAM, such as a 4 K $\times$ 1 bit Intel 2147H, and write a 0 at address 0, and a 1 at address 1 by statically connecting the address, data, and chip select pins to the appropriate levels and changing the write enable signal. Connect all address lines except bit 0 to ground. Connect a TTL pulse source to bit 0 of the address. Observe the output data and the address clock on an oscilloscope and determine the access time. Change the address to a higher value (in the upper end of 4 K) and repeat the above process of memory writing and reading. Verify the data read and compare the access time measured.

INDEX

NAME AND SUBJECT